Environmental Hazards

Series Editor
Thomas A. Birkland

For further volumes:
http://www.springer.com/series/8583

Naim Kapucu • Kuotsai Tom Liou
Editors

Disaster and Development

Examining Global Issues and Cases

 Springer

Editors
Naim Kapucu
School of Public Administration
University of Central Florida
Orlando
Florida
USA

Kuotsai Tom Liou
School of Public Administration
University of Central Florida
Orlando
Florida
USA

ISBN 978-3-319-35665-5   ISBN 978-3-319-04468-2 (eBook)
DOI 10.1007/978-3-319-04468-2
Springer Cham Heidelberg New York Dordrecht London

Springer is part of Springer Science+Business Media (www.springer.com)

Foreword, *Disasters and Development*

The UN declaration of the International Decade for Natural Disaster Reduction (IDNDR) for the 1990s was a major step forward in drawing the attention of various nations to vulnerabilities from natural disasters. Among other things, the IDNDR effort institutionalized high-level efforts within the United Nations and fostered a new strategy for disaster reduction. Over time, these efforts led to a set of resolutions that more closely linked disasters and development. Chief among these was adoption in 2005 of the *Hyogo Framework for Action 2005–2015: Building the Resilience of Nations and Communities to Disasters* following the Indian Ocean Earthquake and Tsunami. That framework singled out as one of three strategic goals more effective integration of disaster risk reduction considerations into sustainable development policies and programs.

The 2010–2011 UN mid-term report about implementation of the Hyogo Framework addressed progress toward the framework goals based on an analysis of reports by various governments. The report noted substantial progress by many nations but an overall pattern of uneven implementation that largely reflected broader economic and institutional differences among regions and countries. One concern was the failure to integrate disaster risk reduction into sustainable development policies and plans. The report noted the mismatch between traditional ways of addressing disasters and development goals in stating: "Handling what is primarily a developmental issue with largely relief and humanitarian mechanisms and instruments, while helpful at the beginning, needs to be reconsidered to ensure that disaster risk reduction plays the role that it must in enabling and safeguarding development gains." More generally, the report noted the challenges governments faced in fostering more holistic implementation of risk reduction efforts.

The chapters in this volume, *Disasters and Development: Examining Global Issues and Cases*, help to elucidate important gaps that are highlighted in the UN mid-term report. Four aspects stand out in this regard. One is gaining a better understanding of the reasons for the unevenness of different national efforts to link disaster risk reduction with development planning and programs. The case discussions of risk reduction efforts by different countries in this volume go beyond the country reports of official UN documents to provide rich depictions of the challenges and possibilities for advancing disaster reduction and sustainable economic development.

Two lessons stand out from these cases. One is that there is no single blueprint for accomplishing risk reduction and development goals given variation in the political, economic and institutional settings. Another lesson is that the differences in ability to bring about change are not captured by simple distinctions between developed and developing countries. The cases here illustrate examples of notable progress in advancing risk reduction, particularly at local levels, in developing countries like China, India, the Philippines, and Turkey. At the same time, other chapters discuss the limits to addressing the forces of economic development that exacerbate disaster vulnerability in more developed countries such as Germany, Japan, and Korea.

A second contribution of chapters in this volume is clarifying some of the key conditions that facilitate and hinder effective progress in linking disaster risk reduction and economic sustainability. Much has been written, as emphasized in the UN international strategy for disaster risk reduction (*A Safer World in the 21st Century: Risk and Disaster Reduction*), about the need for increased capacity in carrying out vulnerability assessments, in gauging the implications of development programs for vulnerability, and for developing meaningful risk reduction and development plans. What stands out in the chapters in this volume, however, is the importance of gaining political commitment—the "buy in" and continuing commitment of governmental leaders at all levels—for both disaster risk reduction programs and for smart development programs that consider the implications of development for hazard vulnerability. This, not surprisingly, is one of the major problems in settings that have pressing economic, political and social problems that make attention to disaster and development issues lower priorities. The case studies in this volume of disaster management in Azerbaijan and Lebanon, and recovery and reconstruction in Haiti illustrate the dilemmas of limited political commitment.

Many chapters in this volume, as a third contribution, highlight the role of supranational institutions, non-governmental organizations, and private entities in building civic capacity for reducing risks. Financing and technical assistance from the World Bank were pivotal for establishment of an earthquake insurance program in Turkey. The private sector is integral to the establishment of housing programs in India. United Nations and other international non-governmental organizations were central providers of aid after flooding in Pakistan and the earthquake in Haiti. While these and other examples in the chapters of this volume make clear the importance of such contributions, they also serve as a reminder of the development-assistance disease. Reliance on such assistance undermines development of central government commitments and capacity building. Chapters about non-governmental organizations and capacity building more generally help to inform an understanding of how civil capacity can be built with inoculation from the development-assistance disease.

A fourth contribution that crosscuts many of the chapters in this volume is the importance of various forms of policy and political learning over time to better design and implement policies and programs that achieve both disaster reduction and economic development goals. Several chapters show the importance of recognition that these need not be incompatible goals as a basis for learning of how to do better. Understanding how the physical features of the environment both constrain and facilitate risk reduction is a central point made in the discussion of flood and hurricane adaptation in Florida. The importance of overcoming the rush to rebuild after catastrophic events and taking time to develop an understanding of community desires were central to rebuilding efforts in Christchurch, New Zealandthat can mitigate the impacts of future earthquakes.

The themes noted in this volume will be invaluable as scholars, policymakers, and others look to future efforts to bridge economic development programs and disaster reduction efforts. There are no quick fixes or easy recipes. If anything, the chapters in this volume underscore the challenges involved. The obstacles are the result of long-standing forces that have made development the engine of national economies, often at the expense of increased vulnerability to natural hazards. Those decisions have made more marginalized citizens the most vulnerable given the location of their housing and livelihoods. A real challenge is overcoming past development decisions that have created an existing hazardous built environment—"killer buildings" that are especially vulnerable to earthquakes, flood-prone shanty areas along low-lying coastal areas and rivers, and construction practices that make residences vulnerable to high winds and other devastating forces. Part of the answer lay in not repeating past mistakes by being smarter about the location of infrastructure and development. Yet, more needs to be done to ameliorate the past mistakes to increase the resilience of the existing built environment to devastating natural hazards.

As the chapters in this volume make clear, the approach to achieving this is as much a political undertaking as it is an economic and social one. The plethora of scholarship about the need for increased resilience, economic sustainability, and enhanced civic capacity gets at the fundamental notion that future efforts need to foster a shared responsibility for risk reduction among citizens, non-governmental, and governmental entities that does not undermine the prospects for economic development. As stated in the mid-term UN report about progress with the Hyogo Framework, more successful programs have created "a social demand for disaster risk reduction so that individuals realize their own share of responsibility in increasing their resilience and in holding governments accountable for the development and implementation of coherent disaster risk reduction plans and investments."

Institutionalizing these commitments requires more than networks of actors working together to overcome adversityas is often the case in the aftermath of a devastating event. The chapters in this volume note the importance of cultivating such networks but also of leadership for setting forth a vision that provides a sense of shared purpose, in mobilizing constituencies in support of that vision, and in establishing institutional structures that channel attention, information flows, and governmental activities. When done well, the end result is a policy regime that serves as the basis for sustained risk reduction and economic development efforts around common goals of resilience for both disasters and the economy. This also provides for a more holistic approach that is too often lacking as efforts are parceled out among various parts of governments and networks of non-governmental actors.

Stronger regimes foster policy legitimacy and enhance implementation prospects because they develop a shared purpose, capitalize on the supportive efforts of key players and supporters, and focus the attention and authority of multiple implementers in support of a common goal. A key concern for any such effort is to avoid backsliding in commitments as the memories of disasters fade, attention turns to other pressing issues, and the demands for economic growth dominate concerns about disaster vulnerability. Although somewhat inevitable, these forces can be resisted when new constituencies—the financial community, leading businesses, civic groups, and others—recognize the potential for catastrophic losses and invest their energies in maintaining efforts to avert them. Fostering such "social demand" for disaster risk reduction is critical for long-run progress in achieving sustained efforts. This volume helps establish a path toward these ends.

University of Washington, Seattle, USA   Peter J. May
                                                        Donald R. Mathews Distinguished
                                                          Professor of American Politics

Preface

It is apparent that there is a need to integrate disaster mitigation and risk reduction into disaster recovery, economic and community development, and environmental policy and management. To reach these goals agencies and disciplines should work together, share knowledge and consider pre-planning strategies with the goal of increasing disaster resiliency and the overall economic health of their community. This book is intended to provide conceptual framework and empirical evidence of the factors contributing to disaster recovery and sustainable economic development. This is a unique feature of the book.

This book offers a systematic, empirical examination of the concepts of disasters and sustainable economic development applied many cases around the world. It will contribute to the literature on emergency management, community sciences, policy and planning, economic development, and environmental management. To date the scholarly literature has tended to approach disaster and development from a case study perspective. Our proposed book approaches research on disaster and development in communities and regions from a comprehensive and integrated manner. The volume provides results that are more generalizable and more widely applicable to a variety of circumstances, disasters, and geographic regions.

Ultimately, this book sheds light on how communities can increase their resiliency in response to and recovering from disasters through policy interventions and governance mechanisms. The book will also advance a scholarly understanding of what influences disaster resiliency and contributes to environmental, economic, and social sustainability of communities and regions.

Key Elements

1. *International and Comparative focus*: Disasters are worldwide. Rather than focusing primarily on disasters in the United States to inform practice and advance the scholarly literature, this book expands the scope of empirical analysis to international contexts. This brings additional contextual factors that improve the general understanding of how communities plan for and manage disasters and economic development.

2. *Interdisciplinary Perspectives*

   a. *Sociological and community perspectives*: This volume also presents chapters that measure vulnerability and resilience at the individual level of analysis. The role of culture, social capital, socio-economic vulnerabilities, and interpersonal social networks provides complementary evidence to the analyses conducted at the larger community and regional scale of disaster planning and management.
   b. *Planning and development perspectives*: The relationships among land-uses, housing decisions, and mitigation strategies to vulnerability of disasters are in some cases clearly evident, such as the recent floods in the Midwest and coastal communities. Chapters in this volume test the relations among these factors, which have implications for planning decisions and policy for both inland and coastal regions. In other cases, the implications are more widely diffuse and long term, but are certainly just as important. The difficulty of coping with the gradual event of sea level rise is addressed in this volume.
   c. *Governance and policy perspectives*: The literature on disaster mitigation, preparedness, recovery, and economic development casts the challenges for communities as creating an integrated disaster preparedness and development efforts. This refocuses attention away from a "silo" approach to a "collaboration" approach in creating disaster resilient and sustainable communities. These perspectives will lead to the development of strategies for improved management in the mitigation, preparation, response, and recovery to/from natural and man-made hazards.

Key Research Questions

The following questions are not intended to be exhaustive of the propositions examined in this volume, but indicate some basic questions that we seek to investigate with the contributing scholars in the field.

1. How can disasters and development be addressed in an integrated manner?
2. How can the concept of resiliency be operationalized/used in a way that is useful as a framework to investigate the conditions that lead to stronger, safer, and more sustainable communities?
3. What explains the resiliency of communities and regions? In other words, what factors account for the variation across jurisdictions and geographic units in the ability to respond and recover from a disaster?
4. What are the various policy interventions and governance mechanisms that can be developed to improve the resiliency and sustainability of communities and reduce their vulnerability to natural disasters?
5. How disaster and development strategies conceptualized, operationalized, and implemented in different parts of the world?

6. How certain catastrophic disasters, as focusing events, impacted policies and practice disaster management and development?
7. What are some of the key differences between developing and developed countries in respect to disaster and development?
8. How does the disaster recovery process impact the social, political, and economic institutions of the disaster stricken communities?
9. How the disaster impacted communities collaborate with multiple stakeholders (local, state, international) during the transition from recovery to development/redevelopment?
10. Can the participatory/collaborative nature of disaster recovery help build resilient communities?

The primary audiences of this book are scholars in emergency and crisis management, planning and policy, disaster response and recovery, and environmental management and policy. This book can also be used as a textbook in graduate and advanced undergraduate programs/courses on disaster management, disaster studies, emergency and crisis management, environmental policy and management, and public policy and administration.

The book includes three parts. Each part has chapters with conceptual and theoretical foci as well as cases. After reading the introductory chapter, each chapter in the book can be read independently. The book has several cases from established disaster management systems around the world as well as emerging disaster management systems. The book covers disaster recovery, disaster resiliency, and sustainability within several chapters. Economic growth aspect was not covered in great detail in the book. The book provides usable knowledge for scholars as well as practitioners. Several of the chapters in the book were presented at the 2013 conference of American Society for Public Administration (ASPA) in New Orleans, USA. We acknowledge guidance and assistance of Thomas A. Birkland, Environmental and Hazards series, and Fritz Schmuhl, Springer editor. We also acknowledge assistance from Judith Terpos during the production phase of the book. The chapters have been blind reviewed by several experts in the field. We acknowledge their contributions to the book as well.

Contents

Contributors

Osman Alhassan Institute of African Studies, University of Ghana, Legon, Accra, Ghana

Simon A. Andrew University of North Texas, Denton, USA

Sudha Arlikatti University of North Texas, Denton, USA

Jesse Sey Ayivor Institute for Environmental and Sanitation Studies, University of Ghana, Legon, Accra, Ghana

Thomas A. Birkland School of Public and International Affairs (SPIA), North Carolina State University, Raleigh, USA

Samuel D. Brody Departments of Marine Sciences, Landscape Architecture and Urban Planning, Texas A&M University, Galveston, Texas, USA

Ralph S. Brower Florida State University, Florida, USA

Kiki Caruson Department of Government and International Affairs, University of South Florida, Tampa, Florida, USA

Louise K. Comfort University of Pittsburgh, Pittsburgh, PA, USA

Fatih Demiroz University of Central Florida, Orlando, USA

Janet Dilling Florida State University, Florida, USA

Frances L. Edwards Mineta Transportation Institute, San Jose State University, San Jose, USA

Fikret Elma Department of Public Administration, Celal Bayar University, Manisa, Turkey

Robin Ersing School of Public Affairs, University of South Florida, Tampa, Florida, USA

Melanie Gall Department of Geography, University of South Carolina, Columbia, SC, USA

N. Emel Ganapati Florida International University, Miami, FL, USA

Vener Garayev Department of Political Science and Public Administration, Gediz University, Izmir, Turkey

Brian J. Gerber School of Public Affairs, University of Colorado Denver, Denver, CO, USA

Daniel C. Goodrich Mineta Transportation Institute, San Jose State University, San Jose, USA

Thomas W. Haase Department of Political Studies and Public Administration, American University of Beirut, Beirut, Lebanon

Christopher Hawkins School of Public Administration, University of Central Florida, Orlando, USA

Qian Hu University of Central Florida, Orlando, FL, USA

Alessandra Jerolleman University of New Orleans, Natural Hazard Mitigation Association, Metairie, USA

David Johnston Joint Centre for Disaster Research, Massey University, Palmerston North, New Zealand; GNS Science, Lower Hutt, New Zealand

Naim Kapucu School of Public Administration, University of Central Florida, Orlando, USA

University of Central Florida (UCF), Orlando, FL, USA

Christine M. Kenney Joint Centre for Disaster Research, Massey University, Palmerston North, New Zealand

Sana Khosa University of Central Florida, Orlando, USA

John J. Kiefer Department of Political Science, University of New Orleans, New Orleans, USA

Claire Connolly Knox School of Public Administration, University of Central Florida, Orlando, USA

Kuotsai Tom Liou School of Public Administration, University of Central Florida, Orlando, USA

University of Central Florida (UCF), Orlando, FL, USA

Cathy Yang Liu Georgia State University, Atlanta, GA, USA

Xiaoli Lu Center for Crisis Management Research, School of Public Policy and Management, Tsinghua University, Beijing, China

Francisco A. Magno De La Salle University, Manila, Philippines

Ljubica Mamula-Seadon University of Auckland, Auckland, New Zealand

Rejina Manandhar Department of Public Administration, University of North Texas, Denton, USA

David A. McEntire Department of Public Administration, University of North Texas, Denton, USA

Daniel Nohrstedt Department of Government, Uppsala University, Uppsala, Sweden

Charles Parker Department of Government, Uppsala University, Uppsala, Sweden

Douglas Paton School of Psychology, University of Tasmania, Tasmania, Australia

Joint Centre for Disaster Research, Massey University, Palmerston North, New Zealand

Abdul Akeem Sadiq Indiana University Purdue University Indianapolis (IUPUI), Indianapolis, USA

Steve Scheinert University of Vermont, Burlington, VT, USA

Rajib Shaw Graduate School of Global Environmental Studies of Kyoto University, Kyoto, Japan

Megan K. Warnement North Carolina State University, Raleigh, USA

William L. Waugh Georgia State University, Atlanta, GA, USA

Ping Xu Department of Political Science, Gender and Women's Studies, University of Rhode Island, Kingston, USA

Dong Keun Yoon School of Urban and Environmental Engineering, Ulsan National Institute of Science and Technology (UNIST), Ulsan, South Korea

Huan Zhang School of Social Development and Public Policy, Beijing Normal University, Beijing, China

Kelvin Zuo Department of Civil and Environmental Engineering, University of Auckland, Auckland, New Zealand

Chapter 1
Disasters and Development: Investigating an Integrated Framework

Naim Kapucu and Kuotsai Tom Liou

1.1 Introduction

Disaster management and economic development have become two major public policies for many developed and developing countries in recent years. To achieve goals of economic development, policymakers and public managers have to design different development policies and programs to seek opportunities for business formation and industry development and to address problems of economic cycles and recessions and consequences of the special financial crisis. While focusing on development goals, the leaders in these countries have to adjust their policies priorities and rearrange valuable resources to deal with occurrences and challenges of a variety of natural, man-made, and technological disasters, which are directly or indirectly related to economic development. For example, the four hurricanes that damaged portions of Florida in 2004, Hurricane Katrina in 2005, the earthquake in Haiti in 2010, the earthquake and resultant tsunami and nuclear power plant accident that struck Japan in 2011, the China's Sichuan earthquake in 2008 (Wenchuan) and again in 2013 (Ya'an) provide unfortunate reminders of the vulnerability of communities to natural disasters. These unfortunate events, like many others, illustrate how disasters impact individuals and communities and affected social-technical systems and economic functions and community lives.

While the impact of disasters on development has been recognized, the complex relationship between disaster and development has not been fully studied by scholars of disaster management and economic development. Disasters and their consequences may produce severe negative effects on economic and social development of communities and interrupt their planned development goals and policies. On the other hand, disaster management, especially disaster recovery, may also provide opportunities for policymakers and community leaders to reconsider their policy priorities and use valuable resources for the consideration of sustainable economic development. The relationship between disaster and development is a dynamic one

N. Kapucu (✉) · K. T. Liou
University of Central Florida (UCF), Orlando, FL, USA
e-mail: kapucu@ucf.edu

N. Kapucu, K. T. Liou (eds.), *Disaster and Development,* Environmental Hazards,   1
DOI 10.1007/978-3-319-04468-2_1, © Springer International Publishing Switzerland 2014

and its implications, positive or negative ones, depends on the unique social and cultural of local communities and the managerial capacity of their disaster and development.

This book is designed to study the dynamic relationship between disaster and development by examining important theoretical and policy issues and reviewing specific cases or examples in different countries. The introductory chapter first introduces major theoretical concerns in the literature of disaster and crisis management and economic development and then offers a conceptual framework to examine factors contributing to disaster recovery and sustainable economic development.

1.2   Conceptual and Theoretical Concerns

In the field of emergency and crisis management, too much emphasis is placed on response to disasters. Research on the economic impacts of disasters and disaster recovery is limited (Chang 1984; Kapucu and Ozerdem 2012; Miller and Rivera 2011; Phillips and Neal 2007). This is due to the fact that researchers and practitioners alike "in disaster and/or development must contend just not with varying disciplinary and political perspectives but also with the tension between academic endeavor and practitioner-led interests" (Fordham 2006, p. 341). In disaster and development, interdisciplinary and inter-agency boundaries are crossed and initiatives to introduce early forms of mitigation tend to be seen as a luxury not a necessity. To add to the problem, not only is the discipline vast but, as noted by Fordham (2006), early research on the subject was carried out due to an environmental divide creating "dissimilarity in subject matter and literature between disaster researchers and development researchers" (p. 337).

Looking at disasters and development as a whole, one must consider the role essential development can play in managing disasters as they affect "society's capacity—both in preparedness and recovery" (Tran et al. 2009, p. 404). Many factors can impede the process of integrating disasters and development. Rapid growth and poor re-development can create challenges and increase disaster risk in vulnerable communities. These issues create a need for a shift from perfecting response and recovery to looking at the benefits of disaster development in the early stages of mitigation and preparedness. Risk reduction and environmental management, in this sense, play a key role in developing policies that sustain communities (Collins 2009; Tran et al. 2009).

1.2.1   Disasters

With disasters on the rise and surmounting threats and impacts of climate change the urgent need presented is how to properly address the stresses and challenges disasters place upon a society. Sanker and Herath (2009) found that there is a relationship between a country's level of vulnerability and how severely it is impacted by an event, revealing the need for strengthening disaster risk reduction. The level of vulnerability is not only determined by the seismic and geological risks and threats in regions but can be exacerbated by the built environment of a community. To explain this Sanker and Herath state, "[i]t is quite evident that the disasters are not 'natural' but disasters are the final effect of the collision of 'natural hazards' with 'vulnerabilities' and 'exposures'" (p. 139). With this information one can then visualize a picture of the built environment constantly under threat from various hazards, both natural and manmade. Thus, as Wisner (2003) puts it, disaster can be equated with failure of human development.

Risk reduction measures and practices are integral for improved disaster response and recovery and overall mitigation of vulnerabilities and risks. Underdeveloped communities increase their susceptibility to hazards and must consider implementing disaster reduction strategies that create "forward-looking policies pertaining to social development and equity, economic growth, justice and environmental quality;" otherwise risks are heightened (Ahrens and Rudolph 2006, p. 208). On the other hand disasters present the opportunity to potentially alter human behavior. Thus, as emphasized earlier, vulnerability to hazards occurs not only as the result of geological conditions but also because of the actions of humans living within a particular environment (Lindell et al. 1997). In that sense, individuals' and local governments' assessment of own vulnerabilities is critical for reducing risk and "determining what mitigation practices can be implemented" (UNISDR 2012, p. 5). Understanding disasters as a two way process, "nonstructural" mitigation, which is the process of amending individual actions, can empower communities to understand the risks they accept regarding land use choices (Kendra and Wachtendorf 2006). These actions will help to create a participatory environment in disaster and mitigation development once individuals are able to identify, in a new light, the risks within their community (Godschalk et al. 1999; Norris et al. 2008; Wisner et al. 2004).

1.2.2 Development

Development is defined by Blakely and Leigh (2010) as local economic development that "is achieved when a community's standard of living can be preserved and increased through a process of human and physical development that is based on principles of equity and sustainability" (p. 75). Unlike disaster responses, development tends to be "forward-focused" (p. 335), reaching for the attainment of long term goals and set on advancing the economic and social environment (Fordham 2006). The relationship between disaster and development is important from the perspective of sustainable development.

The term sustainable development has been referred to as "development that meets the needs of the present without compromising the ability of future generations to meet their own needs" (WCED 1987, p. 43). It has also been defined as "improving the quality of human life while living within the carrying capacity of supporting ecosystems" (IUCN/UNEP/WWF 1991, p. 211). The former definition emphasizes equity issues between the present and future generations and the latter definition addresses the balance between economic development and environment protection. The interest in sustainable development has to do with concerns about negative consequences of environmental and equity problems of economic development. Disasters affect sustainable developments in both generation equity and environment protection because disaster outcomes produce negative effects on local natural and environmental conditions and disaster policies and program affect not only recovery efforts for the current generation, but also development opportunities for future generations. Sustainable development and related quality of life concerns have become major goals of development policies and programs for many local governments (e.g., Greenwood and Holt 2010).

1.2.3   Disaster and Development

Recognizing the importance of disaster and development, researchers have tried to examine key issues or topic about the relationship between disaster and development studies. For example, researchers of the United Nations International Strategy for Disaster Reduction (UNISDR) (Fordham 2006; Tran et al. 2009; UNISDR 2012) have identified three key variables of development in risk reduction and disaster sustainability and stressed the importance of "disaster reduction, social and economic development, and sympathetic environmental management" (Fordham 2006, p. 340). Tran et al. (2009) stress "developmental management, environmental management and disaster risk management" (p. 409) as focus points for sustainable disaster management. In his review of disaster research between 1977 and 1997, Alexander (1997) highlights the same issue and concludes that analysis of disasters should be multi-disciplinary, sophisticated, and comprehensive in terms of study context. These authors all emphasize the importance of managing sustainable development with disaster recovery and management.

Unsustainable development strategies are key reasons for increased cost of disasters (FEMA 2000). Impact of climate change and global environmental challenges necessitate a perspective that integrates disaster and development (Pelling 2003). A need for integrating perspective was also recognized by the United Nations *Millennium Development Goals* by placing sustainable development as critical for disaster risk reduction (UN 2013). To support the connection between disaster and development, researchers (e.g., Kasemir et al. 2003; Paterson 2006) also stress the importance of creating participatory and collaborative processes, allowing civil society to engage in and take responsibility in disaster risk management and sustainable development. For example, Kasemir et al. (2003) promote the participation of citizens and other stakeholders for policy making in sustainable development. They suggest consultation procedures to integrate technical scientific modeling with democratic decision-making processes. Similar concerns are also highlighted by Paterson (2006) in regard to disaster policy.

In order to make progress in the development stages of disaster reduction, interdisciplinary studies must occur to develop better risk reduction strategies and until policies reflect this need, communities will potentially remain vulnerable and weak (Alabaster 2011; Collins 2009; Paterson 2006). It is important to create a participatory processes, allowing civil society to engage in and take responsibility in disaster risk management. This point is highlighted by Paterson (2006) in regard to disaster policy stating that eventually, "law's contribution will be in terms of providing and guaranteeing the processes by which a wide range of actors (state, private sector, and civil society) interact in both the technical and policy aspects of disaster risk management" (p. 73). Participatory actions provide opportunities for a wide range of actors to interact and share knowledge. This process, it is hoped, will bring about connections between environmental knowledge and risk reduction. An example used by Paterson (2006) to highlight this relationship is the need for experts to share earthquake predications in order to inform policy makers of the need to integrate risk reduction.

The holistic approach to integrating disaster with development has been emphasized and practiced by policymakers and public managers of local communities. For example the city of Greensburg Kansas has incorporated multiple aspects into their Sustainable Comprehensive Plan. To complement future rebuilding twelve goals were created to address the "built environment, hazard mitigation, economic development, resource management, housing, transportation, infrastructure, parks and green corridors, and future land use" (Berkebile and Hardy 2010, p. 38). This effort reflects an opportunity presented in the pre-planning stages of disaster management to incorporate not only geological hazards but also social and economic features. The relationship between disaster and development will be further examined in the following section with explanation the integrated framework.

1.3 Integrated Disaster and Development Framework

The idea of integrating disaster and development to understand their relationship is relatively new. The interdependent (and blurred) relationship between disasters and development is extremely complex and requires new strategies to address both (Manyena 2012; UNISDR 2013; UNDP 2004; McEntire et al. 2002; Mileti 1999). Disaster recovery efforts should include sustainable development perspectives with an aim of vulnerability reduction (Bankoff et al. 2004; Berke et al. 1993; McEntire 2004; Stenchion 1997). Disaster and development concepts were not used together with the hazard perspectives before the 1970s. Vulnerability reduction perspective gained some influence in the field during 1980s and 1990s when vulnerability perspectives and disaster development were used together (Manyena 2012). Similar to disaster vulnerability reduction perspectives, disaster resiliency perspectives that focus on sustainable development and disaster risk reduction have also been emphasized in recent years i (Kapucu and Ozerdem 2012; Miller and Rivera 2011).

No one wants any disaster to occur in their communities. However, when they occur, disasters might create opportunity to improve community conditions, reduce risk, and create new economic and community development options (Skidmore and Toya 2002; Waugh and Smith 2006). Webb et al. (2002), in their extensive re-

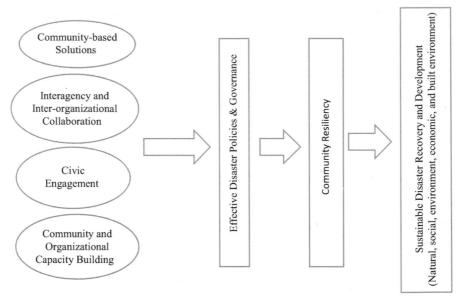

Fig. 1.1 Integrated framework for disaster and development

search on business recovery were unable to find a significant relationship between disasters and long-term recovery. However, they were able to identify a relationship between disasters and short-term recovery and immediate post impact. They also found that the type of business significantly impacts the long-term recovery. The ability of quick adaptation for small business and nonprofit organizations is critical for sustainable economic and community recovery (Boin et al. 2010; Bruneau et al. 2003). Other positive and negative influences included in the study were: age of the business (negative), duration of closure (negative), financial condition of the business (negative), primary market, and business climate (positive). Post-disaster research mostly shows that small businesses and nonprofits are not able to recover, compete, or survive after disasters (Alesch et al. 2001; Ingram et al. 2006; Simo and Bies 2007) since they need assistance from government for long-term recovery.

As depicted in Fig. 1.1, we propose that community based partnerships, networks, social capital with effective disaster risk reduction policies and strategies are important factors in supporting community disaster resiliency and sustainable community recovery. These factors will be briefly reviewed here and will be further examined in different chapters throughout the book. The success of the integrated model is based on such factors as effective disaster policies and governance, community resilience, collaborative capacity, civic engagement and participation, the support of nonprofit and civil society organization.

1.3.1 Disaster Recovery and Sustainable Development

Disasters can be seen as a "window of opportunity" for development and well-planned development can certainly reduce impacts of disasters. However, disasters were not recognized as a potential for development and development agencies were not involved in disaster management (Cuny 1983). This has been the case for both domestic and international development. Literature suggests mixed results on the relationship between disasters and economic recovery and development. Natural disasters might have long term positive impact on human, social, and physical capitals as well as productivity (Chang 1984; Skidmore and Toya 2002; Waugh and Smith 2006). Laura Reese (2006) claims that many of the lessons of disaster response and recovery can be applied to economic recovery as well. Reese indicates important characteristics for an effective disaster recovery and redevelopment, which include commitment, cooperation, creativity, inclusivity, and flexibility (Reese 2006).

1.3.2 Policies and Governance for Hazards and Vulnerability Reduction and Resiliency

Vulnerability and hazards over the past few decades have increased due to poor development practices and policies that lack sustainable outcomes and infrastructure. As a result counties and communities find themselves struggling in the recovery stage to provide both financial and physical resources after a disaster. Using forward thinking development practices implemented after disasters may provide the opportunity and resources needed to enhances sustainable community development, mitigation efforts and decrease future recovery costs (Birkland 1998; FEMA 2000).

Considering the aspect of human involvement and the role individuals play in hazard vulnerability. Kusenbach and Christmann (2013) define vulnerability in a way that incorporates these aspects. The important concept is that an individual's perception of vulnerability is affected by the environment they are in, their ideals and the concerns of others that make up their community provided that individuals do have a choice regarding the level of exposure they choose to live within. When this is considered light is shed on a potential aspect of vulnerability which can then be used to hinder or promote development practices (which include mitigation and the use of vulnerability assessment factors).

A crisis such as a major disaster can become the opportunity for community wide sustainable development. One attainable feature of sustainable development is creating resiliency in the face of catastrophic events (FEMA 2000). It is known that the effects of disasters are linked to poverty and previous degradation of natural land (Alexander 1997). These concepts when considered together highlight the need to create resiliency before a disaster as a way of reducing community vulnerabilities and subsequently, recovery costs after a disaster. There is a dire need for the creation and implementation of structured policies in response to or as mitigative efforts to disasters which regard the physical and natural environment as well as the economic environment. Such policies need to also withstand the pressures of politics at all levels of government.

Resiliency can be created by communities, individuals, and organizations through the execution of choice and action. May (2013) notes that there are many reasons that facilitate the participation and creation of policies in the wake of disasters that can lead to more resilient communities. He discusses two important concepts that can lead to the creation of participation in resiliency efforts. The first is enacting some form of mandate on state and local governments which focuses on the reduction of hazards; while the second is extending planning efforts to identify common recovery goals in partnerships with all levels of the community.

1.3.3   Collaborative Capacity for Development After Disasters

As more citizen service delivery moves to the private sector, it is clear that there is an increased need for collaboration across sectors as well as within organizations and a need for an improvement of current practices. Partnerships, even an unlikely unification of people or services, can enhance response and recovery operations after a disaster through the increase outreach creates. It is said that even the most unlikely partnerships can be of benefit when disaster strikes (Ansell and Gash 2008; Bryson et al. 2006; Gazley 2013; Gray 2007).

Disasters have no discretion or boundaries associated with their paths of destruction. As a result many organizations, jurisdictions, and agencies need to be involved in addressing the impacts. Looking more closely at the local level the resources within a community can belong to and benefit multiple sectors. Consider the infrastructure that may be damaged, and may be public or privately owned and can benefit the whole community or just a few. Thus, the level of participation needed to effectively recover from a disaster must be considered.

Reese (2006) highlights these concerns and the many different players involved in recovering from the East Grand Forks floods. The processes and policies created and followed by this town proved to be efficient and long lasting. The fact that single fund sources were only a small piece to a larger recovery picture illustrating the critical role each organization had in working together as well as towards common goals. In this case partnerships were needed to reach the citizens were imperative to fostering trust in the process (Reese 2006). Partnerships with private companies, faith based organization, community associations, and even schools and local sports organizations extend the reach of local governments creating collaborative capacity while providing a connection to those citizens who may have otherwise been missed.

1.3.4   Civic Engagement and Participation in Recovery and Development

In a study to determine if citizen's evaluation of the processes of civic participation and engagement were perceived as a positive endeavor in recovery efforts, Kweit and Kweit (2004) highlighted the long term success perceived in citizen participation. They examined the recovery efforts of two cities affected in a flood. One used civic participation fully in recovery efforts and the other relied mostly on the guidance of elected officials with some community meetings and discussion forums. Results showed that civic participation was linked to the overall level of trust found within each community. Trust is a critical factor government officials should invest their time and resources in developing prior to a disaster as it enhances recovery processes and creates resiliency. It is in these times that both officials and citizens are vulnerable to making wrong decisions that affect not only themselves but the community as a whole. In some communities members have little control over events that lead to a disaster, but civic participation helps citizens visualize and create the outcomes of disaster recovery which in turn strengthens the social fabric of a community.

It has been found in previous research as well as the aforementioned study that citizen participation provides improved long term recovery efforts (Kweit and Kweit 2004). Considering the creation of policies in the wake of a disaster, it is identified that engaging citizens in this process promotes the success of policies in two ways. These include the acceptance level citizens have towards the policy as well as participation in shaping the policy which will be unique to their community.

1.3.5   The Support of Nonprofit and Civil Society Organizations for Community Based-Solutions

The nonprofit sector has become highly developed and more active in disaster management in more recent times. This was not a direct strategy of the nation but rather the emergence of a response to specific disasters and cultural changes over the past decades (Eikenberry et al. 2007; Ott and Dicke 2012). This emergence has provided a foundation for nonprofit organizations to succeed in the collaborative efforts needed in response to disasters. Kapucu (2007) states that nonprofit organizations can assist in "local, state and national problems through negotiated efforts or partnerships" (p. 552). This idea can also be extended to the processes of international relief.

Foundations established in communities may also play a unique role in disaster recovery. Reese (2006) notes that foundations can provide resources to programs and efforts that are not able to be funded through the government or other nonprofit organizations. This is due to the fact that foundations are semi-private organizations and tend to have less bureaucracy in place and more discretion over spending in occasions of rare instances such as a disaster.

Until recently international aid came mostly in response to disasters and their effects on societies. There are multiple reasons for focus from international relief to center around processes of response and recovery; these include the policies of particular countries and the desires of donors as funding is more expansive when compared to that of preparedness and mitigation (Alexander 1997). In the early 90s international relief organizations started to recognize the link between disasters and underdeveloped communities or what we can refer to as unsustainable development. The turnaround in thinking lead to consideration of the importance of mitigation activities and implementation prior to disasters (Berke 1995).

Community based solutions in the field of emergency management are needed to fully integrate mitigation and preparedness techniques into the creation of development policies. Edwards (2013) contributes this practice to the environment individuals find themselves within, specifically the built environment, the geographic location and the 'natural systems' of the community. It is in this sense that citizens become vulnerable, due to both individual choices and decision made to protect which actually end of creating additional hazards. The example highlighted by Edwards (2013) is the levee breech of New Orleans during Hurricane Katrina as well as dam failures and seawall hazard displacement. Not only does the environment create a need for community based solutions but social demographics in communities may also create unique situations during and after a disaster. Utilizing an approach to address these issues require the mobilization of resources of the entire community.

Edwards (2013) looks at community based solutions as bottom up approaches to managing disaster. She identifies the community actors or organizations as social capital and the benefits in taking stock of them regarding what is unique and a potential benefit to citizens in the event of a disaster. The benefit of social capital and mobilization of additional resources is important because officials can increase community resources by including these organizations in preparedness and planning efforts. Community-based approaches in turn assist in the success of long term recovery efforts such as displacement issues including those of the special needs population, job placement, medical care, etc. It is not only organizations but also individuals that drive recovery efforts after a disaster. Edwards (2013) notes that during the recovery efforts of Hurricane Katrina there were two unique situations where individuals played a leadership role; one situation where an individual was responsible for the re-opening of a community clinic and one incident where an individual single-handedly reopened a local school system.

1.3.6 Capacity Building for Sustainable Development after Disasters

The capacity of an organization or community to respond to and recover from a disaster can be linked to the concept of disaster and development including the concepts of resiliency, policy creation, and governance. Berkes (2007) illustrates this point through defining resiliency as a community concept in which response to a disaster includes adaption and absorption of the impact. Being able to absorb and adapt to the impact of a disaster would mean that a community has the capacity to deal with the event and is resilient. Capacity building takes this concept a step further and integrates preventative and mitigation efforts through the use of effective and feasible disaster development policies (The National Academies 2012). The capacity building perspective requires the creation of a culture of preparedness at individual, family, community, government, and private and nonprofit sector organizations. Capacity building is directly linked to the improvement of managing disasters in ways that development in not compromised.

Capacity building can also be present in the form of institutional memory. After a disaster strikes individuals within an organization that have had experience have a rare opportunity to capture the interest of their peers and share knowledge thus stimulating ideas and increasing the capacity to adapt. Institutional memory is also a factor in self-organization which can be critical in long term recovery efforts and the formation of an adaptive structure within the community (Berkes 2007).

Capacity building in the form of accumulating human resource capacity and other resources creates a better understanding of the situation when disaster management is considered due to the inter-agency and inter-organizational nature of disaster response and recovery (Kapucu et al. 2013; Manyena 2012). The role of these partners is highlighted through the use of their expertise in pre and post disaster efforts. In the aftermath of a disaster situation it is important for officials to come together to fully recognize and anticipate the needs of the community and identify areas where capacity needs to be developed.

1.4 Conclusion

Integrating disaster and development into a framework is valuable in theory and practice and the understanding of the integrated framework will be useful for both scholars and practitioners. Disasters have the ability to both encourage new opportunities for growth or interrupt current or focused development projects. As a result disasters can be either an opportunity for a community or a loss of previous hard work. It is for these reason that disaster and development must be considered together when creating and/or maintaining policies and development procedures concerning local disasters. In this sense disasters may provide an opportunity to reduce community vulnerability, as well as, decrease disruption to future development through increased preparedness and mitigative efforts.

Sustainable development is a process that can be centrally planned or can be the result of local disaster recovery efforts. This is because disasters often present a unique opportunity for communities to not only rebuild but also improve the functions and infrastructure of their community. The only way to fully manage these large scale recovery efforts is with the integration of multiple organizations, and multiple levels of the government, nonprofit sector and even private entities alongside the inclusion of citizens. Civic participation in recovery efforts and policy building as a form of mitigation efforts increases trust, and as a result resiliency in the face of disasters. The multi-level approach required for disaster development relies not only on the decisions of the multiple organizations involved but also the choices and actions of citizens.

The four phases of disaster management each provide opportunities for development. In the preparedness cycle, the need for civic participation in early phases of development will encourage trust between the government and assisting organizations, and to the multiple citizens it is capable of reaching. During the mitigation cycle, the buy-in created in preparedness activities helps to discover and enhance the needs for action within each community, in turn creating a certain level of community backing for mitigation efforts. In the response phase it is important to call on those unique organizations that reach citizens that average government agencies may not have access to. This enhances information exchange and transparency in critical times of disasters. It is found that in the recovery phase these processes then lead to individual self-organization as well as future participation and acceptance of planning and recovery efforts and overall development of sustainable policies. It is also important to monitor success of policies and governance tools in building resilient and sustainable communities. Focusing on different elements of the framework, the chapters in this book provide additional studies to examine theoretical issues and disaster cases among developed and developing countries and communities about the experience and lessons of integrating disaster and development.

References

Ahrens, J., & Rudolph, P. M. (2006). The importance of governance in risk reduction and disaster management. *Journal of Contingencies & Crisis Management, 14*(4), 207–220.

Alabaster, O. (2011) Earthquake response plan vital: U.N. disaster risk expert. *The Daily Star*, October 24. http://www.dailystar.com.lb/News/Local-News/2011/Oct-24/152026-earthquake-response-plan-vital-un-disaster-risk-expert.ashx#ixzz1s K7h10Mg. Accessed 17 April 2012.

Alesch, D. J., Holly, J. N., Mittler, E., & Nagy, R. (2001). *Organizations at risk: What happens when small businesses and not-for-profits encounter natural disasters*. Fairfax: Public Entity Risk Institute.

Alexander, D. (1997). Study of natural disasters, 1977–1997: Some reflections on a changing field of knowledge. *Disasters, 21*(4), 284–304.

Ansell, C., & Gash, A. (2008). Collaborative governance in theory and practice. *Journal of Public Administration Research and Theory, 18*(4), 543–571.

Bankoff, G., Frerks, G., & Hilhorst, D. (Eds.). (2004). *Mapping vulnerability: Disasters, development, and people*. London: Earthscan.

Berke, P. R. (1995). Natural hazard reduction and sustainable development: A global assessment. *Journal of Planning Literature, 9*(4), 370–382.

Berke, P. R., Kartez, J., & Wenger, D. (1993). Recovery after disaster: Achieving sustainable development. *Disasters, 17*(2), 93–108.

Berkes, F. (2007). Understanding uncertainty and reducing vulnerability: Lessons from resilience thinking. *Natural Hazards, 41*(2), 283–295.

Berkebile, R., & Hardy, S. (2010). Moving beyond recovery: Sustainability in rural America. *National Civic Review,* 99(3), 36–40.

Birkland, T. A. (1998). Focusing events, mobilization, and agenda setting. *Journal of Public Policy, 18*(1), 53–74.

Blakely, E. J., & Leigh, N. G. (2010). *Planning local economic development: Theory and practice* (4th ed.). Thousand Oaks: Sage.

Boin, A., Comfort, L. C., & Demchak, C. C. (2010). The rise of resilience. In L. C. Comfort, A. Boin, C., & C. Demchak (Eds.), *Designing resilience: Preparing for extreme events* (1–12). Pittsburgh: University of Pittsburgh Press.

Bruneau, M., Chang, S., Eguichi, R. T., Lee, G. C., O'Rourke, T. D., Reinhorn, A. M., & von Winterfeldt, D. (2003). A framework to quantitatively assess and enhance the seismic resilience of communities. *Earthquake Spectra, 19*(4), 733–752. doi:10.1193/1.1623497.

Bryson, J. M., Crosby, B. C., & Stone, M. M. (2006). The design and implementation of cross-sector collaborations: Propositions from literature. *Public Administration Review, 66*(S1), 44–55 (Special Issue).

Chang, S. (1984). Do disaster areas benefit from disasters? *Growth and Change, 15*(4), 24–31.

Collins, A. E. (2009). *Disaster and development*. London: Routledge.

Cuny, F. C. (1983). *Disasters and development*. Oxford: Oxford University Press.

Edwards, F. (2013). All hazards, whole community, creating resiliency. In N. Kapucu, C. Hawkins, & F. Rivera (Eds.), *Disaster resiliency: Interdisciplinary perspectives* (pp. 21–47). New York: Routledge.

Eikenberry, A. M., Arroyave, V., & Cooper, T. (2007). Administrative failure and the international NGO response to hurricane Katrina. Special Issue. *Public Administration Review, 67,* 160–170.

Federal Emergency Management Agency (FEMA). (2000). *Planning for a sustainable future: The link between hazard mitigation and livability*. http://www.fema.gov/library/viewRecord. do?id=1541. Accessed 10 Aug 2012.

Fordham, M. (2006). Disaster and development research and practice: A necessary eclecticism? In H. Rodriguez, E. L. Quarantelli, & R. R. Dynes (Eds.), *Handbook of disaster research*. New York: Springer.

Gazley, B. (2013). Building collaborative capacity for collaborative capacity. In N. Kapucu, C. Hawkins, & F. Rivera (Eds.), *Disaster resiliency: Interdisciplinary perspectives* (pp. 84–98). New York: Routledge.

Godschalk, D. R., Beatley, T., Berke, P., Brower, D. J., & Kaiser, E. J. (1999). *Natural hazard mitigation: Recasting disaster policy and planning*. Washington D.C: Island Press.

Gray, B. (2007). The process of partnership construction: Anticipating obstacles and enhancing the likelihood of successful partnerships for sustainable development. In P. Glasbergen, F. Biermann, & A. P. J. Mol (Eds.), *Partnerships, governance and sustainable development: Reflections on theory and practice* (pp. 29–48). Northampton: Edward Elgar.

Greenwood, D. T., & Holt, R. P. F. (2010). *Local economic development in the 21st century: Quality of life and sustainability*. Armonk: M.E. Sharpe, Inc.

Ingram, J. C., Franco, G., Rumbaitis-del Rio, C., & Khazai, B. (2006). Post-disaster recovery dilemmas: Challenges in balancing short-term and long-term needs for vulnerability reduction. *Environmental Science and Policy, 9*(7–8), 607–613.

IUCN/UNEP/WWF. (1991). *Caring for the earth: A strategy for sustainable living*. Gland, Switzerland.

Kapucu, N. (2007). Non-profit response to catastrophic disasters. *Disaster Prevention and Management, 16*(4), 551–561.

Kapucu, N., & Ozerdem, A. (2012). *Managing emergencies and crises*. Boston: Jones & Bartlett Publishers.

Kapucu, N., Hawkins, C., & Rivera, F. (Eds.). (2013). *Disaster resiliency: Interdisciplinary perspectives*. New York: Routledge.

Kasemir, B., Jager, J., Jaeger, C. C., & Gardner, M. T. (2003). *Public participation in sustainability science: A handbook*. Cambridge, UK: Cambridge University Press.

Kendra, J., & Wachtendorf, T. (2006). Community innovations and disasters. In H. Rodriguez, E. L. Quarantelli, & R. R. Dynes (Eds.), *Handbook of disaster research* (pp. 316–334). New York: Springer.

Kusenbach, M., & Christmann, G. (2013). Understanding hurricane vulnerability: Lessons from mobile home communities. In N. Kapucu, C. Hawkins, & F. Rivera (Eds.), *Disaster resiliency: Interdisciplinary perspectives* (pp. 61–83). New York: Routledge.

Kweit, M. G., & Kweit, R. W. (2004). Citizen participation and citizen evaluation in disaster recovery. *American Review of Public Administration, 34*(4), 354–373.

Lindell, M. K., Alesch, D., Bolton, P. A., Greene, M. R., Larson, L. A., Lopes, R., May, P. J., Mulilis, J.-P., Nathe, S., Nigg, J. M., Palm, R., Pate, P., Perry, R. W., Pine, J., Tubbesing, S. K., & Whitney, D. J. (1997). Adoption and implantation of hazard adjustments. *International Journal of Mass Emergencies and Disasters, 15,* 327–453 (Special Issue).

Manyena S. B. (2012). Disaster and development paradigms: Too close for comfort? *Development Policy Review, 30*(3), 327–345.

May, P. (2013). Public risk and disaster resilience: Rethinking public and private sector roles. In N. Kapucu, C. Hawkins, & F. Rivera (Eds.), *Disaster resiliency: Interdisciplinary perspectives* (pp. 126–145). New York: Routledge.

McEntire, D. A. (2004). Development, disasters and vulnerability: A discussion of divergent theories and the need for their integration. *Disaster Prevention and Management, 13*(3), 193–198.

McEntire, D. A., Fuller, C., Johnston, C. W., & Weber, R. (2002). A comparison of disaster paradigms: The search for a holistic policy guide. *Public Administration Review, 62*(3), 267–281.

Mileti, D. S. (1999). *Disasters by design: A reassessment of natural hazards in the United States.* Washington, D.C: Joseph Henry Press.

Miller, D. M. S., & Rivera, J. D. (Eds.). (2011). *Community disaster recovery and resiliency: Exploring global opportunities and challenges.* Boca Raton: CRC Press.

Norris, F. H., Stevens, S. P., Pfefferbaum, B., Wyche, K. F., & Pfefferbaum, R. L. (2008). Community resilience as a metaphor, theory, set of capacities and strategy for disaster readiness. *American Journal of Community Psychology, 41*(1–2), 127–150. doi:10.1007/s10464-007-9156-6.

Ott, J., & Dicke, L. (2012). *The nature of the nonprofit sector.* Boulder: Westview Press.

Paterson, J. (2006). A note on georisk, sustainable development and law. *AIP Conference Proceedings, 825*(1), 67–78.

Pelling, M. (Ed.). (2003). *Natural disasters and development in a globalizing world.* New York: Routledge.

Phillips, B. D., & Neal, M. D. (2007). Recovery. In W. L. Waugh, Jr. & K. Tierney (Eds.), *Emergency management: Principles and practice for local government* (2nd ed., pp. 207–233). Washington DC: ICMA.

Reese, L. A. (2006). Economic versus natural disasters: If detroit had a hurricane .... *Economic Development Quarterly, 20*(3), 219–231.

Sanker, S. & Herath, G. (2009). Macroeconomic management and sustainable development. In R. Shaw & R. R. Krishnamurthy (Eds.), *Disaster management: Global challenges and local solutions* (pp. 135–149). Hyderabad: Universities Press.

Shreib, K., Norris, F. H., & Galea, S. (2010). Measuring capacities for community resilience. *Social Indicators Research, 99*(2), 227–247. doi:10.1007/s11205-010-9576-9.

Simo, G., & Bies, A. (2007). The role of nonprofits in disaster response: An expanded model of cross-sector collaboration. *Public Administration Review, 67*(S1), 125–142.

Skidmore, M., & Toya, H. (2002). Do natural disasters promote long-run growth? *Economic Inquiry, 40,* 664–687.

Stenchion, P. (1997). Development and disaster management. *Australian Journal of Emergency Management, 12*(3), 40–44.

The National Academies. (2012). *Disaster resilience: A national imperative.* Washington, DC: The National Academies Press.

Tran, P., Sonak, S., & Shaw, R. (2009). Disaster, environment and development: Opportunities for integration in Asia-Pacific region. In R. Shaw & R. R. Krishnamurthy (Eds.), *Disaster management: Global challenges and local solutions* (pp. 400–423). Hyderabad: Universities Press.

United Nations (UN). (2013). *The millennium development goals report.* New York: UN.

United Nations Development Programme (UNDP). (2004). *Reducing disaster risk: A challenge for development-a global report.* New York: United Nations.

United Nations International Strategy for Disaster Reduction (UNISDR). (2005). *International day for disaster reduction*, United Nation. http://www.unisdr.org/2005/campaign/2005-iddr. htm. Accessed 15 Feb 2012.
United Nations International Strategy for Disaster Reduction (UNISDR). (2012). *Towards a post-2015 framework for disaster risk reduction*. http://www.unisdr.org/files/25129_posthfaconsultationpaperfinal30.pdf. Accessed 15 Feb 2012.
United Nations International Strategy for Disaster Reduction (UNISDR). (2013). *Synthesis report consultations on a post-2015 framework on disaster risk reduction*. Geneva, UN.
Waugh, W. L. Jr., & Smith, B. R. (2006). Economic development and reconstruction on the Gulf after Katrina. *Economic Development Quarterly, 20*(3):211–218.
Webb, G. R., Tierney, K. J., & Dahlhamer, J. M. (2002). Predicting long-term recovery from disasters: A comparison of the Loma Prieta earthquake and hurricane Andrew. *Environmental Hazards, 4*(1), 45–58.
Wisner, B. (2003). Changes in capitalism and global shifts in the distribution of hazard and vulnerability. In M. Pelling (Ed.), *Natural disasters and development in a globalizing world* (pp. 43–56). New York: Routledge.
Wisner, B., Blaikie, P., Cannon, T., & Davis, I. (2004). *At risk: Natural hazards, people's vulnerability and disasters* (2nd ed.). London: Routledge.
World Commission on Environment and Development (WCED). (1987). *Our common future*. Oxford: Oxford University Press.

Part I
Risk Reduction and Policy Learning

This part of the book includes chapters addressing risks, hazards, vulnerabilities and policies in response to these problems and developing resilient and sustainable communities. Rejina Manandhar and David McEntire recognize disasters are rising in frequency and intensity, and explore the importance of policies that promote improved disaster management. In particular, the chapter suggests that all types of vulnerabilities must be reduced while simultaneously enhancing a broad range of emergency management capabilities. Thomas Birkland and Megan Warnement consider how disasters serve as an important element of the agenda setting process in developing countries. These "focusing events" highlight policy failure, and provide opportunities for change and policy learning. This chapter examines the role of disasters in developing nondemocratic countries to see if the principles of focusing events apply to policy change and resilience the way they do in developed democratic countries. Emel Ganapati presents the disaster management reforms that have been introduced after the Marmara earthquake in Turkey in the context of a development and disaster relationship. Vener Garayev and Fikret Elma examine the emergency management system of Republic of Azerbaijan in light of the 2010 Kura river flood. This chapter contributes to the literature on emergency management systems in post-Soviet countries, specifically focusing on the newly created emergency management system in Azerbaijan. John Kiefer and Alessandra Jerolleman provide an overview and assessment of the progress made in mitigating, preparing for, responding to, and recovering from disasters during the period since Hurricane Katrina struck New Orleans, Louisiana on August 29th, 2005. Christopher Hawkins and Claire C. Knox examine policy changes and disaster management system in Florida. Thomas W. Haase reviews Lebanon's disaster management and sustainable development issues, as well as the historical development of the country's civilian disaster management institutions. D.K. Yoon's chapter examines the impacts of natural disasters on disaster management system and policy in Korea and how disaster management systems and policies have been changed and adapted along with the types of disasters and their consequences to increase communities' capacity to cope with disasters and to reduce disaster risk over time.

Chapter 2
Disasters, Development, and Resilience: Exploring the Need for Comprehensive Vulnerability Management

Rejina Manandhar and David A. McEntire

2.1 Introduction

The United Nation defines disaster as "an event or series of events which gives rise to casualties and/or damage or loss of property, infrastructure, essential services or means of livelihood on a scale which is beyond the normal capacity of the affected community's ability to cope with out aid" (DMTP 1994, p. 12). Disasters are a result of hazards that impact human and built environments, thus suggesting that nature, people, buildings and infrastructure interact in complicated ways. Disasters are therefore serious problems as they create devastating short-term and long-term impacts on a community, nation, or region. These events affect human lives, property, employment, infrastructure, and environment (Seneviratne et al. 2010). Furthermore, disasters tend to overwhelm existing resources which compels the affected community to seek outside assistance (Novick 2005).

Several disasters have generated world-wide attention recently. Some become catastrophic, resulting in unimaginable casualties, deaths, and property damage. The impact of these events expands to a much larger population and geographic area, making the recovery effort very difficult, costly, and time consuming. For instance, the Indian Ocean Tsunami traveled as far as 3,000 miles to Africa, affected more than five nations, and caused about 350,000 deaths making it one of the deadliest disasters in history (Resosudarmo and Athukorala 2005). Almost 6 years after Hurricane Katrina, victims are still struggling to recover from the impact of the disaster. Housing and infrastructure systems have not been fully repaired or replaced. The 2010 Haiti earthquake killed over 222,500 people (UNISDR 2011). Making matters worse, health concerns—such as contaminated water and food sources—caused a cholera outbreak, which further added to the number of people affected by the disaster. The 2011 earthquake, tsunami and nuclear reactor leak in Japan were also noteworthy (See Chap. 22 in this volume). They caused about 19,846 deaths, affected 36,8820 people, and generated damage of about

D. A. McEntire (✉) · R. Manandhar
Department of Public Administration, University of North Texas, Denton, USA
e-mail: mcentire@unt.edu

N. Kapucu, K. T. Liou (eds.), *Disaster and Development,* Environmental Hazards,   19
DOI 10.1007/978-3-319-04468-2_2, © Springer International Publishing Switzerland 2014

210,000 billion US $ (EM-DAT 2012). The social and emotional impact of the disaster is also heartbreaking. Physiological impacts such as anxiety, fear, feeling of helplessness, and post-traumatic stress disorder may affect victims self confidence and attitude towards life.

The effects of disasters are even more widespread than recognized. The 2010 British Petroleum oil spill in Gulf of Mexico contaminated numerous water sources, and affected many bird and marine species. Meanwhile, the rising terrorist threat has jeopardized the very notion of "peace and prosperity," and has resulted in societal breakdown and hostility among various social groups (Harding 2007; Perrow 2008). The increasing animosity between a radical minority of the Muslim community and westerners post 9/11 terrorist attacks seems to confirm this claim.

Most importantly, disasters hinder development (Fordham 2007; Harding 2007; Manyena 2012). The economic and infrastructure damage caused by disasters is often unimaginable and may lead a nation into a recession. After a disaster, the focus will mostly divert away from development projects and towards disaster recovery (like restoring damaged infrastructure and uplifting the lost economic condition). The situation will be even worse in developing countries as they often lack resources (financial, educational, workforce, etc.) to effectively deal with disasters (Kusumasari et al. 2010; Pokhrel et al. 2009; McEntire 2011). Hence, disasters are serious problems. They not only hamper physical well-being and disrupt our social activity, but they also hinder a nation's development process. Deaths, injuries, disruption, population displacement, and ecosystem misbalance all erode social livelihood and hinder sustainable development.

With these in mind, the chapter examines the complicated but yet important relationship between disaster and development through the perspective of vulnerability. The chapter begins by discussing disaster trends worldwide and the anticipated impact of future disasters. It then looks into the role of different physical, social, economical, cultural and political factors that play a vital role in disaster risk reduction. Lastly, the chapter stresses on the need for a new kind of development strategies that reverses the social construction of disaster. The chapter concludes by providing implication to practitioners, and direction on future research.

2.2 Rising Disasters

Researchers claim that the number of disaster is increasing rapidly, and anticipate that future disasters will be larger and more destructive due to significant problems like growing urbanization, population growth, and environmental degradation (Sylves 2008). Although the exact number is controversial (Alexander 2006; Collins 2009), statistical records confirm that disasters are increasing in frequency (Alexander 2006; Bouwer et al. 2007; Collins 2009; Perrow 2008; Seneviratne et al. 2010). According to International Disaster Database, the number of natural disasters reported worldwide has increased from about 50 in 1950 to almost 430 in 2010 (EM-DAT 2012). In 2010 alone, 373 natural disasters killed over 296,800 people, making 2010 the deadliest years in at least two decades (UNISDR 2011). Similarly, the number of technological disaster reported worldwide has continuously increased from about 15–20 in 1950 to approximately 360 in 2007 (EM-DAT 2012). Besides natural and technological disasters, the differences in social, cultural, political and religious ideologies and values have resulted in various mass-protests, terrorist attacks, and civil war. The recently observed Arab Spring—including conflict in Syria, Libya, Tunisia, and Egypt—reflect growing differences among the public and the ruling government.

Concomitant with the increase in disaster occurrence, human losses as well as economic and environmental impact are also escalating. The overall number of people affected by disasters has been growing by 6% each year since 1960 and this number is expected to continue in future (UN 2012). Worldwide, the number of deaths between 1991 and 2000 was 939,007, which increased to 1,313,183 between 2001 and 2010. Meanwhile, people affected by disasters have risen from 2,426,321,086 between 1991 and 2000, to 2,676,416,290 between 2001 and 2010. In 2010 alone, reported deaths and the number of people affected worldwide were 304,476 and 304,387,659 respectively (EM-DAT n.d.). Besides the growing physical loss, the economic loss due to disaster has also increased from 54 billion US $ in 1980 to almost 210 billion US $ in 2011 (EM-DAT/UNISDR n.d.).

At present, environmental degradation also poses a threat for all locations irrespective of their proximity to hazards (Thomalla et al. 2006). Although the actual impact of environmental mismanagement is still unknown, many scholars predict that climate change may exacerbate the severity of future disasters, resulting in super-hurricanes and tropical cyclones, long-term sea level rise, extreme temperatures, long term droughts, high levels of precipitation, and large-scale flooding (Alexander 2006; O'Brien et al. 2006; Seneviratne et al. 2010; Thomalla et al. 2006). Bouwer et al. (2007) note that the global costs of weather-related disasters have also increased from an annual average 8.9 billion US $ between 1977 and 1986 to 45.1 billion US $ between 1997 and 2006. The U.S. National Weather Service data also indicates an increase in weather related fatalities. According to the data, there were 138 fatalities in 2010 which increased to 206 in 2011, making the number well above the 10-year average for heat related fatalities (NWS 2012).

It is true that hazards that lead to disaster cannot be completely avoided (O'Brien et al. 2006). However, better understanding of the causes and effects of disasters can provide helpful insights to deal with them more effectively (McEntire 2005, 2011). Knowledge about disaster management strategies, followed by strong policies that support good practices and lessons learned, can help us manage disasters more successfully (McEntire 2005, 2011; Seneviratne et al. 2010).

2.3 Disaster and Development

Contradictory to the term disaster—which has a negative connotation—development often, but not always, has a positive and progressive implication. Development is equated to economic prosperity, technological advancement, poverty reduction, modern amenities, education, freedom, and perhaps even equality. Although the two terms—disaster and development—tend to divert society towards opposite directions, these two concepts are undoubtedly related to each other. According to Fordham (2007) "Many [scholars] (Collins 2009; Manyena 2012; Schilderman 1993; O'Brien et al. 2006) are now realizing that there is strong connection between disaster and development" (p. 345). The Disaster Management Training Programme (1994), McEntire (2004) and Fordham (2007) indicate four ways in which disaster and development support and conflict each other: (1) development increases vulnerability to disaster, (2) development reduces vulnerability to disaster, (3) disaster sets back development, and (4) disaster provides development opportunities. Despite of the growing acceptance of this close relationship between disaster and development, many challenges like political conflict, lack of coordination, and resource inadequacy makes this integration mere rhetoric (Manyena 2012). Nonetheless, development and disaster management activities should be integrated and go hand-in-hand in order to alleviate future impacts of disasters. Development has both advantages and disadvantages which further affects vulnerability to disasters. "Development investments and projects are almost never risk-neutral; they can either increase or reduce vulnerability" (O'Brien et al. 2006, p. 70).

2.4   Development Increases Vulnerability

Poorly planned development interventions can become a source of hazard which further erases the benefits of development investments (O'Brien et al. 2006; McEntire 2004). For instance, the use of low quality construction materials and poor building techniques during development increases vulnerability to disasters. Likewise, development may also be related to other factors that induce disaster impacts such as population growth, urbanization, and human activities that deteriorate the environment. For instance, the economic prosperity and comfort associated with development attracts numerous people to cities resulting in urbanization, population growth, and high density. All of these factors will intensify disaster losses in future (Alexander 2006; Bouwer et al. 2007; Collins 2009). The existence of slums in some mega-cities like Mumbai, Delhi, and Dhaka also increases the chances of post disaster epidemic. Likewise, deforestation and encroachment of agricultural and forest land promotes urban sprawl, which make disaster response activities complicated (McEntire 2011). The advancement in information technology has increased our efficiency in commercial and communication arenas. At the same time, easy access to technology has increased the vulnerability associated with hackers and terrorist activities (Perrow 2008).

Disaster vulnerability is also growing due to global competition among nations; different countries are constructing high-rise buildings, expanding industries and manufacturing, nuclear energy and weapons. Development in construction technology has also contributed to construction of skyscrapers that are vulnerable to earthquake and terrorist attacks. Industrialization and the creation of new chemicals have increased vulnerability to anthropogenic disasters. Similarly, numerous nuclear plants are present around the world, and can produce harmful radiation that have the potential to cause physical disabilities and impairment in multiple future generations; the ecological and environmental harm of this is unthinkable. According to Perrow (2008) "about one-half of U.S. population lives within 50 miles of a nuclear power plant, and with a large release and the right wind orientation it can create lethal impact to the population" , (p. 739). Many countries are now actively developing nuclear weapons, which can threaten the safety of future generations as well as destabilize international geo-political system.

2.5   Development Decreases Vulnerability

While development may indeed create vulnerability, it may decrease vulnerability as well (Collins 2009; Fordham 2007; McEntire 2004). With the advancement in science and technology, we are now able to identify and address hazards. There is much hardware (Ipad, laptop, radio, remote sensing equipment, unmanned vehicles, etc.) and software (Arc GIS, E-Team, worldwide webs, social networking sites, electronic print medias, etc.) that help in all types of emergency management activities. With the aid of GIS, emergency managers are now able to identify places and population that are in risk and adopt protective measures that reduce their vulnerability. Development has also made it possible to use technologies to make buildings and infrastructures that are more resistant to disasters. For example, in Peru, new building techniques along with locally available materials and locally trained work force were used to build earthquake resistant "quincha" houses (Schilderman 1993).

Similarly, development may also foster social and economic opportunities and promote acceptance of new values and cultures. Economic development activities often focus on poverty reduction through job creation. This may reduce vulnerability to disasters. Increased understanding and acceptance of diverse cultures may potentially lower conflicts among various religious and ethnic groups. Development has also expanded educational horizons; many emergency management professionals now have college degrees (Phillips et al. 2012). In United States alone, there are over 150 universities that provide fully developed college degrees (at the Bachelor, Master and PhD levels) in emergency management (Phillips et al. 2012). This could have a profound impact on the management of disasters.

Development also creates public awareness and may emphasize increased political participation for disaster management. Currently there are more special interest groups that advocate for a safe environment and effective disaster management. In the United States, the realization of a need to centralize and streamline emergency management functions has led to the establishment of the Federal Emergency Management Agency (FEMA) (Phillips et al. 2012). Development has also resulted in globalization of emergency management; there are many international organizations (e.g., International Federation of Red Cross, World Bank) that deal with disasters. The International Association of Emergency Mangers (IAEM) brings emergency managers from all over the world together, and provides a platform to address common disaster concerns. There are also numerous international funds and relief organizations for disaster victims worldwide (Phillips et al. 2012). The United Nation is specifically very active in disaster and development issues and has programs (UNISDR, UNDP) that focus on disaster risk reduction and capacity building in developing nations. All of this may produce positive results for the future of emergency management.

2.6 Disaster Sets Back Development

Development is likewise set back by disasters at times (Manyena 2012; Fordham 2007; McEntire 2004). Disasters have social, economic and physical impacts, and hinder individual and community development in the long run (Kulatunga 2010). Hurricane Mitch caused an extensive damage to Honduras and Nicaragua in 1998, setting development back 20 years (O'Brien et al. 2006, p. 60). Likewise, Harding (2007) discusses the impact of massive human rights violation, wars, and political conflicts on Iraq's developmental activities; the social disruption in Iraq has hindered development activities in the country. When a disaster occurs, the resources allocated for development projects will also be diverted to disaster recovery activities (thereby making the community regressive). This, however, does not mean that resources should be diverted away from disaster recovery activities. Of course, resources are crucial to rebound from disasters. Nonetheless consideration should be on incorporating disaster mitigation strategies during development. This will not only provide a disaster resistant structure, but also lowers the amount of resources that will be used aftermath a disaster. For example, retrofitting buildings during construction increases resistance of the building and reduces vulnerability, which can significantly lower the amount of resources that goes into recovery activities in the future.

2.7 Disaster Provides Development Opportunities

Hazards and disasters often have low salience in the political agenda. Nonetheless, disasters also provide development opportunities for communities and nations (Fordham 2007; McEntire 2004). A major disaster can open a "window of opportunity" and help the issue move into the policy agenda (Prater and Lindell 2000; Sylves 2008). Research shows that many communities participate in better construction practices after disaster (Manyena 2012; Schilderman 1993). In the United States, the government became more alert about terrorist activities after the 9/11 terrorist attack Likewise, the 2011 earthquake and Tsunami in Japan raised attention on potential harmful aspects of nuclear power plants. Similarly, the leadership and collaboration failures exhibited during Katrina (Waugh and Tierney 2007) have forced the government to find ways to foster collaboration among sectors and governmental bodies. The importance of an integrated emergency management approach is highly valued after Hurricane Katrina. Communities with previous disaster experience also tend to adopt building technologies that increases the resistance of the buildings (disaster subculture). Likewise a city that was completely destroyed by a disaster may seek to adopt planning strategies that reduce vulnerability to future disasters. Nonetheless, the lack of resources often acts as a barrier to incorporate effective development practices. This is why despite having many devastating experiences with disaster, many of the developing nations are still ill prepared for future hazards.

2.8   The Need to Integrate Perspectives

As can be seen, development and disasters are closely related. A single framework that integrates the two concerns is not only desirable, but is a necessary step for both effective disaster management and good development practices (Collins 2009; Manyena 2012). The method of integrating the two may vary according to context and situation. Despite possible differences, methods of reducing disasters should be the top priority for everyone (Manyena 2012).

The idea of integrating development and disaster management is not new. Efforts like sustainable development or sustainable hazards mitigation have been advocated earlier. NGOs and international organizations (UN, World Bank, International Federation of Red Cross and Red Crescent Societies, Asian Disaster Reduction Center) have proposed numerous development goals (e.g., Millennium Development Goals, Hyogo Framework for Action, World Disaster Reduction Campaign), implemented various programs (e.g., United Nations International Strategy for Disaster Risk Reduction, ProVention Consortium, The International Strategy for Disaster Reduction), and published many reports (World Disasters Report, UNDP's Disaster and Development Report) that focus on integrating disaster and development. The United Nations carries out many developmental programs in developing nations which seeks to reduce disaster vulnerability through capacity building. Collins (2009) and Mileti (1999) have also published much work on disaster and development, yet, the focus seems to be limited only to sustainable development, which has both strengths and limitations (Afedzie and McEntire 2010; McEntire 2005, 2011). For instance, the sustainable development perspective focuses more on natural disasters and hazard mitigation, but ignores other types of disasters and phases of emergency management (Afedzie and McEntire 2010; McEntire 2005, 2011).

Thus, there is a need of a new kind of development that includes, but goes beyond poverty reduction, economic development, and sustainability. A policy that gives necessary considerations to disaster likelihood and reduces vulnerability to future disasters is required. There should be clear understanding of who is most vulnerable and how the interactions between nature and society shape the underlying factors that contribute to vulnerability (Thomalla et al. 2006). There is also a need for management strategies that are capable of handling disaster simultaneously without interrupting development activities. Schilderman (1993) draws special attention to the discrepancy in the amount of aid spent on disaster and vulnerability reduction; he claims that too much is spent to deal with disaster impacts rather than reducing vulnerability (p. 422). Hence, in order to address disasters effectively, a long term solution is vital, which can only be achieved with a change in policy. For this to occur, community based businesses (CBOs), non-governmental organizations (NGOs), and international non-governmental organizations (INGOs) play an important role in challenging the assumptions underlying current relief and reconstruction work, and most importantly advocating for a more development oriented approach to disaster (Schilderman 1993). Furthermore, all emergency management stakeholders (citizens, local, state, federal bodies, private, public and nonprofit sectors) must have an active role in both formulating and implementing such polices.

2.9   Reversing the Social Construction of Disasters

Disasters have been studied from many approaches. For instance, Fordham (2007) discusses the natural hazards, disaster sociology, and vulnerability perspectives. The natural hazard paradigm is correct to concentrate on probable natural hazards. However, this perspective often ignores man-made hazards (technological, terrorism, HAZMAT) (Afedzie and McEntire 2010; McEntire 2005, McEntire et al. 2010). The disaster sociology approach, on the other hand, looks into disasters through organizational and collective behavior lenses (Fordham 2007). Its solitary focus on groups (group behavior and power structure) has often been a potential limitation as it may inadvertently overlook management needs and priorities.

Vulnerability is another perspective that is widely used in disaster studies. Scholars and practitioners have defined the concept of "vulnerability" in different ways (McEntire 2011, p. 295). Cutter (1996) defines vulnerability as "the likelihood that an individual or group will be exposed to and adversely affected by a hazard" (p. 532). Mustafa refers vulnerability as "a state of defenselessness which renders a community powerless to withstand the debilitating effects of events commonly perceived as disaster or natural hazards" (cited in McEntire 2011, p. 295). Kasperson et al. (2001, p. 5) states that "vulnerability is a function of variability and distribution in physical and socio-economic systems and limited human capacity to cope with accumulating hazards and the socio-economic constraints that limits this capacity" (cited in Ibem 2011, p. 29). Other definitions relate vulnerability with potential for loss, risk, damage, stress, state of powerlessness, insecurity, and defenselessness.

Among the three approaches (natural hazards, disaster sociology, and vulnerability), it is the vulnerability perspective which is considered to be most appropriate to understand hazards and disasters due to its ability to explain physical, social, economic, cultural and political variables related to hazards (McEntire 2005, 2011). For example, Alexander (2006) claims that "vulnerability is greater determinant of disaster risk than hazards themselves" (p. 2). Thomalla et al. (2006) believe that vulnerability is an outcome of complex social, economical and environmental interaction that affects the ability of individuals and communities to prepare for, cope with, and recover from, disasters (p. 43). Vulnerability has also been generally studied from sociological perspective where much focus is given to vulnerability to certain groups based on their physical and socio-economic state.

While this view of vulnerability is undoubtedly accurate, it might be somewhat incomplete. David McEntire has worked continuously on this topic for years, and he views vulnerability from the perspective of both liabilities and capabilities. McEntire et al. (2002) advocated a new disaster paradigm—comprehensive vulnerability management—and suggests that the recommendation provides a better understanding of disaster management than the previous disaster paradigms (e.g., comprehensive emergency management, sustainable hazard mitigation, disaster resilient communities, etc.). He argues that comprehensive vulnerability management incorporates all types of disaster agents (natural and manmade), all factors (physical, social, economical, cultural, and political), all stakeholders (private, public, nonprofit, citizens), all disciplines (sociology, geography, planning, political science engineering, psychology, etc.), and an emphasis on all phases of emergency management (mitigation, preparedness, response and recovery). McEntire (2011) refers to vulnerability as "a measure of proneness along with the ability to withstand or react to adverse consequences" (p. 298). According to him, vulnerability has two components—factors that determine proneness (liabilities) as well as other variables associated with the limited capacity (capabilities) (McEntire 2011).

2.10 Liabilities

The various factors that affect disaster vulnerability include physical, technological, social, economic, cultural, and political variables (McEntire 2011). Physical factors include human settlement patterns, building construction, enforcement of laws and regulations. Technological variables include the use of engineering and the impact of computers and other forms of technology both (positive and negative). The socio-economic factor is comprised of variables like social-status or class and ethnicity (Asian, American, Hispanic, etc.). Cultural factors may take account of nonmaterial components (e.g. beliefs, values, languages, family pattern, and networks) (Kulatunga 2010). Meanwhile, political factors focus on issues of power, authority, policies, political systems, leadership, coordination, and collaboration.

It is these and other factors that have a bearing on development and result in socially constructed disasters (Fordham 2007). For instance, prior demographic shifts over the past few decades indicate that the U.S. population has slowly moved towards the seismically active West Coast as well as vulnerable levees and coastal areas in the South (Perrow 2008). These decisions and activities increase the proximity to hazards, making the population vulnerable. Human activities such as unplanned settlements, deforestation, encroachment of the forest-land for cultivation, and building construction are inducing disaster risk (Pokhrel et al. 2009). However, these patterns may be shifting, and can be reversed with other steps to deal with disasters in an effective manner.

2.11 Capacity

Vulnerability is also related capacity building activities in societies (McEntire 2011). Capacity building is defined as "the creation of an enabling environment with appropriate policy and legal frameworks, institutional development, including community participation, human resources development and strengthening of managerial system" (UNDP as cited in Haigh and Amaratunga 2009, p. 83). Capacity building can either focus on providing basic infrastructure facilities or may involve activities that foster development (Haigh and Amaratunga 2009). Formulating programs and policies is not enough; there should be adequate financial, human and organizational resources to make these efforts achievable. Many international organizations provide financial assistance to help poor nations in their developmental efforts. Nonetheless, these types of international aid and development programs are often criticized for creating "dependency syndrome" (Berke 1995). Hence the focus should be on emphasizing local capacity building and resolving long standing physical, social, economic and political problems, so that the communities can be self-reliant. In terms of disaster management, capacity building efforts must focus on a broad array of mitigation, preparedness, response and recovery activities (see Chap. 10 in this volume).

With these in mind, McEntire (2011) indicates four ways to reduce vulnerability: reducing risk, reducing susceptibility, increasing resistance, and increasing resilience. Reducing risk and susceptibility attempt to reduce liabilities. Meanwhile, increasing resistance and resilience are strategies to build capabilities (Fig. 2.1).

2.12 Reducing Risk

Risk reduction focuses on physical variables related to vulnerability—location of buildings in proximity to hazards. Disaster risk can be reduced by implementing policies that encourage safe building practices that move human settlements away from hazards (where possible). Likewise, enforcing standards and regulation (e.g., building codes, regulations on industrial emissions, land use regulations, etc.) and making regular inspection can significantly lower the risk to population as well as the environment (Perrow 2008; Seneviratne et al. 2010). Public education about hazards, vulnerability and disasters can increase awareness among citizen which may ultimately inhibit precarious human activities like deforestation, encroachment of forest land, violation of laws etc.

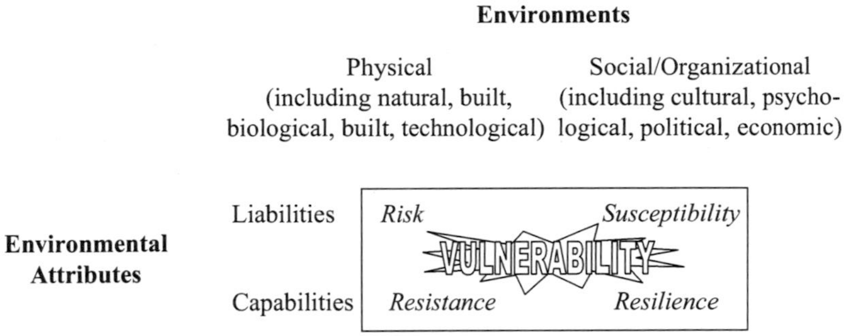

Environments

Physical   Social/Organizational
(including natural, built,   (including cultural, psycho-
biological, built, technological)   logical, political, economic)

Fig. 2.1 Comprehensive vulnerability management. (McEntire 2001)

2.13 Reducing Susceptibility

Unlike risk reduction, the concept of susceptibility looks into social, economic, cultural, and political variables associated with vulnerability (McEntire 2011). Susceptibility also includes the vulnerability of different groups to hazards and disasters. Certain populations like minorities, poor, elderly, immigrants, women, disabled and children may sometimes be vulnerable to disasters as they may lack capabilities to deal with disasters for various reasons (Tobin 1999). Factors such as age and gender can increase vulnerability; the elderly, women and children sometimes have low physical capabilities to deal with disasters. Similarly, women are traditionally viewed as caretakers and spend much time at home, so they might have the additional responsibility of caring for children (in addition to themselves). Social factors like growing economic disparity and inequality in political power also raise the issue of vulnerability among certain groups (like poor and minorities), and act as an obstacle for development (Alexander 2006). Moreover, the World Bank estimates that 20 % of the world's poorest people have some kind of disability (World Bank cited in UN 2006), so policies that bridge the gap between rich and poor are not only desirable, but are a necessity to reduce vulnerability in other ways.

Similar to social and economic variables, cultural values and attitudes can act both as a promoter as well as an obstacle for reducing susceptibility (Kulatunga 2010). Culture provides guidance for individual survival; however, cultural elements also increase vulnerability to disasters. For example, some communities survived the Indian Ocean Tsunami due to their indigenous knowledge regarding Tsunami, but this was not always the case. The Jevanese communities living in Indonesia were subjected to a major threat when they neglected the evacuation order due to their attachment to traditional cultural beliefs. The habitual practice of culture may often result in negative consequences (Kulatunga 2010; Fordham 2007). For instance, the busy lifestyle in the west has resulted in fast-food culture, obesity and massive reliance on big-box stores. These changes and dependencies, along with the rising divorce rate, breakdown of the family and higher dropout rates, may ultimately make society more vulnerable to disaster (McEntire 2011). Thus, a change is culture is important to reduce the potential for disaster (Mileti 1999).

Political factors such as "stove piping" (communicating with others in their own organizations or within the same sector), lack of leadership, poor coordination and communication can also hinder disaster management activities (Kupucu et al. 2010; Phillips et al. 2012; Waugh and Streib 2006). Government lassitude or enthusiasm towards disaster can consequently affect policies and the amount of resources that go into disaster management programs. Similarly, the politicization of disaster events also affects disaster management activities as it diverts the focus from addressing the need of disaster victims to the assignment of blame. Hence addressing social, economic, cultural and political factors will be necessary to reduce vulnerability.

2.14   Increasing Resistance

Building resistance reduces vulnerability by focusing mainly on traditional hazard mitigation approaches. Resistance can be increased by adopting structural mitigation strategies (such as constructing levees, dams, elevating houses in flood prone areas) as well as non-structural mitigation strategies (like careful land use planning, building codes enforcement, and relocating communities where possible). Retrofitting buildings in seismic zones, making wildfire defensible spaces, constructing emergency shelters create an environment that can resists hazard and at the same time with stand harsh disasters. Technology can also play an important role in enhancing these types of mitigation techniques. With the use of new technologies, buildings can use base isolation and vibration damping products that absorb and control seismic motion during earthquakes. Hurricane clips and shutters to strengthen windows and doors further enable houses to be more resistant to high winds. Attention should also be given to eliminating development in hazardous areas, the enforcement of building codes, regular maintenance and building inspection activities. Developing a back-up system for computers can eliminate the threat of losing valuable data after disaster.

2.15   Increasing Resilience

Building resilience is another approach to reduce vulnerability. "Disaster resilience" is commonly referred as the ability to recover or bounce back to normalcy after a disaster (McEntire et al. 2002, p. 269). Nonetheless, many scholars (Berke 1995; McEntire et al. 2002) believe that the focus should be on moving the disaster struck community towards a more stable condition than pre-disaster state. The resilience perspective looks into disasters as anticipated events and emphasizes the act of planning to predict contingencies and reduce the initial shock (Blanke and McGrady 2012, p. 75). Creating hazard mitigation plans, planning evacuation routes, and providing drills and exercises provide emergency management officials and the public guidance on steps to alleviate loss from future disaster. Similarly, allocating sufficient resources for emergency management activities and emphasizing coordination among all levels of government is crucial for increasing disaster resilience (McEntire 2011).

Resilience can also be increased by having pre-existing plans (hazard mitigation plans, emergency operation plans, and recovery plans). However just having a pre-existing plan does not guarantee the safety of the community, it is crucial for local officials to implement those plans, and improvise according to the situation. Particularly for private firms, creating a business continuity and disaster recovery plan (BCDRP) enables corporations to effectively cope with the disaster and at the same time helps to promptly resume normal operations (Blanke and McGrady 2012). For instance, the Volunteering Nursing Association (VNA) headquarter building in Texas was destroyed by fire; however, its business continuity and disaster recovery plan (BCDRP) helped the organization fully recover from the fire, without interruption of its service (Blanke and McGrady 2012). Likewise, the acquisition of grants and the implementation of policies are also crucial in managing disasters. The accessibility to grants and good polices enables disaster stricken community to rebound quickly, effectively and efficiently after disaster.

2.16 Implications

As can be seen, a proactive approach that focuses on future disasters (rather than presuming disasters are low probability events and retaining a reactive stance) should be adopted. Rising disasters indicate that emphasis should be on new and holistic form of development that reduces the potential for disaster and augments capacity (Collins 2009). Manyena (2012) provides an example of Ethiopia's Institutional and Support Project (ISP); the project made efforts to integrate development and disaster concerns by using risk reducing and capacity building strategies. The project adopted an early warning system which made information accessible to remote areas. However, the project lacked effective use of local knowledge and values so the impact was not as extensive as it could be (Manyena 2012).

Thus, it is true that the focus on any one of these approaches—reducing risk, reducing susceptibility, building resistance, and building resilience—can help to reduce vulnerability to some extent. Nevertheless, it is the combination of all four strategies that is essential to address disaster vulnerability in the long term. For example, locating a population further away from ocean shore may decrease the impact of tsunamis. Nonetheless, this action does not lower the need for policies relating to anthropogenic hazards (which may result due to unsafe human practices such as deforestation, failure to follow laws or safety protocols, incomplete planning or incorrect decision making). All of the strategies mentioned here (reducing risk, reducing susceptibility, increasing resistance, and increasing resilience) should thus be integrated together for successful disaster management activities (McEntire 2011).

Scholars and researchers have consistently highlighted the importance of incorporating broad disaster management principles during development activities. However, not enough is known about how the integration can be achieved. For instance, are there communities that are making strides in each area? Why are some jurisdictions and nations more successful than others? What can be done to move all countries and communities forward?

Regardless of our lack of knowledge, emphasis should be on an integration that will reduce vulnerability and that provide development opportunities. Bouwer et al. (2007) suggest that development through implementation of innovative financial mechanism (like catastrophe risk insurance and deficit rainfall insurance) can support disaster reduction. Schilderman (1993) recommends the use of local materials, community involvement and policy integration for disaster and development consolidation. O'Brien et al. (2006, p. 65) argue that long-term development actions can be threatened by environmental concerns, and hence the consensus and planning approaches that link development and disaster should extend in this area.

Of course, it should be recognized that programs that focus on disaster and development are bound to be impacted by various challenges such as political conflict, lack of resources, leadership failures, and coordination issues that create obstacles (Manyena 2012). For example, Harding (2007) posits that the political conflict between the Middle East and the United States has at times hindered developmental work in Iraq. In addition, most of the third world countries are incapable of integrating disaster and development programs due to lack of resources so they must rely heavily on international aid. In Ethiopia, nonprofit organizations did play a major role in initiating and supporting development projects aftermath of a disaster (Manyena 2012). Although this type of NGOs and INGOs' involvement is helpful for development, foreign aid may also result in dependency syndrome (Berke 1995). Hence, future research should focus on successful ways to merge disaster and development activities while also seeking to lower an over-reliance on foreign aid and outside resources.

Likewise, focus should also be given to risk and capability assessments, and applying research findings and technologies (monitoring tools) to enable practitioners to identify vulnerable populations (Collins 2009). For example, foreseeing the possible impact of an event can help the government to allocate resources more carefully which makes response and recovery activities more effective and less problematic. Steps should also be made to streamline the documentation of all types of disasters, and also to make more accurate prediction of environmental degradation and future disasters (Thomalla et al. 2006). Knowledge about the actual impact of climate change on disasters could help practitioners carry out necessary activities to reduce the severity of such events.

Practitioners, on the other hand, should incorporate valuable research findings in their professions. Being better able to determine the hazards and disasters that are specific to the community can help practitioners adopt disaster management activities that are required and realistic. Similarly, understanding best practices related to preparedness, training and exercises can provide practitioners with valuable knowledge that can be utilized in time of disasters. Furthermore, existing research also makes participating organizations more aware about their weaknesses and strengths. The knowledge of local ordinances, federal/state regulations and recommendations on how to work in the political realm is also equally important for practitioners (Waugh and Tierney 2007). Studies likewise reveal that emergency managers should make efforts to bridge the gap between all levels of government (federal, state, and local) and all sectors (private, public, and non-profit), by creating bridges that promote mutual understanding, collaboration, and coordination. Schneider (2002) recommends numerous ways to link hazard mitigation with sustainable community development, all of which are also applicable in integrating development and emergency management systems. For instance, it is important to educate all public managers and planners about the relationship between disaster management and development. Emphasis should be on stressing the value of mitigating hazards during early phases of development. Similarly, policy makers should follow prior research findings to link emergency management systems and priorities with policies related to development. For example, policies on land use planning and transportation should be tied with disaster management activities. Since community based emergency management activities empower the public and enhance their sense of responsibility for disaster management activities (Kulatunga 2010), practitioners should also incorporate citizen inputs during formulation and implementation of various disaster policies and be receptive to people's suggestions and feedback. Likewise, research has revealed that practitioners should consider a vulnerable population as active resources instead of viewing as helpless victims (Berke 1995; Schilderman 1993).

Regardless of diversity of policy recommendations and methods for practical implementation, the integration of disaster and development activities should always consider all types of threats, all phases of emergency management, all variables relating to vulnerability, all actors involved in disasters, and all relevant disciplines (McEntire et al. 2002). Many disaster paradigms (eg. sustainable hazard mitigation, disaster-resistance community, and disaster-resilience community) focus more on natural hazards and ignore other types of threats (technological, civil, terrorist, biological). This should not be the case because most types of hazards can impact a community. Secondly, it is crucial to give attention to all phases of disaster cycle (mitigation, preparedness, response and recovery). Historically, the response phase seems to dominate disaster research and the agenda of practitioners (Phillips et al. 2012). Hence, researchers and practitioners should also consider other phases of emergency management while focusing on proactive, rather than reactive efforts.

It is also vital to take into account the various factors that affect the degree of vulnerability. Communities vary based on their demographics, social and cultural contexts, access to resources and political realities. A hazard that can be particularly deadly to one community might just be a minor event for another jurisdiction. Hence, all types of variables matter and efforts should therefore be made to identify and reduce the broad array of vulnerabilities.

Another consideration should be the involvement of all actors (private, public, nonprofit, public, government bodies and citizens) related to disasters and emergency management. Local officials, community planners and the general population mistakenly assume that emergency management is not their responsibility or concern (Schneider 2002). Similarly, it is a myth that only some organizations (like fire department, police department, EMS) have active role in disaster response. Disaster management problems cannot be solved in isolation (Mileti and Gailus 2005), so it is important to change this kind of mindset. Effective management of disaster is only possible when all sectors, all levels of government and citizens support and contribute their efforts to address hazard and disaster vulnerabilities. This requires a shift in emergency management system (i.e. moving from a command and control system to a more network based system that fosters communication, collaboration and cooperation among all actors) (Kapucu et al. 2010; Waugh and Streib 2006).

Finally, disaster management is and should be multidisciplinary in nature; scholars and practitioners from numerous disciplines (engineering, public administration, geography, climatology, sociology, psychology, economics, city planning, information technology, political science, and anthropology, etc.) must contribute their knowledge and provide valuable insights on disaster management (Phillips et al. 2012). Disaster and development studies must be approached from different perspectives such as public health, environmental management, physical/geography, urban planning, social and behavioral studies, emergency management (Collins 2009). Nonetheless, a holistic understanding of the relationship between disaster and development can only be achieved if it follows an interdisciplinary perspective that moves beyond individual fields of interest.

Hence, there is a need for a new policy that concentrates on reversing the social construction of disasters by assessing all types of liabilities, assessing all types of capabilities, and then closing the gap between the two. A holistic policy that encourages collaboration across hazards, variables, phases, actors and disciplines should be the top priority. Since it is important for researchers and practitioners to view disasters and emergencies from a holistic perspective, the concept of "comprehensive vulnerability management" may prove advantageous.

2.17   Conclusion

Disasters are increasing worldwide and their impact on society is also escalating to new heights. Among the various consequences of disaster, its impact on development is noticeable. Disaster and development are clearly related. Poor development can intensify vulnerability and cause devastating impact, but at the same time wise development practices may reduce vulnerability. Disasters also hinder developmental works, but can concurrently provide development opportunities as well. Hence, there is a need for a new kind of development that reverses the social construction of disaster, and fosters the positive aspects of the "disaster and development" relationship.

Likewise, consideration should also be on reducing all types of vulnerabilities while simultaneously enhancing a broad range of emergency management capabilities. For this to occur, future research should focus on ways to reduce risk and susceptibility, and at the same time find ways to increase resistance and resilience. Concurrently, practitioners should also play an important role for overcoming the challenges associated with integrating disaster and development by focusing on policies that address all hazards, disaster variables, emergency management phases, disaster participants and academic disciplines. By taking a more aggressive and holistic approach we will be better able to reverse the rising trend of disasters.

References

Afedzie, R., & McEntire, D. A. (2010). Rethinking disasters by design. *Disaster Prevention and Management, 19*(1), 48–58.

Alexander, D. (2006). Globalization of disaster: Trends, problems and dilemmas. *Journal of International Affairs-Columbia University, 59*(2), 1.

Berke, P. R. (1995). Natural-hazard reduction and sustainable development: A global assessment. *Journal of Planning Literature, 9*(4), 370–382.

Blanke, S. J., & McGrady, E. (2012). From hot ashes to a cool recovery reducing risk by acting on business continuity and disaster recovery lessons learned. *Home Health Care Management & Practice, 24*(2), 73–80.

Bouwer, L. M., Crompton, R. P., Faust, E., Höppe, P., & Pielke, R. A. Jr. (2007). Confronting disaster losses. *Science-New York then Washington-, 318*(5851), 753.

Collins, A. E. (2009). Disaster and development. *Routledge perspective on development.* Taylor & Francis: New York.

Cutter, S. L. (1996). Vulnerability to environmental hazards. *Progress in Human Geography, 20*(4), 529–539.

Disaster Management Training Programme (DMTP). (1994). *Disasters and Development (2nd Edition).* Retrieved June 20, 2012 from http://www.untj.org/docs/Disaster_Management/Resources%20Page/disaster_development.pdf.

EM-DAT. (n.d.). Disaster Data [Table illustration for Total number of people reported killed and affected by disasters by country and territory (1991 to 2000; 2001 to 2010; and 2010]. *Access from World Disasters Report.2011.* Retrieved from http://www.ifrc.org/PageFiles/89755/Photos/307000-WDR-2011-FINAL-email-1.pdf.

EM-DAT. (2012). Disaster Trends [Data File]. *Disaster Trends Data Available from EM-DAT: The OFDA/CRED International Disaster Database–University of Louvain Belgium.* Retrieved June 23, 2012 from http://www.emdat.be/disaster-trends.

EM-DAT/UNISDR. (n.d.). [Graph illustration for Annual reported economic damages from natural disasters: 1980-2011]. *2011 Disasters in numbers–Disaster profiles Access from the PreventionWeb.* Retrieved June 13, 2012 from http://www.preventionweb.net/files/24697_24692 2011disasterstats1.pdf.

Fordham, M. (2007). Disaster and development research and practice: A necessary eclecticism? In Handbook of disaster research (pp. 335–346). New York: Springer New York.

Haigh, R., & Amaratunga, D. (2009). Capacity building for post disaster infrastructure development and management. *International Journal of Strategic Property Management, 13*(2), 83–86.

Harding, S. (2007). Man-made disaster and development the case of Iraq. *International Social Work, 50*(3), 295–306.

Ibem, E. O. (2011). Challenges of disaster vulnerability reduction in Lagos megacity area, Nigeria. *Disaster Prevention and Management, 20*(1), 27–40.

Kapucu, N., Arslan, T., & Demiroz, F. (2010). Collaborative emergency management and national emergency management network. *Disaster Prevention and Management, 19*(4), 452–468.

Kasperson, R. E., Kasperson, J. X., Dow, K., Ezcurra, E., Liverman, D. M., Mitchell, J. K., Samuel, J. R., O'Riordan, T. & Timmerman, P. (2001). *"Introduction: global environmental risk and society"*, in Kasperson, J. X. and Kasperson, R. E. (Eds), Global Environmental Risk, United Nations University Press, Tokyo, pp.1–48.

Kulatunga, U. (2010). Impact of culture towards disaster risk reduction. *International Journal of Strategic Property Management, 14*(4), 304–313.

Kusumasari, B., Alam, Q., & Siddiqui, K. (2010). Resource capability for local government in managing disaster. *Disaster Prevention and Management, 19*(4), 438–451.

Manyena, S. B. (2012). Disaster and development paradigms: Too close for comfort? *Development Policy Review, 30*(3), 327–345.

McEntire, D. A. (2001). Triggering agents, vulnerabilities and disaster reduction: Towards a holistic paradigm. *Disaster prevention and Management, 10*(3), 189–196.

McEntire, D. A. (2004). Development, disasters and vulnerability: A discussion of divergent theories and the need for their integration. *Disaster Prevention and Management, 13*(3), 193–198.

McEntire, D. A. (2005). The history, meaning and policy recommendations of sustainable development: A review essay. *International Journal of Environment and Sustainable Development, 4*(2), 106–118.

McEntire, D. A. (2011). Understanding and reducing vulnerability: From the approach of liabilities and capabilities. *Disaster Prevention and Management, 20*(3), 294–313.

McEntire, D. A., Fuller, C., Johnston, C. W., & Weber, R. (2002). A comparison of disaster paradigms: The search for a holistic policy guide. *Public Administration Review, 62*(3), 267–281.

McEntire, D. A., Crocker, C. G., & Peters, E. (2010). Addressing vulnerability through an integrated approach. *International Journal of Disaster Resilience in the Built Environment, 1*(1), 50–64.

Mileti, D. S. (1999). *Disasters by design: A reassessment of natural hazards in the United States.* Washington, D.C: Joseph Henry Press.

Mileti, D., & Gailus, J. (2005). Sustainable development and hazards mitigation in the United States: Disasters by design revisited. In C. Haque (Ed.), *Mitigation of natural hazards and disasters: International perspectives* (pp. 159–172). Springer Netherlands.

National Weather Services (NWS). (2012). *2011 Heat Related Fatalities* [Data File]. Retrieved June 21, 2012 from http://www.nws.noaa.gov/os/hazstats/heat11.pdf.

Novick, L. F. (2005). Epidemiologic approaches to disasters: Reducing our vulnerability. *American Journal of Epidemiology, 162*(1), 1–2.

O'Brien, G., O'Keefe, P., Rose, J., & Wisner, B. (2006). Climate change and disaster management. *Disasters, 30*(1), 64–80.

Perrow, C. (2008). Disasters evermore? reducing our vulnerabilities to natural, industrial, and terrorist disasters. *Social Research: An International Quarterly, 75*(3), 733–752.

Phillips, B. D., Neal, D. M., & Webb, G. R. (2012). *Introduction to emergency management.* Boca Raton: CRC Press.

Pokhrel, D., Bhandari, B. S., & Viraraghavan, T. (2009). Natural hazards and environmental implications in Nepal. *Disaster Prevention and Management, 18*(5), 478–489.

Prater, C. S., & Lindell, M. K. (2000). Politics of hazard mitigation. *Natural Hazards Review, 1*(2), 73–82.

Resosudarmo, B. P., & Athukorala, P. (2005). The Indian Ocean Tsunami: Economic impact, disaster management, and lessons. *Asian Economic Papers, 4*(1), 1–39.

Schilderman, T. (1993). Disasters and development. A case study from Peru. *Journal of International Development, 5*(4), 415–423.

Schneider, R. O. (2002). Hazard mitigation and sustainable community development. *Disaster Prevention and Management, 11*(2), 141–147.

Seneviratne, K., Baldry, D., & Pathirage, C. (2010). Disaster knowledge factors in managing disasters successfully. *International Journal of Strategic Property Management, 14*(4), 376–390.

Sylves, R. (2008). *Disaster policy and politics: Emergency management and homeland security* (1st ed.). Washington, DC: CQ Press.

Thomalla, F., Downing, T., Spanger-Siegfried, E., Han, G., & Rockström, J. (2006). Reducing hazard vulnerability: Towards a common approach between disaster risk reduction and climate adaptation. *Disasters, 30*(1), 39–48.

Tobin, G. A. (1999). Sustainability and community resilience: The holy grail of hazards planning? *Global Environmental Change Part B: Environmental Hazards, 1*(1), 13–25.

United Nations (UN). (2006). International Convention on the Rights of Persons with Disabilities. [Facts Sheet]. Retrieved July 10, 2012 from http://www.un.org/disabilities/convention/pdfs/factsheet.pdf.

United Nations (UN). (2012). *Indicators of Sustainable Development: Guidelines and Methodologies (Third Edition)*. Retrieved August 2, 2012 from http://www.un.org/esa/sustdev/natlinfo/indicators/methodology_sheets.pdf.

United Nations International Strategy for Disaster Reduction (UNISDR). (2011). [Press Release, 24 January 2011]. Retrieved June 20, 2012 from http://cred.be/sites/default/files/Press_Release_UNISDR2011_03.pdf.

Waugh, W. L. Jr., & Streib, G. (2006). Collaboration and leadership for effective emergency management. *Public Administration Review, 66,* 131–140.

Waugh, W. L. Jr., & Tierney, K. (Eds.). (2007). *Emergency management: Principles and practice for local government* (2nd ed.). ICMA Press.

Chapter 3
Focusing Events in Disasters and Development

Thomas A. Birkland and Megan K. Warnement

3.1 Focusing Events in Disasters and Development

Disasters, crises and catastrophes are important political events that can significantly alter the course of politics and of public policy. In the United States, events such as the September 11th attacks and Hurricane Katrina led to intensive but short-lived public interest in the wide range of problems highlighted by these events. Similar events, such as the 2004 South Asia Tsunami, or the 2011 Fukushima nuclear disaster, itself triggered by a severely damaging tsunami, attract worldwide attention. These kinds of events often trigger considerable effort to adopt potential solutions to the revealed problems.

Such events are also important to scholars of the public policy process who wish to understand the role of sudden events in the processes by which people and political institutions understand and address the problems revealed by these disasters. While it is likely that political actors have long known that crises provide opportunity for change, the agenda setting literature did not begin to develop a sense of the role of these events until the 1970s. Cobb and Elder spoke of "circumstantial reactors" that fulfill a similar function (1983, p. 83). John Kingdon (2003, in earlier versions of the book) most notably adopted the term "focusing events" to describe events that "focus" attention on problems and solutions. Later, Birkland (1997) took up the question: What factors make an event more or less influential on the agenda? The fundamental question that such studies engage is whether and to what extent policy change is the result of "lessons learned" about policy failures that made the event more damaging than it "should" have been, from a normative perspective.

Most studies of agenda setting have focused on democratic systems in the United States and Europe. While world trends point toward increasing democratization,

T. A. Birkland (✉)
School of Public and International Affairs (SPIA), North Carolina State University, Raleigh, USA
e-mail: tom_birkland@ncsu.edu

M. K. Warnement
North Carolina State University, Raleigh, USA
e-mail: meg.warnement@gmail.com

N. Kapucu, K. T. Liou (eds.), *Disaster and Development,* Environmental Hazards,
DOI 10.1007/978-3-319-04468-2_3, © Springer International Publishing Switzerland 2014

many developing countries were or are characterized by low levels of democracy and low policymaking capacity. In such systems, the politics of disasters will differ from disaster politics in democratic states, as citizens and policy advocates are not able to use the familiar democratic means to mobilize support for better policy.

In this chapter, we use three case studies in this discussion to apply the concept of "focusing events" that have struck three different developing nations: Nicaragua, China, and Haiti. We then raise two central questions: what is the role of a focusing event in a non-democratic society, both in terms of domestic politics and in international development assistance, in improving resilience in nondemocratic contexts?

3.2   An Overview of the Idea of "Focusing Events"

Focusing events are an element of the agenda-setting process, in which some issues gain and others lose attention among policy makers and the public. Agenda setting is interesting for political scientists because "the definition of the alternatives is the supreme instrument of power" (Schattschneider 1975, p. 66), where alternatives can mean issues, events, problems, and solutions. Groups, or agglomerations of such groups, called *advocacy coalitions* (Sabatier et al. 1996; Sabatier and Jenkins-Smith 1999), engage in rhetorical battles, in different venues, to elevate these issues on the agenda while attempting to deny agenda access to other actors (Cobb and Ross 1997). Group competition is fierce because the agenda space is limited by individual and organizational constraints on information processing, so that no system can accommodate all issues and ideas (Walker 1977; Baumgartner and Jones 1993; Cobb and Elder 1983). The competition in agenda setting is not simply about raising an issue on the agenda; it is focused more on propagating the preferred story about how a bad condition came to be (Stone 2002), and a story of how the problem revealed by an event might be prevented or mitigated in the future.

This competition is over both which problems are most important, and over what causes and solutions surround any particular problem (Hilgartner and Bosk 1988; Lawrence and Birkland 2004; Birkland and Lawrence 2009). The agenda setting process is therefore a system of sifting issues, problems and ideas and implicitly assigning priorities to these issues. John Kingdon (2003) argues that agenda change is driven by two broad phenomena: changes in indicators of underlying problems, which lead to debates over whether and to what extent a problem exists and is worthy of action; and *focusing events*, or sudden shocks to policy systems that lead to attention and potential policy change.

In Kingdon's "streams metaphor" of the policy process, three conceptual streams describe the agenda-setting phase of the policy process. These are the *problem* stream, which contains ideas about various problems in society to which public policy might be applied; the *politics* stream, containing the ebb and flow of electoral politics, public opinion, and the like; and the *policy* stream, which contains a set of ideas about how problems could be addressed. In all three streams, problems and solutions, and the means by which to implement solutions to solve problems, are not self-evident, but are subject to considerable debate and conflict. A window of opportunity opens when two or more streams come together at a moment in time where problems are matched with solutions, and where politics aligns in such a way to make this matching more likely. Often, the opening of these windows is driven and promoted by the efforts of policy entrepreneurs to join favorable politics with problem definitions and ideas for solutions. A focusing event, therefore, opens these windows because they provide an urgent, symbol-rich example of what many would argue is obvious policy failure, the result of which is what pro-change forces would argue is the necessity for rapid policy change to prevent the recurrence of the recent disaster. Of course, other interests may argue that an event is atypical, that existing systems can and will bring the acute problems under control, and that sweeping change is not necessary. Regardless of the ultimate outcome, we can argue that pressure for policy change does increase in most focusing events and that debate over ideas increases, both among policy makers and in the attentive public, to the extent that an attentive public exists. Focusing events may trigger greater attention to problems and solutions because they increase the likelihood of more influential and powerful actors entering the conflict on the side of policy change (Schattschneider 1975; Baumgartner and Jones 1993). These actors can use the event to make claims of policy failure and to press for a more active search for solutions, leading to a greater likelihood of policy change (May 1992; Birkland 2006).

John Kingdon used the term "focusing event" within a general discussion of "Focusing Events, Crises, and Symbols" (2003, pp. 94–100). Kingdon calls focusing events a "little push" "like a crisis or disaster that comes along to call attention to the problem, a powerful symbol that catches on, or the personal experience of a policy maker." Kingdon argues that focusing events gain their agenda-setting power by aggregating their harms in one place and time: one plane crash that kills 200 people will get more attention than 200 single fatal accidents; one big terrorist attack in one place will get more attention than several hundred smaller events, particularly if they are far away. Traditionally, sudden events were thought to "simply bowl over everything standing in the way of prominence on the agenda" (Kingdon 2003, p. 96). Birkland's work is based on this "bowling over" effect, and he assumes that the event so dominates the agenda that talk of policy change—that is, learning the "lessons" of the recent event—is the normatively desirable result (Birkland 2006) because the event itself often reveals policy failures from which we normatively expect to learn (May 1992).

Birkland (1997, 1998) refined Kingdon's definition of a focusing event, defining *potential focusing event* as an event that is

> sudden, relatively rare, can be reasonably defined as harmful or revealing the possibility of potentially greater future harms, inflicts harms or suggests potential harms that are or could be concentrated on a definable geographical area or community of interest, and that is known to policy makers and the public virtually simultaneously. (1997, p. 22)

The term *potential* means that an event may be of a type that, based on historical experience, we would believe would gain a great deal of attention. But it is difficult to know in advance how "focal" an event will be. Will the event make a more significant and discernible difference on the agenda than competing events and issues?

Will that event have any discernible influence on policy changes? Will it have a greater influence on the agenda, and on policy, than did similar events in the past?

When we look retrospectively, we can see that an actual focusing event, by definition, is accompanied by a great deal of attention to a public issue or problem. Baumgartner and Jones (1993) note that increased attention to a policy problem is usually *negative* attention, which often motivates further political debate, driving moving issues closer to potential policy changes. However, political elites do not always resist change, even if policy monopolies are reorganized. Focusing events, particularly in policy domains characterized as "policies without publics," (May 1990) yield "internal mobilization" efforts to promote the change that policy elites or experts prefer (Cobb and Elder 1983). Indeed, Best (2010) cites Molotch and Lester's (1975) claim that "actor-promoted events" (APEs, in her term) are more likely to generate attention when the actors are elites with which news media already have steady contacts, and whose actions are considered important by virtue of the actors' position in society. The messages these actors seek to convey are generally pro-status quo, or at least pro-elite, to the extent that elites sometimes desire policy change, particularly in "policies without publics" (May 1990). These findings may be relevant to studies of nondemocratic developing states, in which decision making is largely confined to elites with little public participation. Policy change, under such conditions, are often characterized by "internal mobilization" or by technocracy, which has implications for the nature and substance of policies enacted (or not enacted, as we will see) to address the most recent crisis.

3.3   Disasters as Focusing Events in Non-Democratic Systems: Three Cases

We assume that, in the regular functioning of focusing-event politics in democracies, focusing events are able to attract attention (that is, the event is not secret), they are able to gain critical or negative attention with respect to the status quo (interest groups and a free press are able to operate), and there are consequences for unpopular decision making, such as failure to win reelection. Furthermore, we assume that voters expect their leaders to be problem solvers who engage in debate over the nature of problems and their potential solutions. We also assume that a nation has a functioning state administrative apparatus to react to events and to manage policy change. And we assume that political debate is free, open, and pluralistic.

But the history of development in much of the world has also been a history of varying degrees of autocracy and suppression of democratic movements. We have selected three countries that experienced damaging earthquakes to illustrate the role of focusing events in what we consider "developing countries." The term is, of course, controversial, but we define developing countries in two ways. First, we define a developing country as being at or below the level of "moderate" human development according to the UN's Human Development Index (HDI) as of 2010. Second, we consider a country to be developing if it has only begun, or has not yet made, the transition to democracy or full democracy as defined by the Polity IV dataset, a project headed by Monty Marshall and Keith Jaggers at the University of Colorado. Their website, http://www.systemicpeace.org/polity/polity4.htm, is our source for our data on democratization, and we will refer to and quote their country reports in this chapter; the reader is directed to this web site to view these materials.

The Polity IV dataset provides a measure of the degree to which a state is democratic, autocratic, or mixed ("anocracy"), on a scale ranging from -10 (fully autocratic) to $+10$ (fully democratic). The data have been collected since 1980, and the Polity scores calculated back to 1946, allowing for comparative and time-based comparisons of democratic development. To get a sense of the scaling, the United States has a Polity IV score of 10, and has had this score since 1946; by contrast, North Korea has had a score of -9 since the early 1970s, and has rated no higher than -6, suggesting a lengthy history of repression. States with negative polity scores are, we argue, less likely to see focusing events as important to normal political discourse.

We selected three case studies, all earthquakes, which illustrate the workings of focusing events in developing countries.

- *The December 23, 1972 Nicaragua earthquake*, an M6.2[1] event which severely damaged Managua, the capital city. Recovery was greatly slowed by the corruption of the authoritarian regime of dictator Anastasio Somoza Debayle's junta. Much of the external aid delivered to the country was spent corruptly or stolen, and downtown Managua was never effectively rebuilt. Nicaragua's polity score was -8 until the Sandinista revolution of 1979, at which time the polity score began to move more toward democracy, eventually settling at $+9$ in the late 2000s.
- *The May 12, 2008 Sichuan earthquake in China,* M7.9, which revealed crucial shortcomings in the construction of schools, in particular. This led to a period of popular protest and state repression of that protest, but it also appears to have elevated earthquake safety and seismic construction standards to the policy elite's agenda. China has long been an authoritarian regime, with a current Polity score of -7, which has held steady since 1975, but which has been as low as -9, during the most repressive days post-Mao.
- *The 10 January 2010 Haiti earthquake*, a M7.0 event that killed as many as 300,000 Haitians, and which did catastrophic damage to the capital, Port au Prince, in the poorest nation in the Western Hemisphere. Haiti's Polity score reflects its severely repressive nature during the worst years of the Jean Claude "Baby Doc" Duvalier dictatorship, falling as far as -10 in the mid-1970s. Since the end of the Duvalier regime, Haiti has been characterized by considerable political instability, with the Polity score now at about $+5$ as democracy becomes more institutionalized. Still, Haiti remains vulnerable to extralegal regime change, and this instability, combined with the overall poverty of the nation, has made developing an effective state administrative structure extremely difficult, despite years of international assistance. As we note later, the degree of political repression in Haiti is far less important in its recovery from the earthquake than is the near total absence of a functioning state administration.

---

[1] Any references to the magnitude of earthquakes in this chapter is to the Moment Magnitude Scale (M or M_w), which is a refinement of the largely replaced Richter scale.

To what extent did these disasters yield increased attention to the earthquake hazard, understandings of policy failure, and attempts to learn and to change policy? How do the political systems in these countries shape the role of these events? These case studies will show that focusing events can be important, but not always in the way that theories about democratic systems would expect.

3.4   The Nicaragua Earthquake

The Nicaragua earthquake struck on December 23, 1972, at about 12:30 am local time, when many people were at home and asleep. It killed 5,000 people and severely damaged the national economy. This earthquake gained particular attention in the United States because a widely admired baseball player, Roberto Clemente, was killed when the plane he chartered to bring aid supplies soon after the quake crashed after takeoff from Miami (Maraniss 2006).

Nicaragua at that time was ruled by Anastasio Somoza Debayle, whose family led a military dictatorship for 49 years, ending with the "Sandinista" revolution of 1979. From 1979 to about 1983, Nicaragua went through a period of leftist revolutionary change, although during that time moved away from authoritarianism. Between 1979 and 1983, Nicaragua's Polity IV score moved from −9 to −5. By 1991, after competitive democratic elections Nicaragua's Policy IV score moved to +6 and, with the consolidation of democratic institutions, stands at +9. At the time of the earthquake, and until the 1979 revolution, it was a truly repressive state, with an autocracy score of −8.

In any disaster, deeper thinking about the causes of the destruction and possible solutions are subordinate to disaster relief and recovery efforts in the immediate term (Birkland 1997; Haas et al. 1977). However, it is well known that actions taken during relief will considerably influence the relative success of recovery. In the Nicaragua case, the recovery was poorly managed, and the recovery from this disaster was, at best, incomplete. Much of the aid and capital made available for reconstruction was stolen from aid donors and from the state, either as a matter of direct theft or through its being channeled through corrupt enterprises managed by the Somoza family (Lernoux 1977; George 2012).

It is tempting to say that the earthquake was the "cause" of the 1979 Sandinista Revolution. Indeed, even as Penny Lernoux was writing in *The Nation* on the corruption of the Somozas' reaction to the earthquake, she was arguing that the Sandinista movement posed little threat to the regime. On the other hand, the Sandinista movement (*Frente Sandinista de Liberación Nacional*, or FSLN) was founded in the early 1960s, and began to gain momentum until the early 1970s. The 1972 earthquake actually slowed revolutionary activity for many of the reasons that all sectors of society were affected: the vast damage done to the nation's infrastructure, the temporary creation of national solidarity to rebuild, and so on. But this solidarity was extremely short lived, as it became clear that the Somoza regime had begun to steal or to misallocate relief and recovery assistance. While Lernoux argues that most of the FSLN's leadership was killed, captured, or driven underground by 1977, the movement was also on the cusp of significant revolutionary change because the FSLN's opposition to the dictatorship was joined by other members of the business, religious, and intellectual elite that was motivated by continued dismay with the regime's mishandling of earthquake relief and other national challenges that were greatly exacerbated by the earthquake. Ultimately, the FLSN took power in 1979, but it also ran a repressive regime until peace and democratization agreements led to the seeds of a democratic system in the 1990s.

In this case, Kingdon's and Birkland's notions of focusing events cannot be said to have led to a period of introspection, learning, and event-driven policy change centered on seismic safety, The earthquake and the events that followed influenced fundamental changes in the national order, but changed very about earthquake hazards: it did not engender, for example, a reexamination of building code provisions or enforcement; as George (2012) notes, "procedural" corruption in Nicaragua made such enforcement unlikely. There was no meaningful post-disaster recovery planning, or careful urban planning for Managua that would take into account seismic safety or quality of life, and where there was once an identifiable "downtown" Managua, there is now a sprawling, largely unplanned city without an urban center.

But the earthquake—or, more properly, the response to it—revealed the corruption of the Somoza regime. The mismanagement of earthquake recovery sufficiently angered a large portion of the Nicaraguan middle class, including, perhaps most notably, Joaquin Chamorro, the editor of *La Prensa,* the major newspaper in Nicaragua that took an increasingly hostile tone to the Somoza regime. Chamorro was assassinated in 1978, which led to further anger with Somoza, who was assumed responsible for the assassination. Chamorro's death was the catalyst for the end stage of the Nicaraguan Revolution, but the earthquake can be said to have started this process by driving a wedge between Somoza and the middle class.

We can say then that the "problem solving" orientation that we assume is adopted by policy actors after some focusing events was not evident by the government in Nicaragua which, given its level of repression, was not open to debate over something as esoteric as seismic safety. But we can say that Kingdon's "streams" responded to the earthquake. Substantial changes happened in the political stream, where normal institutions of democracy did not exist, but where extra-state or even extra-legal means were employed to express popular disappointment with not only the poor response to the earthquake, but to overall corruption. The problem stream, therefore, was dominated by the problem of the repression and corruption of the regime itself, and its contribution to wealth inequality and poverty. The policy stream did not generate ideas about how to improve seismic safety, because the usual institutions that would develop such ideas—think tanks, universities, civil society groups, and a free press—did not exist in a meaningful way, either because of repression or because there was really no excess room for such esoterica compared with the rather more pressing problem of political repression and kleptocracy. Without these democratic institutions, when significant change comes, it comes in the form of revolt or rebellion, not in the form of change following pluralistic political debate. And, even then, by the time the Sandinista revolution took firm control, the urban form and function of Managua cannot be said to have "recovered" to pre-quake functionality.

Of course, many disasters have significantly altered the physical form of cities. Fire codes have changed after great urban fires (London 1666, Chicago 1871, Seattle 1889, San Francisco 1906), and after major earthquakes (Long Beach 1932, Anchorage 1964, Los Angeles 1971, 1994 San Francisco 1989). Floods have done considerable damage but were addressed, in some cases, by exploiting opportunities to improve urban function, such as the creation of riverside parks as flood buffers in Grand Forks, ND, Cedar Rapids, IA, and Tulsa, OK, among other communities. Most of these changes to urban form and function are grounded in the belief, at least, that these changes would yield hazard mitigation and improve resilience. By contrast, the 1972 earthquake did not yield thoughtful consideration of Managua's urban form. Instead, the city center was badly damaged and not rebuilt, growth resumed in the outlying areas, and rendered Managua a city without a central district, save for some squatter settlements. The city therefore functions poorly, and there are stark differences in the lives of the poor and the well-to-do that continue to define the city (Rodgers 2004).

3.5   The Sichuan Earthquake

The Sichuan earthquake struck on May 12, 2008, in the afternoon local time (please also see Chap. 24 in this volume). It was an M7.9 earthquake, a very large event, and damage and destruction was widespread. The distinguishing feature of this earthquake was the widespread destruction of school buildings. A large number of school buildings failed catastrophically, killing at least 70,000 people, including at least 5,300 students (Branigan 2009). Because of China's well-known "one child" policy, this aspect of the disaster was particularly tragic for many families, many of whom lost their only child in these collapses. Ultimately, to address these parents' mounting anger, the Chinese government eased this policy for families who lost children in the disaster (Spencer 2008). Often, schools—called "tofu buildings" (Chengpeng 2012; Menefee and Nordtveit 2012) for their propensity to collapse—failed even as nearby public buildings and private structures performed much better. Ultimately the reason for many of the schools' collapses was laid to overall corruption in the construction of schools, which led to substandard construction regardless of the reasonably modern building codes under which they were to have been built, and to corruption, in which public officials would claim to pay for more expensive construction materials and techniques, but would buy lower-quality substitutes, and pocket the difference (Wang 2008). Indeed, this is not a matter of design deficiencies—Chinese engineers are well trained in seismic design—but is a matter of the proper enforcement of existing design standards.

Despite the remarkable gains in economic output and in national and personal income, mostly concentrated in the urban centers in the east of the country, the People's Republic of China cannot be considered a free country. Its Polity IV score is − 7, classifying the PRC as an "authoritarian regime." But China is, in some ways, opening up to debate, if not dissent. As the Polity IV 2010 country report noted

On the other hand, rapid economic growth has begun to create a new class of economic entrepreneurs who are amassing the resources necessary to create functional autonomy from government control. It is also increasing pressures on traditional groups and urban workers who are witnessing a growing gap in consumption patterns and a growing pressure on political intermediaries and local officials who are challenged by the rising influence and autonomy of the entrepreneurial class and increasing opportunities for official corruption. *Local and grassroots protests have been increasing as affected people react to corrupt officials and/or policies that threaten traditional livelihoods, allow deterioration of environment, or seize land for developers.* (My emphasis) http://www.systemicpeace. org/polity/China2010.pdf

While China is not a "free" country, its leaders are more sensitive to popular sentiment than was the Somoza regime. China's response to the Sichuan earthquake stood in remarkable contrast with government responses to the Nicaragua earthquake. The initial response to the disaster was not as disorganized and, ultimately, as corrupt as was the Nicaraguan regime's efforts. Relief efforts—aided by the Chinese People's Liberation Army—were well organized and efficient, and there was a considerable amount of civil society and NGO participation in the relief effort, on a scale rare in China (The Economist 2008)

The Chinese government, while acknowledging the need for change, sought to suppress popular anger over the corruption and shoddy building that exacerbated this disaster. It is well known that the Chinese government tolerates or even supports the emergence of social media and news on the internet, but it is equally well known that China will move very quickly to suppress popular dissent when it fails to serve regime interests (King et al. 2013). This repression is direct, as when the government suppressed a demonstration of grieving parents complaining about the performance of the shoddily built "tofu buildings" (and prevented memorial services), or when the government censored Internet outlets that sought to describe the event and its causes.

The 2008 Sichuan earthquake was indeed a focusing event for the Chinese government and Chinese people. Because of changes in information technology and a slight increase in government tolerance of popular discussion of national problems, the earthquake motivated significant numbers of citizens and elite policy makers to call for change. But we cannot say that it led to the sort of interest group and media mobilization and attention that such an event often generates in the United States or other democracies. On the other hand, the Chinese government was strongly motivated to respond to the earthquake, provide relief, and to investigate and correct the errors in construction that so angered the Chinese people and that garnered worldwide criticism, much of it negative.

However, we cannot simply dismiss whatever resulted from the Sichuan earthquake as a natural outgrowth of a totalitarian regime. Rather, we can also explain the nature of earthquake policy in any state as being a "policy without publics," which is a category of public policy in which there is relatively little popular mobilization about problem identification and the search for solutions (May 1990). In the school collapses in China, we see a greater level of nascent popular anger than in the United States after large earthquakes, likely because current construction standards for such structures are sufficiently strict that they are less likely to fail. Furthermore, the United States is fortunate to have never had a catastrophic earthquake strike during school hours during an era in which school construction standards were poor. The Long Beach, California, earthquake of 1933 destroyed a number of poorly built brick school buildings, but did so early in the morning. Still, the results were sufficiently shocking that they marked the first truly serious effort in California to improve building codes for public buildings (Geschwind 2001; Fatemi and James 1997).

In China, then, while there was great popular anger over the collapse of the schools, and significant efforts to suppress this anger, it is also true that the engineering community and the central government took note of the disaster, understood the reasons why schools performed so poorly, and took steps to remedy the problem. In particular, more stringent design and construction standards were adopted (Yujia 2009). It is, however, unclear whether these requirements are effectively enforced. Building code enforcement is a challenge in all countries subject to natural hazards, and the best codes are only as good as their enforcement regimes.

3.6   The Haiti Earthquake of 2010

Historically, Haiti has been a repressive autocracy, and, unlike Nicaragua after the fall of the Somoza regime, Haiti has not transitioned from one-man rule to stable democracy. Since the end of the Duvalier regime, Haitian politics have been unstable, and while democratization was begun in the 1990s, this project is incomplete. But what we do know about Haiti is that it is highly dependent on NGOs for economic and social support, such that, when Haiti was struck by its most destructive disaster ever, it lacked the social and political institutions that usually respond to such events and that seek to learn from disaster to improve on future performance.

On January 12, 2010, just before 5 pm local time, southwestern Haiti was struck by a catastrophic M7 earthquake that crippled Haiti's capital, Port au Prince (please also see Chap. 23 in this volume). The death toll from this disaster was substantial, although the range of estimates of the death toll is between 40,000 and over 300,000 people. The Haitian government has always produced the highest estimate (Reuters 2011), based on unclear methods, while the United States has argued that the toll was much lower (Associated Press 2011). Regardless of the death toll, this earthquake was catastrophic, and required a major outpouring of international relief and recovery assistance, because it rendered an already weak state unable to effectively respond to the earthquake.

This need for assistance is outlined by the extent of the damage done by the earthquake. Because Haiti is a poor nation, its construction and design standards were insufficient to mitigate earthquake damage, and traditional materials and methods yielded structures that were not designed to withstand even moderate ground shaking, such that a field team found that "90 % of the [reinforced concrete] structures surveyed in Haiti would have been classified as seismically vulnerable before the earthquake" (O'Brien et al. 2011). Vulnerability was compounded by poor infrastructure, such as poor roads and water supply, as well extremely high levels of poverty and political instability.

To the extent that democratic governance and professional public administration were developing in Haiti, these efforts were badly set back by the destruction of government buildings, and by the actual loss of important members of government and civil society in Haiti. But, for all this damage and loss of life, the major reason for Haiti's vulnerability—and its failure to bounce back in any meaningful way—is a result of uncoordinated aid delivered by an NGO establishment that has rarely invested in capacity building to create an effective state sector or civil society in Haiti. Rather, the NGO and foreign aid community continued to pursue particular and uncoordinated both before and after the earthquake and, indeed, simply accelerated its pre-earthquake efforts rather than orienting its efforts to appropriately helping the Haitian people and government to recover and to manage risk. The NGOs have largely provided things that have little to do with earthquake recovery and in building state or community-based resilience (Klarreich and Polman 2012). Indeed, due to continued political instability and partisan bickering in Haiti, which limited the ability of the government to develop the capacity to act that its citizens demanded, the NGOs actively avoided interactions with the government under the belief that the government, to the extent it existed, was an impediment to their work and to *their*, not the government's or the people's priorities (Kushner 2012; Wilentz 2011). In any case, Haiti has long been dependent on NGO and international development assistance (Booth 2011; Klarreich and Polman 2012), even well before the earthquake, and needs are largely assessed and addressed in the absence of government or popular participation or input. After the earthquake, this trend continued, even to the extent that, as a New York Times article noted, earthquake relief was provided "where Haiti wasn't broken" (Sontag 2012).

Curiously, then, from a policy perspective it is difficult to claim that the 2010 earthquake in Haiti was a focusing event with respect to domestic public policy related to natural disasters. Haiti had, and still lacks, a state structure that could design and enforce an advanced seismic safety building code, and the NGO community has done little to build government capacity toward this end (Klarreich and Polman 2012). The nation remains poor, its needs to create housing remain immediate even 3 years after the earthquake (NPR 2013), and pressure for progress is, as in so many places, likely to overcome any efforts to greatly improve design and construction practices. Ultimately, if we say that focusing events are focal because they attract attention to problems and thereby lead to learning opportunities, it is not clear that this earthquake served as a focusing event in that sense. As Klarreich and Polman's (2012) searing indictment of Haiti's "NGO Republic" makes clear, the earthquake revealed more about the shortcomings of Haiti's dependence on NGOs that have conflicting agendas. In that respect, it appears that the NGOs have learned little from this event, thereby denying the idea that any actor treated this disaster as a learning opportunity.

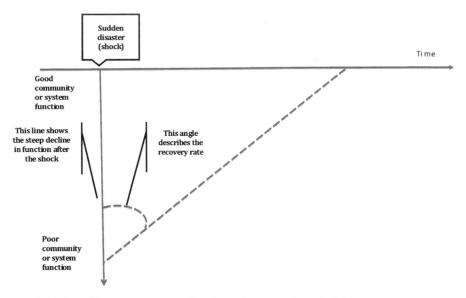

Fig. 3.1 The resilience delta. (Source line: MCEER, University at Buffalo)

3.7   Focusing Events and Resilience[2]

In recent years, it has become common to describe a nation's or a community's "resilience" to disasters. The idea of "resilience" is appealing, because it implies that some sort of system—social, ecological, engineered—can bend in the face of strain, but not break. In engineering, resilience stands in contrast with the idea of brittleness and robustness: a system can be strongly engineered to resist a great deal of force, but when that force exceeds the system's ability to withstand the force, the system fails catastrophically. Resilient systems are more effective because they "fail gracefully," do not fail catastrophically, and contain some degree of rebound capacity. De Bruijne et al. further explain the concept of disaster resilience by explaining that resilience within a crisis context includes the ability to both absorb the shock of the event and have the capacity to recover from the event (De Bruijne et al. 2010; Janssen et al. 2006).

Researchers at MCEER (formerly the Multidisciplinary Center for Earthquake Engineering Research) at SUNY Buffalo have developed the idea of a "resilience delta" (MCEER 2008) that reflects three dimensions of resilience: pre-disaster functionality of an infrastructure system, the extent of damage to the infrastructure system, and the speed of recovery of that system. Their vision of resilience is reflected in Fig. 3.1, which suggests that resilience is both a function of the extent of damage done by an event and the time it takes to recover pre-event functionality after the event.

---

[2] Parts of this section draw heavily from Birkland and Waterman 2008.

Their "R4" notion of resilience includes[3]

- Robustness—strength, or the ability of elements, systems, and other units of analysis to withstand a given level of stress or demand without suffering degradation or loss of function;
- Redundancy—the extent to which elements, systems, or other units of analysis exist that are substitutable, i.e., capable of satisfying functional requirements in the event of disruption, degradation, or loss of function;
- Resourcefulness—the capacity to identify problems, establish priorities, and mobilize resources when conditions exist that threaten to disrupt some element, system, or other unit of analysis (resourcefulness can be further conceptualized as consisting of the ability to supply material—i.e., monetary, physical, technological, and informational—and human resources to meet established priorities and achieve goals); and
- Rapidity—the capacity to meet priorities and achieve goals in a timely manner in order to contain losses and avoid future disruption.

These features of a resilient system are contained in what MCEER calls the four dimensions of resilience:

- Technical—the ability of physical systems (including all interconnected components) to perform to acceptable/desired levels when subject to disaster;
- Organizational—the capacity of organizations—especially those managing critical facilities and disaster-related functions—to make decisions and take actions that contribute to resilience;
- Social—consisting of measures specifically designed to lessen the extent to which disaster-stricken communities and governmental jurisdictions suffer negative consequences due to loss of critical services due to disaster; and
- Economic—the capacity to reduce both direct and indirect economic losses resulting from disasters.

In a similar vein, Aguirre notes that communities have different resilience profiles based on "physical, biological, psychological, social, and cultural systems" (Aguirre 2006, p. 1). Similar thinking has informed research and programs on "disaster resistant" communities, although it is important to note that 100 % *resistance* to a disaster is not usually achievable, from economic, political, or engineering perspectives. Rather, a resilient community is one that may not be entirely resistant—that is, things may break during the disaster—but that has the capacity to recover quickly, both because vulnerability is reduced and resilience is promoted. If we conceive of a community faced with a known hazard—high probability of earthquakes, for example—we would expect that a learning and resilient community would be one that experiences a series of disasters that become less disruptive, presumably due to learning and the application of experience (Birkland 2006). Of course, such learning may be more common—and more culturally embedded—in areas that suffer

---

[3] The language in lists that follow is directly quoted from MCEER 2006.

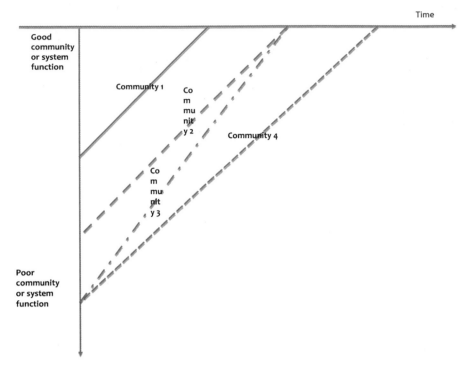

Fig. 3.2 Different Communities and their Resilience Deltas. (Source: Draws on work at MCEER, SUNY at Buffalo)

from repeated and seasonal threats, such as coastal and riverine areas prone to hurricanes, cyclones, or monsoons.

The disaster delta is an important concept because it contains three key aspects of resilience: the extent of initial damage to a system, the *rate* at which recovery occurs, and the level of functionality to which the community returns. MCEER originally applied this concept to physical infrastructure, but we can apply these concepts to damage to community functionality of the community: that is, to what extent does the community work as a community, with people, organizations, infrastructure, and government supporting the activities that characterized the community pre-disaster? The horizontal line along the top extends from the 100% functionality of the community, and, if successful, the line joining the depth of the vertical axis—the extent of damage—describes the rate at which a community can recover. In simplest terms, given these simplified linear models, the angle between described by this line is a representation of the rate of recovery to the prior level of community functionality.

In Fig. 3.2, the basic model is altered in some substantial ways:

- The line representing Community 1 is the basis for comparison.
- Community 2 is less resilient because the length of time it takes to recover from the same shock as Community 1 is considerably greater, for whatever reason.

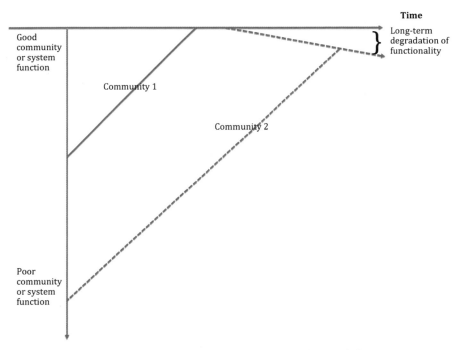

Fig. 3.3 Resilience deltas when community function is permanently degraded. (Source: Draws on work at MCEER, SUNY at Buffalo)

- Community 3 is also less resilient, because, although rapidity and resourcefulness are great, the community is less resilient overall because initial damage is so great. In other words, the costs and challenges of resilience are represented in all these cases by the area of the delta, not merely the depth of damage or the extent of recovery time.
- Community 4 is the least resilient community because, even though the slope of the recovery line is the same as that in community 1, damage is more severe, and recovery time takes longer.

Figure 3.3 shows two communities, one that suffers less overall damage than the other does. In both communities, the rate of recovery is roughly the same, but since community 2 was so profoundly harmed, even at the same recovery rate as community 1, it recovers only to a point that represents some degradation of pre-disaster functionality. We can conceive of community 2 as something like Managua, where some recovery happened but because of decisions (or inaction) made by the central government, the overall function of Managua-or, perhaps, the expectations for its function–degraded as the response plodded along.

When viewed this way, it is clear that there are three key features of resilience: the prevention of damage, the speed of recovery, and the prevention of substantial decay in the functionality of the community. Community functionality is not the same as a return to the *status quo ante*, which often replicates the very vulnerabilities

Table 3.1 Performance in the 4Rs of disaster resilience

	Nicaragua 1972	Sichuan 2008	Haiti 2010
Robustness of engineered and social systems	Not robust—buildings failed under stress of earthquake	Moderately robust—many modern buildings performed well	Not robust—Catastrophic damage to buildings and social systems
Redundancy	Not redundant. Poorly performing infrastructure. Poorly operating state institutions. Little room for emergent organization in an autocratic system (short of, ultimately, a revolution)	Moderately redundant	Not redundant. Poorly performing infrastructure that was barely adequate even before the earthquake. Poorly operating state institutions. Little room for emergent organization in an autocratic system (short of, ultimately, a revolution)
Resourcefulness	Not resourceful. Characterized by high degrees of political and economic corruption and autocracy	Resourceful. China is a large and economically powerful country. The military and other assets were used to rush aid to stricken areas	Not resourceful. Poorly performing state institutions that are very heavily dependent on NGOs for support (*New York Times*, 2012)
Rapidity	Very slow. Managua never recovered its social or physical form. Failure of the state to plan for post-earthquake reconstruction yielded a hollowed out and sprawling Managua	Rapid. China has the resources to respond to an event of this size	Very slow. Few structures have been built and there is an extreme housing shortage, with survivors still living in tent cities (*New York Times*, 2012)

that lead to poor resilience in the first instance. Rather, the community would, it is hoped, return to its previous level of social and economic function while reducing vulnerability. These goals are often viewed as incompatible, particularly if people express a strong desire to return to "normalcy," if normalcy also reproduces preexisting vulnerabilities.

These resilience deltas are highly simplified, of course. Communities are composed of many systems—infrastructure systems, health care, schools, private industry, and the like, each with their own resilience deltas that, when summed together, provide a fuller picture of community resilience, particularly when we note that these systems are tightly coupled and highly interdependent.

With this in mind, we can consider how the communities struck by the 1972, 2008, and 2012 earthquakes compare along the dimensions outlined by MCEER. In this case, we consider "communities" very broadly, at the scale of the region or the entire state. We also combine these elements with the idea of focusing events as a

Table 3.2 Performance in the technical aspects of recovery

	Nicaragua 1972	Sichuan 2008	Haiti 2010
Technical	Very poor. Infrastructure and the built environment not built to contemporary standards for seismic hazards	Moderately good. Urban areas were quickly rebuilt, although the quality of construction is unclear	Very poor. Infrastructure systems did not work well, and were poorer after the disaster than they were before; these systems were far below world standards before the earthquake
Organizational	Very poor. Little chance for civil society to develop under the regime. State organizations were corrupt and diverted aid to the Somoza family enterprises. Contractors used to promote recovery were corrupt and often owned by the Somozas	Good. The Chinese state is well organized, although corruption is deeply problematic	Very poor. There is nearly no state establishment in Haiti, and the nation is highly dependent on NGOs for support. The NGOs tend to promote their own particular strengths, and are not well coordinated
Social	Very poor. Massive numbers of casualties and displaced people	Fair: Emergent organization did occur, but was often suppressed. Protest movements designed to press for better construction practices were powerfully suppressed	Poor for this earthquake. Haitian people are not well integrated into their own recovery process[a]. Extremely poor
Economic	Poor	Strong. China has a large and resilient economy	Very poor. Much of the productivity capacity of Haiti was in and around Port au Prince, and was all destroyed

[a] "Haiti after the earthquake: Civil society perspectives on Haitian reconstruction and Dominican-Haitian bilateral relations," *Reliefweb*, September 10, 2010. http://reliefweb.int/report/haiti/haiti-after-earthquake-civil-society-perspectives-haitian-reconstruction-and-dominican, accessed January 29, 2012

driver of political or policy change. To simplify, we arrange the key themes of the case studies in the following two tables (Tables 3.1 and 3.2).

The resilience delta provides a depiction of the dimensions of resilience. In addition to these dimensions, tensions also exist in terms of trying to understand whether a community is "resilient." Boin et al. (2010) describe three such tensions; speedy recovery vs. timely adaptation, severity of disturbance, and state of return following disaster. Speedy recovery vs. timely adaptation refers to the notion of where resilience occurs. Resilience that occurs before the disturbance is when a community is able to scan the environment and be better prepared. This differs from resilience that is situated after the disturbance and is seen as "the last line of defense separating a stricken community from structural demise or extinction" (Boin et al. 2010, p. 8)

Another tension is the severity of disturbance. Is resilience the capacity to deal with the rare, extreme events or is resilience dealing with everything, even expected events? The third tension involves the ability for a community to "bounce back" following an event. How is this concept of "bounce back" defined? Is it when the community returns to the state it was before the event or is it when it is minimally functioning again? Events such as earthquakes provide the opportunity to not only return to "normal," but also to mitigate hazards so that the level of disruption does not recur. Indeed, if this happens, we can say that this is prima facie evidence of "policy learning" (Birkland 2006). Therefore, is resilience when the community returns to a "stronger" state and how is this determined without the presence of a similar event?

Based on these tensions, Boin et al. formulate the following definition of resilience:

> Resilience is the capacity of a social system (e.g., an organization, city or society) to proactively adapt to and recover from disturbances that are perceived within the system to fall outside the range of normal and expected disturbances. (Boin et al. 2010, p. 9)

A key feature of this definition is the *proactive* nature of adaptation. A focusing event may be about the immediate event, but it also provides a learning opportunity to prevent equally bad outcomes in the future. Since we are looking at the concept of resilience through the lens of focusing events, we depart slightly with Boin et al.'s application of the tensions of resilience. Based on our definition of a focusing event as being sudden, rare, and harmful, we have a more stringent view of resilience. Our definition of resilience would situate the concept of resilience after the "shock" of the event (tension 1), the severity of disturbance would define resilience as dealing with rare and devastating events (tension 2) and this influences the state of return causing it to take longer to return to its pre-shock state if that is even feasible—the new goal being functioning in the face of a serious disturbance (tension 3).

This view can be seen in the cases of the Nicaragua, Sichuan and Haiti earthquakes. These three communities were not able to "scan the environment" to be better prepared for the earthquakes. The earthquakes were rare and devastating events, and, at least in Haiti and Nicaragua, little pre-disaster efforts to be resilient were evident. Nor have the Nicaragua and Haiti earthquakes proven learning opportunities in these countries; in Nicaragua, the regime simply had no incentive to learn from these disasters and to properly recover. The earthquake was just another opportunity for elite theft, this time from aid donors.

One element of learning after a focusing event, we argue, is adopting a problem-solving orientation and hastening the pace of recovery. There is some evidence of this in China, where the regime and the Chinese people learned of the very real costs of corrupt building practices. But there is no apparent problem-solving orientation in Nicaragua, where the regime was not designed to solve the peoples' problems, nor in Haiti, where the state did not have or develop the capacity to learn. In the latter case, the NGOs apparently have learned little as well, while failing to undertake efforts to build Haitian capacity to learn and respond to disasters.

The Chinese government did have the capacity to immediately and effectively respond to the disaster, and, in the year following, enacted stronger building codes and standards. But before the earthquake, a combination of corruption and overall poor building practices created a degree of vulnerability that surprised many senior Chinese government officials. By contrast, we cannot say that focusing events provided an opportunity for resilience-oriented learning in Haiti because of its poverty, poor infrastructure, and nonfunctional state sector. In short, to the extent that the Haiti earthquake was a learning opportunity, it is likely to be a less effective lesson because of the poor capacity of the Haitian political system, both before and after the earthquake, to observe, learn, and implement lessons that might otherwise be learned after a disaster.

As we have noted, to the extent services are provided in Haiti, they are provided by NGOs, but these NGOs did not use the earthquake as an opportunity to learn about how to make Haiti more resilient; rather, they used the focusing event as a warrant to increase their already disjointed, mission-driven efforts. These missions were not changed by the earthquake, but, rather, were intensified, so that the "solution" to Haiti's problems were "more" of what was being done already: more microcredit, more food aid, more education, more health care services. Certainly, these things were needed, but few, if any, of these efforts were designed to improve state capacity or community resilience in a way that the nation could use its resources, however limited, to respond to a future disaster.

3.8 Conclusion

Developing countries provide an interesting background to study the concept of resilience since they are more vulnerable than developed countries. Mirza (2003) argued that climate change would increase the frequency and magnitude of extreme events, thus increase the vulnerability of developing countries since they do not have the plans and infrastructure in place to prepare for such events. Cummins (2009) found that in economic terms, disasters have a greater impact on the proportional GDP of developing countries than developed countries. Our discussion of resilience is a useful way of considering the interactions between politics, policy, communities, the built environment, and the hazard itself. But resilience is not the main story, because resilience is not a *policy* principle in our cases. Rather, it is an organizing principle for understanding why some communities fare better than do others before or after disasters.

We do not undertake, in this chapter, the question of whether democratic institutions yield better development outcomes in normal times. Rather, we seek to illustrate how sudden events can provide opportunities for problem solving in "developing" countries. The term "developing" in this context means political and economic development, together. Because our theories of agenda setting and focusing events are predicated on notions of democracy, we certainly take the normative position that "democracy" is better than autocratic or authoritarian government. But we come to this conclusion for a practical reason: that when a disaster strikes, democratic systems are better able to register, process, and address popular demands for change. Those popular demands may not always understand the technical issues involved in seismic safety, nor should we expect these demands to be couched in technical language. Rather, the mitigation of hazards and the analysis of risk is often left to experts, but works best in pluralistic systems where there are opportunities for debate about whether and to what extent more or less stringent measures should be taken to prepare for, respond to, and mitigate natural hazards.

References

Aguirre, B. E. (2006). *On the concept of resilience*. Newark: University of Delaware, Disaster Research Center.

Associated Press. (2011). Haitian earthquake death and homeless figures questioned in US report. *The Guardian*. May 31. http://www.guardian.co.uk/world/2011/may/31/haitian-earthquake-death-toll-question. Accessed 12 Jan 2013.

Baumgartner, F. R., & Jones, B. D. (1993). *Agendas and instability in American politics*. Chicago: University of Chicago Press.

Best, R. (2010). Situation or social problem: The influence of events on media coverage of homelessness. *Social Problems, 57*(1), 74–91.

Birkland, T. A. (1997). *After disaster*. Washington: Georgetown University Press.

Birkland, T. A. (1998). Focusing events, mobilization, and agenda setting. *Journal of Public Policy, 18*(1), 53–74 (April).

Birkland, T. A. (2006). *Lessons of disaster*. Washington, D. C.: Georgetown University Press.

Birkland, T. A., & Lawrence, R. G. (2009). Media framing and policy change after Columbine. *American Behavioral Scientist, 52*(10), 1405–1425.

Birkland, T. A., & Waterman. S. (2009). The politics and policy challenges of disaster resilience. In *Resilience engineering perspectives. Volume 2: Preparation and Restoration*, ed. C. P. Nemeth, E. Hollnagel and S. Dekker. Burlington, VT: Ashgate.

Boin, A., Comfort, L. K., & Demchak, C. C. (2010). The rise of resilience. In L. K. Comfort, A. Boin, & C. C. Demchak (Eds.), *Designing resilience: Preparing for extreme events*. Pittsburgh: University of Pittsburgh Press.

Booth, W. (2 February 2011). NGOs in Haiti face new questions about effectiveness. *The Washington Post*, sec. World. http://www.washingtonpost.com/wp-dyn/content/article/2011/02/01/AR2011020102030.html. Accessed 20 Jan 2013.

Branigan, T. (2009). China releases earthquake death toll of children. *The Guardian*. http://www.guardian.co.uk/world/2009/may/07/china-earthquake-anniversary-death-toll. Accessed 20 Jan 2013.

Chengpeng, L. (25 May 2012). Patriotism with Chinese characteristics. *The New York Times*, sec. Opinion. http://www.nytimes.com/2012/05/26/opinion/patriotism-with-chinese-characteristics.html. Accessed 19 Feb 2014.

Cobb, R. W., & Elder, C. D. (1983). *Participation in American politics: The dynamics of agenda-building*. Baltimore: Johns Hopkins University Press.

Cobb, R. W., & Ross, M. H. (1997). *Cultural strategies of agenda denial: Avoidance, attack, and redefinition*. Lawrence: University Press of Kansas.

Cummins, J. D. (2009). *Catastrophe risk financing in developing countries: Principles for public intervention*. Washington, D. C.: World Bank.

De Bruijne, M., Boin, A., & Van Eeten, M. (2010). Resilience: Exploring the concept and its meanings. In L. K. Comfort, A. Boin, & C. C. Demchak (Eds.), *Designing resilience: Preparing for extreme events*. Pittsburgh: University of Pittsburgh Press.

Fatemi, S., & James, C. (1997). *The long beach earthquake of 1933*. National Information Service for Earthquake Engineering, University of California, Berkeley. http://nisee.berkeley.edu/long_beach/long_beach.html. Accessed 20 Jan 2013.

George, S. (2012). And then the earthquake: In Central America, earthquakes topple more than buildings. *Bertelsmann Future Challenges*. http://futurechallenges.org/local/and-then-the-earthquake-in-central-america-earthquakes-topple-more-than-buildings/. Accessed 20 Jan 2013.

Geschwind, C. H. (2001). *California earthquakes: Science, risk, and the politics of hazard mitigation*. Baltimore: Johns Hopkins University Press.

Haas, J. E., Kates, R., & Bowdon, M. (1977). *Reconstruction following disaster*. Cambridge: MIT Press.

Hilgartner, J., & Bosk, C. (1988). The rise and fall of social problems: A public Arenas model. *American Journal of Sociology, 94*(1), 53–78.

Janssen, M. A., Schoon, M. L., Ke, W., & Börner, K. (2006). Scholarly networks on resilience, vulnerability and adaptation within the human dimensions of global environmental change. *Global Environmental Change, 16*(3), 240–252 (August).

King, G., Pan, J., & Roberts, M. (2013). How censorship in China allows government criticism but silences collective expression. *American Political Science Review, 107*(2), 1–18. http://gking.harvard.edu/publications/how-censorship-china-allows-government-criticism-silences-collective-expression. Accessed 20 Feb 2014.

Kingdon, J. W. (2003). *Agendas, alternatives, and public policies 2nd ed. Longman classics in political science*. New York: Longman.

Klarreich, K., & Polman, L. (31 October 2012). The NGO republic of Haiti. *The Nation*. http://www.thenation.com/article/170929/ngo-republic-haiti#. Accessed 19 Feb 2014.

Kushner, J. (2012). Haiti's politics of blame. *GlobalPost*. http://www.globalpost.com/dispatch/news/regions/americas/haiti/120112/haiti-earthquake-politics-michel-martelly. Accessed 20 Jan 2013.

Lawrence, R. G., & Birkland, T. A. (2004). Guns, hollywood, and school safely: Defining the school-shooting problem across public arenas. *Social Science Quarterly, 85*(5), 1193.

Lernoux, P. (1977). The Somozas of Nicaragua. *The Nation, 225*(3), 72–77.

Maraniss, D. (2006). The meaning of Roberto Clemente. *Sports Illustrated, 104*(15), 56. Master-FILE complete, EBSCOhost. Accessed 20 Jan 2013.

May, P. J. (1990). Reconsidering policy design: Policies and publics. *Journal of Public Policy, 11*(2), 187–206.

May, P. J. (1992). Policy learning and failure. *Journal of Public Policy, 12*(4), 331–354 (December).

MCEER (formerly Multidisciplinary Center for Earthquake Engineering Research). (2008). *Engineering resilience solutions: From earthquake engineering to extreme events*. Buffalo: MCEER, University at Buffalo, SUNY. http://mceer.buffalo.edu/about_MCEER/MCEERbrochure-2-26-09.pdf. Accessed 10 Aug 2013.

Menefee, T., & Nordtveit, B. H. (2012). Disaster, civil society and education in China: A case study of an independent non-government organization working in the aftermath of the Wenchuan earthquake. *International Journal of Educational Development, 32*(4), 600–607. doi:10.1016/j.ijedudev.2011.10.002 (July).

Mirza, M. M. Q. (2003). Climate change and extreme weather events: Can developing countries adapt. *Climate Policy, 3*(3), 233–248.

Molotch, H., & Lester, M. (1975). Accidental news: The great oil spill as local occurrence and national event. *The American Journal of Sociology, 81*(2), 235–260 (September).

National Public Radio (NPR). (2013). Effects of 2010 earthquake still Mar Haiti. *Weekend Edition Saturday*. Washington, D. C.: NPR. http://www.npr.org/2013/01/12/169209923/effects-of-2010-earthquake-still-mar-haiti. Accessed 19 Feb 2014.

O'Brien, P., Eberhard, M., Haraldsson, O., Irfanoglu, A., Lattanzi, D., Lauer, S., & Pujol, S. (2011). Measures of the seismic vulnerability of reinforced concrete buildings in Haiti. *Earthquake Spectra, 27*(S1), S373–S386. doi:10.1193/1.3637034. (October 1).

Reuters. (12 January 2011). *Haiti revises quake death toll up to over 316, 000*. http://www.reuters.com/article/2011/01/12/haiti-quake-toll-idUSN1223196420110112. Accessed 20 Jan 2013.

Rodgers, D. (2004). 'Disembedding' the city: Crime, insecurity and spatial organization in Managua, Nicaragua. *Environment and Urbanization, 16*(2), 113–124. doi:10.1177/095624780401600202. (October 1).

Sabatier, P. A., & Jenkins-Smith, H. (1999). The advocacy coalition framework: An assessment. In P. A. Sabatier, ed. *Theories of the policy process*. Boulder: Westview.

Sabatier, P. A., Jenkins-Smith, H. C., & Lawlor, E. F. (1996). Policy change and learning: An advocacy coalition approach. *Journal of Policy Analysis and Management, 15*(1), 110.

Schattschneider, E. E. (1975). The semisovereign people. Hinsdale, Ill.: The Dryden Press.

Sontag, D. (5 July 2012). Earthquake relief where Haiti wasn't broken. *The New York Times.* http://www.nytimes.com/2012/07/06/world/americas/earthquake-relief-where-haiti-wasnt-broken.html. Accessed 19 Feb 2014.

Spencer, R. (26 May 2008). China to drop one-child policy for earthquake parents. *Telegraph.co.uk.* http://www.telegraph.co.uk/news/worldnews/asia/china/2032924/China-earthquake-China-to-drop-one-child-police-for-earthquake-parents.html. Accessed 19 Feb 2014.

Stone, D. A. (2002). *Policy paradox: The art of political decision making.* New York: Norton.

The Economist. (2008). China's earthquake: Days of disaster. *The Economist.* Accessed 15 May. http://www.economist.com/node/11376935. Accessed 20 Jan 2013.

Walker, J. L. (1977). Setting the agenda in the U.S. senate: A theory of problem selection. *British Journal of Political Science, 7,* 423–445.

Wang, S. (2008). Changing models of China's policy agenda setting. *Modern China, 34*(1), 56–87.

Wilentz, A. (16 January 2011). Haiti's political earthquake. *Los Angeles Times.* http://articles.latimes.com/2011/jan/16/opinion/la-oe-wilentz-haiti-20110116. Accessed 19 Feb 2014.

Yujia, M. (2009). Report reveals reasons behind school buildings collapse in Sichuan. china.org.cn. http://www.china.org.cn/china/news/2009-05-26/content_17839434.htm. Accessed 19 Feb 2014.

Chapter 4
Linking Development to Disasters in Turkey: Moving Forward After the Marmara Earthquake

N. Emel Ganapati

4.1 Introduction

On August 17, 1999, the northwestern part of Turkey was rocked for by a 7.4 magnitude earthquake. The earthquake—referred to as the Marmara earthquake—killed more than 17,000 people and affected 1,358,953 people (CRED 2012). The epicenter of the earthquake was the district Golcuk in Kocaeli province neighboring Istanbul. The earthquake damaged more than 210,000 housing units, the most number of housing units damaged by any earthquake in Turkey (PMO-CMC 2000). The impact of the earthquake on Turkey's economy was extensive, estimated at around USD 20 billion (CRED 2012).

The Marmara earthquake was the most important *focusing event* (Birkland 1997) in the history of Turkey. It displayed the country's weaknesses in terms of preparing for, responding to, and recovering from disasters. Several studies in the literature have studied the changes that have taken place in Turkey's disaster management structure in the aftermath of the Marmara earthquake and proposed measures to be undertaken (Ozerdem and Barakat 2000; Balamir 2002; Corbacioglu and Kapucu 2006; Celik 2007; Ganapati 2008; Unlu et al. 2010). Among these, Ozerdem and Barakat's (2000) was the only study that touched upon the link between disasters and development in Turkey.

Defining development as a process of reducing vulnerability to disasters (Anderson 1985), this chapter approaches disasters from a developmental perspective. Its main contributions to the literature are twofold. First is to provide a long-term perspective to still ongoing disaster management reforms in Turkey. Ozerdem and Barakat's (2000) study was conducted in the immediate aftermath of the earthquake. Hence, it mainly discussed the limitations of Turkish disaster management structure in terms of mitigation and preparedness in the context of a disaster and development relationship. Since then, many organizational and legislative reforms (e.g., establishment of a focal agency and an obligatory earthquake insurance scheme) have

N. E. Ganapati (✉)
Florida International University, Miami, FL, USA
e-mail: ganapat@fiu.edu

N. Kapucu, K. T. Liou (eds.), *Disaster and Development,* Environmental Hazards,
DOI 10.1007/978-3-319-04468-2_4, © Springer International Publishing Switzerland 2014

Table 4.1 Impact of disasters in Turkey (1900–2012). (Source: CRED 2012)

Disaster type	# of events	Killed	Total affected	Damage USD (in thousands)
Earthquake	76	89,236	6,924,005	24,685,400
Flood	39	1,342	1,778,520	2,195,500
Landslide	9	286	13,481	26,000
Storm	9	100	13,639	2,200
Avalanche	3	407	1,075	–
TOTAL	136	91,371	8,730,720	26,909,100

been introduced in the country. Although the main focus of these reforms were not vulnerability reduction per se, these reforms are likely to have an impact on vulnerability reduction in the long run in Turkey. Second is to examine country-level plans and programs on disasters (e.g., National Earthquake Strategy and Action Plan) and development (e.g., Five Year Development Plans) for the first time to better understand the linkages between the two.

The sections that follow provide a brief disaster profile of Turkey. This is followed by a discussion on the Marmara earthquake as a focusing event and the major reforms introduced after the earthquake. The chapter concludes with a list of broader strategies that still need to be addressed in Turkey in the context of a development and disaster relationship. It suggests that the government of Turkey needs to put disasters back in national-level Five Year Development Plans, integrate social vulnerability to the National Earthquake Strategy and Action Plan and give a voice to local actors (e.g., local governments, community based organizations) for vulnerability reduction.

4.2   Turkey's Disaster Profile and the Marmara Earthquake

Turkey is located on the Alpide belt, which stretches from Indonesia through the Himalaya mountain range in Asia, southern Europe, and out into the Atlantic Ocean. This belt is second most seismic region in the world after the Pacific Ring of Fire. According to the official Earthquake Hazard Zoning Map of Turkey dated 1996—still in effect—, sixty six percent of Turkey's land are in the first and second degree seismic zones (JICA 2004, p. 25). These zones not only house 71 % of the country's population but also a significant majority of its industries and dams (76 and 69 %, respectively).

Although Turkey remains vulnerable to several other natural hazards (e.g., floods, landslides, storms and avalanches) due to its topography and meteorological conditions, earthquakes cause the most damage in the country in terms of death toll, number of people affected, and financial damage. As shown in Table 4.1, from 1900 to 2012, Turkey was struck by 76 major earthquakes. Together these earthquakes claimed 90,000 lives, which correspond to 97 % of all disaster related deaths in the country. They affected 7 million people and caused damages in excess of USD 24 Billion.

Floods are the second costly disasters in Turkey in terms of human suffering and financial damages. Between 1900 and 2012, there have been 39 major flood occurrences in the country, causing 1,342 deaths and displacing 1.7 million people. The financial burden of the floods was around USD 2.2 Billion. Landslides, storms and avalanches also cause damages in Turkey. However, they affect a relatively small percentage of the population and cause lesser damages compared with the earthquakes and the floods.

Although many disasters that preceded the Marmara earthquake were pivotal events in the evolution of disaster management in Turkey (e.g., 1939 Erzincan earthquake), the Marmara earthquake was unique in its size and importance. It perhaps brought the most attention to the problems of Turkey in terms of preparing for, responding to and recovering from disasters, serving as a *focusing event* (Birkland 1997).

According to Rubin (2007), a focusing event exhibits most of the following characteristics: large magnitude, high impact, eligibility for disaster declaration, high visibility, unusual location, a unique threat agent, and surprise. The Marmara earthquake meets all the criteria outlined by Rubin but one (the unusual location) as explained below.

Large Magnitude The earthquake affected eight provinces which housed one fourth of the country's population and accounted for approximately 35% of the Gross Domestic Product prior to the earthquake (SIS 2000).

High Impact The earthquake officially claimed 17,480 people and damaged approximately 244,383 housing and business units (PMO-CMC 2000).

Eligibility for Disaster Declaration The Marmara earthquake was immediately declared as a disaster as the number of damaged buildings far exceeded the set criteria for disaster declaration (at least 50 in settlements with more than 50,000 residents).

High Visibility The earthquake had high visibility, partly because the area affected by the earthquake was heavily urbanized and industrialized (e.g., some neighborhoods in Istanbul) and partly because the Turkish TV stations and international media outlets were broadcasting from the earthquake zone day and night.

Unique Threat Agent A unique threat agent linked to the Marmara earthquake was the fire which started at the TUPRAS Refinery (Turkish Petroleum Refineries A.S.) in Izmit, the largest refinery in Turkey. The fire created a huge panic in the quake zone and led to evacuation orders for the refinery personnel and the nearby disaster survivors (Milliyet 1999a).

Surprise Turkey is no stranger to powerful earthquakes as mentioned earlier. What caught many by surprise, however, was the state's lack of preparedness to the Marmara earthquake. The Turkish government's search and rescue teams and the Turkish Red Crescent Society reportedly arrived in the earthquake zone after some foreign search and rescue teams and the Red Cross (Milliyet 1999c, d). When the earthquake hit, the earthquake fund, one of the three major disaster funds in Turkey, had the equivalent of USD 2 (1 Million TL) in it (Milliyet 1999b).

4.3   Post-Marmara Earthquake Changes in Turkey's Disaster Management Structure

The changes that were introduced in Turkey following the Marmara earthquake have dealt with all four phases of disasters—preparedness, response, recovery, and mitigation. This section details the major changes below, including the establishment of one focal agency for disaster management, establishment of an earthquake council, preparation of a national-level strategy on earthquakes, safer construction practices, compulsory earthquake insurance system, and enhanced search and rescue capacity.

4.3.1   Establishment of One Focal Agency: Disaster and Emergency Management Presidency

Prior to the Marmara earthquake, there was no focal agency in Turkey that had a legal mandate to coordinate preparedness, response, recovery and mitigation activities at the national level. Lack of such agency contributed to the bureaucratic confusion in the aftermath of the Marmara earthquake. In November 1999, the General Directorate of Emergency Management was established under the auspices of the Prime Minister's Office to provide top-level coordination during response and rescue and to oversee disaster mitigation measures undertaken by government at the central level. However, the responsibilities of this directorate overlapped with the responsibilities of other governmental agencies and added to the confusion on which government agencies were responsible for which phase of disasters (Ganapati 2008). The directorate also lacked the necessary resources (e.g., in terms of staff and funding) to undertake its coordinative tasks on disaster mitigation (JICA 2004).

In light of these criticisms, the government passed Law No. 5902 in 2009, establishing the Disaster and Emergency Management Presidency (DEMP) (*Afet ve Acil Durum Yonetimi Baskanligi*) under the auspices of the Prime Minister's Office. DEMP helped unite fragmented organizations that dealt with disasters in the county by absorbing several such organizations, including the General Directorate of Emergency Management, the General Directorate of Disaster Affairs, and the General Directorate of the Civil Defense. DEMP now functions as an all hazards, focal agency at the national level, addressing all phases of emergency management, including preparedness, response and rescue, recovery and mitigation (similar to the Federal Emergency Management Agency in the U.S. prior to its placement under the Department of Homeland Security).

At the provincial level, DEMP undertakes its responsibilities, including those that relate to vulnerability reduction, through Provincial Disaster and State of Emergency Directorates and Civil Defense Search and Rescue Team Directorates. All provincial governments in Turkey include the former, whose responsibilities range from preparing provincial level civil defense plans and ensuring disaster-related coordination among the government agencies to accrediting civil society organizations involved in disasters (e.g., search and rescue teams), identifying risks within their jurisdictions, and educating the public. Only selected provinces house the latter, the civil defense arm of Provincial Disaster and State of Emergency Directorates (Adana, Afyon, Ankara, Bursa, Diyarbakir, Erzurum, Istanbul, Izmir, Sakarya, Samsun, and Van).

4.3.2 Science-Government Collaboration: Earthquake Advisory Council

Another important post-earthquake development was enhanced collaboration between the scientific community and the government. Such collaboration started in March 2000 with establishment of the National Earthquake Council (Ulusal Deprem Konseyi), an independent, advisory organization to provide policy and research guidance on disaster mitigation in the country, among other things. Consisted of academicians, the Council was established under the Scientific and Technical Research Council of Turkey (Turkiye Bilimsel ve Teknolojik Arastirma Kurulu [TUBITAK]). One of the main contributions of the Council in terms of vulnerability reduction was to issue a national earthquake mitigation strategy report in 2002. The Council could not conduct its duties as expected and effectively ceased to exist due to changes made in TUBITAK (JICA 2004, p. 48). In 2007, the Council was abolished by the Prime Minister's Office (Hurriyet Daily News 2007), reportedly due to politics (e.g., the chairman of the Council belonging to the opposition party) (Turkish Politics Updates 2011).

Following the establishment of DEMP, a new advisory body called the Earthquake Advisory Board (Deprem Danisma Kurulu) was formed in 2009. The Board's main tasks include presenting alternative policies for mitigating earthquake damages and determining priority areas for earthquake related research. Headed by the Director of DEMP, the Board brings different multi-stakeholders on earthquakes together, including academicians and representatives of governmental agencies and civil society organizations. It has thirteen members, five of whom are directors or representatives of agencies that deal with earthquakes. The remaining eight members are appointed by the Director of DEMP. These include five academicians (nominated by the Higher Education Council or Yükseköğretim Kurulu [YOK]) and three representatives from civil society organizations in the country. The Board has a mandate to meet at least four times a year.

4.3.3   Preparation of the First National Plan: National Earthquake Strategy and Action Plan

Before the Marmara earthquake, Turkey did not have a national-level plan which detailed the structure of coordination between different levels of government, non-governmental organizations, and the private sector in times of disasters (similar to the National Response and Disaster Recovery Frameworks in the U.S.). With the help of the Earthquake Advisory Board, DEMP launched Turkey's first national level plan on earthquakes, entitled the National Earthquake Strategy and Action Plan (NESAP 2012–2023), in August 2011. The plan touches upon preparedness, response, recovery, and mitigation, is implemented by DEMP. Its main objective is to prevent or reduce the effects of earthquakes in the "physical, economic, social, environmental and political" realm by creating "earthquake-resistant, safe, well prepared and sustainable settlements" (DEMP 2011, p. 7).

NESAP is structured around three principal themes, namely learning about earthquakes, earthquake safe settlement and construction, and coping with the consequences of earthquakes. It includes 87 action items total with respect to the principal themes. For each action item, NESAP specifies the period during which that particular action item should be undertaken as well as the types of action necessary to undertake (e.g., cooperation and coordination, legislative arrangements, organizational structuring, and capacity enhancement). In addition, the document identifies which agencies will be responsible for ensuring coordination at the national level and which agencies will play supportive roles in undertaking that particular action item.

NESAP was approved by the Disaster and Emergency High Board, established in 2009 along with DEMP. This Board is headed by the Prime Minister or his/her appointee. It consists of several ministers, including the Ministers of National Defense, Interior, Foreign Affairs, Finance, National Education, Environment and Urbanization, Health, and Forestry and Water Affairs as well as Transport, Maritime, and Communication. The main role of this Board is to approve plan, program, and reports on disasters and emergencies in the country.

4.3.4   A New Financing Mechanism: Compulsory Earthquake Insurance System

Prior to the Marmara earthquake, the Turkish government replaced all property damaged or destroyed from disasters in accordance with the Disaster Law No. 7296. Due to assurances from the state, homeowners did not have an incentive to purchase private disaster insurance. Only three percent of residential buildings carried private earthquake insurance prior to the Marmara Earthquake (GFDRR 2011). With technical and financial assistance from the World Bank, the Turkish government introduced a compulsory earthquake scheme in 2000 to reduce its fiscal exposure to disasters as costs of government-funded post-disaster reconstruction after the earthquake were extensive.

NO OF INSURANCE POLICIES (000)

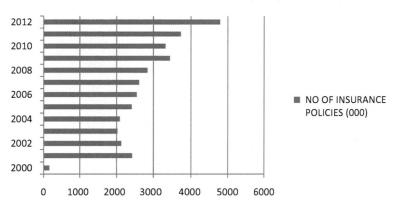

Fig. 4.1 Number of obligatory earthquake policies sold (in thousands) by Year (2000–2012). (Source: Natural disaster insurance institution 2013)

The compulsory earthquake scheme requires that all homeowners in urban areas insure their property against earthquake disasters to be eligible for post-disaster housing aid from the state—either in the form of permanent housing or subsidies for housing repair and retrofitting. The annual premiums are affordable (mainly due to country-wide pooling of earthquake risks) but not subsidized by the government (GFDRR 2011). They are collected under the Turkish Catastrophe Insurance Pool (TCIP). This pool is "sufficient to withstand a 1-in-350 year earthquake" (GFDRR 2011, p. 1). The government's role in TCIP is limited to covering damages that exceed the claims-playing capacity of the TCIP (Gurenko et al. 2006; GFDRR 2011).

The Compulsory Earthquake Insurance is managed by the Natural Disaster Insurance Institution (NDII) under the Under Secretariat of Treasury. NDII is directed by the Board of Directors of the TCIP, which include high level public officials (five) and representatives from the private sector (one) and academia (one). NDII's most operations are outsourced to private companies, indicating close collaboration between the government and the insurance sector.

According to the NDII website, as of 23 February 2013, the coverage ratio of the obligatory earthquake scheme in Turkey was 30%. Coverage in some provinces were as high as 51% (Bolu) or as low as 7% (Sirnak). As shown in Fig. 4.1, the number of earthquake insurance policies sold doubled from 2.4 million in 2001 (the first full year after introduction of the obligatory insurance) to 4.8 million in 2012.

To ensure deeper market penetration of earthquake insurance in Turkey, the Turkish government passed new legislation (No. 6305), which restricts power and water connections in uninsured homes (Milliyet 2012). This law took effect in August 2012 and is more likely to impact homes that are rented out. Despite this law, many homeowners choose not to purchase this obligatory insurance with the expectation that the Turkish government will eventually pay for damages to their homes regardless of whether or not they have the obligatory coverage.

With respect to disaster risk reduction, the NDII plays a dual role. First, it provides incentives (e.g., discounts based on building performance appraisals) for contractors to comply with the building codes, and it does not provide insurance coverage to buildings that do not have the necessary permits (Gurenko et al. 2006; Asian Development Bank 2008). Second, it helps promote public awareness of earthquake risks (e.g., in school textbooks) (Gurenko et al. 2006).

4.3.5 A Path toward Safer Buildings and Settlements: Changes in Construction Supervision

The high death toll from the Marmara earthquake underlined the problems in construction practices in the country. Prior to the earthquake, local governments enforced the relevant building codes and other standards related to urban development and planning—including regulations on earthquake resistant design—within their boundaries. The task of enforcing the building codes and standards outside the municipal boundaries was left to provincial governments. The local and provincial governments often lacked qualified personnel to conduct proper building inspections at the construction site.

The Turkish government has undertaken several initiatives after the Marmara earthquake to ensure safer buildings and settlements in the country. With respect to the safety of new building structures, the government introduced privately-owned *Building Inspection Firms (BIFs)* as one of the key players in vulnerability reduction in the country (Decree No. 595) in April 2000, within eight months of the earthquake. BIFs inspect construction in all new buildings (except public buildings) and report to local or provincial governments for issuance of construction and occupancy permits. They are held responsible for the structural defects of the buildings up to 15 years and for non-structural defects up to 2 years (e.g., through payments for damage compensation). BIF activities are closely monitored by the government. Although BIFs were involved in construction inspection in selected pilot provinces initially, their coverage area was expanded to all provinces in 2010. As of 27 February 2013, there were 1,499 BIFs registered in the country (Merkez Yapi Denetim Komisyonu Baskanligi 2013).

In addition to the measures taken to ensure the safety of new building structures, the Turkish government has taken steps towards retrofitting and reconstructing the seismically vulnerable existing building stock in the country. First, it has started to carry out a large scale disaster risk reduction project in Istanbul. Istanbul is not only located on the North Anatolian fault but it is also the most populous (13.8 million people) and dense city (2,666 persons per square meter) in Turkey (Turkish Statistical Institute 2013) with a high concentration of industrial facilities. According to a JICA estimate, a seismic event of the same magnitude as the Marmara earthquake in Istanbul could damage 350,000 public and private buildings and lead to a death toll of 87,000 people and injuries of 135,000 people (World Bank 2012a).

To address Istanbul's seismic vulnerabilities, the Turkish government borrowed loans from the World Bank, the European Investment Bank, and the Council of Europe Development Bank to implement a project entitled Istanbul Seismic Risk Mitigation and Emergency Preparedness Project (ISMEP) (World Bank 2012b). The main goal of ISMEP is to save lives and reduce the impact of future earthquakes through seismic risk mitigation for public facilities, enhancing emergency preparedness, and enforcement of building codes (World Bank 2013).

According to the World Bank (2012a), some of the major accomplishments of the ISMEP project to date include: (a) retrofitting or reconstruction of 701 public buildings (e.g., schools and hospitals) and places of historical interest (e.g., the Topkapi palace); (b) improved systems for issuance of building permits in the pilot municipalities of Pendik and Bagcilar; (c) training of 3,630 engineers in the seismic retrofitting code; and (d) training of 450,000 neighborhood volunteers through the Public Awareness and Neighborhood Community Volunteers Programs.

Second, the Turkish government has recently passed legislation on regeneration of areas for the purposes of disaster risk reduction. The two main pieces of legislation regulating this matter were passed in 2012, and they include the Law on the Regeneration of Areas under Disaster Risk (Law No. 6306), commonly known as the Urban Regeneration Law, passed by the Council of Ministers; and the Regulation on the Implementation of the Law on the Regeneration of Areas under Disaster Risk issued by the Ministry of Environment and Urbanization.

The new legislation on urban regeneration allows for large-scale demolishment of risky building structures and/or buildings on risky areas and construction of safer ones. According to a Ministry of Environment and Urban Planning estimate, "approximately 10,000,000 buildings nationwide to fall within the scope of the urban regeneration plan, making Turkey the biggest construction site on earth" (Ozeke 2012). The legislation defines "areas under risk" as areas where disasters could lead to loss of life and property.

The risky areas are identified by the Ministry of Environment and Urbanization, Housing Development Administration of Turkey (Toplu Konut Dairesi Baskanligi), or municipalities, with input from DEMP. Property owners could also initiate the identification process by requesting the Ministry of Environment and Urbanization or municipalities to determine whether or not their area could be classified as a risky one. Upon proposal of the Ministry of Environment and Urbanization, the risky areas are decided by the Council of Ministers.

4.3.6 Broader Role for the Civil Society: Search and Rescue Teams

Prior to the earthquake, there were only three governmental search and rescue (SAR) teams in Turkey (based in Ankara, Istanbul and Erzurum) (Aydogdu and Altintas 2011). They were part of the General Directorate of Civil Defense under the auspices of the Ministry of Interior. Following the Marmara earthquake, the country's SAR capacity has increased tremendously, both at the governmental and non-governmental levels. At the governmental level, the Directorate of Civil Defense, which is now housed within DEMP, established SAR teams in eight additional provinces (Adana, Afyonkarahisar, Bursa, Diyarbakir, Izmir, Sakarya, Samsun, and Van) a year after the earthquake. These teams currently operate as regional teams, serving several provinces in the country. The Armed Forces and several municipalities (e.g., Istanbul, Ankara) have also set up their own SAR teams.

At the non-governmental level, voluntary SAR teams have been established throughout the country. These teams respond to a variety of disasters (e.g., earthquakes, floods, train or helicopter accidents) in Turkey and abroad. Although the Directorate of Civil Defense had no formal partnership with voluntary SAR teams prior to the Marmara earthquake, it has started to collaborate with them while undertaking training activities and drills and responding to disasters.

Although some SAR teams in the country focus exclusively on search and rescue, others take on additional activities that relate to vulnerability reduction, such as helping the general public better prepare for disasters. Below are some notable examples to such teams dedicated to both search and rescue and disaster risk reduction: (1) *GESOTIM (Golcuk SAR)*: This team was established in the city of Golcuk, the epicenter of the Marmara earthquake, after the earthquake with a mission to "save a soul for life." GESOTIM is now one of the eight SAR teams representing Turkey in International Search and Rescue Advisory Group (INSARAG), a global network of countries and organizations dedicated to SAR under the United Nations umbrella. Since its establishment, the team has been deployed following several national (e.g., 2011 Van earthquake) and international disasters (e.g., 2003 Iran-Bam earthquake). In addition to its search and rescue activities, GESOTIM publishes newsletters to raise disaster awareness. (2) *The Neighborhood Disaster Volunteer System:* The Neighborhood Disaster Volunteer System was an initiative that was launched as a project entitled "Neighborhood Disaster Support Project" by the Swiss Agency for Development and Cooperation in 2000 in the Kocaeli province. This project aimed at strengthening the capacity of neighborhoods so that the residents can: (a) respond to disasters in a timely manner; and (b) develop a high level of disaster awareness. The project was then extended to other provinces in the country. Neighborhoods participating in this initiative set up two structures in their areas: SAR teams called Neighborhood Disaster Volunteers (MAG) and Neighborhood Disaster Committees (MAK). The former includes volunteers who participate in a SAR training program and are expected to respond to a disaster in a timely manner and assist professional search and rescue teams after their arrival in the neighborhood. Upon the completion of the training program, the neighborhood establishes what is called the "Crisis Center," a container with equipment (e.g., emergency communication equipment) to be used by the volunteers. There are currently 105 neighborhoods with 4,960 MAG members in the provinces of Istanbul, Kocaeli, Bursa, Yalova and Izmir (Mahalle Afet Gonulluleri 2013). The latter are led by *muhtars*, the heads of neighborhoods. Their main tasks are to assess disaster threats and to raise disaster awareness and response capacity in the neighborhood. In addition to the *muhtars*, these committees consist of individuals who undertake the tasks of logistics, risk and damage assessment, and coordination of volunteers. MAK members are elected by MAG members but they do not have to be a member of MAG per se.

4.4 Challenges Ahead: Linking Development and Disasters

The post-Marmara earthquake changes in disaster management detailed above were major steps taken towards disaster risk reduction in Turkey. There is still more to do with respect to each, however. The Disaster and Emergency Management Presidency's politically appointed directors lack the necessary professional expertise in disaster management (Hurriyet Daily News 2011). The country-wide coverage rate of the obligatory earthquake insurance is still low, especially in the eastern provinces, despite recent attempts to restrict power and water connections in homes that do not carry the earthquake insurance. Many in Turkey believe that the urban regeneration legislation was passed for political and economic reasons rather than for benefiting the public (Ozeke 2012). Furthermore, similar urban regeneration projects undertaken in Istanbul ended up producing homes that are not culturally appropriate for their residents (Karaman 2013). The agency's governmental search and rescue teams also failed to respond to the recent Van earthquake in an effective manner (Hurriyet Daily News 2011).

Although a lot has been written on what is or what is not working with respect to the post-earthquake developments in terms of disaster management in Turkey (e.g., Balamir 2002; Corbacioglu and Kapucu 2006; Celik 2007; Ganapati 2008; Unlu et al. 2010), we still do not know what else needs to be done in Turkey in the context of a development and disaster relationship. This section suggests three strategies for this purpose: putting disasters back in development (e.g., in the case of five year development plans), integrating social vulnerability to the national earthquake strategy and action plan, and giving voice to local actors.

4.4.1 Putting Disasters Back in Development

The national discourse on development in Turkey is shaped by five year development plans that set macroeconomic and social policies that are to be followed by all government agencies. In the past, these plans were prepared by the State Planning Organization. In 2011, this organization was replaced by the Ministry of Development, whose main responsibilities include providing advice to the government on development, preparing national level development strategies, and coordinating development agencies at the regional level (total of 26).

The First Five Year Plan was prepared for 1963–1967. Since then, five year plans have been issued continuously (T.C. Kalkinma Bakanligi 2013). These plans typically provide an assessment of the country from a development perspective and out-

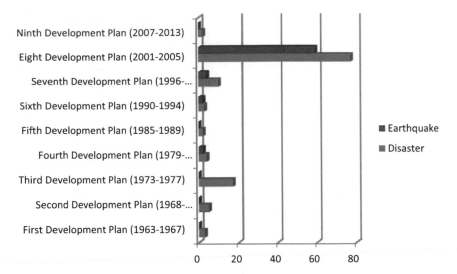

Fig. 4.2 No. of times the words "Disaster" and earthquake mentioned in five year development plans. (Source: Data compiled from T.C. Kalkinma Bakanligi 2013)

line basic targets, strategies, principles and policies to be followed during the five year plan period. The plans have historically paid little or no attention to disasters in general and to disaster risk reduction in specific with one significant exception as explained below: the Eight Five Year Development Plan. According to a word frequency search conducted using the NVivo qualitative software program, most plans used the term "disaster" five times or less (six out of nine plans) and failed to use the term "earthquake" even once (five out of nine plans) (See Fig. 4.2 for the frequency of the terms "disaster" and "earthquake" in Five Year Development Plans). One common theme under which disasters were mentioned (seven out of nine plans) was housing projections, specifically how many housing units will be needed in urban and rural areas during the plan period due to potential earthquakes.

The attention paid to disaster risk reduction has evolved over time. The initial three plans did not deal with disaster risk reduction issues at all. The Fourth Five Year Plan was the first plan to mention disaster risk reduction although it simply called for different standards to be followed in new buildings and retrofitting of existing buildings in seismically vulnerable areas. The consequent two plans also touched upon disaster risk reduction albeit briefly. The Fifth Five Year Plan had one sentence suggesting rehabilitation of rural housing in villages, especially in villages that are located on first degree seismic zones. The Sixth Five Year Plan mentioned disaster risk reduction in the context of housing and environmental problems, with respect to the need to identify and utilize earthquake resistant technologies in buildings located on seismic zones and the need to mitigate the impacts of such disasters as deforestation, desertification, soil erosion, and floods.

The Seventh Five Year Plan paid relatively more attention to disaster risk reduction than its predecessors. It explicitly acknowledged the earthquake related risks and recommended the following: (a) to prepare earthquake maps to be used in regional and urban planning efforts; (b) to take the necessary precautions to minimize natural hazard risks in general and earthquake risks in specific; (c) to revise the outdated disaster law; and (d) to ensure coordination among agencies involved in different aspects of disasters prior to, during and after disasters.

Prepared immediately after the Marmara Earthquake, the Eight Five Year Plan paid the most attention to disaster risk reduction issues in Turkey. This plan detailed how the 1999 Marmara and Bolu-Duzce earthquakes affected the economy in a negative way and led to an increase in unemployment and a decline in tourism. At the same time, the plan acknowledged that post-disaster housing reconstruction activities (e.g., by the World Bank) had a positive impact on the construction sector in the country. It indeed had a subsection devoted to natural hazards, under the section on "Enhancement of Efficiency in Public Services." Furthermore, the plan outlined "Basic Targets, Principles and Policies" for natural hazards.

The Eight Five Year Plan highlighted the importance of disaster risk reduction activities for the country and called for allocation of funds prior to disasters for such activities. Some of the disaster risk reduction activities proposed by this plan focused on a comprehensive overhaul of the disaster management structure in the country (e.g., through revising legislation as well as restructuring the organizations involved in different phases of disasters and ensuring coordination among the organizations). Others targeted regional and urban planning efforts that take disaster risks into account (e.g., through selection of sites for settlements, dealing with illegal housing stock). Still others dealt with supervising construction in new buildings and retrofitting the existing building stock in high risk areas. The Eight Five Year Plan even mentioned the need for specialized education of engineers on disasters and professional ethics as well as for a national database on disasters and training of the public.

Another interesting aspect of the Eight Five Year Development Plan was its acknowledgement of social vulnerabilities in the context of disasters. This plan, for instance, mentioned the need to ensure food security for the poor as they may become poorer following disasters. It noted that the country's social assistance and social service institutions did not meet the needs following disasters. It called for vocational training in disaster-stricken areas. Moreover, it highlighted the need to improve the socio-economic conditions of the villagers as a way to prevent deforestation.

The most recent plan, the Ninth Five Year Development Plan, was a step back from its predecessor. This plan did not go beyond highlighting the confusion at the organizational level with respect to disaster management (prior to the establishment of Disaster and Emergency Management Presidency) and calling for regional development in high risk rural areas. The Ninth Five Year Development Plan's lack of attention to disasters in general and disaster risk reduction in specific is perhaps an indication of the belief among the government circles that the necessary reforms on disaster management have already taken place. Alternatively, the lack of attention is a continuing proof of disconnect between the development discourse and the disaster discourse unless the country is shaken by another catastrophe similar to the magnitude of the Marmara earthquake.

4.4.2 *Integrating Social Vulnerability to the National Earthquake Strategy and Action Plan*

As mentioned earlier, Turkey's first country-wide level plan on earthquakes, the National Earthquake Strategy and Action Plan (NESAP 2012–2023), was launched in 2011. The NESAP indeed does a better job of linking disasters to development than the Five Year Development Plans, perhaps because representatives from the Ministry of Development were involved in preparation of the plan. It acknowledges the role of government agencies that shape development in the country—i.e., the Ministry of Development and the regional-level Development Agencies—in creating seismically safe settlements and buildings, one of the three goals of the NESAP.

The NESAP specifically states that, under the leadership of the Ministry of Development, "Development Agencies shall take into account earthquake hazards and risks within their domains, and conduct their activities such that these risks will not be increased, or actually reduced" (p. 30). In addition, the plan mandates that the Provincial Special Administrations and Municipalities determine the hazards and risks in their province prior to preparing provincial development and environmental plans and develop strategies for disaster risk reduction, especially in urban areas (p. 31). Furthermore, the NESAP lists the Ministry of Development as the lead agency for formulating earthquake related Research and Development that could contribute to seismically safer construction practices and uninterrupted supply of infrastructure services in times of disasters (e.g., natural gas, power, communication systems) in the country. Overall, the plan mentions the name of the Ministry of Development in one-fifth of all Action Items (18 out of 87; two as a lead agency and 16 as a supportive agency).

Although the NESAP's effort to link disasters to development is noteworthy, the plan does a poor job of addressing social vulnerabilities, an important component of disaster risk reduction. A content analysis of the plan shows that the plan's overall focus as far as vulnerability is concerned is on the seismic vulnerability of settlements and buildings. The plan mentions socially vulnerable groups (elderly, disabled persons, women and children) only briefly.

Furthermore, the NESAP does not include renters as one of the socially vulnerable groups. As noted by disaster scholars (Comerio 1998; Mukherji 2010; Ganapati 2013), renters are often neglected during governmental disaster assistance and face additional barriers while recovering from disasters. In the aftermath of the Marmara earthquake, for instance, only homeowners whose property was destroyed or heavily damaged from the earthquake were the beneficiaries of large scale permanent housing projects that were undertaken in the country. Renters were pushed to live in buildings that were damaged by the earthquake after removal of the prefabricated housing units. The damaged buildings were retrofitted after the earthquake. Yet, there were fears among the public that they were still unsafe. The people of Golcuk, the epicenter of the earthquake, referred to such buildings as "cement tombs" due to fears that these buildings will collapse in the next earthquake.

The NESAP's oversight with respect to socially vulnerable groups in general and renters in specific could be explained by the composition of the working groups that participated in the preparation of the plan. According to the Union of Chambers of Turkish Engineers and Architects (TMMOB 2011), the backgrounds of scientists who were members of the working groups through working groups were mainly geology and engineering, and therefore, half of the topics that should have been included in the plan were excluded.

4.4.3 Giving Voice to Local Actors

The post-disaster recovery period presents policymakers with windows of opportunities for fostering the longer-term process of development as part of recovery (Cuny 1983; Anderson and Woodrow 1989), for putting in place mitigation strategies (e.g., building codes) to lessen future vulnerabilities to disasters (Ingram et al. 2006) and for building resilient (Vale and Campanella 2005; Guo 2012) communities. The challenge for the policymakers is how to achieve all these through a participatory approach, however.

In Turkey, the recovery processes are handled in a top-down manner. Local governments are indeed the key players in urban planning at the local level in Turkey. However, Article 9 of the urban development law allows the Ministry of Environment and Urbanization to intervene in the preparation and revision of local plans in times of disasters. Following the Marmara earthquake, the Ministry of Public Works and Settlement, the predecessor of the Ministry of Environment and Urbanization, invoked this article within 6 days of the earthquake and announced that it would prepare plans for the permanent housing areas in the earthquake zone.

The Ministry's decision meant that local actors were divorced from the planning processes of permanent housing areas, especially the local governments and civic organizations of those affected by the disaster. In the city of Golcuk, for instance, the mayor had to visit Ankara, located more than 270 km away, 181 times over one and a half years since key recovery decisions were given in the capital city by central government officials (Ganapati and Ganapati 2009). The major's staff felt powerless in recovery decisions as all they could do was to give tours of the area to central government officials who were trying to decide where to build the permanent homes. Their opinions were not asked. Similarly, civic organizations which were established after the earthquake to have a say in the recovery of their city and in disaster risk reduction (e.g., Gölcük 17 August Association) were left out from the planning processes despite their repeated pleas from the central government. What is needed is not only to give a voice to key local actors like the local governments and civic organizations but also to assess the capacities of civic organizations that are dedicated to disaster risk reduction, help build their capacity and collaborate with them.

4.5   Conclusion

This chapter presented the Marmara earthquake as a focusing event (Birkland 1996, 1997) and highlighted the disaster management reforms that have been introduced in Turkey after the Marmara earthquake. Although some of the reforms covered in this chapter have been implemented for more than a decade, others have been introduced recently. The country's national level agency in charge of all disaster phases, the Disaster and Emergency Management Presidency, was established only in 2009. As a relatively new organization, this agency is yet to be tested in an earthquake with the magnitude of the Marmara earthquake. According to many, this agency failed to ensure coordination in response to the 2011 Van earthquake, which claimed more than 600 lives in the country (Hurriyet Daily News 2011). Turkey's first national level plan on earthquakes, the National Earthquake Strategy and Action Plan, was released in August 2011. It is also too early to tell to what extent the strategy would contribute to disaster risk reduction. Similarly, the legislation on the regeneration of urban areas was passed in 2012, and its effects on disaster risk reduction are yet to be seen.

Overall the disaster management reforms that have been introduced in Turkey were positive steps towards reducing the vulnerabilities in the country. However, more needs to be done in terms of linking development to disasters, both at the national and local levels. At the national level, the five year development plans prepared by the Ministry of Development need to pay more attention to disasters. Although the plan which was issued immediately after the Marmara earthquake has done that well, the most recent plan barely mentions disasters and earthquakes. Similarly, at the national level, there is a need to put more emphasis on social vulnerabilities in the National Earthquake Strategy and Action Plan. Despite the fact that this plan makes a more explicit effort to link disasters to development than the Five Year Development Plans, its focus on vulnerability is rather limited to the seismic vulnerability of settlements and buildings rather than the vulnerabilities of the people (elderly, disabled persons, women and children). Lastly, at the local level, it is essential to acknowledge that the capacities exist for disaster risk reduction and ensure that the local governments and the civic organizations have a voice in decision making processes while preparing for, responding to and recovering from disasters.

References

Anderson, M. B. (1985). A reconsideration of the linkages between disasters and development. *Disasters*. Harvard Supplement.

Anderson, M. B., & Woodrow, P. J. (1989). *Rising from the Ashes: Development strategies in times of disaster*. Boulder: Westview Press.

Asian Development Bank. (2008). Earthquake insurance: Lessons from international experience and key issues for developing earthquake insurance in the PRC. http://reliefweb.int/sites/reliefweb.int/files/resources/F_R.pdf.pdf. Accessed 15 Feb 2013.

Aydogdu, S., & Altintas, K. H. (2011). Characteristics of civil defense search and rescue units, Turkey, 2008–2009. *Iranian Red Crescent Medical Journal, 13*(9), 651–659.

Balamir, M. (2002). Painful steps of progress from crisis planning to contingency planning: Changes for disaster preparedness in Turkey. *Journal of Contingencies and Crisis Management, 10*(1), 39–49.

Birkland, T. A. (1996). Natural Disasters as Focusing Events: Policy Communities and Political Response. *International Journal of Mass Emergencies and Disasters, 14*(2), 221–243.

Birkland, T. A. (1997). *After disaster: Agenda setting, public policy and focusing events.* Washington, D.C.: Georgetown University Press.

Celik, S. (2007). Towards local and protective Turkish disaster management system. *Dumlupinar Universitesi Sosyal Bilimler Dergisi, 19,* 95–116.

Centre for Research on the Epidemiology of Disasters (CRED). (2012). Country profile for Turkey. http://www.emdat.be/country-profile. Accessed 12 Dec 2012.

Comerio, M. C. (1998). *Disaster hits home: New policy for urban housing recovery.* Berkeley: University of California Press.

Corbacioglu, S., & Kapucu, N. (2006). Organizational learning and self-adaptation in dynamic disaster environments. *Disasters, 30*(2), 212–233.

Cuny, F. C. (1983). *Disasters and Development.* New York: Oxford University Press.

Disaster and Emergency Management Presidency (DEMP). (2011). *National Earthquake Strategy and Action Plan 2012–2023.* Ankara: DEMP.

Ganapati, N. E., & Ganapati, S. (2009). Enabling participatory planning in post-disaster contexts: A case study of World Bank's housing reconstruction in Turkey. *Journal of the American Planning Association, 75*(1), 41–59.

Ganapati, N. E. (2008). Disaster management structure in Turkey: Away from a reactive and paternalistic approach? In J. Pinkowski (Ed.), *Disaster management handbook* (pp. 281–320). CRC Press.

Ganapati, N. E. (2013). Measuring the processes and outcomes of post-disaster housing recovery: Lessons from Gölcük, Turkey. *Natural Hazards, 65*(3), 1783–1799.

Global Facility for Disaster Reduction and Recovery (GFDRR). (2011). Turkish catastrophe insurance pool. http://www.gfdrr.org/sites/gfdrr.org/files/documents/DFI_TCIP_Jan11.pdf. Accessed 1 March 2013.

Guo, Y. (2012). Urban resilience in post-disaster reconstruction: Towards a resilient development in Sichuan, China. *International Journal of Disaster Risk Science, 3*(1), 45–55.

Gurenko, E., Lester, R., Mahul, O., & Gonulal, S. O. (2006). *Earthquake Insurance in Turkey: History of the Turkish Catastrophe Insurance Pool.* Washington, DC: World Bank.

Hurriyet Daily News. (2007). National earthquake council abolished, 2 March. http://www.hurriyetdailynews.com. Accessed 1 March 2013.

Hurriyet Daily News. (2011). "Turkey's disaster agency struggles in Van quake test," 14 November. http://www.hurriyetdailynews.com. Accessed 1 March 2013.

Ingram, J. C., Franco, G., Rio, C. R.-d., & Khazai, B. (2006). Post-disaster recovery dilemmas: Challenges in balancing short-term and long-term needs for vulnerability reduction. *Environmental Science and Policy, 9*(7), 607–613.

Japan International Cooperation Agency (JICA). (2004). *Country strategy paper for natural disasters in Turkey.* Ankara: JICA.

Karaman, O. (2013). Urban renewal in Istanbul: Reconfigured spaces, robotic lives. *International Journal of Urban and Regional Research, 37*(2), 715–733.

Mahalle Afet Gonulluleri (MAG). (2013). Anasayfa. http://www.mag.org.tr/tur/mag.asp. Accessed 1 March 2013.

Merkez Yapi Denetim Komisyonu Baskanligi. (2013). Türkiye'de Özel Sektör Yapı Denetimi. http://www.yds.gov.tr/index.php. Accessed 27 Feb 2013.

Milliyet. (1999a). Katrilyonluk tesis alev alev, 18 August. http://www.milliyet.com.tr/1999/08/18/ekonomi/eko00.html. Accessed 23 Dec 2013.

Milliyet. (1999b). Afet fonlarıyla borç ödeniyor! 19 August. http://www.milliyet.com.tr/1999/08/19/siyaset/siy02.html. Accessed 23 Dec 2013.

Milliyet. (1999c). Kızılay yüz kızartıyor, 26 August. http://www.milliyet.com.tr/1999/08/26/ haber/hab000.html. Accessed 23 Dec 2013.

Milliyet. (1999d). CHP, afet raporu hazırladı, 28 August. http://www.milliyet.com.tr/1999/08/28/ siyaset/siy03.html. Accessed 23 Dec 2013.

Milliyet. (2012). Deprem sigortası için son 4 gün!, 13 August. http://ekonomi.milliyet.com.tr/ deprem-sigortasi-icin-son-4-gun-/ekonomi/ekonomidetay/13.08.2012/1580366/default.htm. Accessed 15 Feb 2013.

Mukherji, A. (2010). Post-earthquake housing recovery in Bachhau, India: The homeowner, the renter, and the squatter. *Earthquake Spectra, 26*(4), 1085–1100.

Natural Disaster Insurance Institution (2013). Compulsory earthquake insurance. http://www.tcip. gov.tr/. Accessed 23 Feb 2013.

Ozeke, H. B. (2012). Turkey: The biggest construction site on earth: Turkey what to expect from the urban regeneration law. http://www.mondaq.com. Accessed 1 March 2013.

Özerdem, A., & Barakat, S. (2000). After the Marmara earthquake: Lessons for avoiding short cuts to disasters. *Third World Quarterly, 21*(3), 425–439.

Prime Minister's Office-Crisis Management Center (PMO-CMC) (T.C. Basbakanlik Kriz Yonetim Merkezi). (2000). *Depremler 1999*. Ankara: PMO–CMC.

Rubin, C. (2007). *Emergency management: The American experience, 1900–2005*. Fairfax: Public Entity Risk Institute.

State Institute of Statistics (SIS) (Devlet Istatistik Enstitusu). (2000). *Statistical yearbook of Turkey 1999*. Ankara: SIS.

T. C. Bakanligi, Kalkinma (2013). Kalkinma Planlari 1–9. http://ekutup.dpt.gov.tr/plan. Accessed 5 Jan 2013.

Turk Muhendis ve Mimar Odalari Birligi (TMMOB). (2011). Ulusal Deprem Stratejisi Ve Eylem Plani 2012–2023 Taslak Metni Hakkindaki Tmmob Şehir Plancilari Odasi Afet Ve Risk Komisyonu Görüşü. http://www.spo.org.tr. Accessed 1 March 2013.

Turkish Politics Updates. (2011). The week in Turkish politics, 26 October 2011. http://turkishpoliticsupdates.wordpress.com. Accessed 1 March 2013.

Turkish Statistical Institute. (2013). 28 Ocak 2013 Haber Bulteni. http://www.tuik.gov.tr/PreHaberBultenleri.do?id=13425. Accessed 1 Feb 2013.

Unlu, A., Kapucu, N., & Sahin, B. (2010). Disaster and crisis management in Turkey: a need for a unified crisis management system. *Disaster Prevention and Management, 19*(2), 155–174.

Vale, L., & Campanella, T. J. (Eds.). (2005). *The resilient city: How modern cities recover from disaster*. New York: Oxford University Press.

World Bank. (2012a). World Bank—Turkey partnership: Country program snapshot Sept 2012. http://www.worldbank.org.tr/. Accessed 4 Jan 2013.

World Bank. (2012b). Improving the assessment of disaster risks to strengthen financial resilience. Washington, DC: World Bank. http://www.gfdrr.org/G20DRM. Accessed 4 Jan 2013.

World Bank. (2013). Istanbul Seismic Risk Mitigation Project (ISMEP). http://www.worldbank. org.tr. Accessed 9 March 2013.

Chapter 5
Disaster Management System in Azerbaijan: The Case of 2010 Kura River Flood

Vener Garayev and Fikret Elma

5.1 Introduction

It is clear today that emergencies, disasters and crises are much more destructive and consequential than ever. In this regard, communities face natural and man-made disasters of unprecedented nature, scope and impact. Despite the efforts to minimize their possible negative consequences, respective human vulnerability to disasters across the world has, generally speaking, not been properly minimized or adequately addressed. Dealing with disasters has mostly been partial, short-run and reactive in nature, thus, postponing required solutions until next disasters strike (Kapucu and Ozerdem 2011).

One of the important related problems still to be tackled is the way communities respond to disasters and the way they deal with the post-disaster stage, especially at the economic development level. Bringing life back to normal is not only a problem of single agency, but requires a set of inter-related policies and reforms targeted at long-term mitigation and sustainable development in all spheres of the community affected by the disasters. The approach to above-mentioned issues, however, is determined by several factors, including capacity, economy, culture, and politics-related issues. This chapter intends to analyze the post-disaster stages, namely recovery and development measures, in light of a recent flood disaster in Azerbaijan.

Azerbaijan is one of the post-Soviet countries trying to reframe the structures and policies inherited from the Soviet period. Being somehow disconnected from development policies, disaster management is one of the fields in the country that needs reforms and policy adjustments. To date, most of the disaster response efforts in Azerbaijan have been reactive in nature, while development policies require

V. Garayev (✉)
Department of Political Science and Public Administration, Gediz University, Izmir, Turkey
e-mail: vener.garayev@gediz.edu.tr

F. Elma
Department of Public Administration, Celal Bayar University, Manisa, Turkey
e-mail: fikret.elma@cbu.edu.tr

N. Kapucu, K. T. Liou (eds.), *Disaster and Development*, Environmental Hazards,   79
DOI 10.1007/978-3-319-04468-2_5, © Springer International Publishing Switzerland 2014

a more systemic, sustainable and wider approach. This chapter analyzes the case of Kura river flood of 2010 in Azerbaijan, and tries to understand the connection between disaster response/recovery/mitigation and subsequent regional development and community planning policies in the aftermath of the disaster (See Chap. 12 in this volume). The government's post-disaster approach presents an opportunity to identify how state links recovery efforts with sustainable development of the region. The main questions addressed in this chapter are: Has Azerbaijani state been effective in disaster response and recovery? Have disaster response/recovery efforts been integrated into long-run mitigation and economic development of the region? What are the future trends and opportunities for the Azerbaijani state in the field of disaster management and economic development? The chapter contributes to the discussion of disaster management and economic policies that have been deemed as separate fields from each other, and contextualizes the topic in the case of a developing post-Soviet country. Analysis of the case presents a comparative perspective for other post-Soviet or regional countries with similar institutional and structural legacies.

5.2 Emergency Management and Development

Disasters are mostly unexpected events that disrupt the functionality of communities bringing about human, social, material, economic and environmental losses (Ahrens and Rudolph 2006). It is not a surprise today that the natural and man-made disasters have changed in many aspects. They have changed in frequency, scope and severity, all of which have only increased over the past decades. Regardless of geography or nation, communities today face an unprecedented nature of destructiveness and consequences brought by disasters. Whether an earthquake, a hurricane, a flood, or a drought—all of the disasters and emergencies have blatantly shown the high level of vulnerability that communities face and should tackle with. In other words, disaster/emergency/crisis management encounters an overwhelming job of creating resilient communities that can minimize, if not eliminate, the impacts of unexpected and unwanted disasters. Such a job, without doubt, requires a comprehensive emergency management framework encompassing an all-hazards approach to dealing with disasters.

Traditional emergency management cycle comprises four phases, namely mitigation, preparedness, response and recovery. While mitigation is a process focusing on continuous efforts to eliminate or reduce the communities' vulnerability to disasters, preparedness focuses on activities with the purpose of getting ready for an upcoming disaster. Response, in turn, is the phase of implementing risk-aversion strategies to settings being impacted by the disasters, while recovery is a post-disaster stage aiming to restore the pre-disaster conditions. It is important to note that while preparedness and response stages are generally relatively short-run and reactive, recovery and mitigation stages entail relatively long-run and proactive strategies and programs.

The way emergency management cycle functions is very much dependent on the nature of disasters. In this regard, as opposed to slow-onset disasters (hurricanes, drought, etc.) that entail all four phases of the emergency management cycle, the rapid-onset or unexpected and sudden disasters (earthquakes, landslides, floods, etc.) entail mostly recovery and mitigation stages. Such a distinction, without doubt, brings about different disaster management tools and strategies varying in terms of time, resources, and human capital. In both cases, however, the post-disaster measures generally focus on immediate actions to restore the "normal life" conditions of the communities under impact. Recovery phase of the emergency management cycle per se is the stage presenting mostly short-run solutions, while mitigation entails further measures to avoid future similar cases. In essence, while recovery stage is mostly about post-disaster measures, mitigation is and should be a stage encompassing the whole emergency management cycle without interruptions. In other words, mitigation phase targets at identification and elimination/minimization of possible man-made and natural risks to communities (Schwab et al. 2007).

There are several benefits of mitigation to the community, and the most important of them is that it creates disaster-resilient communities through protection of lives and property, minimizes financial impact on the communities, speeds up the recovery process, and eliminates/minimizes uncertainty about disasters and their possible consequences (FEMA 2012). A report by Multihazard Mitigation Council (2005) states that every dollar spent on mitigation policies and measures saves $ 4 in terms of future benefits to the community. Despite its widely-accepted benefits, however, mitigation is mostly ignored, whether by communities or respective governments, due to its invisibility or absence of immediate results. The impact of mitigative policies is mostly seen after several years and only when a disaster tests community's resilience again. Therefore, emergency management is not only a matter of need and necessity to protect communities, but also of political will to implement long-run risk-aversion and hazard mitigation strategies.

The mitigation tools may be classified as *structural* ("hard" engineering measures like building levees or relocating buildings) or *non-structural* ("soft" engineering measures like using forests to prevent land sliding) (Sylves 2008). They can also be grouped into *project* (physical efforts to minimize risk) and *process* (policy and regulatory adjustments) measures (McCarthy and Keegan 2009). Regardless of the approach, however, communities would not become disaster-resilient unless post-disaster recovery and, most importantly, long-run mitigation efforts are linked to development policies (Godschalk 2007). In other words, mitigation measures should be implemented along with development programs in a coordinated way (Godschalk and Brower 1985). It is only through integration of disaster management policies and community planning that disaster-resilient communities emerge (Pierce 2003). Having noted this, it is important to say that traditional approach is generally characterized by no link between emergency management practices and community development programs. In essence, the former deals with elimination and minimization of mostly immediate risks/impacts of man-made and natural disasters to the communities, while the latter focuses on sustained efforts to improve economic conditions of the same communities. When the two are disconnected, however, the long-run benefits of the latter are under the risk of being vanished due to some disaster consequences that might be avoided if coordination between the emergency managers and community planners is in place. Godschalk and Brower (1985) state this issue in the following way:

> Development management programs use government powers and resources to guide and influence the location, type, amount, density, quality, and rate of development within a local or regional jurisdiction. Development management tools include an area's plans, regulations, public facilities programs, land acquisition, and taxation measures. With the addition of hazard reduction elements, these tools can also serve the objectives of the hazard mitigation strategy. (p. 65)

Therefore, communities that combine development measures with risk minimization measures would produce more resilient societies. The way the authors define a hazard mitigation strategy is as following:

> A hazard mitigation strategy is a coordinated and consistent set of goals, policies, and tools for reducing or minimizing human and property losses from hazards and resulting disasters. Once a jurisdiction has completed its risk and vulnerability analyses and estimated its probable economic losses, then it must identify measures for reducing expected losses and combine them into an effective hazard mitigation strategy. Such a strategy should be directed toward those selected implementation actions that will eliminate key hazard problems, both before and after disaster strikes. An advantage of an integrated mitigation approach is its possible power to increase community acceptance of mitigation actions by demonstrating the full scope of hazard threats and potential economic losses. (Godschalk and Brower 1985, p. 69)

In addition to having an integrated recovery/mitigation strategy as a result of coordination between the emergency managers and community planners, however, there is also a need to include community members into decision-making process. Such an approach not only provides a chance to present reasonable solutions, but also makes them acceptable and legitimate in the eyes of the community (Pierce 2003). The process involving emergency managers, community planners and the citizens at the same table, in turn, is a strong guarantor of achieving resilient and sustainable communities in the long run (Godschalk 2007), the basis of which is the development programs and measures implemented in accordance with recognition of and commitment to minimize the risks of hazards in the community (Rubin and Barbee 1985). Moreover, Ahrens and Rudolph (2006) argue that there is a cyclic relationship between the level and quality of disaster management capacity and sustainable economic development of the communities, both of which reinforce each other. The main goal in connecting the two is to make the society resilient and sustainable so that disasters don't keep undermining the functionality of life conditions (Kapucu and Garayev 2013).

5.3   Context

This section of the chapter describes the context of the study, namely, the structure and functions of the emergency management system in Azerbaijan, and the Kura River flood of 2010, with an account of governmental response and recovery efforts.

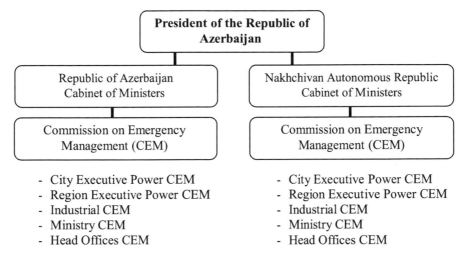

Fig. 5.1 Emergency management system of Azerbaijan (1992–2005). (ARFHK 1992)

5.3.1 Emergency Management System in Azerbaijan

Republic of Azerbaijan is a South Caucasian country with a population of 9 million and area of 86.6 km^2 (33.4 mi^2). The country was part of the Soviet Union until its collapse in 1991. The post-Soviet Azerbaijan is an oil-rich developing country with a strategic geo-political location in the region. As in the case of most post-Soviet countries that faced a transition period in early 1990s, Azerbaijan encountered several political, economic, and social problems. The period was characterized by systematic administrative and structural ineffectiveness and inefficiency in many crucial fields such as governance, education, health, social security, environment and emergency management. With gradual progress in terms of political stability in subsequent years, the government gained a chance to focus on the above-mentioned issues that led to several gradual reforms.

One of the fields requiring immediate reforms was the field of emergency management. The reason for such a need was the fact that, as a result of the collapse of the Soviet Union, the previously existing related agency, namely the Soviet-period Council of Ministers Commission on Emergency Situations, became essentially dysfunctional. The Commission was re-designed in 1992 through a Statute that created a Commission on Emergency Situations under the Cabinet of Ministers. The Commission was headed by one of the deputies of the Prime Minister and was responsible for emergency management at all governmental and administrative levels including economic and industrial spheres (Fig. 5.1).

With several legal and structural adjustments attempting to develop and improve the new emergency management system, a series of key legislative arrangements followed in subsequent years, some of which are summarized below (ARFHN 2013a):1997 presidential decree "About Fire Safety;" 1998 presidential decree "About Civil Defense;" 2000 presidential decree "About Technical

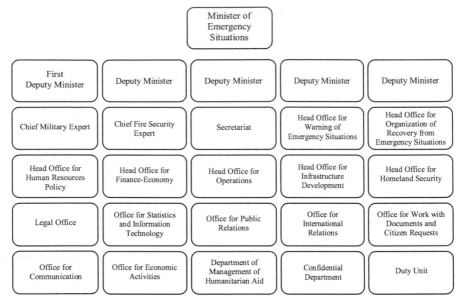

Fig. 5.2 Structure of ministry of emergency situations of Azerbaijan (2005–present). (ARFHN 2013b)

Safety;"2000 updated presidential decree "About Civil Defense;" 2001 executive order about "Framework Convention on Civil Defense Assistance;" and 2002 presidential decree "About Enhancement of Rules of Issue of Special Permission (license) for Some Types of Activity."

Despite the system in place, there was a need for a more comprehensive and sophisticated emergency management system to deal with natural (geological, climatic, etc.) and man-made (industry, urbanization, etc.) complexities. Such a need pushed for related reforms targeting at creation of specialized agency that would deal with potential and imminent hazards threatening the critical infrastructure, and the lives and properties of the population. The expected reform occurred in 2005, when the President Ilham Aliyev signed a presidential decree that resulted in creation of the Ministry of Emergency Situations, responsible for prevention of and response to the natural, technological and man-made disasters as well as coordination of related activities under single agency. It is important to note that, while previous structure was just a Commission under the Cabinet of Ministers, the newly-created structure was an autonomous ministry specifically responsible for management of all types of emergencies. As shown in Fig. 5.2, the Ministry is headed by a Minister, who has five deputies responsible for administration of several offices.

The above-mentioned Ministry structure encompasses a number of specialized agencies, some of which are: Crisis Management Center, Special Risky Rescue Service, State Service for Fire Protection, State Agency for Construction Safety Control, State Agency for Regulation of Nuclear and Radiological Activities, and Academy of Ministry of Emergency Situations. In addition, the emergency management system was divided into nine geographically-arranged regional centers, namely, Baku Regional Center, Northern Regional Center, Southern Regional Center, Northwestern Regional Center, Sumgait Regional Center, Ganja Regional Center, Karabakh Regional Center, Aran Regional Center, and Mughan Regional Center (ARFHN 2013c).

The main goal of the Ministry of Emergency Situations was to provide a comprehensive framework for civil defense functions, especially for the response and recovery stages of the emergency management cycle. These functions were to be performed under a single umbrella agency with the main purpose of coordinated action. While being still far from a perfect emergency management system that covers mitigation, preparedness, response and recovery stages of emergency management cycle, the above-mentioned attempt to create an independent agency with all-hazard approach to emergency management was a crucial development for the country. The subsequent years were characterized by legislative reforms aiming at development and improvement of critical infrastructure protection, and ministry structure and employee standards. The main developments were related to such issues as telecommunication, environment, construction codes, water resources, fire and radiological safety, and oil-gas export pipelines protection. In addition to legal and structural improvements, the government signed several cooperation and mutual-aid agreements with Russia, Turkey, France, Germany, South Korea, Ukraine and Belarus.

5.3.2   2010 Kura River Flood

Being a geographically diverse and an industrially developed country, Azerbaijan faces a threat of several natural and man-made disasters. While both types of disasters are possible, the past experience shows that natural disasters prevail. In this regard, the country is mostly affected by such natural disasters as earthquakes, floods, landslides, volcano eruption and erosion, while man-made disasters are possible due to the industrial activities related to oil and natural gas production as well as unplanned, and unregulated urbanization. Floods, however, are among the most frequent disaster types, thus, requiring specific attention and address. Almost every year Azerbaijan faces floods of different scope and severity, causing damage to residential and agricultural areas.

One of the recent floods the country has experienced is the flood of Kura and Araz rivers of 2010. In May and April months of 2010 Azerbaijan was struck by one of the destructive floods in recent decades. The provinces around the major rivers Kura and Araz were severely impacted by the flood that destroyed or damaged thousands of residential homes, public facilities, major infrastructure, and plantation areas (Fig. 5.3).

The impacted population was forced to leave their homes and transferred to public facilities or shelters during evacuation. While the consequences in some of the river-bank-close provinces were characterized by devastated plantations, in others they were characterized by more severe results such as damaged houses and infrastructure. Despite controversial claims by different sources, 5 people were reported to be killed during the disaster (Gülalıyev 2010).

Fig. 5.3 Sight from impacted area after Kura river flood. (Photo: Courtesy of Maarif Chingizoglu, RFE/RL reporter)

While there are many possible institutional, structural, economic, or even political explanations for the causes and/or consequences of the Kura river flood, it would be also useful to look at the issue from disaster management perspective. Mammadov and Verdiyev (2009) state that Azerbaijan does have an experience of dealing with floods especially through structural and engineering measures (water basins, bank construction, dams, etc.). Nevertheless, the authors argue, little attention is paid to development of non-structural measures such as flood insurance, flood forecasting, and early-warning systems.

5.3.2.1 Response and Recovery

Despite significant efforts to base the content of this chapter on as many reliable sources as possible, the account of the disaster and its consequences, apart from popular media, appears to be limited, both in terms of governmental and non-governmental sources. Two major sources were extensively used for the purposes of this chapter, namely, the information and reports provided on the official website of the Ministry of Emergency Situations, and the independent reports provided by a group of experts representing different nonprofit organizations called "Kura" Civil Society Headquarter (KCSH). The latter was specifically established with the aim to monitor the emergency response and recovery efforts and identify issues for consideration and improvement. Both of the sources mentioned were utilized in order to get a balanced and objective perspective on the issue.

Based on the information from the Ministry's website, the flood impacted 11 provinces/cities, namely Sabirabad, Salyan, Saatli, Imishli, Zardab, Kurdamir, Hajigabul, Neftchala, Beylagan, Fuzuli, and Shirvan. The number of affected population is reported to be around 4,200, with the largest impact being in Sabirabad (ARFHN 2013c). According to Kura Civil Society Headquarter, on the other hand, the number of affected provinces/cities is reported as 40. Some 30,000 residential homes were damaged, and around 1,10,000 ha of plantation areas were flooded (Gülalıyev 2010). As a result, thousands of people were forced to leave their homes and be transferred to shelters and public facilities until recovery process is over. Most of the critical infrastructure and public facilities like schools and hospitals as well as private businesses were damaged and became dysfunctional for a long period.

The first response by the government was to establish a State Commission that would deal with and monitor disaster management efforts. The government paid due attention to the disaster immediately in the first days after the flood struck, and the President visited the region to monitor the situation. In addition, the regional headquarter of the Ministry of Emergency Situations was transferred to Imishli and Saatli for better and more effective management of the disaster. The Cabinet of Ministers, in addition, held its meetings in Shirvan, one of the impacted cities, with the aim to have a full control of the situation and make proper decisions. Accordingly, the local and regional resources were mobilized to evacuate the impacted population, which was consequential in terms of minimizing human life loss. Food, clothes, sheltering and medical resources were immediately provided to the affected people, with further efforts to re-establish the functionality of the main infrastructure (Gülalıyev 2010).

In addition to the immediate response, the government took additional step to identify the scope of the damage caused by flood. On July 9 of 2010 the Cabinet of Ministers passed the 197s order that envisioned creation of commissions comprising representatives from different ministries in order to observe and monitor the consequences of the disaster in the field (ARFHN 2013c). The order projected allocation of 300 million manats ($ 240 million) for disaster recovery purposes, that would be distributed to the Ministry of Emergency Situations, the Ministry of Transportation, and the Irrigation and Water Management Joint Stock Company (Gülalıyev 2010).

As a result, the government started construction of new houses in provinces and cities that were drastically impacted by the flood (Fig. 5.4). Construction was envisioned for citizens whose houses were not possible to be restored. Five new residential districts were constructed in five provinces/cities, namely Sabirabad, Saatli, Salyan, Imishli and Zardab. The new districts included such public facilities as school, kindergarten, medical facility, library and post office.

The main infrastructure utilities such as water, electric, natural gas, and heating systems were also established, along with a total of 1,12,500 m^2 asphalt within the districts. As of April of 2013, out of 95 public facilities that were envisioned for recovery purposes, 90 had been already completed, among which 49 were restored, 40 were newly-built, and 1 was strengthened. Beyond residential houses, the restoration activities included such facilities as mosques, government buildings, elementary schools, and maternity hospitals (ARFHN 2013c).

As of April of 2013, a total of 4,217 residential houses had been completed and inhabited. In addition to restoration of damaged and construction of new houses, the disaster victims received one-time payments depending on the scope of damage to their properties or lands. Accordingly, based on order 71s that was passed on March 4 of 2011, 14,254 families received approximately 20 million manats ($ 16 million). Likewise, based on order 231s that was passed on August 8 of 2011, a total of 13,178 families received a one-time payment in the amount of 6,005 manats ($ 4,804) (ARFHN 2013c).

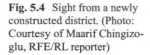

Fig. 5.4 Sight from a newly constructed district. (Photo: Courtesy of Maarif Chingizo-glu, RFE/RL reporter)

In order to get additional insight about the above-mentioned issues, some of the disaster victims were interviewed in June of 2013. Indeed, some of the victims were satisfied with the situation and the level of compensation/aid they had received. Accordingly, the government provided timely response and sufficient resources to eliminate the negative results of the disaster.

In addition to restoration activities related to residential areas and public facilities, the government implemented environmental measures to mitigate the impact of possible future floods on Kura and Araz rivers. For these purposes, a new channel was opened to Araz river in order to alleviate the speed and amount of water flow. In addition, the damaged dams and levees around Kura and Araz rivers were repaired and strengthened. Some related activities also included river-engineering techniques targeted at widening of the banks and river-depth control. A total of 48 million manats ($ 38.4 million) was allocated and expended for the purposes of the above-mentioned disaster recovery activities (ARFHN 2013c).

Lastly, following the disaster, the Ministry of Emergency Situations to some extent continued its efforts beyond structural measures. Accordingly, the Ministry signed agreements with two private companies from Netherlands, namely Deltares and Royal Haskoning, specializing in water management and engineering consultancy respectively. The joint efforts were targeted at development of forecasts and river basin management models, identification of potential risks and related solutions, and management of water reservoirs along Kura and Araz rivers. Appropriate suggestions were made and prepared for implementation through government programs (ARFHN 2013c). Related structural and engineering measures such as riverbed deepening (FHN Qəzeti 2012) and river bank strengthening (FHN Qəzeti 2013) in Kura river have been implemented over the course of the last 3 years.

5.3.2.2 Critiques

While government efforts to recover from the disaster were substantial, several non-governmental sources reported instances of mismanagement and misuse related especially to distribution and expenditure of resources allocated for recovery purposes (Mirqədirov 2010). One of the significant related reports is the "Kura" Civil Society Headquarter's report of 2010. The report mainly mentions several mismanagement problems as well as discrepancies between projected and reported expenditures of allocated state budget, some of which are:

Public Awareness and Communication: Most of the activities performed during recovery stage have been poorly reported in media, with little reference to the nature and scope of the activities. In addition, the initial order 197s has never been published on the Ministry's website or anywhere publicly.

Damage Assessment: The State Commission created to manage the disaster situation performed no, partial, or non-objective damage assessment. Damage assessment was also unprofessional in terms of methodology and techniques.

Compensation and Aid: The monetary compensation and/or aid envisioned to be delivered once a week was delivered in many instances with delays. In addition, the amount of money delivered was not compatible with real needs and minimum wage standards. There were many cases in which no explanation or receipts were given to the receivers of compensation and/or aid. Lastly, the amount of money given to victims as a compensation for house restoration purposes was significantly below required amounts. Again, no explanation was given in terms of what standards or methodology was utilized to identify the amount of compensation.

Aid Distribution: Several instances of corruption and appropriation were determined during aid distribution process. In this regard, some of the victims sued related public officials and got compensated appropriately.

Environmental Issues: Address of environmental issues following the disaster has been limited. Erosion of lands and trees as well as congestion of poisonous water around residential areas are among issues for consideration (Gülalıyev 2010).

The above-mentioned issues were also confirmed by some of the interviewed victims that were dissatisfied with how government responded to disaster and subsequent measures of recovery. The main emphasize of the victims was the fact that compensation/aid provided to them was not sufficient enough to repair their houses or build new ones. In addition, some claimed that it was government that was responsible for not properly monitoring the pre-disaster conditions, thus, leading to disaster. Yet others mention insufficient attention by government officials towards disaster victims in specific areas impacted by the flood.

Perhaps the most important critique of the way the government managed the disaster was about lack of comprehensive and sophisticated program to respond and recover from the disaster. In other words, the way government response was reactive in nature, thus, lacking connection with long-run policies. For example, opening a new channel from Araz river in order to alleviate the scope and speed of the river flow was decided and performed without proper analysis of and linkage to environmental issues. Moving victims from impacted areas to newly-constructed districts, likewise, was an immediate decision disregarding demographic and economic policies of the region (Həsənov 2010).

5.4   Linking Disaster to Development

Disasters are events that test societies' level of resilience, thus, presenting them an opportunity for change if the current system is ineffective (Boin et al. 2008). They are "focusing events" that are unusual in nature and scope, and have an ability to shape post-disaster policies (Birkland 1997). The Kura river flood of 2010 in Azerbaijan is one of such cases that presents an opportunity for change in disaster management as well as social and economic policies. While the country does have experience of dealing with disasters like floods, most of the practices focused on structural measures such as building dams or widening of the river banks rather than on non-structural measures that focus on long-run projects and policies (Mammadov and Verdiyev 2009). Such a case, in turn, creates systemic gaps between desired economically sustainable resilient communities and current situation characterized by reactive approach. Every time a disaster strikes, thus, community faces social and economic shocks that prevent it from long-run undisrupted development.

In case of disaster management in Azerbaijan, proactive approach that links mitigation with community planning is one of the issues to be addressed and institutionalized at the governmental level. The claims by non-governmental entities and media shows that post-disaster recovery is not performed in a professional and/ or systematic way (Gülalıyev 2010). First of all, despite government's efforts and some best practices during post-disaster situation, there have been areas and communities that have been paid little or no attention. This is mostly due to irregular and non-standard approach to risk and damage assessment. Secondly, there was a problem with standards and criteria regarding how and which areas, victims, and issues were addressed, showing a methodological concern in dealing with post-disaster situation. Thirdly, the recovery process has been reactive in nature, thus, characterized by government expenditures "for the sake of expenditures", rather than eliminating systemic problems. Lastly, the recovery measures have been implemented with little, or, in some instances, no connection to human, social, economic and environmental issues.

Accordingly, construction of new houses in newly established districts, displacing disaster victims forcing them to leave the agricultural areas, opening a new channel from Araz river with no substantial ecologic analysis, compensations in negligibly small amounts are among issues that depict the situation, which is considerably different from what Ahrens and Rudolph (2006) see as part of sustainable development process in which "the exploitation of resources, the direction of investments, the orientation of technological development and institutional change are all in harmony and enhance both current and future potential to meet human needs and aspirations" (p. 208). The lack of systematic and professional disaster management measures as well as missing connection to community planning and development make it imperative to develop a more robust and long-run-based policies, programs and projects.

One should note that because Azerbaijan is a country with a relatively short history of political and economic independence, and, accordingly, with a short history of emergency management system, the government's performance and management of pre-disaster, disaster, and post-disaster situations deserves substantial credit. While not outstanding and sufficient in nature and scope, the government's efforts are a good example of political will, management and leadership every community needs in times of disasters. Considering the fact that Azerbaijan faces disasters like floods quite frequently, on the other hand, the government should improve the way it manages disasters throughout all stages of a disaster cycle.

5.5   Conclusion

This chapter covered the case of the 2010 flood disaster in a post-Soviet country Republic of Azerbaijan. While the country's emergency management system is relatively new, disasters have been effective in shaping the way emergencies and disasters are tackled. While the system is partially effective in terms of immediate response, as reflected in the government data, independent reports and media, it is impossible to conclude in the same way about long-run measures and policies.

The main finding appears to be the fact that Azerbaijan's emergency management system is reactive, rather than proactive, in nature, thus, imposing limits on sustainable economic and social development. This, in turn, means the government elaborates on structural or engineering measures of mitigation and recovery, rather than on regulatory or systemic policies or projects integrated into emergency management system. The system may be improved by investing into non-structural measures and long-run policies that seek connection between mitigation and community planning. Specific adjustments are required in terms of provision of professional, expertise-based, methodology-educated and fair process during emergency response stage.

This study has limitations, specifically in terms of obtaining relevant government reports about post-flood government activities. The scarcity of literature on the country's emergency management system was another hindrance to effective analysis of structural and institutional characteristics of the system. Further research on different types of disasters or a different post-Soviet country might give additional insight about the nature and effectiveness of emergency management systems in the region.

References

Ahrens, J., & Rudolph, P. M. (2006). The importance of governance in risk reduction and disaster management. *Journal of Contingencies and Crisis Management, 14*(4), 207–220.

ARFHK. (1992). *Azərbaycan Respublikası Nazirlər Kabinetinin fövqəladə hallar üzrə Komissiyası haqqında Əsasnamə. Azərbaycan* Respublikasının Fövqəladə Hallar Nazirliyi: http://www.fhn.gov.az/uploads/legislation/aze/rmerkez/d%C3%B6vlet_sistemi1992.pdf. Accessed 1 April 2013.

ARFHN. (2013a). *Azərbaycan Respublikası Prezidentinin fərman və sərəncamları*. Azərbaycan Respublikasının Fövqəladə Hallar Nazirliyi: http://www.fhn.gov.az/?aze/menu/18. Accessed 1 April 2013.

ARFHN. (2013b). *Nazirliyin strukturu*. Azərbaycan Respublikasının Fövqəladə Hallar Nazirliyi: http://www.fhn.gov.az/?aze/menu/12. Accessed 1 April 2013.

ARFHN. (2013c). *Təbii və texnogen mənşəli fövqəladə halların aradan qaldırılması*. Azərbaycan Respublikasının Fövqəladə Hallar Nazirliyi: http://www.fhn.gov.az/index.php?aze/pages/33. Accessed 13 July 2013.

Birkland, T. A. (1997). *After disaster: Agenda setting, public policy, and focusing events*. Washington, D.C.: Georgetown University Press.

Boin, A., McConnell, A., & 't Hart, P. (2008). Governing after crisis. In A. Boin, A. McConnell, & P. 't Hart (Eds.), *Governing after crisis*. Cambridge: Cambridge University Press.

FEMA, F. E. (2012). *Mitigation's value to society*. Federal Emergency Management Agency: http://www.fema.gov/library/viewRecord.do?id=3031. Accessed 20 Jan 2013.

FHN Qəzeti. (2012, April 12). *Fövqəladə hallar naziri Kür boyunca dib-dərinləşdirmə işlərinin təşkili üçün Neftçalada inşa edilən əsas bazada olmuşdur*. Fövqəladə Hallar Nazirliyinin Qəzeti: http://www.fhn.gov.az/newspaper/?type=view_news & news_id=2329. Accessed 1 April 2013.

FHN Qəzeti. (2013, January 12). *FHN Kür çayında dibdərinləşdirmə işləri aparır*. Fövqəladə Hallar Nazirliyinin Qəzeti: http://www.fhn.gov.az/newspaper/?type=view_news & news_id=2850. Accessed 1 April 2013.

Godschalk, D. R. (2007). Mitigation. In W. W. Waugh & K. Tierney (Eds.), *Emergency management: Principles and practice for local government* (2nd ed.). Washington, D.C.: ICMA.

Godschalk, D. R., & Brower, D. J. (1985). Mitigation strategies and integrated emergency management. *Public Administration Review, 45*(Special Issue), 64–71.

Gülalıyev, O. (2010). *Kür daşqınlarının yaratdığı problemlərlə bağlı fəlakət bölgəsində aparılmış monitorinqin nəticələrinə dair aralıq hesabat*. "Kür" Vətəndaş Cəmiyyəti Qərargahı. http://www.kur-az.com/jurnal/xburaxilis.pdf. Accessed 1 April 2013.

Həsənov, S. (2010). *Kür daşqınları; səbəb, nəticə və çıxış yolları*. "Kür" Vətəndaş Cəmiyyəti Qərargahı. http://www.kur-az.com/jurnal/xburaxilis.pdf. Accessed 1 April 2013.

Kapucu, N., & Garayev, V. (2013). Mitigation and emergency management. In A. Jerolleman & J. Kiefer (Eds.), *Natural hazard mitigation: A guidebook for practitioners and academics* (pp. 19–42). Boca Raton: CRC Press.

Kapucu, N., & Ozerdem, A. (2011). *Managing emergencies and crises*. Boston: Jones & Bartlett Publishers.

Mammadov, R., & Verdiyev, R. (2009). *Integrated water resources management as basis for flood prevention in the Kura river basin*. United Nations Economic Commission for Europe: http://www.unece.org/fileadmin/DAM/env/water/meetings/flood/workshop%202009/presentations/session%203/Mammadov_Verdiyev_article.pdf. Accessed 1 April 2013.

McCarthy, F. X., & Keegan, N. (2009). *FEMA's pre-disaster mitigation program: Overview and issues*. Washington, D.C.: Congressional Research Service.

Mirqədirov, R. (2010). *Məmur məsuliyyətsizliyi*. "Kür" Vətəndaş Cəmiyyəti Qərargahı. http://www.kur-az.com/jurnal/xburaxilis.pdf. Accessed 1 April 2013.

Multihazard Mitigation Council. (2005). *Natural hazard mitigation saves: An independent study to assess the future savings from mitigation activities*. Washington, D.C.: National Institute of Building Sciences.

Pierce, L. (2003). Disaster management and community planning, and public participation: How to achieve sustainable hazard mitigation. *Natural Hazards, 28*, 211–228.

Rubin, C. B., & Barbee, D. G. (1985). Disaster recovery and hazard mitigation: Bridging the intergovernmental gap. *Public Administration Review, 45*(Special Issue), 57–63.

Schwab, A. K., Eschelbach, K., & Brower, D. J. (2007). *Hazard mitigation and preparedness*. Hoboken: Wiley.

Sylves, R. T. (2008). *Disaster policy and politics: Emergency management and homeland security*. Washington, D.C.: CQ Press.

Chapter 6
What Has Been Done? Louisiana After Katrina

John J. Kiefer and Alessandra Jerolleman

6.1 Introduction

This chapter provides an overview and assessment of the progress made in mitigating, preparing for, responding to, and recovering from disasters during the period since Hurricane Katrina struck New Orleans, Louisiana on August 29th, 2005. While the focus of the chapter will be Louisiana, many of the changes that are examined extend far beyond the state and will therefore be included in this chapter. Indeed, a disaster of this magnitude engendered a re-examination of the nation's readiness that extended far beyond the region.

The chapter begins by describing the state of readiness of Louisiana in the years that lead up to Hurricane Katrina, including the role that development decisions played in exacerbating existing vulnerabilities. By any yardstick, people in south Louisiana were highly vulnerable to disaster, some due to economic circumstances, some not understanding the risk posed by hazards, and many by both. Levees and pumping stations were also at risk due to lack of maintenance and a high degree of reliance on human interaction during a disaster. Federal, state, and local legislation and policies were in place that crippled response and recovery; from an inflexible Stafford Act to state legislation that hampered the recovery with cumbersome and inefficient property acquisition to hastily enact "not in my backyard" laws that affected displaced populations. Federal policies also played a role in promoting unwise development decisions, both in terms of housing and economic development, which affected the extent of damages. Using the 5-year period before Katrina as a baseline, the assessment is made of the progress in three major areas: social, technical, and political systems. These areas have been generally accepted as important yardsticks for community resilience, and it is often where two or more of these

J. J. Kiefer (✉)
Department of Political Science, University of New Orleans, New Orleans, USA
e-mail: jkiefer@uno.edu

A. Jerolleman
University of New Orleans, Natural Hazard Mitigation Association, Metairie, USA

N. Kapucu, K. T. Liou (eds.), *Disaster and Development,* Environmental Hazards, 93
DOI 10.1007/978-3-319-04468-2_6, © Springer International Publishing Switzerland 2014

systems interact that major vulnerabilities in a community's ability to deal effectively with a disaster will occur (Leavitt and Kiefer 2006).

6.2   Land Use and Development Trends—Subsidizing Risky Behavior

Current federal policies serve to create subsidies for risky development decisions. These subsidies impact development and redevelopment, and include: federal grants for infrastructure without stringent land use and building code requirements for the areas which may then be developed; Stafford Act loans and grants following disasters with insufficient requirements to mitigate against likely risk; and, tax deductions for mortgage interest and losses (Conrad and Thomas 2013). Across the United States, development has continued to increase in at risk areas, leading to ever-increasing property exposure to risk from hazards such as flooding[1]. Although the National Flood Insurance Program (NFIP) does require certain standards, it also heavily subsidizes flood insurance premiums. Some researchers have even argued that the NFIP has had the unintended consequence of increasing development within the nation's floodplains by making insurance available (Jerolleman 2013).

Recent efforts have been made to reform the NFIP, particularly the Biggert-Waters Act of 2012 that begins to eliminate some of the subsidies, such as those for second homes; but the legislation faces many challenges due to the immediate hardships it imposes upon property owners in at risk areas. This is particularly challenging in Louisiana, where the premium increases will be sufficiently high as to potentially render some homeowners unable to pay their mortgages. Local political leaders, including several Parish presidents have been lobbying for changes to the Act, in particular for the return of the grandfather clause. They argue that residents who have built to the standards they were given are now being unfairly penalized due to re-mapping (Linderman 2013). Louisiana, and other Gulf Coast state, residents have already been impacted by significantly increased rates for homeowner's insurance following Hurricane Katrina.

However, without significant changes to current land use and development practices, the United States can expect disaster spending to continue to increase tremendously (Thomas and Bowen 2009). Under present policies, the federal government covers a great deal of the loss, thereby eliminating a significant disincentive to continuing to develop at risk areas, or continue to redevelop at risk areas without significantly higher standards. The federal government provides funding for redevelopment through dozens of programs among multiple agencies, much of which is not clearly tied to stringent hazard mitigation requirements. As a result, rebuilding following a disaster often results in the recreation of pre-disaster conditions (Godschalk et al. 1999; Thomas et al. 2011).

---

[1] Some communities have been successful at pursuing development while making sound mitigation decisions. These communities illustrate the fact that development is not inherently risky or unwise, it simply unwise to undertake development which does not fully take the natural environment into account.

A recent report released by the Brookings Institute in 2013 recommended the following three categories of reform to eliminate the extensive federal subsidies: (1) further reforms to the NFIP beyond Biggert-Waters; (2) greater local and private cost-sharing of disaster costs; (3) incentives for higher development standards including the implementation of higher standards for any type of federal support of construction, including reconstruction and infrastructure (Conrad and Thomas 2013).

These same factors, which have influenced development decisions across the United States, have helped to shape the New Orleans area. Although the location of New Orleans, within a deltaic plain at the foot of the Mississippi River, is inherently at a certain level of risk; it is also a very necessary location. However, the expansion of the city of New Orleans into the East and other at risk areas, while abandoning the historic practice of building elevated home, is a key example of development decisions increasing risk significantly for residents long after the developers have profited and left(Baxter unpublished).

In a recent case study of the development of New Orleans East, an area that was experienced some of the very worst impacts from Hurricane Katrina; Baxter (unpublished) found that comprehensive approaches to flood control were abandoned in favor of approaches such as levees that required less land. In other words, sound mitigation decisions and concerns were overlooked in the rush to make a profit. The developer effectively lobbied to have the city bear the costs of developing drainage infrastructure in order to be able to market the ideal of a single-family slab home. These homes were constructed on land that local engineers had described as uninhabitable, cost-prohibitive to develop, and bound to flood. In fact, the homes began to experience slab issues, due in part to the use of insufficient fill and other construction practices, within 5 years.

Another example of unwise decisions justified for the purposes of potential economic development is the Mississippi River Gulf Outlet (MRGO), a navigation canal connecting New Orleans to the Gulf of Mexico. Initial efforts to obtain funding for the MRGO focused on the economic benefits of the connection and attempted to downplay the significant environmental concerns (Freudenburg et al. 2008). The MRGO devastated wetlands and exacerbated the impacts of Hurricane Katrina, becoming a key factor in the death and destruction that ensured (Freudenburg et al. 2008). Unlike the earlier example, in which development focused on housing, this example focuses on the concept of economic development and growth of industry. In both cases, there were very real impacts to residents and significantly increased risk. This same tension between land-use, housing decisions and mitigation is present in the post-Katrina recovery landscape, as will be described further below.

6.3   "The Big One"—Louisiana and Katrina

Hurricanes are complex events, from the warnings to the evacuation, response, and recovery. The impacts of the losses caused by Katrina are significant. Thousands of homes and businesses and even entire neighborhoods in the New Orleans metropolitan area were destroyed by flood. Strong winds also caused damage in the

New Orleans area, including downtown where windows in some high-rise buildings were blown out and the roof of the Louisiana Superdome was partially peeled away. Many other structures in Louisiana that avoided surge or fresh water floods, including some areas well inland, were damaged by strong winds and tornadoes (Leavitt and Kiefer 2006).

The economic and environmental ramifications of Katrina have been widespread and long lasting due to impacts on large population and tourism centers, the oil and gas industry, and transportation. The hurricane severely affected or destroyed workplaces in New Orleans and other heavily populated areas of the northern Gulf coast, resulting in thousands of lost jobs and millions of dollars in lost tax revenues for the impacted communities and states. A significant percentage of United States oil refining capacity was disrupted after the storm due to flooded refineries, crippled pipelines, and several oil rigs and platforms damaged, adrift or capsized. Several million gallons of oil were spilled from damaged facilities scattered throughout southeastern Louisiana. Key transportation arteries were disrupted or cut off by the hurricane. Traffic along the Mississippi River was below normal capacity for at least 2 weeks following the storm. Major highways into and through New Orleans were blocked by floods. Major bridges along the northern Gulf coast were destroyed, including the Interstate 10 Twin Span Bridge connecting New Orleans and Slidell, Louisiana. The American Insurance Services Group (AISG) estimated that Katrina was responsible for $ 40.6 billion of insured losses in the United States. A preliminary estimate of the total damage cost of Katrina in the United States is assumed to be roughly twice the insured losses, or about $ 81 billion. This figure makes Katrina was by far the costliest hurricane in United States history. Even after adjusting for inflation, the estimated total damage cost of Katrina is roughly double that of Hurricane Andrew (ISO Property 2010).

6.4   Are Our Citizens Ready?

After the levees around New Orleans failed and floodwaters inundated most of the City of New Orleans, evidence of social vulnerability was nowhere more evident than in the faces of the thousands who flocked to the Louisiana Superdome to seek safety. The scene was played out thousands of times over network television and in printed media. Vulnerable populations living in the city either did not heed the evacuation warnings, did not trust the warnings, or did not have the resources to evacuate (Kiefer and Montjoy 2006). The reasons for the existence of the rather large population of vulnerable citizens in south Louisiana before Katrina struck are many and complex. Some of the most important factors include: An economy largely based on low-wage service jobs; Relatively low levels of attainment of public school students, especially in the City of New Orleans; The region's history of racially discriminatory practices; Minority populations' distrust of their government; A culture of government dependency; Underfunded and mismanaged public housing; and Public health care that was often difficult to access.

It is difficult to assess the degree to which our social systems have recovered since Katrina. There has been a large shift in the region's population, based to some degree on differences in socio-economic status. Populations in some of the region's most vulnerable communities are now significantly lower than they were before Katrina struck. St. Bernard Parish, just outside New Orleans, struggles to meet Parish operating costs. Much of Plaquemines Parish and eastern New Orleans remain devoid of businesses. Many commercial properties remain out-of-commerce. Despite the valiant and impressive efforts of the "Make-it-Right" foundation (spearheaded by Brad Pitt and Angelina Jolie) and Musicians' Village (spearheaded by Branford Marsalis & Harry Connick Jr.) in the 9th Ward, only about one-fifth of the 9th Ward's population has returned (Liu and Plyer 2008).

Even though our country moved through the challenges of the recession, south Louisiana has fortunately been largely sheltered from the recession's impact due to the influx of jobs related to the long-term recovery process. The result was that the region's net loss of jobs was significantly less than the national average loss. At the same time, average wages post-Katrina grew at an impressive 14 %, and the number of schools that meet state performance standards has increased by 44 % (REIS 2010).

6.5   An Evaluation of Technical Readiness

It will take several more major hurricanes to truly test any improvements in the technical readiness of south Louisiana's technical systems such as pumping stations and levees. Many pumping stations have been reinforced to withstand sustained hurricane winds, and facilities have been hardened to ensure the safety of pump operators during the worst weather. A competency-based framework has, due to public outcry, replaced the old system of using a system of patronage in appointments to local levee boards.

Evacuation Planning; Hurricane Katrina clearly demonstrated very poor planning, and the results shook the nation. The then-mayor, Ray Nagin, announced that the Louisiana Superdome, home to the National Football League's New Orleans Saints, would open as a "shelter of last resort." As multiple levees failed at the 17th Street, London Avenue, and Industrial Canals, water began to rise in the streets of New Orleans. Many living in heavily flooded areas crawled to their roofs and waited in desperation for rescue. Others drowned in the floodwaters. Emergency teams rescued over 60,000 people along the Gulf Coast from rooftops and flooded homes, according to the United States Coast Guard. Yet almost 25,000 people were stranded at the Superdome and the Ernest N. Morial Convention Center in days after the storm (Dwyer and Drew 2005).

The New Orleans Office of Homeland Security and Emergency Preparedness (NOHSEP) developed, wrote, and coordinated the execution of the City Assisted Evacuation Plan (CAEP). The CAEP services citizens who need to evacuate during an emergency, but who lack the capability to self-evacuate. It is an evacuation method of last resort, and only for those citizens who have no other means to leave, or have physical limitations that prohibit self-evacuation. Upon the declaration of a mandatory evacuation by the Mayor of the City of New Orleans, emergency managers activate the CAEP. Under the operations plan, the CAEP has approximately 42 h to collect residents at 17 neighborhood pick-up points throughout the city of New Orleans. Four of these pick-up points are specifically equipped to handle senior citizens and the disabled. Buses begin to pick people up at neighborhood pick up points 54 h before tropical storm force winds hit Louisiana coastline (H-hour) and can continue until 12 h prior to landfall. The CAEP provides evacuees with a round trip ticket to and from a state or regional shelter out of harm's way during the storm.

Emergency managers implemented the CAEP for the first time on Saturday, August 30, 2008, in advance of Hurricane Gustav. The CAEP evacuated over 20,000 citizens, more than 300 pets and over 14,000 visitors from New Orleans in less than 35 h. Residents reported to the 17 pick-up points, and New Orleans Emergency Medical Services (NOEMS), in association with the Regional Transit Authority, provided door-to-door service with paratransit buses for residents who identified themselves as physically unable to get to a pick-up point. The successful implementation of the CAEP resulted from mechanisms coordinating and deploying the assets of various local, state, and federal agencies and organizations. The CAEP is compliant with National Incident Management System (NIMS) and uses Incident Command Structure (ICS) structure, currently the national approach to disaster management. NIMS provides a systematic, proactive approach to guide departments and agencies at all levels of government, nongovernmental organizations, and the private sector, to work seamlessly to prevent, protect against, respond to, recover from, and mitigate effects of disasters, regardless of cause, size, location, or complexity (FEMA 2010).

In April of 2009, the University of New Orleans's Center for Hazard Assessment, Response and Technology (CHART) program surveyed 700 people who utilized the CAEP during Hurricane Gustav to gauge the participants' satisfaction with the CAEP. Results showed that citizens generally regard the CAEP as a successful strategy. Nearly three-quarters of evacuees reported satisfaction with their experience, and would use the CAEP again. Almost 70 % of participants rated their re-entry experience as "good" or better and over half of the participants rated transportation out of the city as "good" or better. Study findings showed evacuees were generally willing to listen to government officials, cooperate with them and contribute to the effectiveness of evacuation efforts. A majority of these evacuees indicated that evacuation preparedness had improved considerably since Hurricane Katrina. While there is work to do to improve the CAEP, Hurricane Gustav's evacuation and resulting research concluded that compared to Hurricane Katrina, the CAEP allowed a much more orderly, pre-emptive evacuation of New Orleans' residents without reliable transportation (Kiefer and Jenkins 2009).

Communications The most notable flaw during Hurricane Katrina evacuation efforts was the breakdown in emergency communication. Vertical communication failures occurred between local, state and federal authorities; horizontal failures continued between regions and within municipalities. In New Orleans, storm surges destroyed one communications tower, while floodwaters damaged another two causing a catastrophic failure in communications throughout the city. In

Plaquemines Parish, the storm destroyed the only communications tower and the Emergency Operations Center (EOC). As a result, all 911 communications failed for 3 weeks. In St. Bernard Parish, winds destroyed communications towers, antennas, and communications buildings; both sheriff and fire department personnel had to be evacuated. Additionally, the Sheriff's Office in Jefferson Parish lost its primary communications tower.

Prior to Katrina, "stovepipe" systems in place allowed first responders to communicate within their own jurisdiction or agency, but not with others. Police could not talk to the fire department and ambulances could not coordinate with police. This poorly managed communications infrastructure led to ineffective coordination of resources, lack of situational awareness, and put first responders, military personnel, and civilians in harm's way (NOUASI 2007).

After Katrina, emergency managers recognized and placed much greater emphasis on completing the interoperable communications system. The strategic plan called for an effective, redundant, voice and data communications systems that shared law enforcement and first responder information across parish and jurisdictional boundaries during hazardous incidents and day-to-day operations (New Orleans Urban Area Security Initiative 2007). The Office of Emergency Communications (OEC), under the Department of Homeland Security that was created out of the Post-Katrina Emergency Management Reform Act, provides technical assistance, coordinates regional emergency communication efforts, and conducts outreach to all grant funded areas. State and federal partners developed a Tactical Interoperable Communications Plan (TICP).

Nursing Homes Significant improvement has been made in the region's nursing home readiness. During Katrina, twenty-one nursing homes were evacuated out of sixty nursing homes affected. Twenty-two residents died at the Lafon Nursing Home of the Holy Family (Hull and Struck 2005), and thirty-four perished at St. Rita's nursing home. The owners of St. Rita's were charged, and acquitted, of thirty-four counts of negligent homicide after residents perished without being evacuated (Associated Press 2007).

During Hurricane Katrina, nursing homes who evacuated reported a wide range of significant problems with the evacuation. Some of the most common problems included transportation contracts that were not always honored, that evacuation travel took longer than expected, that medication needs complicated travel, that host facilities were unavailable or inadequately prepared, that facilities could not maintain adequate staff, that food and water shortages occurred or were narrowly averted, and that prompt return of residents to facilities was difficult (Fink 2012).

Perhaps one of the most difficult problems to overcome in evacuating nursing home residents is the significant increase in mortality during and in the weeks after evacuation. Since Katrina, all nursing homes that receive federal funding for Medicare and Medicaid must have an emergency and evacuation plan. Compliance with the planning requirements has been viewed as successful, with compliance rates for emergency planning and training at 94 and 80 % respectively (Fink 2012).

Housing Many of the homes that were rebuilt following Hurricane Katrina were required to be rebuilt to a higher elevation due to changes in the base flood elevation coupled with the level of damage, and the NFIP's substantial damage requirements. However, the base flood elevation as designated is below the flood of record in many areas of the city. In other words, were New Orleans to experience another catastrophic levee failure, these homes would again flood. These homes are now less likely to flood from drainage events or from hurricanes that do not overwhelm the levee protection or pumping capacity. Additionally, home elevation may reduce the potential for flooding, but does not necessarily eliminate the need for evacuation.

Many homeowners sought to avoid elevating their homes by seeking to have their properties identified as not having experienced substantial damage. This was due to a desire to rebuild quickly, as well as the very challenge of funding home elevation. Although mitigation funding was made available to many homeowners, the application process was lengthy and the funds available were often insufficient; particularly as the market for home elevations increased and the costs rose accordingly. Additionally, a great deal of the home elevations were managed through the Road Home Program, a program that raised significant controversy due to issues with the appeals process and contractor (Jerolleman 2013).

6.6   Policies that Promote Resiliency

Pets: Several important new policies that promote resiliency have been implemented post-Katrina. At the federal level, the Pet Evacuation and Transportation Standards Act of 2006 (Pets Evacuation and Transportation Standards Act of 2006) ensures that emergency preparedness operational plans take into account the needs of individuals with household pets and service animals, before, during, and after a major disaster. Much of the impetus for this new policy resulted from the media images in the aftermath of Katrina—those of animals being left behind or not allowed onto evacuation buses.

At the local level, the City of New Orleans Evacuation plan now has specific language for pets in the evacuation system. For example, if an evacuee has small pet (about 15 pounds) that can be held on the lap or can fit in a lap carrier, it can ride along with the evacuee on the bus. When an evacuee has a pet that is not allowed on an evacuation bus, a police officer at the pick-up location will call the SPCA to dispatch a vehicle that can bring the pet to the appropriate pick-up location (NORTA 2013). These pet-oriented policy changes have already been related to improved evacuation outcomes. In an analysis of the evacuation for Hurricane Gustav, evacuees (Kiefer and Jenkins 2009) reported not one pet-related concern or incident.

Collaborative Planning The Post Katrina Emergency Management Reform Act was the most significant Federal law passed in the aftermath of Katrina to change the way the Federal government responded to emergencies. Key highlights of this Act include the implementation of a "Whole Community" plan of emergency response that involves government, private, private non-profit, and citizens in response and recovery effects (US House of representatives 2011). The Act created thirteen regional and three national full time Incident Management Assistance Teams, greatly expanding the coverage of Federal response teams. Prior to Katrina, FEMA had Emergency Response Teams that had primary day-to-day duties in other areas, precluding their effective response to disasters.

A New Mass Care Strategy The Act also provided a National Mass Care Strategy, "Providing a framework to strengthen and expand resources available to help shelter, feed, and provide other mass care services by pooling expertise and identifying partnership opportunities" (US House of representatives 2011). This strategy called for both FEMA and the American Red Cross to lead the mass care portion of Emergency Support Function (ESF) #6 to facilitate the planning and coordination of mass care services.

The National Disaster Recovery Framework In September of 2011, the National Disaster Recovery Framework was implemented to "provide the overarching interagency coordination structure for the recovery phase for Stafford Act incidents, and elements of the framework may also be used for significant non-Stafford Act incidents" (FEMA 2011). The Framework Lays out a recovery continuum to show where and where it does not overlap with the National Response Framework (NRF) and defined roles in recovery that previously had not been included such as: Core recovery principles; Roles and responsibilities of recovery coordinators and other stakeholders; Coordinating structure to facilitate communication and collaboration among all stakeholders, guidance for pre and post disaster recovery planning; The overall process by which communities can capitalize on opportunities to rebuild stronger, smarter, and safer(FEMA 2011)

The National Disaster Recovery Framework introduced six new recovery support functions Community planning and capacity building; Economic, health and social services; Housing; Infrastructure systems; Natural Resources; and Cultural Resources. It also established a National Emergency Child Locator Center that is responsible for establishing and maintaining a toll-free hotline and website to receive reports and provide information about displaced children and to provide the public with additional resources. This was necessary because "in the few months following Katrina and Rita, 5,192 children had been confirmed as either missing or displaced" (US House of representatives 2011).

Louisiana and City Policies Prior to and during Hurricane Katrina, no separate, customized plan existed to evacuate tourists and visitors in New Orleans. While some hotels offered guests the option to stay at the hotel during the course of the storm (Mowbray 2005), others forced guests to leave the hotel with no evacuation plan (Superdome evacuation 2005). With no other viable option, some hotel guests ended up at the Superdome and the Convention Center. Like everyone living at these shelters, they suffered lack of food, running water, and highly unsanitary, and unsafe conditions. Visitors able to remain at their hotels had significantly better experiences than those at shelters of last resort, but they too still experienced dangerous scenarios. Glass windows broke, looters attempted to gain access, and water flooded the first level of many hotels (Superdome evacuation 2005).

After Hurricane Katrina, the government received heavy criticism for its poor response in evacuating hotel guests. Additionally foreign tourists reported first responders and rescue crews gave preferential treatment to American citizens (Townsend 2005). Hurricane Katrina exposed major shortcomings in visitor evacuation planning. According to the Convention and Visitors Bureau, there are 30,000 hotel rooms that may be inhabited during hurricane season and there must be a plan for guests to safely evacuate the city.

After Katrina, state and local governments recognized the crucial need for a separate evacuation strategy focused exclusively on tourists and visitors. Together they established the New Orleans Hotel and Lodging Visitor Evacuation Plan (NOHLVED) in 2006. The plan activates either when the mayor's office initiates the CAEP or when a convention organization requests assistance in evacuating their attendees. Under this plan, visitors have three options. Anyone with prearranged air or ground transportation plans that does not require city evacuation assistance will be asked to leave the city immediately. Those whose departure dates do not coincide with the evacuation schedule will be told to contact their airline company or rental car company to change their date of departure in an effort to secure their own transportation out of the city. Finally, the hotel staff will direct those unable to do either to staging centers for bus transportation to Louis Armstrong International Airport.

To arrange for emergency evacuation plans, NOHLVEP keeps constant communications with commercial airline companies and emergency planners. Airlines ensure all visitors have an opportunity to leave the city via plane prior to the airport's closing, regardless of whether the visitor obtained a ticket for departure. In the event of an evacuation, many airlines will send more planes, while others may choose to send charter planes. The contribution made by airlines varies from company to company. Regardless, all visitors receive the chance to procure a seat on a plane leaving the city. The New Orleans Hotel and Lodging Visitor Evacuation Plan was viewed as necessary because of lessons learned during the Katrina evacuation.

Upgraded Emergency Operations Facilities In 2011, New Orleans implemented a much-needed upgrade to their EOC. New Orleans' City Hall is located in the Central Business District of New Orleans and opened for business in 1957. The New Orleans Office of Homeland Security and Emergency Preparedness (NOHSEP) is currently located on the ninth floor of this building. In times of crisis, the Emergency Operations Center (EOC) is made operational. During Hurricane Katrina, the City of New Orleans' EOC was ill equipped to handle the magnitude of the disaster and operational continuity suffered during the immediate crisis. Today, the New Orleans' EOC occupies a state-of-the art all-inclusive space designed to withstand a major hurricane and to remain operational during all-hazards events. Opened in 2009, the new EOC offers major improvements in square footage, communications capabilities and physical layout. Prior to the new EOC, FEMA consistently rated the City of New Orleans' EOC "poor" in terms of its physical ability to handle a major event pre-and post-Katrina. As late as 2008, before the new EOC opened, FEMA wrote that the New Orleans EOC lacks space to support interrelated functions adequately in a common area. The limited space hinders the ability of even a select few ESFs to coordinate agency and multi-agency activities effectively and efficiently (FEMA 2010).

Discussions surrounding a permanent EOC began while emergency responders worked inside the Hyatt hotel. From concept to opening, the City's new EOC took 4 years to complete. The new EOC measures 10,000 square feet and has a dedicated communications room with a satellite push to talk phone as well as a HAM radio system, National Warning System, and UHF-VHF radios. It also includes Broadband Global Area Network (BGAN) satellite dishes for loss of Internet, office cubicles for day-to-day staff activity and the ability to expand and house 96 people in the event of activation.

Sheltering The Stafford Act authorized FEMA assistance during the first 2 years post-Katrina. In 2009, the Senate Subcommittee on Housing reported that the federal housing response was inadequate, resulting in the needs of hundreds of thousands of citizens not being met. It went on to report that FEMA had engaged in an ad hoc response to the catastrophic housing needs on New Orleans's residents because FEMA lacked a fully developed response plan (U.S. Congress 2009).

FEMA's ad hoc response called for extensive use of manufactured houses—mostly trailers—that proved to be an inadequate shelter option especially for long-term housing. Over 80 % of housing trailers were placed on private property for individuals to live in as they rebuilt their own homes. Unfortunately, this ad hoc response created significant problems for local communities. The travel trailers provided by FEMA created additional vulnerability. Indeed, in the months that followed Katrina, local emergency managers had to deal with significant numbers of highly vulnerable trailers in their communities.

Nearly all of FEMA's housing assistance went to homeowners, with those renting being largely left out of the recovery process. While the Section 403 hotel shelter program was repeatedly extended for short-terms for renters, the result was intense uncertainty among non-home owning disaster victims who were not sure when they would have to leave their hotel rooms. Proposals to assist renters was blocked by FEMA's Office of General Counsel on grounds that the Stafford Act prohibited repairs of rental property.

Trailers were also extremely expensive. FEMA estimated that it cost $ 59,000 to provide and operate a trailer for 18 months, and FEMA spent $ 5.5 billion on the trailer program. Moreover, this estimate was made before the discovery of high levels of formaldehyde in the trailers. At this time, the full extent of the health consequences is unknown, but high levels of respiratory infections were reported in children living in the homes.

To attempt to alleviate these failures, the Post-Katrina Emergency Management Reform Act mandated nine provisions on FEMA housing policies. These included: Developing a coordinated housing strategy with federal, state, and local governments; Identifying the most efficient and cost effective federal programs to meet short- and long-term housing needs; Clearly defining the role, programs, authorities, and responsibilities of all levels of government; Describing in detail the programs offered by the different responsible organizations; Considering the methods of providing housing assistance where employment and other living resources are available; Describing programs directed to meet the special needs and low income populations and that of ensuring sufficient housing units for the disabled; Describing plans for operations of housing clusters; Describing plans for promoting the rehabilitation of existing rental housing; and Describe additional authorities needed to effectively implement the new plan.

Yet FEMA was slow to create the required plan, and the plan they did finally submit did not comply with the Act. FEMA finally submitted a compliant plan on the last day of the Bush administration in January 2009, although it still lacked operational plans. Nonetheless, it did lay out an approved strategy. Still, critics claim that FEMA still does not have adequate policies in place to meet temporary housing needs during disasters. LaCheen (2010) notes that FEMA still does not have a unified approach to temporary housing. He says this is especially important when required to meet the needs of citizens with disabilities and the special needs of children.

Unfortunately, when Hurricane Sandy hit the Northeast on the evening of October 29th, 2012, many of the problems seemed similar to those experienced after Katrina. Those initially displaced by Sandy were placed in shelters and hotels around the region. According to one estimate, about 34,000 people went to hotels and motels (Castellano 2012), costing over 300 million for temporary housing, mostly in the form of rental assistance (Serna et al. 2012). On November 8th FEMA started shipping in trailers for temporary housing. The initial shipment was a relatively small one hundred units because FEMA did not have an estimate of need (Heyboer and McGlone 2012).

Enacting Higher Codes and Standards In 2006, the Louisiana Legislature mandated the use of nationally recognized codes and standards as part of the Uniform Construction Code. These codes, which went into effect in early 2007, increased the building standards for much of the State. In addition to the adoption of a new code, much of Louisiana has received updated Flood Insurance Rate Maps from FEMA—another key component in ensuring that future construction is built to a more appropriate elevation and better able to withstand high winds. These changes are a positive step forward, but many land use decisions are local and are difficult to impact with federal or state policy.

Several groups have been active in promoting sound land use practices in Louisiana. Among these groups is the Center for Planning Excellence (2013), which has published *Best Practices for Development Coastal Louisiana Manual and Toolkit* as well as the *Louisiana Land Use Toolkit*. These two resources provide strategies for reducing the risks of flooding and protecting the natural environment. The Land Use Toolkit includes model development codes, which can be adapted by local government officials.

Increased Awareness of Hazard Mitigation: Many agencies and organizations have devoted significant resources to educating the public about hazard mitigation and risk. FEMA provided funding to support community outreach and education through the Hazard Mitigation Grant Program. These funds covered workshops, publications, and other key outreach components all of which contribute to an increase in understanding of risk and the options available to reduce that risk. A wide range of funding was granted to the University of New Orleans' Center for Hazard Assessment, Response and Technology (UNO-CHART). UNO-CHART is an applied social science hazards research center that collaborates with and supports Louisiana communities in efforts to achieve disaster resilience. The objectives of CHART projects are to assist residents and local and state officials in reducing risk to natural hazards, especially hurricane and climate hazards, and to help them gain a better understanding of their risk and what they can do to protect themselves from these hazards. Some examples of the projects and sources of funding follow:

Coastal Restoration Companies and the RESTORE Act (Environmental Defense Fund). Louisiana and the other Gulf Coast states are expected to receive significant new funding to support coastal restoration work through the Deepwater Horizon payments and settlements. We seek to collect information from businesses that will inform us on the issues relating to the economic impact, long-term success and implementation of projects to sustain sensitive coastal areas.

The Socioeconomic Effects of the BP Deepwater Horizon Oil Spill on Gulf Coast Communities (Oxfam). UNO-CHART is examining the impact of the BP Deepwater Horizon Macondo oil well blowout and disaster on the affected Gulf Coast communities. The goal is to show the array of impacts and portray a sense of the severity of each type o impact experienced by residents.

Blending Science and Traditional Ecological Knowledge (Sci-TEK) to Support Ecosystem Restoration (NOAA-CREST, MS/AL Sea Grant, LA Office of Coastal Protection and Restoration-OCPR). Coastal erosion is challenging the very existence of communities in Louisiana. Commercial fishermen and navigation experts have partnered with physical scientists, engineers and government agency staff to better inform coastal restoration decision making. A UNO team of a physical scientist, community representatives and social scientists from UNO-CHART developed a mapping 'tool' to support this partnership: Sci-TEK Collaborative Mapping. This tool translates TEK into a mappable format so that it can be incorporated in the modeling process already utilized by coastal scientists.

Louisiana Community Education & Outreach Program (FEMA/GOHSEP).(one of the authors serves as co-principal investigator for this project). UNO-CHART is implementing mitigation outreach and education projects intended to inform citizens, business owners, non-profit organizations, and local officials about the risks to which they are vulnerable and ways in which those risks can be reduced through multiple mitigation methods. The education and outreach for these projects reaches across a variety of communities from decision makers to the most vulnerable populations. Each of the following five projects work towards increasing the overall awareness of our ability to address potential disasters before they happen:

Continuity Planning for Community Organizations Relying on national best practices and information gathered from stakeholder focus groups, the project team is holding community continuity workshops that focus on mitigation, risk assessment, identification of resources, communications, information technology, etc. Workshops are being conducted throughout the State (on site and via video conferences) and a manual is near completion that will allow others to conduct the initial training. The project team also continues to review and provide feedback on multiple participant continuity plans.

Hazards Resiliency Curriculum The goal of this project is to develop a multidisciplinary instructional program in disaster resilience studies throughout the various colleges and departments within the University that will produce community/regional leaders with strong professional capacities to support statewide mitigation. To date, the project team has developed: A *Hazard Policy Specialization* within Public Administration Program; A *Hazard Mitigation Planning Specialization* within Master of Urban & Regional Planning Program; A *Minor in Disaster Resilience Studies* within the College of Liberal Arts; and *New/Revised Courses* with a Focus on Resilience.

Risk Literacy This project focuses on teaching concepts of disaster resilience through literacy programs and literacy through disaster resilience. Based on current practice and research on the issues surrounding literacy and disaster resilience, the project team has developed materials for literacy programs relying on curricula from the few programs that exist nationally and internationally. Stakeholder meetings were held to outline content and models for training. One segment of the curricula, a manual "Preparing for Storms in Louisiana" has been created by the team and reviewed by multiple groups. [2]The related facilitator's guide has also been prepared. Both the manual and facilitator's guide are available in print and via the Internet for use by a wide variety of agencies and programs across the State and the country. Currently, the team is working on curricula geared towards other reading levels and languages.

Disaster Resilient University Statewide Conference UNO-CHART hosted the Disaster Resistant University Workshop 2011—Building Partnerships in Mitigation, February 16–18, 2011, on its main campus. More than 100 participants representing 25 universities/colleges from around the country and Canada attended the Workshop. A second Workshop, held March 20–22, 2013, included over 130 attendees representing 36 universities and colleges from around the country, plus Canada, Japan, and South Africa. Presentation topics included but were not limited to mitigation program assessment, campus violence, risk communication, mitigation planning, risk assessment, floodproofing, and instructional continuity.

Community Executives Program in Risk Management In 2011, the CHART team facilitated Executive Symposia in Resilience and Risk Management intended to enhance the capacity of community and business leaders to make strategic decisions related to risk management based on sound, scientific principles of resilience. Additional meetings will be held in 2013.

Repetitive Floodloss Reduction Project for the States of LA/TX (FEMA). This project focuses on working with communities to reduce their numbers of repetitive

---

[2] The manual, one of the first of its kind, can be found online at http://www.chart.uno.edu/docs/RiskLiteracyManual.pdf.

flood losses through three major deliverables: (1) website and web-based data portal, (2) area analyses, (3) education and outreach.

Community Rating System (CRS) User Group Support. The CRS, a voluntary program available to National Flood Insurance Program (NFIP) participating communities, provides incentives for communities to go beyond the minimum floodplain management regulations established by the NFIP to minimize risk. The purpose of a CRS User's Group is to serve as a support and educational resource for local communities who participate in the CRS. UNO-CHART facilitates two separate CRS User's Groups in Louisiana; one around the Lake Pontchartrain area, and one in the Baton Rouge area. These two groups are comprised of only 16 of the 42 CRS participating communities in of Louisiana, but make up almost half of the total CRS discounts in the State. Groups are comprised of local officials who have been designated as CRS Coordinator for their communities. UNO-CHART provides support through facilitation of meetings and research.

Reconsidering the "New Normal": Impact of Trauma on Urban Ecological and Social Diversity (NSF—subcontract to Tulane University). In collaboration with Tulane and Xavier Universities, CHART conducted an NSF-funded interdisciplinary research project on the impacts of traumatic events on social and ecological diversity, with New Orleans utilized as a case study. CHART's role on this project was to collect neighborhood-level data in the Lower 9th Ward, Hollygrove, and Pontchartrain Park neighborhoods, to trace the social networks, and to conduct comparative analyses of pre- and post-event diversity.

Mitigation Funding and Homeowner Assistance (Solutient). UNO-CHART is assisting Solutient in providing homeowner consultations regarding individual mitigation programs and related funding. This project currently serves residents of lower Jefferson Parish.

SOARS (UCAR). In partnership with SOARS, UNO-CHART engages university students from around the country in a participatory action research model (PAR) in ethnically diverse, at risk, coastal communities in Southeast Louisiana.

Resident-Driven Action Planning for Community Resiliency in Plaquemines Parish (Greater New Orleans Foundation). In partnership with residents of Plaquemines Parish, the UNO-CHART project team utilizes Participatory Action Research methods to work towards increased citizen involvement in the planning process for the Parish. The team works directly with the residents to ensure that local knowledge and the community's own vision are included in the multiple planning efforts that will have direct effects on Plaquemines Parish.

The BP Horizon Oil Disaster: Media Accounts and Community Impacts (NSF). UNO-CHART partnered with the University of Louisiana-Lafayette to address the following research question: What is the impact of the media accounting on the ways that communities come to understand the BP Deepwater Horizon Oil Disaster and are there intervening factors that mitigate community adoption of the media accounting?

Floodplain Management Policy & Program Evaluation (ASFPM). UNO-CHART collaborated with ASFPM to assist the Federal Interagency Floodplain Management Task Force FIFM-TF in the task of identifying impacts and barriers of federal programs with regard to achieving the goals of floodplain management, and those policies and programs that promote and support sound floodplain management.

Hazard Mitigation Project Scoping & Plan Amendment (DHS-FEMA/GOHSEP). Through this grant, UNO-CHART assisted the University in identifying, evaluating, prioritizing, and scoping potential mitigation projects for our campus.

6.7 Conclusion

It is almost impossible to make a clear and definite assessment of the readiness of south Louisiana post-Katrina. If we look that the social, political and technical systems discussed in this chapter, the results are mixed. Perhaps first, the reader must ask, "ready...ready for what?" Remember that our baseline, Hurricane Katrina, was not so much a natural disaster as it was a technical one. During the late evening of August 29th and early morning of August 30th, 2005, many emergency managers reported they were "breathing a sigh of relief" at having been spared by the worst of Hurricane Katrina (personal accounts to the author). Things were winding down, Katrina had moved to the east. Then the waters began to rise. Some first responders reported increasing flooding, but did not know where the water was coming from.

The social fabric of the region has changed. It is less poor, less minority. Housing is more expensive; primarily due to both the long-term shortages of housing stock and the need to rebuild to ensure compliance with updated building codes, (building a structure higher above base flood elevation is more expensive). In many ways, the steps undertaken to reduce the vulnerability of the built environment are having short-term negative impacts on the social environment. The pre-Katrina stock of public housing has not been restored, devastated neighborhoods lack investment in grocery stores to spur resettlement, and the public health system is still not adequate.

The tremendous investment by the Federal Government in reinforcing and re-evaluating the technical systems that protect the region has been expensive, expansive, and impressive. Some of this technical infrastructure was tested during Hurricane Isaac. While the levee systems seemed to fare well and provided enhanced protection to some communities, some have argued that the reinforced levee systems and new flood barriers have had a displacing effect, causing detrimental flooding to communities that were only marginally affected by Hurricane Katrina. Additional non-structural measures, such as further home elevations, floodproofing, and investment in coastal restoration, are still needed to reduce the residual risk that remains behind the levees.

While there are undoubtedly areas for improvement, this is perhaps the most promising area for meaningful change. Important portions of the Stafford Act have been revised. The NFIP has also been revised to attempt to reduce the subsidies for risky development, and to encourage hazard mitigation. FEMA has more trained personnel ready to respond in the aftermath of a disaster. Local and state leaders have more experience in the complex intergovernmental relationships that are essential to ensuring adequate mitigation, preparedness, response and recovery.

References

Associated Press. (2007, September 7). *Katrina nursing home owners acquitted.* http://www.msnbc.msn.com/id/20649744/ns/us_news-crime_and_courts/t/katrina-nursing-home-owners-acquitted/#.UPMIX28c1EI. Accessed 15 Dec 2012.

Baxter, V. (unpublished). *Rent, real estate, and flood mitigation in New Orleans East.*

Castellano, A. (2012, November 6). *Superstorm sandy: FEMA trailers may be used to house homeless.* http://abcnews.go.com/US/superstorm-sandy-fema-trailers-house-homeless-noreaster-approaches/story?id=17649727. Accessed 15 Dec 2012.

Center for Planning Excellence. (2013). *Best practices for development in coastal Louisiana manual and toolkit.* Available at: http://coastal.cpex.org/.

Conrad, D. R., & Thomas, E. A. (2013). *Proposal 2: Reforming federal support for risky development.* The Hamilton Project. Brookings Institute.

Dwyer, J., & Drew, C. (2005, September 29). Fear exceeded crime's reality in New Orleans. *The New York Times.* http://www.nytimes.com/2005/09/29/national/nationalspecial/29crime.html?ei=5090&en=1ba20914f5888e10&ex=1285646400&adxnnl=1&partner=rssuserland&emc=rss&adxnnlx=1127998837-dUb23oxvthQ0MrMgl9neEg&pagewanted=print. Accessed 12 Oct 2013.

Federal Emergency Management Agency. (2010). *Incident Command System (ICS).* http://www.fema.gov/emergency/nims/IncidentCommandSystem.shtm. Accessed 10 July 2010.

Federal Emergency Management Agency (FEMA). (2011). *National disaster recovery framework.* Washington, D.C.

Fink, S. (2012, August 28). *Superstorm sandy: Hospitals, nursing homes are better prepared for hurricane Isaac than earlier storms.* http://www.nola.com/hurricane/index.ssf/2012/08/hospitals_nursing_homes_better.html. Accessed 15 Dec 2012.

Freudenburg, W. R., Gramling, R., Laska, S., & Erikson, K. (2008). Organizing hazards, engineering disasters? Improving the recognition of policial-economic factors in the creation of disasters. *Social Forces, 87*(2), 1015–1038.

Godschalk, D. R., Beatley, T., Berke, P., Brower, D. J., & Kaiser, E. J. (1999). *Natural hazard mitigation: Recasting disaster policy and planning.* Washington, D.C.: Island Press.

Heyboer, K., & McGlone, P. (2012, November 8). *Post-Sandy notebook: FEMA sending mobile housing units for victims in metropolitan area.* http://www.nj.com/news/index.ssf/2012/11/post-sandy_notebook_fema_sendi.html. Accessed 15 Dec 2012.

Hull, A., & Struck, D. (2005, September 23). *At nursing home, Katrina dealt only the first blow.* http://www.washingtonpost.com/wp-dyn/content/article/2005/09/22/AR2005092202263.html. Accessed 15 Dec 2012.

ISO Property. (2010). *AIR worldwide estimates total property damage from hurricane Katrina's storm surge and flood at $ 44 billion.* http://www.isopropertyresources.com/Press/2005-Press-Releases/AIR-Worldwide-Estimates-Total-Property-Damage-from-Hurricane-Katrina-s-Storm-Surge-and-Flood-a.html. Accessed 12 Oct 2013.

Jerolleman, A. (2013). *The privatization of hazard mitigation: A case study of the creation and implementation of a Federal Program.* (Dissertation).

Kiefer, J., & Jenkins, P. (2009). *"An analysis of the hurricane Gustav city-assisted evacuation," for the city of New Orleans, 2009.*

Kiefer, J., & Montjoy, R. (2006). Incrementalism before the storm: Network performance for the evacuation of New Orleans. *Public Administration Review, 66,* 122–130. (Special issue).

LaCheen, C. (2010). *More than five years after Katrina, FEMA is not prepared to meet the temporary housing needs of people with disabilities and others after a disaster.* National Center for Law and Economic Justice.

Leavitt, W., & Kiefer, J. (2006). Infrastructure interdependency and the creation of a normal disaster: The case of hurricane Katrina and the city of New Orleans. *Public Works Management & Policy, 10*(4), 306–314.

Linderman, J. (2013, May 17). Parish presidents band together to fight against Biggert-Waters, flood insurance rate hikes. *Times-Picayune.*

Liu, A., & Plyer, A. (2008, August). *The New Orleans index: Tracking the recovery of New Orleans & the metro area: Anniversary edition: Three years after Katrina.* Brookings Institution Metropolitan Policy Program & Greater New Orleans Community Data Center.

LSU Ag Center. (2008). *The international building codes in Lousiana.* www.lsu.edu/sglegal/. Accessed 12 Oct 2013.

Mowbray, R. (2005). Evacuation to hotels comes with own set of hazards. *The Times Picayune.* http://www.nola.com/hurricane/t-p/katrina.ssf?/hurricane/katrina/stories/083005_a15_hotels. html. Accessed 12 Oct 2013.

New Orleans Regional Transport Authority (NORTA). (2013). *You're on your way!* http://www. norta.com/getting_around/Evacuation_Plans/index.html. Accessed 15 Dec 2012.

New Orleans Urban Area Security Initiative (NOUASI). (2007). *Louisiana region one communications network FY COPS technology grant.* New Orleans.

Pets Evacuation and Transportation Standards Act of 2006, Public Law 109-308, 42 USC 5121.

Regional Economic Information System (REIS). (2010). Bureau of economic analysis, U. S. Department of Commerce.

Serna, J., Hennessy-Fisk, M., & Bengali, S. (2012, November 9). *FEMA sending mobile homes for superstorm sandy victims.* http://www.latimes.com/news/nation/nationnow/la-na-nn-fema-mobile-homes-hurricane-sandy-victims-20121109,0,6521856.story?page=1. Accessed 15 Dec 2012.

Superdome Evacuation Completed. (2005). http://www.msnbc.msn.com/id/9175611/. Accessed 15 Dec 2012.

Thomas, E. A., & Bowen, S. (2009, November). Preventing human caused disasters. *Natural Hazards Observer, 34*(2), 1–9.

Thomas, E. A., Jerolleman, A., Turner, T., Punchard, D., & Bowen, S. (2011). *Planning and building, safe & sustainable communities: The patchwork quilt approach.* Natural Hazard Mitigation Association.

Townsend, M. (2005). *You're on your own, British victims told. The Guardian.* http://www.abc.net. au/am/con10t/2005/s1455687.htm. Accessed 12 Oct 2013.

U. S. Congress. (2009). Committee on Homeland Security, Ad Hoc Subcommittee on disaster recovery, special report—Far from home: Deficiencies in federal disaster housing assistance after hurricanes Katrina and Rita and recommendations for improvement, 111th Cong., 1st sess., (Washington: GPO). http://www.gpoaccess.gov/congress/index.html. Accessed 12 Oct 2013.

U. S. House of Representatives. (2011). *Five years later: An assessment of the post Katrina emergency management reform act, hearing before the House Committee on Homeland Security, Subcommittee on Emergency Preparedness, Response, and Communications, (testimony of Craig Fugate).* http://www.fema.gov/pdf/about/programs/legislative/testimony/2011/10_25_2011_five_years_later_assessment_of_pkemra.pdf. Accessed 12 Oct 2013.

Chapter 7
Disaster Events and Policy Change in Florida

Christopher Hawkins and Claire Connolly Knox

7.1 Introduction

There is a wide range of hazard mitigation policies local governments have adopted
to reduce the environmental, economic, and social impacts from natural disasters.
One stream of research has focused on explaining the adoption of mitigation techniques, policy change, and the interorganizational relations that are established to
prepare for and respond to disasters. Local and regional characteristics (Brody and
Gunn 2012; Kapucu et al. 2012), planning processes (Hawkins 2012), and the configuration of policy networks (Demirez et al. 2012) are found to influence disaster
mitigation policy and plans aimed at reducing vulnerabilities.

This chapter primarily focuses on contributing to this research base by examining
proposed solutions to collective problems that emerge after focusing events in the
context of natural disasters (See Chap. 3/Birkland in this volume)). The challenges
in emergency management are difficult given the diffuse nature of hazards. There
are, however, "focusing events" that can draw the attention of policy makers and
influence the policy making process (Birkland 1997, 2006). The focusing events
that occurred in South Florida in 1992 (Hurricane Andrew) and Central Florida in
2004 (hurricanes Charley, Frances, and Jeanne) that resulted in policy changes at
the county level in hazard mitigation practices are the focus of this chapter.

As noted by Nohrstedt and Weible (2010), the theoretical and empirical efforts
in the public policy literature have predominately focused on whether an external
event caused a change in policy while less careful attention has been paid to the nature of the event, the type of change, and the causal mechanisms linking the external
event and change. We seek to address these questions by examining the proposed
policy changes contained in the 1993 Lewis Report and the after action reports
prepared by three counties in Central Florida in 2005. Because many emergency

C. Hawkins (✉) · C. C. Knox
School of Public Administration, University of Central Florida, Orlando, USA
e-mail: christopher.hawkins@ucf.edu

C. C. Knox
e-mail: claire.knox@ucf.edu

N. Kapucu, K. T. Liou (eds.), *Disaster and Development,* Environmental Hazards,   111
DOI 10.1007/978-3-319-04468-2_7, © Springer International Publishing Switzerland 2014

management policy and planning changes prior to 2004 were based on the Lewis Report recommendations, we compare these documents to demonstrate how county governments respond to focusing events by proposing changes to existing policy that is based on a process of instrumental learning. We begin by briefly reviewing the literature and then provide background information on hurricanes Andrew, Charley, Frances, and Jeanne. We then discuss our sample of counties and data collection, and then present our findings and conclusion.

7.2   Disaster Events

Development is conceptualized as the production of our built environment. The development process can range from the installation of infrastructure, such as roads and utilities, to the construction of housing and commercial buildings and the ancillary services that support these uses, such as parks and fire stations. The potential for a disaster and the development patterns of a community, particularly in urban areas, are invariably linked. This nexus is particularly strong when one considers the wide range of structural mitigation techniques, such as levees, floodwalls, and fills that modify the built environment in order to protect property and lives from flooding (Birkland et al. 2003; Brody et al. 2010; Burby et al. 2000). However, structural measures such as these have been critiqued for their limited capacity to adequately retain flooding (Burby et al. 1999; Larson and Plasencia 2001; Stein et al. 2000); their expense and potential for neglect (Brody et al. 2010); and in some cases their adverse environmental impacts on natural ecosystems because they encourage development in inappropriate locations (Abell 2000; Birkland et al. 2003; Burby 2006). Nonetheless, structural mechanisms will continue to represent one approach to mitigating potential disaster events.

To complement traditional structural mechanisms, greater emphasis has been placed on developing strong inter-organizational arrangements and governance systems that aim to not only reduce vulnerably to disasters through planning, but also support the effective response following a disaster (Birkland et al. 2003; Birkland and Waterman 2008; Kapucu et al. 2012). Studies have shown that the damage from disaster events are significantly reduced when appropriate land-use planning and policy efforts are implemented (Burby 2005, 2006; Nelson and French 2002) and when organizations work collaboratively in the planning and response stages of a disaster (Kapucu et al. 2008). In essence, communities that invest the time to specify emergency management and hazard planning policies and create administrative structures that enable effective management, are more likely to reduce loss and implement more effective disaster response strategies (Burby et al. 1999; Nelson and French 2002; Schwab et al. 1998).

Cash strapped counties, of course, must prioritize proposed policy changes. Moreover, limited staff and resource capacity, low risk perception, and a manager's partial understanding of their emergency management role oftentimes pushes many emergency policy items to the proverbial "back burner" (Grant 1996). It is well documented in the planning and hazards literature that this occurs even when policy changes are necessary to save lives, protect property, and improve the overall disaster resiliency of a community (Godschalk 2003; Somers & Svara 2009).

Also influencing government action are the characteristics of disasters. Some disasters pose what May (2012) refers to as "public risks"—low probability but potentially high consequence—in which there are few relevant publics who are advocating for risk reduction efforts. As opposed to private risks, the broadly distributed, diffused, and temporally remote impacts of disasters makes it difficult to establish a public constituency for policy change outside of specialized agencies. This difficulty is compounded where the multi-level governance structure of disaster planning and response creates a commitment problem and undermines investments in disaster mitigation efforts (Burby and May 1998).

Yet there are circumstances were local governments have made progress in advancing policy agendas, formulating alternative mechanisms, and deciding on policy direction to mitigate hazards. Plans are changed, policies are adjusted, resources reallocated, and organizational structures rearranged. Why? In the following section, we briefly review the literature on agenda setting and focusing events that is used as a lens to answer these questions.

7.2.1 Agenda Setting and Focusing Events

Agenda setting is the process by which problems and alternative solutions gain or lose public attention and the attention of elected officials (Birkland 1997, 2006). The agenda can either be a government agenda, which is a list of subjects that officials are paying serious attention to at any given time, or a decision agenda, which is a short list of those subjects from the government agenda that have survived competition and are moving into position for some definitive decision (Kingdon 1995). In either case, the agenda represents problems, the cause of the problems, and proposed solutions (Birkland 2006). Kingdon (1995) suggests the heightened public interest on an issue creates a "window of opportunity" that makes it easier for policy makers to support and adopt a controversial policy once it is placed on the agenda.

Research on agenda setting suggests that significant events are crucial for creating the conditions necessary for government policy intervention. In the issue attention cycle proposed by Downs (1972), for example, events trigger media scrutiny and focus public attention on a particular problem. As press coverage and public opinion supporting change increases, an issue is placed high on the government's agenda. Baumgartner and Jones (1993) characterize public policy making as experiencing long terms of stability, which is punctuated by periods of abrupt change. This process entails some sort of trigger event, such as a natural disaster, which raises issue salience and the need for government action to address a public problem.

Birkland (2006) uses the term "focusing event" to describe sudden, unpredictable events that influence the public policy making process or trigger that can lead to policy change. Specifically, a focusing event is a "rare, harmful, sudden event that becomes known to the mass public and policy elites virtually simultaneously"

(Birkland 2006, p. 3). These events can be essential factors in agenda setting and policy change. As opposed to subtle changes that are less likely to be viewed as "events," such as sea level rise, it is much easier to identify a dramatic and visible event as a cause of agenda change at a particular moment. Stressors and enablers include natural or man-made disasters such as floods, hurricanes, and oil spills. A focusing event is important because it draws attention to a problem and helps open a "window of opportunity" (Kingdon 1995). The results can be major changes in the core aspects of governmental programs (Sabatier and Jenkins-Smith 1999; Nohrstedt and Weible 2010).

In emergency management, focusing events are often considered an exogenous impact on a policy subsystem, such as a hurricane on the transportation, economic, environmental, or built systems of a city or region (Nohrstedt and Weible 2010). External shocks may be necessary at times to ignite or foster learning by changing what is known about a group's goals, functions, or outcomes (Birkland 2006; Kingdon 1995). Heclo (1974) suggests for example that "learning can be taken to mean a relatively enduring alteration in behavior that result from experience; usually, this alteration is conceptualized as a change in response made in reaction to some perceived stimulus" (p. 306). May refines the policy learning concept by suggesting that instrumental learning "entails lessons about the viability of policy instruments and implementation designs" (May 1992, p. 350). Instrumental learning consists of obtaining new information from evaluations and implementing new policy instruments, such as funding; penalties or enforcement; training, educational, or technical assistance; administrative arrangements; and actions from specific rules.[1] The external events, or shocks, in 1992 and 2004 that set in motion policy learning and resulted in policy changes to hazard mitigation practices in Florida is the focus of this chapter. In the following section, we describe Hurricane Andrew and the Lewis Report, which provides the foundation for many current state and county level policy changes, and Hurricanes Charley, Francis, and Jeanne that damaged large portions of Central Florida in 2004.

7.3 Study Context

7.3.1 Hurricane Andrew and the Lewis Report

Hurricane Andrew made landfall on August 24, 1992, south of Miami as a relatively dry, fast moving, Category 4 storm. Experts estimate damages at $ 30 billion and 52 people died (Godschalk et al. 1999). After the devastating effect of Hurricane Andrew and Florida's inadequate response, Governor Chiles created the Governor's Disaster Planning and Response Review Committee (Executive Order 92–242) to

---

[1] A second form of policy learning "entails lessons from the social construction of policy problems, the scope of the policy, or policy goals" (May 1992, p. 350), and includes changes in the scope, goals, and/or objectives of the policy; paradigm shifts; and "new or reaffirmed social construction of a policy by the policy elites in a given policy domain" (May 1992, pp. 350–337).

review the preparedness, response, recovery, and mitigation efforts before and after Hurricane Andrew. The Committee, led by Chairman Philip D. Lewis, met eight times between October 12, 1992 and January 6, 1993. The Committee's final report (hereinafter Lewis Report) concluded with 94 recommendations for public, private, and nonprofit sectors. The initial goal of the report was that fully implementing the recommendations would provide Florida with the best emergency management system in the U. S. (The Governor's Disaster 1993).

The Lewis Report outlined numerous required and recommended policy and management changes; Coastal High Hazard Areas, Hurricane Vulnerability Zones, and building codes (i.e., specifically the lack of enforcement of existing codes) were among the more high profile policies that were modified (Birkland 2006; Godschalk et al. 1999). Subsequent to the issuance of the Lewis Report, Florida created the Emergency Management Preparedness and Assistance Trust Fund, established the Division of Emergency Management under the Department of Community Affairs, and passed the Emergency Management Act (EMA) (Chap. 252 of the Florida Statutes). The EMA specifically mandates mutual aid agreements between municipalities, counties, and state governments, an emergency operations plan by all local governments, and alignment of plans and operations with county agencies (Kory 1998; Wilson and Oyola-Yemaiel 2001). Researchers have recognized Hurricane Andrew as a turning point in emergency management from a command and control to a more networked approach leading to faster response by public, private, and nonprofit sectors (Waugh and Streib 2006; Sahlin 1992), as well as use the Lewis Report in other studies (e.g., Knox 2013; Maher 1992; Mittler 1997; Wilson and Oyola-Yemaiel 2001).

7.3.2   Hurricanes Charley, Frances, and Jeanne

The 2004 hurricane season was the first major test of Florida's new emergency management policy and administrative changes since Hurricane Andrew in 1992. Because hurricanes occur relatively frequently across many months and across a wide region, Birkland (2006) suggests they generate attention, but not as an individual focusing event. Hurricanes are not unexpected per say, but the severity of three hurricanes on one area is usually unexpected. This for all purposes is rare. When aggregated over a short time and within a relatively small geographic area, however, we suggest they can become a focusing event and set in motion an agenda setting process.[2] Map 1 shows the areas affected by these hurricanes. Collectively the 2004 hurricanes discussed below caused approximately $ 45 billion in damages

---

[2] While three hurricanes affecting one area are unusual, it is also important to note there were only four hurricanes affecting central Florida since 1950: King in 1950, Donna in 1960, David in 1979, and Erin in 1995. Although four hurricanes affected Florida during the 2004 hurricane season, this study focuses on the three directly affecting central Florida: Charley, Frances, and Jeanne. These three storms "necessitated the full emergency response and disaster recovery operations from the local jurisdictions" (Volusia County 2005, p. 2). Hurricane Wilma in 2005 is excluded from this analysis because the focus of this chapter is the 2004 hurricane season, which was the first test of the policy and institutional changes created after the implementation of the Lewis Report recommendations.

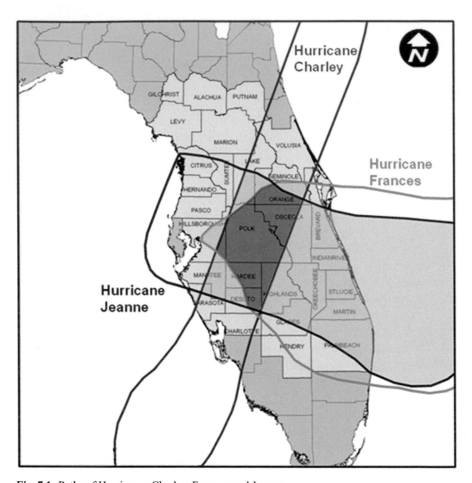

Fig. 7.1 Paths of Hurricanes Charley, Frances, and Jeanne

and killed 114 people (Kapucu et al. 2008). In this section, we provide background information on these events, and then turn to our data collection and analysis that examines the relationship between these focusing events and proposed policy changes (Fig. 7.1).

Hurricane Charley made landfall on August 13, 2004, as a Category 4 hurricane on Florida's southwest coast. The eye passed over Punta Gorda and Port Charlotte with devastating results. Charley continued over the central Florida peninsula downgrading to a Category 1 when it passed through Orange County in Central Florida. The storm left the state near Daytona Beach on Florida's northeast coast (Pasch et al. 2011; Orange County 2005). Estimated damages were approximately between $ 13 and 15 billion (Seminole County 2005).

Hurricane Frances made landfall on September 5, 2004, about 3 weeks after Hurricane Charley as a Category 2 hurricane near Seawall's Point, FL. Unlike Hurricane Charley, Frances was a very slow moving storm, which required the public to stay in homes and shelters longer. A major disaster declaration was issue on September 4, 2004. Estimated damages were approximately $ 8 billion; Frances was responsible for approximately 24 deaths (Seminole County 2005).

Hurricane Jeanne made landfall on September 25, 2004, about 3 weeks after Hurricane Frances as a Category 3 hurricane on the east coast of Florida. The eye crossed the coast at the southern end of Hutchinson Island just east of City of Stuart (Lawrence and Cobb 2005). As the fourth storm affecting Florida in 6 weeks, local waterways were at or over capacity; flooding became a significant issue with Hurricane Jeanne. The estimated total damages ranged between $ 8 and 16 billion (Seminole County 2005).

7.4   Selection of Counties and Data Collection

In Florida, county government is responsible for emergency management activities, including coordinating municipal government within county jurisdictional borders as well collaborating with county governments within the region. Moreover, they work closely with state departments in implementing policy directives and mandates that come from state legislatures and executive level departments. Because of their legal authority over development, county governments are heavily involved in decision-making processes, including setting planning goals and developing alternative strategies for land development. Under state statute, they are also responsible for the administration and management of planning policy in unincorporated areas and work directly with cities within their jurisdictional borders in the approval process for land development projects.

Orange, Seminole, and Volusia counties were selected for the analysis of proposed policy changes because they are relatively diverse in population and economic activities. For example, whereas Orange County has a population in excess of 1 million, both Seminole and Volusia County have fewer than 500,000 residents. Although Orange County is the largest of the three counties, it has much fewer homeowners and has a smaller white population. Seminole and Volusia counties are generally close in total population, yet Volusia County has almost twice the population of elderly and nearly 14 % fewer residents with a Bachelor degree. Because Volusia County is located on the eastern coast of Florida, it emphasizes disaster planning and management along its nearly 50 miles of vulnerable coastal zone. While not a coastal county, Orange County has large tourist attractions in some areas, while in others it is relatively rural. Table 7.1 provides socio-economic characteristics for the three counties.

Per Chap. 252 of Florida Statute, every county in Florida must prepare a Comprehensive Emergency Management Plan (CEMP). The CEMP is an operations-oriented document that establishes the framework for an effective planning and response system to ensure that counties will be adequately prepared to deal with the occurrence of emergencies and disasters. The plan outlines the roles and responsibilities of local, state, and federal agencies and private and nonprofit organizations within each county and the region. The roles of these organizations are organized

Table 7.1 Study area characteristics. (Source: 2010 U.S. Census)

	Orange	Seminole	Volusia
Population	1,145,956	422,718	494,597
Percent over age 65	9.9	12.4	21.5
Percent white	70	81.6	85.2
Percent with bachelor degree	30	34	20.8
Percent homeowner	59.6	71.5	75
Median household income	$ 49,731	$ 58,908	$ 44,169
Population density	1,269	1,367	449.2

on the Emergency Support Functions (ESFs). A lead or primary agency is selected to head each ESF based on its expertise, authority, resources, and capabilities in the functional area. Within each ESF, numerous support agencies and organizations work cooperatively with the primary agency to provide a comprehensive approach to the mitigation, planning, protection, response, and recovery from hazards and emergencies.

We distinguish in our analysis between two avenues for policy change aimed at improving disaster mitigation and emergency management at the local level. One avenue originates from state or national levels, as represented in the Lewis Report discussed above. Policy changes may also originate from the local governing unit that is primarily responsible for integrating planning activities into local operations and overseeing implementation mechanisms. This approach is represented by the voluntary after action reports prepared by Orange, Seminole, and Volusia counties. After action reports capture observations made during and immediately after an exercise, training, or disaster incident. While there is no set format for these reports, report writers commonly organized it around ESFs and contain recommendations for policy, planning, and administrative improvements. There is no requirement by state or federal law to prepare an after action report by county governments after a disaster. The Homeland Security Exercise and Evaluation Program only mandates after action reports/improvement plans as a final product of an exercise or training.

In both cases, changes are assumed to be proposed and adopted in order to address a need that becomes apparent after a focusing event. Bottom-up and top-down policy changes are not completely independent. In many cases, local governing units seek changes in policy because higher-level governments do not adequately address an issue that is of immediate concern to local officials and stakeholders. In other cases, state government may establish new protocols, mandates, or incentives that constrain or provide more opportunities for local officials to make policy changes. For example state-level policy changes that occurred after the 2004 hurricane season included sweeping changes to residential property insurance coverage and building codes (Mang and Santurri 2009; Newman 2006). These changes stemmed from the massive residential property damage in coastal and inland counties from the three hurricanes affecting central Florida.

Table 7.2 Differences between Federal and Florida Emergency Support Functions

ESF Number	Federal	Florida
1	Transportation	Transportation
2	Communications	Communications
3	Public works and engineering	Public works
4	Firefighting	Firefighting
5	Emergency management	Info and planning
6	Mass care, emergency assistance, housing, and human services	Mass care
7	Logistics management and resource support	Unified logistics
8	Public health and medical services	Health and medical
9	Search and rescue	Search and rescue
10	Oil and hazardous materials response	Hazmat
11	Agriculture and natural resources	Food and water
12	Energy	Energy
13	Public safety and security	Military support
14	Long-term community recovery	Public information
15	External affairs	Volunteers and donations
16		Law enforcement
17		Animal services
18		Business, industry and economic stabilization
19		Damage assessment (Orange County)
20		Utilities (Orange County)

7.5 Analysis of After Action Reports

To conduct our analysis of proposed policy changes, we organized the Lewis Report and the three after action reports recommendations based on Florida's 18 Emergency Support Functions. Differences between Federal and Florida ESFs are detailed in Table 7.2. It should be noted that some counties modify the ESFs, which are reflected in Table 7.2. Additionally, the after action reports included recommendations outside the ESFs which were not included in the analysis; these categories included executive policy group, office of emergency management, emergency operations center (EOC), and Radio Amateur Civil Emergency Service (R.A.C.E.S.).

Table 7.3 displays the number of recommended policy changes of the Lewis Report and the after action reports of three counties based on Florida's ESF. First, Orange County's after action report provided 93 general and ESF specific recommendations. The majority of recommendations (61/93 or 66%) fell under one of the 18 ESFs. The remaining 34% (or 27 items) were organized under executive policy group, office of emergency management, ESF/emergency coordinating officers joint recommendations, and emergency coordinating officers.

Seminole County's after action report identified seven major issue areas and provided 84 strategic action items organized around the ESF system. The seven major issue areas were special needs, sheltering, water/ice/food, emergency operations center design, public information, mitigation/public education, and disaster related

Table 7.3 Comparison of proposed policy changes

ESF	Lewis Report	After action reports		
		Orange County	Seminole County	Volusia County
Transportation	9	2	1	0
Communications	18	7	14	15
Public works	0	1	3	0
Firefighting	0	3	6	0
Information and planning	23	2	3	38
Mass care	13	3	6	10
Unified logistics	8	9	11	8
Health and medical	7	8	8	8
Search and rescue	1	2	0	0
Hazmat	0	2	0	0
Food and water	0	2	4	1
Energy	0	3	3	9
Military support	1	0	2	0
Public information	3	10	6	13
Volunteers and donations	6	3	2	0
Law enforcement	5	2	2	1
Animal services	0	1	1	1
Economic stabilization	0	1	0	0

stress. The majority of strategic action items (63/84 or 75 %) fell under one of the 18 ESFs. The remaining 25 % (or 21) items were organized under EOC, municipalities/organizations, R.A.C.E.S., and miscellaneous.

Volusia County's after action report identified nine areas of improvement and provided 105 recommendations. The majority (27/105 or 26 %) of recommendations fell under planning. The remaining recommendations were organized under training, coordination, facilities, evacuation and sheltering, public information, energy, communications, and staffing.

A noticeable change between the Lewis Report recommendations and the three after action reports is the increase in the number of proposed policy changes for agencies, organizations, and specific ESFs to work together in the planning and mitigation phases. For example, Orange County recommends:

ESFs 4 (firefighting), 9 (urban search and rescue), and 10 (hazardous materials) should work in concert with ESF 16 (law enforcement and security) and the county Office of Emergency Management to enhance security in the Emergency Operations Center and on the building's perimeter (Orange County 2005, p. R-13).

Orange County also recommends ESFs 1 (transportation), 2 (communications), 6 (mass care), 8 (health and medical), 11 (food and water), and 14 (public information) participate in all EOC shelter meetings.

The largest number of proposed policy changes in the Lewis Report corresponds to Information and Planning (23) and Communications (18). With regards to the Communications ESF, there is relatively large number of related proposed policy changes in the Seminole (14) and Volusia (15) after action reports. Compared to the Lewis Report recommendations, there is seemingly a continued need to improve communications across agencies that plan for and respond to emergencies.

Table 7.4 Proposed policy changes, Lewis Report and Orange County. (Source: Hawkins 2012)

ESF categories	Recommended policy changes	
	Lewis Report	After action report
Operations		
Firefighting, search and rescue, hazmat	0	7
Law enforcement and military	6	2
Health and medical, food and water	7	10
Logistics		
Mass care and animal services	13	4
Volunteers and donations management	6	3
Unified logistics	8	9
Planning and information		
Transportation, public works, and energy	9	3
Communications	18	7
Public information, and information and planning	26	12
Economic stabilization	0	1

The decreased number of proposed policy changes in Orange County suggests that perhaps emergency management efforts after the Lewis Report have focused on establishing strong interorganizational networks and collaboration. Specifically facilitating access to information and resources, and providing a mechanism to overcome collective action problems (Hawkins 2012). Demiroz et al. (2012), in fact, find the emergency management network in Orange County is relatively decentralized. The county emergency management office has established an extensive communication network with organizations involved in planning for disaster events. With regards to the Information and Planning ESF, Volusia County had 38 proposed policy changes in the 2004 after action report while Orange and Seminole Counties had far less than the number of changes recommended in the Lewis Report.

Table 7.4 provides supporting evidence related to Orange County. It shows the total number of proposed policy changes grouped into three categories for Orange County: operations, logistics, and planning and information (Hawkins 2012). Fewer proposed policy changes were oriented to ESFs that focus primarily on managing the exchange relations among planning and response organizations (Communications), disseminating information to key organizations and planning for disasters (Information and Planning), as well as organizations and agencies that have responsibility for public infrastructure (Transportation).

Figure 7.2 shows the differences in the proposed policy changes graphically for all three counties and the Lewis Report. Across the majority of ESFs, proposed policy changes appear to address new needs identified after the 2004 hurricane season. For example, in 1993, the Lewis Report focused primarily on information and planning, communication, mass care, transportation, and unified logistics. These policy changes emerged after a comprehensive needs assessment of disaster mitigation and the capacity for county and state government to respond to disasters, particularly through enhanced coordination. In 2004, however, the focus of

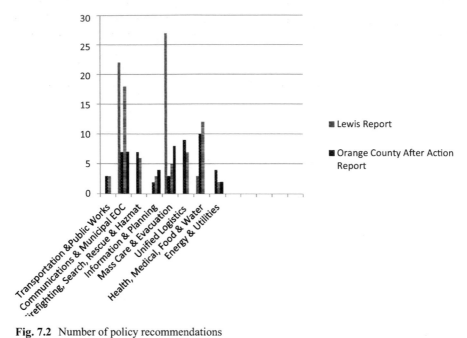

Fig. 7.2 Number of policy recommendations

proposed policy changes shifted to ESFs that are responsible for meeting the basic needs of people (i.e., health and medical, mass care, food and water, and public information). The differences also suggest that counties proposed changes in the after action report related to operations that the Lewis Report did not emphasize, including public information and food and water, as well as basic infrastructure such as energy, firefighting, and public works functions of county government.

Since the hurricanes damaged the same counties within a short period, some of the recommendations are unique to this case study. The primary group of recommendations focused on inadequate staffing and training in the EOCs. By the time Hurricane Jeanne stuck central Florida, staffing reserves had been diminished. Recommendations by all three counties emphasized the basic needs of EOC staff and to cross train the ESF staff. Volusia County specified the need to add more staff for ESF #5 and to keep ESF #11 staffed at all times. Staffing and training issues were not limited to the EOC, but across the counties at the shelters. Specifically, there was a need, lightly addressed in the Lewis Report (recommendation #34), regarding sheltering for special needs populations. Seminole County had the most recommendations for this population that focused on special needs sheltering, staffing, discharge plans, registration, and communication.

There were needs across the three counties to not only network more with outside agencies and organizations, but to bring them into the EOC during an incident. Seminole County recommended Lynx (city bus transportation) and other transportation providers have a representative "invited to the EOC during exercises and actual events" (2005, p. 18). Orange County recommended Orange Utilities

Commission have a representative in the EOC and a Public Information Office in the 3-1-1 Center. Volusia County had the most recommendations for bringing more representatives to the EOC. Specifically, they recommended having a school board representative at ESF #6 (Mass Care), an EMS/EVAC liaison at ESF #8 (Health and Medical), a utility liaison at ESF #12 (Energy), and a municipal liaison in the county EOC.

7.6 Discussion of Findings and Conclusion

In this study, the focusing events in 1992 and 2004 created a window of opportunity for policy changes to be crafted by local officials and presented to policy makers and relative stakeholder groups. In 1992, Hurricane Andrew raised the visibility of disaster planning and emergency management among the public and local officials, particularly the vulnerability of the built environment and the need to devote resources to adequately plan for, mitigate, and respond to potential disasters. The Lewis Report's exhaustive list of recommendations institutionally changed emergency management and planning efforts in Florida, and influenced related changes throughout the United States. Specifically, Florida and 10 other states (e.g., North Carolina, Arizona, and California) require communities to have a comprehensive plan that includes strategies for natural hazards and more stringent building codes (Birkland 2006). The 2004 focusing event of three major hurricanes within 6 weeks provided another window of opportunity to change what we know about local functions, governance structures, and capacity to mitigation and plan for disasters (Birkland 2006). The aftermath of hurricanes Charley, Frances, and Jeanne spurred local officials to reflect on what went right in terms of planning and response, and what could be improved to more adequately manage a disaster event.

Comparison of the proposed policy changes in the Lewis Report and those contained within the after action reports suggests that a process of instrumental learning—obtaining new information from evaluations and implementing new policy instruments—occurred after focusing events in 1992 and 2004. For instance, an emphasis on communications, information and planning, mass care, and transportation in the 1993 Lewis Report indicates that these operations were inadequate given the type of disaster that could occur (e.g., Hurricane Andrew). The observable change in proposed policies following the 2004 hurricanes emphasized new priority areas for disaster planning and emergency management. For example, 7 of the 18 ESFs did not have any corresponding recommended actions in the Lewis Report but were identified as areas needing to be address following the 2004 hurricanes. Another example is the noticeable decrease in the number of proposed policies in the 2004 after action reports that correspond to the Transportation ESF compared to the Lewis Report. While the Lewis Report had nine transportation-related recommendations, there are only two proposed changes in Orange County, one in Seminole County, and none in Volusia County.

One important finding from our analysis is that although communication systems that are critical for managing disaster events was a priority area that needed to be addressed (based on the recommendations contained in the Lewis Report), it continued to be a major focus of new policy proposals after 2004 for Seminole and Volusia counties. On the other hand, the number of proposed policy changes affecting information and planning noticeably decreases, particularly for Orange and Seminole counties. This suggests that since implementation of many Lewis Report recommendations, interorganizational networks became more established and institutionalized as a feature of emergency preparation.

The comparison also reveals a change in the focus of emergency management efforts to improving the overall care of the local population. Whereas the Lewis Report emphasized the establishment of a robust interorganizational system, the after action reports indicated the need to provide basic services to residents (Food and Water, Energy) as well as response efforts (Firefighting, Hazmat). Other ESFs, however, continued to receive attention after the focusing events, including Health and Medical, Mass Care, and Volunteers and Donations.

Although we suggest that changes in policy indicate learning, this is by no means the only reason to explain these changes. Previous research demonstrates the wide range of factors that influence changes in disaster management policy, such as support from senior leadership (Kartez and Kelly 1988; Scanlon 1996; Somers and Svara 2009). Our analysis is also based on hurricane events that occurred in South and Central Florida, which are quite different than other natural or man-made disasters. Regardless of the type of disaster, essential community functions can be disrupted, transportation networks can be damaged, energy systems can be compromised, built environments damaged, and adequate shelter, food, and water can be inaccessible to residents and businesses. In 1992 and in 2004, Florida experienced events that turned these possibilities into reality.

The findings presented in this chapter suggest that focusing events are important for communities to critically evaluate emergency management policy, learn from this evaluation, and leverage the opportunities for agenda setting to propose policy that will address emergency management needs. Specifically, we cannot ignore the value of creating and implementing the recommendations from the Lewis Report and the 2004 after action reports that resulted in various local, state, and national policy, planning, and administrative changes. It could be argued that without the new information from these evaluation reports, instrumental learning on emergency management policy and planning in Florida and throughout the U.S. would have been hampered. Therefore, we recommend after action reports be mandatory after every disaster to facilitate instrumental policy learning at all levels of government.

References

Abell, R. A. (2000). *Freshwater ecoregions of North America: A conservation assessment*. Washington, DC: Island Press.

Baumgartner, F. R., & Jones, B. D. (1993). *Agendas and instability in American politics*. Chicago: The University of Chicago Press.

Birkland, T. A. (1997). *After disaster: Agenda setting, public policy, and focusing events*. Washington, DC: Georgetown University Press.

Birkland, T. A. (2006). *Lessons of disaster: Policy change after catastrophic events.* Washington, DC: Georgetown University Press.

Birkland, T., & Waterman, S. (2008). Is federalism the reason for policy failure in Hurricane Katina? *Publius: The Journal of Federalism, 38,* 692–714.

Birkland, T. A., Burby, R. J., Conrad, D., Cortner, H., & Michener, W. K. (2003). River ecology and flood hazard mitigation. *Natural Hazards Review, 4*(1), 46–54.

Brody, S. D., & Gunn, J. R. (2012). Examining environmental factors contributing to community resilience along the Gulf of Mexico coast. In N. Kapucu, H. Hawkins, & F. Rivera (Eds.), *Disaster resiliency: Interdisciplinary perspectives.* New York: Routledge.

Brody, S. D., Kang, J. E., & Bernhardt, S. (2010). Identifying factors influencing flood mitigation at the local level in Texas and Florida: The role of organizational capacity. *Natural Hazards, 52,* 167–184.

Burby, R. J. (2005). Have state comprehensive planning mandates reduced insured losses from natural disasters? *Natural Hazards Review, 6*(2), 67–81.

Burby, R. J. (2006). Hurricane Katrina and the paradoxes of government disaster policy: Bringing about wise Governmental decisions for hazardous areas. *Annals of the American Academy of Political and Social Science, 604*(1), 171–191.

Burby, R. J., & May, P. J. (1998). Intergovernmental environmental planning: Addressing the commitment conundrum. *Journal of Environmental Planning and Management, 41*(1), 95–110.

Burby, R. J., Berke, P. R., Doyle, R. E., French, S. F., Godschalk, D. R., Kaiser, E. J., Kartez, J. D., May, P. J., Olshansky, R., Paterson, R. G., & Platt, R. H. (1999). Unleashing the power of planning to create disaster-resistant communities. *Journal of the American Planning Association, 65*(3), 247–258.

Burby, R. J., Deyle, R. E., Godschalk, D. R., & Olshansky, R. B. (2000). Creating hazard resilient communities through land-use planning. *Natural Hazards Review, 1*(2), 99–106.

Demiroz, F., Kapucu, N., & Dodson, R. (2012). Community capacity and interorganiozational networks for disaster resiliency: Comparison of rural and urban counties. In N. Kapucu, H. Hawkins, & F. Rivera (Eds.), *Disaster resiliency: Interdisciplinary perspectives.* New York: Routledge.

Downs, A. (1972). Up and down with ecology: The issue attention cycle. *Public interest, 28*(1), 38–50.

Godschalk, D. R. (2003). Urban hazard mitigation: Creating resilient cities. *Natural Hazards Review, 4*(3), 136–143.

Godschalk, D. R., Beatley, T., Berke, P., Brower, D. J., & Kaiser, E. J. (1999). *Natural hazard mitigation: Recasting disaster policy and planning.* Washington, DC: Island Press.

Grant, N. K. (1996). Emergency management training and education for public administration. In R. T. Sylves & W. L. Waugh, Jr. (Eds.), *Disaster management in the U.S. and Canada: The politics, policymaking, administration and analysis of emergency management* (pp. 313–326). Springfield: Charles C. Thomas.

Hawkins, C. V. (2012). Networks, collaborative planning, and perceived quality of comprehensive plans. In N. Kapucu, C. Hawkins, & F. Rivera (Eds.), *Disaster resiliency: Interdisciplinary perspectives.* New York: Routledge.

Heclo, H. (1974). *Modern social policies in Britain and Sweden: From relief to income maintenance.* New Haven: Yale University Press.

Kapucu, N., Berman, E. M., & Wang, X. H. (2008). Emergency information management and public disaster preparedness: Lessons from the 2004 Florida hurricane season. *International Journal of Mass Emergencies and Disasters, 26*(3), 169–196.

Kapucu, N., Hawkins, C. V., & Rivera, F. (2012). Disaster resiliency: Interdisciplinary perspectives. In N. Kapucu, C. Hawkins, & F. Rivera (Eds.), *Disaster resiliency: Interdisciplinary perspectives.* New York: Routledge.

Kartez, J. D., & Kelley, W. J. (1988). Research-based disaster planning: Conditions and implications. In L. K. Comfort (Ed.), *Managing disasters: Strategies and policy perspectives* (pp. 126–146). Durham: Duke University Press.

Kingdon, J. W. (1995). *Agendas, alternatives, and public policies: Second edition*. New York: Harper Collins College Publishers.

Knox, C. C. (2013). Analyzing after action reports from Hurricanes Andrew and Katrina: Repeated, modified, and newly created recommendations. *Journal of Emergency Management, 11*(2), 160–168.

Kory, D. N. (1998). Coordinating intergovernmental policies on emergency management in a multi-centered metropolis. *International Journal of Mass Emergencies and Disasters, 16*(1), 45–54.

Larson, L., & Plasencia, D. (2001). No adverse impact: A new direction in floodplain management policy. *Natural Hazard Review, 2*(A), 167–181.

Lawrence, M. B., & Cobb, H. D. (2005). Tropical cyclone report: Hurricane Jeanne. National Hurricane Center. http://www.nhc.noaa.gov/2004jeanne.shtml. Accessed 1 Aug 2013.

Maher, S. T. (1992). Emergency decision making during the state of Florida's response to Hurricane Andrew. *Nova Law Review, 17,* 1009.

Mang, D. A., & Santurri, R. J. (2009). The winds of change: How wind storms changed the course of regulation in Florida. *Federation of Regulatory Counsel Journal, 20*(4), 1–5.

May, P. (1992). Policy learning and failure. *Journal of Public Policy, 12,* 331–54.

May, P. (2012). Public risks and disaster resilience: Rethinking public and private sector roles. In N. Kapucu, C. Hawkins, & F. Rivera (Eds.), *Disaster resiliency: Interdisciplinary perspectives*. New York: Routledge.

Mittler, E. (1997). A case study of Florida's emergency management since Hurricane Andrew. Natural Hazards Research and Applications Information Center, Institute of Behavior Science, University of Colorado.

Nelson, A., & French, S. (2002). Plan quality and mitigating damage from natural disasters: A case study of the Northridge earthquake with planning policy considerations. *Journal of the American Planning Association, 68*(2), 194–207.

Newman, J. W. (January 2006). Law and ordinance coverage. Government Report: Florida Office of Insurance Regulation.

Nohrstedt, D., & Weible, C. (2010). The logic of policy change after crisis: Proximity and subsystem interaction. *Risk, Hazards and Crisis in Public Policy, 1*(2), 1–32.

Orange County. (2005). After action report. Orange County Government.

Pasch, R. J., Brown, D. P., & Blake, E. S. (2011). Tropical cyclone report: Hurricane Charley. National Hurricane Center. http://www.nhc.noaa.gov/pdf/TCRAL032004_Charley.pdf. Accessed 1 Aug 2013.

Sabatier, P., & Jenkins-Smith, H. (1999). The advocacy coalition framework: An assessment. In P. Sabatier (Ed.), *Theories of the policy process* (pp. 117–166). Boulder: Westview Press.

Sahlin, M. (1992). Paradigm shift. Speech to the Conference of the National Volunteer Organizations Active in Disaster (NVOAD). www.nvoad.org/articles/paradigm.php. Accessed 1 Aug 2013.

Scanlon, J. (1996). The crucial role of the mayor in Canadian emergency management. In R. T. Sylves & W. L. Waugh, Jr. (Eds.), *Disaster management in the U.S. and Canada: The politics, policymaking, administration and analysis of emergency management* (pp. 294–310). Springfield: Charles C. Thomas.

Schwab, J., Topping, K. C., Eadie, C. C., Deyle, R. E., & Smith, R. A. (1998). *Planning for post disaster recovery and reconstruction*. Chicago: American Planning Association Publication.

Seminole County. (2005). After action report. Seminole County Government.

Somers, S., & Svara, J. H. (2009). Assessing and managing environmental risk: Connecting local government management with emergency management. *Public Administration Review, 69*(2), 181–193.

Stein, J., Moreno, P., Conrad, D., & Ellis, S. (2000). *Troubled waters: Congress, corps of engineers, and wasteful water projects*. Washington, DC: Taxpayers for Common Sense and National Wildlife Federation.

The Governor's Disaster. (1993). Planning and Response Review Committee Final Report. Tallahassee, FL. http://www.floridadisaster.org/documents/Lewis%20Report%201992.pdf.

Volusia County. (2005). After action report. Volusia County Government.

Waugh, W. L., Jr., & Streib, G. (2006). Collaboration and leadership for effective emergency management. *Public Administration Review, 66*(Special Issue), 131–140.

Wilson, J., & Oyola-Yemaiel, A. (2001). The evolution of emergency management and the advancement towards a profession in the United States and Florida. *Safety Science, 39*(1–2), 117–131.

Chapter 8
Beyond the Hyogo Framework: Disaster Management in the Republic of Lebanon

Thomas W. Haase

8.1 Introduction

Located at the crossroads of Europe, Africa and Asia, the Republic of Lebanon is a country whose history has been shaped by external invasion, colonization, civil war and political instability. While the political tensions that define the lives of those who presently live within Lebanon may give rise to concerns about the resurgence of civil war, not all of Lebanon's disasters are political in nature. In recent years, a variety of natural and man-made risks have emerged to threaten Lebanon's social, economic and political stability.

These risks represent the full spectrum of natural and man-made hazards. In 2004, for example, Lebanon's central government deployed 10,000 response personnel to manage the consequences of a snowstorm that disrupted much of the country (Rare Snow 2005). During the summer of 2006, the central government delivered relief assistance to those displaced by a conflict between Hezbollah and Israel (Shearer and Pickup 2007). In 2008, Ethiopian Airlines Flight 409 crashed into the Mediterranean after taking off from Beirut-Rafic Hariri International Airport, killing all 90 passengers (Ethiopian Airlines Crash 2010). Although governmental and non-governmental institutions managed to respond to these events, problems with coordination, power sharing and information management raised serious questions about Lebanon's disaster management capacities.

Those familiar with Lebanon would not be surprised to learn that the central government has failed to improve of its disaster management capacities since the crash of Ethiopian Airlines Flight 409. This is because Lebanon's political and administrative institutions have become paralyzed, the result of the political deadlock that has largely existed since the conclusion of the civil war. This political deadlock has two interdependent and mutually reinforcing consequences. With respect to its governmental institutions, Lebanon appears to have a fully

T. W. Haase (✉)
Department of Political Studies and Public Administration,
American University of Beirut, Beirut, Lebanon
e-mail: th30@aub.edu.lb

N. Kapucu, K. T. Liou (eds.), *Disaster and Development,* Environmental Hazards,
DOI 10.1007/978-3-319-04468-2_8, © Springer International Publishing Switzerland 2014

functioning government, as social conflict is, at the present time, largely contained within, or restrained by, public institutions. Just below the surface, the public sector is constrained by problems such as corruption, the reluctance of the government to fill civil services vacancies, which is upwards of 60% in some sectors, deficiencies in substantive and professional expertise, the lack of financial resources, and an inadequate technological and communication infrastructure. These problems have not only undermined the capacity of Lebanon's public sector to function, they have encouraged the steady erosion of the rule of law, so much so, that even basic traffic regulations are difficult for the government to enforce.

The political situation inhibits the Lebanese government's ability to address problems related to economic and social development. For instance, large portions of the country presently receive less than 12 hours of electricity a day. There also exist increasing concerns about the availability of clean water, which has become scarce due to industrial and agricultural pollution, poor sanitation, and environmental degradation. Likewise, deficiencies in urban planning have lead to the growth of densely populated areas that lack the appropriate lifeline systems, as well as transportation infrastructures that do not reduce congestion or improve public safety. For many who live in Lebanon, avoiding the consequences of these problems has become a daily affair.

The present situation seems to suggest that Lebanese policy makers have matters to address that are more important than disaster management. While this may or may not be the case, the point of this chapter is that Lebanese communities need not choose to live under the shadow of risk. More importantly, Lebanon's local governments and civil society organizations can promote sustainable development and disaster management, even in the absence of support and guidance from the central government. Within this context, this chapter begins with an overview of the relationship between disaster and development, and how this relationship came to be endorsed by the signatories of the *Hyogo Framework for Action*, an international agreement that promotes the reduction of global disaster risk. The second section of the chapter shifts the substantive focus to Lebanon with a description of the disaster and development challenges that exist throughout the country. The subsequent section reviews the historical development of Lebanon's civilian disaster management institutions, focusing on the evolution of their responsibilities and the extent to which they are prepared for disaster events. The fourth and final section of this chapter reviews some of the challenges that have prevented Lebanon from making progress on issues of disaster management and sustainable development. In doing so, the section will identify some general steps that can be taken to promote progress in these domains without having to rely on Lebanon's central government. While these recommendations will not replace a comprehensive national strategy for disaster management and sustainable development, they are intended to help local governments and civil society organizations to identify the means and resources needed to reduce risk without the support and guidance of the central government.

8.2 Towards Hyogo: Disaster Management and Development

This chapter is guided by the assumption that disaster events can undermine the scope and progression of development, and if not undertaken properly, development can increase the frequency and severity of disaster events. This is the line of reasoning advanced by Dennis S. Mileti (1999), who argued that the consequences of disaster are often the result of adverse interactions between natural systems, social systems and constructed systems. Of the diversity of conclusions that can be drawn from Mileti's argument, two are fundamental to this chapter's thesis. First, Mileti argues that humans cause the consequences generated by disaster events, which are the result of poor development planning, unwise perceptions about hazards and disasters, and bold assumptions about the relationship between the environment, science, and technology. Second, Mileti argues that disaster consequences can be reduced through the intelligent design of social and constructed systems. This means that communities can take steps to reduce disaster risk by engaging in developmental activities that focus on economic and social development, as well as hazard mitigation and prevention. This perspective is exemplified by the United Nations, which has adopted the perspective that, "[h]uman development can contribute to a serious reduction in disaster risk" (United Nations Development Program [UNDP] 2004, p. 1).

The international community embraced the relationship between disasters and development in 2005, when 168 states adopted the *Hyogo Framework for Action*, which promoted the reduction of disaster risk in five areas of priority (International Strategy for Disaster Reduction [ISDR] 2005, p. 6). First, risk reduction would become an issue of national prominence, meaning that participating governments would develop the institutional capacities needed to implement disaster risk reduction strategies. To this end, governments would develop the policy frameworks necessary for the reduction of disaster risks as well as the mechanisms needed to monitor and evaluate the reduction of risk. Second, member states would undertake hazard and vulnerability studies that would enable them to identify, assess, and monitor disaster risks. The findings generated by these studies would be used to improve early warning capacities. Third, there would also be attempts to improve community resilience through the development of relevant forms of knowledge and the use of education to improve awareness and preparedness in at risk communities. Fourth, states would develop polices and projects to reduce the underlying risk factors related to disaster events. Finally, participating states would undertake reforms that would strengthen their disaster response capacities at all jurisdictional levels.

The signatories, including all of the members of the League of Arab States, committed themselves to taking steps towards sustainable development practices (ISDR 2005, p. 10). These steps required countries such as Lebanon to undertake development planning that protected the environment and reduced the risks associated to climate variability and climate change. They further agreed to adopt social and economic development practices that would strengthen infrastructure and prepare vulnerable populations for disasters through education, food security, health care strategies and risk-sharing mechanisms. To this end, states agreed to update their land-use strategies by improving urban planning techniques, reducing poverty, revising building codes, and including disaster risk considerations in infrastructure projects.

The international community reiterated their commitment to disaster management and sustainable development during the *Global Platform for Disaster Risk Reduction* (IDSR 2009a, p. 26). That same year, the League of Arab States released a report on their progress towards the implementation of the *Hyogo Framework* (ISDR 2009b). While the report confirmed the Leagues' commitment to the goals of the *Hyogo Framework*, it also disclosed:

> The commitment does not translate in most cases to operational capacities nor commitment of resources to effectively implement the strategies or integration risk reduction development plans. There is weak progress on education and public awareness as well as on collection, availability and accessibility of data and information risk and vulnerabilities, tools and methodologies for disaster risk reduction and multi-risk and multi-hazard approaches. (ISDR 2009b, p. 4)

The League's lack of policy progress is of particular concern, especially considering that many of its member countries have growing urban populations and limited natural resources, for example, potable water and arable land. Following this line of reasoning, the *Arab Human Development Report* indicated that the lack of emphasis on sustainable development may become problematic for Arab countries (Elasha 2010). After a comprehensive review of the impacts of climate change, both observed and anticipated, the report concluded that climate change: "will act as a threat multiplier that is likely to exacerbate the existing vulnerability of the region to current climatic and non-climatic stresses, leading to large scare instability with severe environmental, economic, political, and security implications" (Elasha 2010, p. 35). Even if only partially true, the consequences predicted by the *Arab Human Development Report* need not occur, as there remain numerous opportunities for action (Collins 2009).

8.3   Lebanon's False Choice: Disaster or Development

While Lebanon has recently experienced a period of relative political stability, uncontrolled development activities have begun to facilitate the emergence of complex issues that, if left unresolved, will eventually come to threaten the country's long-term security and prosperity. From the development perspective, these issues are substantial in scope and rapidly expanding. Despite the fact that Lebanon is ranked 72nd in the *Human Development Index* (UNDP 2013), which suggests that the country has had some success with respect to human development, reality is slightly more complicated. For instance, poverty remains rampant, with 28.5 % of the population living under the poverty line (El Laithy et al. 2008). Expenditures on education represent just 7 % of public spending, one of the lowest rates in the region (World Bank 2011). Graduation rates are equally problematic: one in ten students fail to graduate from primary school and one in four students fail to graduate from secondary school (Tzannatos 2012). Additionally, the lack of environmental regulations and administrative oversight have led to widespread water and air pollution, which has created conditions ripe for a public health crisis. Furthermore, much of Lebanon's physical infrastructure is inadequate. This is clearly illustrated by the problems that exist in the country's electricity sector. Lebanon's demand for electricity exceeds 2,400 MW per day, while production facilities can only generate 1,500 MW per day. This means that parts of the country are subjected to electrical blackouts that last anywhere from 3 to 20 h (Sassine 2012). Not only do these development issues represent a nuisance, the lack of central oversight means that these issues will eventually have detrimental impacts on Lebanon's economy and the health of its citizens.

Those who live in Lebanon must also confront a variety of disaster hazards. The most substantial of these hazards relate to geological instability (UNDP 2012). Given Lebanon's location along a major fault zone, particularly the Mount Lebanon Thrust, which runs from Sidon to Tripoli, the country is highly susceptible to earthquakes (Huijer et al. 2011). The Romans were well aware of this danger, as an earthquake destroyed the city of Beirut and the temples of Heliopolis, a city now known as Baalbek, in the year 551. This earthquake also generated a massive tsunami, which destroyed several coastal cities and killed tens of thousands of people. While tsunami may not be the first thing that comes to mind when one thinks about the hazards confronted by modern-day Lebanon, a substantial portion of the country's population lives near the coastal cities of Tyre, Sidon, Beirut, and Tripoli. Fortunately, Lebanon has not experienced a sizeable earthquake in recent years. The last major seismic event occurred in 1956, when an earthquake shook more than 400 villages in the Chouf region. The earthquake destroyed 6,000 homes, damaged 17,000 more, and killed 136 people (Brazee and Cloud 1956, p. 50). Subsequent earthquakes could be catastrophic, especially for Beirut, which has yet to implement policies to mitigate seismic risk.

Lebanon's other hazards include forest fires, floods, landslides, and drought. Unlike earthquakes, these hazards tend to occur on a more regular basis. In October 2007, the Lebanese's government faced a crisis brought about by the simultaneous burning of more than 240 forest fires. These fires destroyed large tracks of forest, as well as homes and personal property, and pushed the government's response capacity to its limit. Of equal concern are the hazards related to flooding. Although located in the Middle East, Lebanon's annual rainfall levels create ample opportunities for flooding, as rivers tend to overtop their banks and inundate the heavily populated and poorly developed coastal areas. In January 2013, for example, a 4-day rainstorm generated floods that killed seven people and caused substantial damage to homes and livelihoods of the communities located in the flood zones (Lakkis and Lutz 2013).

Finally, there are the man-made disasters, many of which are caused by Lebanon's poorly planned development activities and limited administrative capacities. Familiar to anyone who has braved the Lebanon's highways, the country has one of the highest rates of traffic deaths in the entire world; the result of unsafe road conditions, poor emphasis on driver safety, and the lack of properly implemented transportation regulations (Choueiri et al. 2010). Other risks include the unregulated release of human and industrial waste, which has contaminated the country's coastlines with "fecal coliform and high levels of toxic metals" (Korfali and Jurdi 2012, p. 351). Concerns also exist about buildings constructed without government oversight. In January 2012, the collapse of a six-story building killed 27 and injured 11 in Ashrafieh, an upscale neighborhood of Beirut (Search at Collapsed 2012). The consequences caused by building collapses are likely to increase. While Beirut's municipal government condemned 25 buildings as structurally unsafe in the 12 months after the 2012 collapse, the government had not been able to completely vacate these buildings (Strum 2013). Other man-made hazards reflect the consequences of decades of ill-planned development and include air pollution, soil erosion, poor sanitation, the loss of biodiversity, and the exhaustion of natural resources.

Discussions about Lebanon's man-made hazards cannot occur without reference to military conflict. During Lebanon's 34 day conflict with Israel in 2006, for example, the Israelis attacked oil facilitates located next to Lebanon's Jiyeh power plant. This attack caused the release of 10,000–15,000 t of heavy fuel oil, which inundated 150 km of coastline and affected ports, private marinas, fisherman's wharves, and public and private beaches (Ministry of Environment 2006). This conflict also created a substantial humanitarian crisis. According to the United Nations Development Program (2007, p. 12):

> Over one million persons (a quarter of the population of Lebanon) were displaced; 1,200 persons, mostly civilians, one-third of them children, lost their lives; 5,000 people were injured, many permanently; more than 500,000 people lost their homes; and several thousands lost their jobs or sources of livelihoods across all sectors of economic activity; agriculture, industry, services. An estimated 100,000 people, mainly youth, emigrated.

Lebanon's governmental institutions, civil society and international actors managed to formulate a humanitarian response to this conflict. Fortunately, the response was substantial enough to avoid an even larger humanitarian crisis. The lessons learned during the crisis have been applied during subsequent events, albeit less publicly. For instance, as of November 14, 2013, after 2 years of instability in neighboring Syria, the United Nations reported that 816,811 Syrians had registered themselves as refugees that were living in Lebanon (UNHCR 2013). The Lebanese government estimated that the number of Syrians present in Lebanon at that time had exceeded 1,000,000. The concern is that the constant flow of Syrian refugees will place additional pressures on governmental institutions already stressed by limited capacities, and if action is not taken, the refugee situation may come to threaten Lebanon's existence.

This review of Lebanon's hazards indicates that unmanaged economic development and the lack of risk awareness have placed the country in a dangerous position. A single disaster could eliminate most of the progress that Lebanon has made during the last quarter century. While the Lebanese must begin to prepare for disasters such as earthquakes, this would not be sufficient. The Lebanese must also be concerned about the compound disasters that could materialize when two or more hazards occur simultaneously or in close succession. In light of the political and administrative constraints that exist within Lebanon's central government, the challenge is to identify steps that local governments and civil society organizations might take to reduce risk on their own. Before contemplating how this might be accomplished, this chapter turns to a brief review of Lebanon's civilian disaster management institutions.

8.4 The Evolution of Disaster Management in Lebanon

Other than a small number of governmental and non-governmental experts, relatively few people in Lebanon are concerned about disasters. This lack of concern is reflected in the research that exists on the subject, which tends to fall into one of two categories. The initial category relates to an expanding, but disconnected, body of research on issues such as tobacco, pollution, seismic risk, sanitation, traffic safety, and urban planning. The next category, which is of importance for the present chapter, is comprised of a collection of governmental and non-governmental reports that focus on issues of disaster management. These reports indicate that Lebanon's civilian disaster institutions developed in four distinct stages (UNDP 2010).

8.4.1 Stage One: The 1970's

Lebanon's civilian disaster management system was first established in December of 1976, when the central government passed Law Number 35/1 (UNDP 2010). This legislation created the High Relief Committee (HRC), the first governmental institution within Lebanon that possessed the mandate to undertake disaster response activities. The HRC was presided over by the Prime Minister, and comprised of government ministers, director generals and representatives of the Lebanese Army and the Internal Security Forces. The HRC was charged with several important responsibilities and functions. First, the committee would accept donations of food and materials given to the Lebanese government to relieve populations affected by disaster events. Second, the committee would establish the procedures needed to "receive and distribute" these donations. Third, and beyond the development of the necessary administrative procedures needed to fulfill its mandate, the committee would manage the logistical tasks related to the receipt, transportation, storage, and delivery of relief donations. Fourth, the committee received the authority to undertake information collection activities necessary to complete its responsibilities. Lastly, the committee could ask for assistance from public institutions during disaster situations. The creation of the High Relief Committee represented a significant step in the development of Lebanon's disaster management institutions. The civil war that disrupted the country from 1975 until 1989, however, temporarily undermined the potential for additional progress.

8.4.2   Stage Two: The 1990's

The Lebanese government returned to the issue of disaster management following the signing of the *Ta'if Accord*. This renewed attention was driven by the experiences of the government during the conflict, which killed or displaced more than a quarter of a million people, as well as the country's subsequent reconstruction. There were two major disaster management developments that occurred in Lebanon during the 1990's. These developments expanded the authority of the High Relief Commission through the modification of Law Number 35/1 of 1976. The first expansion occurred with the passage of Law Number 30 in 1993. In this instance, Lebanese policy makers expanded the mandate of the HCR to include the acceptance "donations of all kinds given to the Lebanese state by international, regional, national and local communities, organizations and individuals to [relieve] the affected population together with other substances provided to it by the council of ministers" (UNDP 2010, p. 18). Until the adoption of this amendment, the HRC was not authorized to accept financial donations for disaster management activities.

The second expansion of HCR authority occurred with the adoption of Law Number 4 in 1994. This legislation expanded the HRC's mandate to include disaster prevention activities, provided the HRC with the authority to assign disaster management tasks to appropriate line ministries and director generals, and allowed the HRC to request relief assistance from public and private institutions (UNDP 2010, p. 19). Up until this point, the HRC only had the authority to request assistance from public institutions during disaster situations. This expansion also represented a fundamental shift in the mandate of the HRC, as the committee was allowed, at least on paper, to take a more proactive role in disaster management activities. Unfortunately, the United Nations indicates that the HRC has not fulfilled its expanded mandate, as the activities of the HRC have remained restricted to disaster relief (UNDP 2010, p. 19).

8.4.3   Stage Three: The 2000's

During the first decade of the twenty-first century, the Lebanese government adopted legislation and strategies directed towards improving Lebanon's ability to protect its infrastructure against disaster risks (UNDP 2010). These activities fell into three general categories. The first category focused on the prevention of environmental disasters. For instance, Law 444 of 2002 mandated that the National Council for the Environment (NEC) would oversee the development of policies necessary to protect Lebanon's natural environment. This mandate included the oversight of issues related to environmental monitoring and assessment, as well as reducing the risks related to hazardous materials. The NEC would further identify the "protective measures that should be implemented to mitigate against environmental disasters due to natural causes, incidents, or war acts" (UNDP 2010, p. 31).

One of the consequences of the environmental legislation was that the Lebanese government had created the situation whereby two separate organizations had mandates related to disaster management activities. This development gave rise to serious questions about authority, jurisdictional boundaries and the appropriate distribution of political power within the central government. These questions become even more complicated as the Lebanese Armed Forces, Lebanon's largest and most widely distributed governmental institution, began to develop plans related to disaster response and relief activities. Done without strategic guidance or support from the central government, the Lebanese Armed Forces, with the support of the Lebanese Civil Defense, which is situated within the Ministry of Interior and Municipalities, and the Lebanese Red Cross, have become Lebanon's primary disaster response actors.

The second category of activities strengthened Lebanon's regulations with respect to construction and public safety. The intent behind the adoption of this legislation was two-fold: to address almost two decades of illegal and unregulated construction, especially that which occurred during the civil war, and to mitigate the potential consequences of disasters caused by earthquakes and fires. With the passage of Law 646 of 2004 and Public Safety Decree 14293 of 2005, the government mandated that all buildings over three stories in height must be certified to have earthquake resistant designs. To reduce the consequences of fires and explosions, the government used Decree 14293 of 2005 to mandate that all buildings install fire resistant doors of a certain minimum thickness. This concern about safety extended to the areas immediately surrounding buildings, especially those under construction, by outlining the safety precautions that construction companies needed to take in order to protect the public from harm.

The third category of activities represented Lebanon's attempt to moderate the risks and consequences generated by forest fires. Motivated by the series of devastating fires that swept through the country in 2007, civil society organizations and government ministries came together to formulate what would eventually become *Lebanon's National Strategy for Forest Fire Management* (Mitri 2009). This strategy, which was adopted by Lebanon's Council of Ministers, was designed to reduce the number and intensity of forest fires, and equally important, to improve the government's capacity to respond to forest fire events. The central government sought to accomplish these goals by following a strategy that emphasized five pillars. First, to conduct research that would help to facilitate the development of an information system that would facilitate the analysis of forest fire data. Second, to moderate risks related to forest fires through land management and community awareness. Third, to procure fire-fighting equipment and develop an early warning system that would reduce response times. Fourth, to improve coordination among the stakeholders involved in forest fire response activities. Finally, the government would set out to strengthen its recovery activities, particularly with respect to reforestation and community recovery.

8.4.4   Stage Four: Beyond the Hyogo Framework

With the ratification of the *Hyogo Framework,* Lebanon positioned itself to become one of the first Middle Eastern countries to implement a comprehensive national disaster management policy. Indeed, the legislative accomplishments that occurred up until 2006, while imperfect, indicated that Lebanese policy makers possessed the desire and momentum necessary to effectuate significant policy change. A subsequent resurgence of political instability, namely a series of political assassinations, one of which took the life of Prime Minister Rafic Hariri, and the subsequent conflict with Israel, pulled the attention of national policy makers away from issues of disaster management. In the years since, other than the passage of the *National Fire Management Strategy* and the ad hoc emergence of the Lebanese Armed Forces as the primary disaster response and relief actors, the Lebanese government has not managed to improve its disaster management institutions and procedures.

This problem has been discussed by a variety of experts. David Shearer and Francine Pickup (2007), United Nations officials who participated in the international humanitarian response to the 2006 conflict, provide insights into capabilities of the High Relief Commission. Their conclusions indicate that the High Relief Commission, while useful for providing information and serving as liaison between political parties and key decision-makers, was unable meet its full potential (Shearer and Pickup 2007). In their analysis of the Lebanese government's response to the conflict, they noted that the activities of the HRC were "weakened by the absence of solid support from key ministries" (Shearer and Pickup 2007, p. 345). Referring to the criticisms discussed in the media, they indicated that there appeared to be management issues related to politics, and the "lack of transparency, corruption and [the] favoring of one community over another" (Shearer and Pickup 2007, p. 345). In contrast to these problems, Shearer and Pickup noted that Lebanon's civil society organizations managed to develop working relationships with local governments and overcame the deficiencies present in the national response system. This success appears to be limited, however, as a World Vision study concluded that communities in five of Lebanon's most vulnerable regions had yet to make significant progress in any of the *Hyogo Framework's* five areas of priority (Maalouf 2009).

The Lebanese government presented a more balanced opinion with respect to its progress towards meeting the goals of the *Hyogo Framework* (Fawaz 2011). According to the central government, progress has been made on the collection of data related to various hazards and risks that exist within the country. This includes, first, the development of a risk assessment profile that includes the establishment of a national risk database, the mapping of hazards and risks, and the assessment of critical infrastructure (Fawaz 2011, p. 8). Second, although still incomplete, the government indicated that it has encouraged the development of early warning systems, especially for severe weather events and forest fires (Fawaz 2011, p. 9, 13). Third, the government reported progress in natural hazards research, referring to studies undertaken on topics such as earthquakes, tsunamis, floods, and forest fires. While not necessary funded by the public, the government referred to research undertaken by several Lebanese universities and research institutions (Fawaz 2011, p. 12). Fourth, the government indicated that it had finalized the development of a national strategy to promote awareness about disaster risk reduction to the general public. This strategy outlined the processes that will be used to issue public warnings and educate citizens about the measures they can take to prevent disasters in their communities (Fawaz 2011, p. 13). Finally, the government indicated that the municipalities of

Beirut, Tripoli, Sidon, Byblos, Baalbek, and Tyre had joined the *Resilient Cities Campaign*, a United Nations program that had been established to empower local communities to develop their resilience to disaster events.

To its credit, the Lebanese government did acknowledge that it had largely failed to meet the aspirations of the *Hyogo Framework*. Most significantly, the government revealed that, while institutional support for the development of a national disaster management policy may have been achieved, its achievements have been neither comprehensive nor substantial. The government elaborated on why this was the case. One explanation was that the central government lacked the capacity to effectively implement its laws, especially when it comes to the previously mentioned construction codes and public safety regulations (Fawaz 2011, p. 14, 25). The government continued by citing the lack of budgetary support and expertise as the primary reasons for these shortcomings. Relatedly, the government indicated that it had yet to develop training and educational programs and curricula needed to educate its citizens about the risks of disaster (Fawaz 2011, p. 11). Another explanation was that government does not know how to coordinate the stakeholders that participate in the response and recovery activities that follow a disaster event (Fawaz 2011, p. 22). Finally, and perhaps most importantly, the government explained that political instability limited its ability to focus on relief and response activities (Fawaz 2011, p. 3).

Nevertheless, the Lebanese government stressed that it remained committed to the development of a comprehensive national disaster management policy. To this end, the government indicated that national policy must be "amended in a way that appoints the HRC as the only national decision making authority for disaster risk management. Such a role should encompass preparedness and mitigation, relief and response, rehabilitation and recovery" (Fawaz 2011, p. 7). The attainment of this objective, the government argued, would help to resolve many of the planning and coordination issues that have plagued the government since 2005. Equally important, the Lebanese government recommended that it should "take disaster prevention, mitigation preparedness, and vulnerability reduction into account in its already existing strategy for sustainable development" (Fawaz 2011, p. 3).

Lebanon appears to be formally committed to the goals of the *Hyogo Framework*. Unfortunately, since 2005, the central government has accomplished little of practical significance to facilitate the systematic and comprehensive reform of Lebanon's national disaster management policy. While acknowledgment and commitment on the part of the Lebanese government are necessary to reduce disaster risk, neither is sufficient to bring about policy reform, recognizing that this says nothing whether these reforms could be successfully implemented. To be sure, Lebanon's central government has taken advantage of periods of relative political stability to create some of the foundations needed to reduce the risks of disaster. The government's ability to build upon these foundations, however, has been undermined by political instability, technical complications, resource and staffing constraints, and problems related to organizational coordination. To further illustrate this point, there are no fewer than three governmental organizations currently responsible for disaster management activities in Lebanon: the National Council for the Environment, the High

Relief Commission, and the Lebanese Armed Forces. As a result of these problems, and as an added layer of complexity to the situation, there are several independent and overlapping disaster management activities that are undertaken by local governments, municipal unions, and nongovernmental organizations throughout the country.

8.5 Promoting Community Preparedness and Resiliency

A variety of tasks must be undertaken to reduce the disaster risks that are present in Lebanon. These tasks read like an extensive to-do list: secure financial resources, protect the environment, implement current laws and regulations, conduct risk assessments, improve partnerships with private and non-governmental entities, protect national resources, and most importantly, develop community awareness. While it might be tempting to suggest that policy makers should focus their energy on the development of a national disaster management plan, practical problems prevent Lebanon from taking constructive policy steps. Many of these problems cannot be resolved given the present political situation. There are important opportunities, however, which if taken advantage of, could improve the resilience of Lebanon's communities, even in the absence of national consensus or comprehensive legislation.

8.5.1 Secure Independent Financial Support

Communities committed to strengthening their disaster resilience must undertake tasks that might be beyond their respective administrative and financial capabilities. Lebanon's local policy makers confront several constraints when it comes to securing financial resources related to disaster management activities. In its progress report on the implementation of the *Hyogo Framework*, the Lebanese government admitted that it does not consider disaster management to be a funding priority (Fawaz 2011, p. 5). This means that Lebanon does not allocate any national funds towards disaster risk reduction, and makes almost no financial investments in hazard reduction, the development of disaster management institutions, the completion of risk assessments, or the development of early warning systems. When national funds do get allocated, they are provided to the HRC to facilitate a response to a specific emergency situation. Even when this happens, the funds are appropriated from the central government's emergency budget. It is hard to identify the specific reason why the financial support for disaster management has been so limited. The Lebanese government suggests that funding levels are low because they have other policy priorities, for example, education and healthcare. Alternatively, funding levels may be low due to the lack of trust that pervades Lebanon's highly centralized political and administrative systems (Haase and Antoun 2014).

When local officials and organizations find themselves in need of funds to support their activities, they can turn to actors from outside their jurisdiction for assistance. The concern is that these external actors, in many instances the central government, suffer from their own financial constraints. Rather than committing their own resources, both national and local governments could take better advantage of the resources provided by the development organizations that already operate in their jurisdictions. In its strategy for disaster risk reduction in the Arab world, the Council of Arab Ministers (2010, p. 18) reinforced this idea:

> In the Arab region, funding remains the main challenge that faces national and local authorities, civil society organizations and humanitarian workers implementing disaster risk reduction measures targeting communities at risk. In keeping with emerging global commitments, the League of Arab States encourages its members to dedicate at least 1 % of national development funding and development assistance towards disaster risk reduction measures.

In situations where national sources of funding are limited, or the national government decides not to allocate 1 % of its development funds to risk reduction, the Council of Arab Ministers suggested that governments could take advantage of two alternative funding options. The first option consists of investment vehicles such as sovereign wealth funds, which could be employed to promote resilience in local communities through the funding of sustainable development projects. The second option would be to apply for additional financial assistance from humanitarian and development organizations. This would require local governments and civil society organizations to expand their grant writing and financial management capacities. Regardless of the option selected, the critical point is that alternative sources of financial resources exist; local officials and civil society organizations simply need to take advantage of existing funding mechanisms and determine how to best direct the financial resources that are secured towards disaster management issues.

8.5.2 Promote Foundational Research

The reduction of risk requires that local officials and decision makers develop the capacities needed to respond to complex disaster situations. The effective management of such situations requires knowledge and actionable information derived from empirical research conducted in at-risk communities. A critical problem, especially for the states that surround the Mediterranean, relates to what Shaikh and Musani (2006) refer to as the research deficit. In their own words (2006, p. S56): "a critical factor that further compounds the lack of knowledge, especially in complex humanitarian emergencies and large scale national disasters, is the partial or total collapse of systems for routine information collection and analysis in the event of structural, social and political instability." It is not just that there is not enough research that is problematic; the quality of the research results is equally important. The concern is that bad information can lead to bad decisions. Shaikh and Musani (2006, p. S56) continue, indicating that "[t]he information that is available is too often derived from a variety of sources using non-standardized methods. [The research that does exist] inherently lacks consistency and is of poor reliability and validity, and is arguably of limited use for establishing baselines, making comparisons or tracking trends." Their essential point is that, regardless of the strength of a country's intent, the success of risk reduction and sustainable development policies depends, in large part, upon the availability of information that is derived from high quality research.

The lack of foundational research severely constrains disaster management activities in Lebanon. According to a report on risk assessment released by the United Nations Development Program (UNDP 2010), Lebanon also suffers from a deficit in risk assessment research. While a variety of risk assessment studies have been undertaken throughout the country, the methodologies used to identify and quantify risk have not been disclosed (UNDP 2010, p. 43). This makes it difficult for those interested in risk reduction to compare and evaluate the findings generated by these assessment studies. To make progress, the stakeholders involved in risk assessment research must come to a consensus on which methodologies are acceptable for use in disaster research. Furthermore, these stakeholders must commit to research that identifies risk, identifies how to reduce risk, and more importantly, identifies the extent to which risks have been reduced. The development of linkages between the researchers who work in Lebanon would help with the distribution of knowledge and expertise among government institutions, academic institutions, and non-governmental organizations. Moving forward, Lebanon's local governments and civil society organizations could encourage researchers to validate their data, and help researchers to create hazard intensity and hazard zoning maps that can actually assist decision-makers (UNDP 2010, p. 44).

8.5.3 Develop Administrative Capacity and Professional Expertise

The next issue that prevents action on disaster management and sustainable development issues relates to the lack of professional expertise and professional capacity (See Chap. 10 in this Volume). During the Fourth Islamic Conference of Environmental Ministers, the participants acknowledged that no country is immune to the threats of disaster (Draft Strategy 2010, p. 39). The participants indicated that Arab governments could protect their populations and infrastructure by improving their administrative capacities and professional expertise. To do so, Arab governments were encouraged to make some critical investments. First, they should develop a cadre of professionals who have the capacity to develop and maintain information networks and databases. Second, they should ensure that disaster management professionals are qualified, and possess the levels of training and the knowledge needed to fulfill their responsibilities. Finally, they should promote cooperation and information sharing amongst critical stakeholders, including public, private, and academic organizations.

For Lebanon to develop the professional expertise and capacities necessary to promote community resilience, the advice outlined in the preceding paragraphs is instructive. These steps should not be considered sufficient, especially considering heavy constraints and limitations placed upon Lebanon's national institutions. In the absence of national leadership, local governments can take independent action, as least to the extent allowed under Lebanese law (Haase and Antoun 2014). First, there is a need to enhance the skills of the local officials, especially those that already possess disaster management responsibilities. Second, there is a need to address crucial shortcomings in current capacities. For example, Lebanon's communities lack expertise related to decision-making, crisis management, geology, and geophysics.

Over the short-term, these deficiencies can be addressed by expanding the pool of individuals that possess the required expertise, for instance, by working with non-governmental organizations. These individuals, if hired in a consultative capacity, could be used to support risk assessment and hazard mapping projects. Over the long-term, efforts should be directed towards expanding the educational opportunities made available to Lebanese citizens. An initial step would be to develop intensive training programs for the professionals that already work in the field. These training programs, which could be funded by non-governmental organizations, would provide local professionals with the opportunity to update and expand their current capacities. A subsequent step would be to expand undergraduate and graduate training at local universities by offering courses on subjects such as infrastructure analysis, earthquake modeling, flood modeling, climate change and hazard management (UNDP 2010, pp. 56–57). Taking university training one step further, these courses could be combined to create the first academic or certification programs related to disaster management in the Arab region.

8.5.4   Facilitate Information Collection and Dissemination

The reduction of disaster risk requires more than the generation of high quality research and the development of local disaster management expertise and capacities. While these factors are certainly important contributors to risk reduction, it is also necessary to educate individual citizens so that they acquire the capacity to take action on their own, independent of the capacities and interests of their respective national and local governments. To promote such self-organization, citizens must be educated about the risks that exist in their communities, as well as the actions that they could take to reduce their vulnerability to those risks. For the Arab region, this requires governmental and non-governmental organizations to emphasize disaster management education and public awareness and to develop disaster management products that are appropriate for children, schools, universities, general public, and the media (ISDR 2009b, p. 8). The success of these activities depends on the availability and dissemination of quality information, which has not been a priority for governments in the region (ISDR 2009b, p. 4).

Those who live in Lebanon remain largely unaware of the risks that exist in their communities. To the extent that they are aware of these risks, they are unlikely to know how to respond after the occurrence of a catastrophic event. According to the United Nations, Lebanon could develop its capacity to disseminate information by updating its administrative mechanisms and procedures, thereby improving its ability to communicate directly, and in real-time, with its citizens (UNDP 2010, p. 44). For instance, the Lebanese government could make its risk assessment reports and hazard maps available to the general public via public websites and online databases. The government could also make use of its telecommunication system to issue emergency warnings, alerts, and instructions via SMS directly to cell phone users. To accomplish these tasks, the government must increase the availability of bandwidth, reduce the cost of Internet access, reduce computer illiteracy, and make regular and systematic updates to public websites. If the use of the national communication systems is impractical, local actors could, upon holding discussions with relevant stakeholders, determine whether they could make use of the public announcement systems employed by mosques, churches, and local communities. Given the prevalence of these systems in Lebanon, they could provide communities with low cost platform for the dissemination of disaster management information. Improvements could be made in other areas as well. Shaikh and Musani (2006, p. S57), for example, noted that, "the mechanisms (scientific journals) through which foundational research are disseminated to end-users of research results, especially those in developing counties, are … unsuitable." Whether the intended audience is academics, policy makers, or the public, the mechanisms used to disseminate research must ensure that the relevant findings, outputs, and recommendations make their way to the end-users, whether though academic journals, newspapers, websites, public announcements or public workshops.

8.6   Conclusion

When Lebanon became a signatory to the *Hyogo Framework*, the country had made substantial progress towards developing the institutional foundations necessary to promote the reduction of disaster risk within its communities. In a twisted sense of irony, this process began when the Lebanese government established the High Relief Committee during the early years of the civil war. The High Relief Committee was given the authority to manage disaster response activities and became Lebanon's first civilian disaster management institution. Although the HRC was initially responsible for the distribution of relief supplies, its authority has expanded to include disaster prevention and mitigation activities, as well as the collection and distribution of disaster relief funds. After the civil war, the Lebanese government passed legislation with the intent to protect the natural environment, enhance public safety, reduce forest fires, and regulate construction. Given this progress, Lebanon was positioned to become a regional leader in disaster management policy. The ratification of the *Hyogo Framework*, however, seems to be a high water mark of sorts, as Lebanon has subsequently failed to adopt or implement policies that would further enhance its resilience. Consequently, Lebanon's disaster management system remains underfunded, short-staffed, uncoordinated, and principally focused on ad hoc response activities. What disaster response capacity that exists has been developed through the actions of the Lebanese Armed Forces, the Lebanese Civil Defense, and the Lebanese Red Cross.

Lebanon is not unique in its predicament. Many countries have failed to meet the full objectives of the *Hyogo Framework*. The fact that issues of sustainable development and disaster management do not fall on the institutional policy agendas of many of these countries, means that this lack of attention will likely to continue for the foreseeable future. For such countries, the question is not whether a comprehensive national disaster management policy should be developed. Rather, the question is what, if anything, can done to strengthen community resilience in the absence of guidance and administrative support from the central government.

One might answer this question by thinking strategically about how issues related to sustainable development and disaster management might be placed on the institutional policy agendas of countries that do not, or cannot, take such issues seriously. While there is guidance on how this might be accomplished, for example, through the use of symbols, policy arguments and focusing of media attention (Stone 1989; Kingdon 1995), little is known about the policy processes of non-western countries. More importantly, this approach is problematic in that it depends upon the decision-making and policy processes of the national government.

An alternative approach might be to wait for the occurrence of a focusing event, for example a massive earthquake (Birkland 1997). Although the consequences would be devastating, such an event would certainly direct substantial amounts of attention towards the issues of disaster management and sustainable development. Such a strategy, however, is ill advised. On the one hand, acceptance of this strategy would not only embrace the idea that the consequences of disaster are outside of human influence, it reflects the idea that it is appropriate to gamble with the lives of those who live under conditions of great risk. On the other hand, the strategy suggests that nothing can be done to reduce the risks of disaster until a solution is found for the problems that have rendered Lebanon's national institutions ineffective. In either case, it could take decades, if not longer, before policy makers can undertake the political and administrative reforms that are needed at the national level.

The answer to the question posed above is less about the formulation of comprehensive disaster management policies, undertaking national administrative reforms, or influencing the institutional policy agenda. The answer requires the identification of practical solutions to the problems of administration and governance that prevent Lebanese communities from taking action to reduce their risk and vulnerabilities. The answer also requires that these practical solutions circumvent the political deadlock that has paralyzed the central government. Instead of focusing attention on the decision makers and institutions at the top of the national hierarchy, individual administrators, local government officials, as well as members of nongovernmental and civil society organizations should identify actions that they can take to reduce risk that are independent of the control and oversight of the central government.

There are indications that this is already occurring in many of Lebanon's communities, as evidenced by the *Resilient Cities Campaign*. One of the central themes of this chapter, however, is that communities should become even more proactive in such endeavors. Local actors can begin by taking steps to secure the necessary financial resources. While it may not be possible to secure budgetary commitments from Lebanon's central government, other revenue sources are already available, including sovereign wealth funds and the grant programs of development organizations. Furthermore, success means moving beyond the retroactive funding of disaster relief activities towards the funding of mitigation and prevention activities linked to sustainable development. Likewise, steps can be taken to improve the level of local expertise on matters of sustainable development and disaster management. This could be accomplished, for example, by having local governmental and non-governmental organizations train their personnel on matters related to disaster management, or more ambitiously, to support the creation of academic programs in emergency or disaster management in Lebanon's institutions of higher education. Moreover, foundational research must be undertaken that improves our understandings of the presence and levels of risk. Even descriptive research that reviews the demographic composition and the capacities within Lebanon's communities would go a long way to improve current levels of preparedness. Whether considered individually or collectively, these steps provide communities the opportunity to improve their capacity to respond to the risks and vulnerabilities in the absence of support and guidance from the central government.

Acknowledgments   This chapter would not have been possible without the support of numerous individuals, especially my graduate students at American University of Beirut, Eric J. Economy, Ayman Hussein and Anna Nersesyants. I am also indebted to my colleagues George Bitar-Ghanem and Wen-Jiun Wang for comments on earlier versions of the chapter. Finally, I would also like to acknowledge the thoughtful contributions of an anonymous reviewer. Any error or admission is the sole responsibly of the author.

References

Birkland, T. A. (1997). *After disaster: Agenda setting, public policy, and focusing events*. Washington, D.C.: Georgetown University Press.

Brazee, R. J., & Cloud, W. K. (1956). *United States Earthquakes, 1956*. Washington D.C. UNT Digital Library. http://digital.library.unt.edu/ark:/67531/metadc40349/. Accessed 3 Feb 2014.

Choueiri, E. M., Choueiri, G. M., & Choueiri, B. M. (June 2010). Analysis of accident patterns in Lebanon. Paper presented at the 4th International Symposium on Highway geometric Design, Valencia, Spain. www.4ishgd.valencia.upv.es/index_archivos/62.pdf.

Collins, A. E. (2009). *Disaster and development* (Vol. 5). New York: Routledge Press.

Council of Arab Ministers. (2010). The Arab strategy for disaster risk reduction 2020. Adopted by the Council of Arab Ministers Responsible for the Environment. Resolution: #345. 22nd Session of the League of Arab States. http://www.unisdr.org/we/inform/publications/18903.

Draft strategy on management of disaster risks and climate change implications in the Islamic world. (2010). Fourth Islamic Conference of Environmental Ministers. Tunisian Republic. http://www.sesric.org/imgs/news/image/icme_3.2%20Disaster%20Strategy.pdf.

El Laithy, H., Abu-Ismail, K., & Hamdan, K. (2008). *Poverty, growth and income distribution in Lebanon* (No. 13). United Nations Development Program: International Policy Centre for Inclusive Growth. http://www.ipc-undp.org/pub/IPCCountryStudy13.pdf.

Elasha, B. O. (2010). *Mapping of climate change threats and human development impacts in the Arab region*. (UNDP Arab Development Report-Research Paper Series). United Nations Development Program, Regional Bureau for the Arab States. www.arab-hdr.org/publications/other/ahdrps/paper02-en.pdf.

Ethiopian Airlines Crash. (24. January 2010). Plane with 90 passengers caught fire, fell into the Mediterranean. Associated Press. www.aparchive.com.

Fawaz, F. (2011). *Lebanon: National progress report on the implementation of the Hyogo framework for action (2009–2011)*. Presidency of Council of Ministers: Beirut, Lebanon. http://www.preventionweb.net/files/19726_lbn_NationalHFAprogress_2009-11.pdf.

Haase, T. W., & Antoun, R. (2014). Administrative decentralization in the Republic of Lebanon: Contemplating the path towards reform [Working paper], American University of Beirut. Beirut, Lebanon.

Huijer, C., Harajli, M., & Sadek, S. (2011). Upgrading the seismic hazard of Lebanon in light of the recent discovery of the offshore thrust fault system. *Lebanese Science Journal, 12*(2), 67.

International Strategy for Disaster Reduction (ISDR). (2005). Hyogo framework for action 2005–2015: Building the resilience of nations and communities to disaster. www.unisdr.org/wcdr.

International Strategy for Disaster Reduction (ISDR). (2009a). Proceedings for the global platform for disaster risk reduction, second session, Geneva, Switzerland: Creating Linkages for a better tomorrow. www.unisdr.org/wcdr.

International Strategy for Disaster Reduction (ISDR). (2009b). Progress in reducing disaster risk and implementing Hyogo framework for action in the Arab region. www.unisdr.org/wcdr.

Kingdon, J. W. (1995). *Agendas, alternatives and public policies*. New York: Longman Publishers.

Korfali, S. I., & Jurdi, M. (2012). Chemical profile of Lebanon's potential contaminated coastal water. *Journal of Environmental Science and Engineering, 1*(3), 351–363.

Lakkis, H., & Lutz, M. (10. January 2013). Government OK's $ 2 mln for storm relief. The Daily Star. http://www.dailystar.com.lb/News/Local-News/2013/Jan-10/201608-government-oks-$2-mln-for-storm-relief.ashx#ixzz2HaHKDcOQ.

Maalouf, S. (2009). *Views from the frontline country report for Lebanon*. Beirut: World Vision. www.preventionweb.net/files/10263_Lebanoncountryreport.pdf.

Mileti, D. S. (1999). *Disasters by design: A reassessment of natural hazards in the United States*. Washington, D.C.: Joseph Henry Press.

Ministry of Environment. (2006). *Lebanon oil spill crisis*. Beirut. http://www.moe.gov.lb/oilspill2006/index.htm.

Mitri, G. (Ed.). (2009). *Lebanon's national strategy for forest fire management*. Beirut: Ministry of Environment Publication.

Rare Snow Whitens. (15. February 2004). USA today. http://usatoday30.usatoday.com/weather/news/2004-02-15-middleeast-snow_x.htm.

Sassine, G. P. (2012). Powering the grid: Lebanon's electricity sector between regulation and decentralization. Executive Magazine. http://www.executive-magazine.com/economics-and-policy/Powering-the-grid/5378.

Search at Collapsed Building Called Off, 27 Dead. (17. January 2012). The Daily Star. http://www.dailystar.com.lb/News/Politics/2012/Jan-17/160129-search-for-survivors-of-collapsed-building-called-off-charbel.ashx#axzz2Hkboywql.

Shaikh, I. A., & Musani, A. (2006). Emergency preparedness and humanitarian action: The research deficit. Eastern mediterranean region perspective. *Eastern Mediterranean Health Journal, 12*(2), S55.

Shearer, D., & Pickup, F. (2007). Still falling short: Protection and partnerships in the Lebanon emergency response. *Disasters, 31*(4), 336–352.

Stone, D. A. (1989). Causal stories and the formation of policy agendas. *Political Science Quarterly, 104*(2), 281–300.

Strum, B. (10. January 2013). Beirut struggles to vacate condemned buildings. *The Daily Star*. http://www.dailystar.com.lb/News/Local-News/2013/Jan-10/201596-beirut-struggles-to-vacate-condemned-buildings.ashx#ixzz2HmXYMfUf.

Tzannatos, Z. (2. October 2012). Unemployment in Lebanon: Lack of skills or lack of skilled jobs? *Executive Magazine*. http://www.executive-magazine.com/banking-and-finance/Unemployment-in-Lebanon/5237.

United Nations Development Program. (2004). *Reducing disaster risk: A challenge for development— A global report*. New York: United Nations. http://www.undp.org/content/undp/en/home/librarypage/crisis-prevention-and-recovery/reducing-disaster-risk-a-challenge-for-development. html.

United Nations Development Program. (2007). UNDP's participation in Lebanon's recovery in the aftermath of the July 2006 war. www.undp.org.lb/proforma.pdf.

United Nations Development Program. (2010). Disaster risk assessments in Lebanon: A comprehensive country situation analysis. Geneva, Switzerland. http://www.gripweb.org/gripweb/sites/default/files/documents_publications/Lebanon_CSA_2012_04_25_0.pdf.

United Nations Development Program. (2012). Strengthening disaster risk management capabilities in Lebanon. http://www.undp.org/content/dam/rbas/doc/Crisis%20prevention/Lebanon_Strengthening%20Disaster%20Risk%20Management%20Capacities.pdf.

United Nations Development Program. (2013). National human development report for Lebanon. http://hdrstats.undp.org/en/countries/profiles/LBN.html.

United Nations High Commissioner for Refugees. (2013). Syria regional refugee response: Information sharing portal. http://data.unhcr.org/syrianrefugees/regional.php.

World Bank. (2011) World development indicators: Lebanon. http://data.worldbank.org/data-catalog/world-development-indicators.

Chapter 9
Disaster Policies and Emergency Management in Korea

Dong Keun Yoon

9.1 Introduction

Though it is almost impossible to prevent natural hazards and the damage they cause, it is possible to alleviate their negative effects on human lives, infrastructure, and property. In the past, most disaster-related efforts have focused primarily on response and recovery (Alexander 2000; Kapucu and Ozerdem 2013; NRC 1994). However, as natural disaster losses have mounted, disaster managers and other decision makers have recognized that sole reliance on the strategies of response and recovery portend continuously escalating costs along with the attendant disruptions associated with natural disasters. Although response and recovery are essential in disaster management, they must be accompanied by increasing attention to reducing losses through effective mitigation programs (NRC 1994). With the paradigm shift from reactive disaster management to disaster risk reduction, there is an increasing emphasis on proactive pre-disaster interventions: mitigation and preparedness.

To better understand the paradigm shift in disaster management in Korea, this chapter examines how disaster management systems and policies have been changed and adapted to reduce disaster risk over time. This chapter consists of three parts. The first part provides an overview of natural disasters in Korea. This part examines characteristics of disasters in terms of disaster types, and their consequences (economic and human losses). The second part of this chapter examines the disaster management system in Korea. The national and local governments' organizational structure for disaster management are examined. The roles and functions of organizations at national and local levels in disaster management are also reviewed. The third part of the chapter examines the characteristics of disaster policies, including laws and regulations, in Korea. This part overviews the history of disaster policies and discusses their implementation's impact on disaster risk reduction. This part also compares natural disaster policies in terms of disaster mitigation perspective in

D. K. Yoon (✉)
School of Urban and Environmental Engineering, Ulsan National Institute
of Science and Technology (UNIST), Ulsan, South Korea
e-mail: dkyoon@unist.ac.kr; dkyoon26@gmail.com

N. Kapucu, K. T. Liou (eds.), *Disaster and Development,* Environmental Hazards,   149
DOI 10.1007/978-3-319-04468-2_9, © Springer International Publishing Switzerland 2014

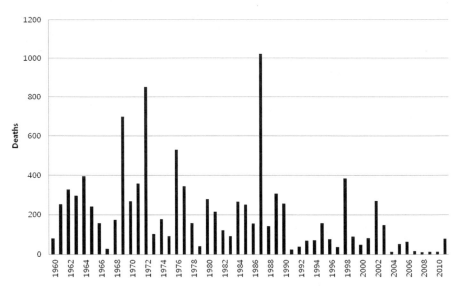

Fig. 9.1 Trends in deaths from all hazards, 1960–2011. (Source: National Disaster Information Center (NDIC))

Korea and discusses challenges and issues relating to current disaster management in practice to increase a community's capacity to deal effectively with future disasters. The fourth part of the chapter discusses the cause and effect relationship between disasters and development through a case example of the 2011 Seoul Flood.

9.2   Natural Disasters in Korea

Korea experiences floods, typhoons, droughts, landslides, snowstorms, earthquakes, and tsunami. The most common types of natural hazards affecting Korea are floods and typhoons. Two thirds of these natural disasters occur between June and September each year. In terms of frequency, torrential rain, storms, and typhoons make up 78 % of the natural disasters, accounting for 84 % of the total damage. According to the Nation Emergency Management Agency (NEMA) of Korea, there are seven incidents of flooding as well as two or three typhoons each year. Natural hazards caused over \$ 41 billion in property damage and nearly 10,000 deaths during the 1960–2011 period (NDIC 2012).

The total number of deaths averaged slightly more than 200 per year during the period from 1960 through 2011 (Fig. 9.1). The deadliest year was 1987, attributable to around 1,000 deaths resulting from Typhoon Thelma and floods (NDIC 2012). Unlike the trend in deaths from natural hazards, economic damage, including property and crop damage, has climbed over time as a direct result of large-scale events, with new loss records set in 2002 and 2003 (Fig. 9.2). Typhoon Rusa hit Korea in

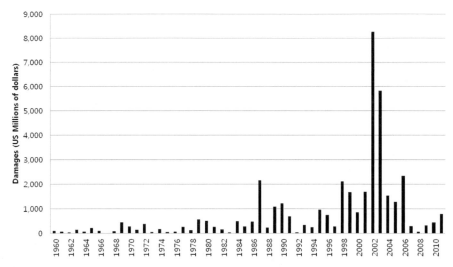

Fig. 9.2 Trends in damages from all hazards, 1960–2011. (Source: National Disaster Information Center (NDIC))

2002, with 246 casualties and an estimated loss of $ 6 billion (Sohn et al. 2005). In 2003, Typhoon Maemi hit the southeastern coast of Korea, the most powerful typhoon to hit the Korean peninsula. Typhoon Maemi caused $ 4.8 billion in economic damage (NDIC 2012). These two devastating typhoons and the Daegu subway fire in 2003, which caused 193 deaths initiated changes in attitudes toward disaster management in Korea. These "focusing events" brought considerable attention to the demand for a comprehensive and all hazards approach to disaster management (Birkland 1997). In 2004, with the passage of the Disaster and Safety Management Basic Act, NEMA was established as a single cabinet-level agency to comprehensively manage natural and man-made disasters.

9.3   Disaster Management System and Structure in Korea

The Constitution of Korea makes it clear that government should have the responsibility to manage all types of disasters, to prevent disasters, and to protect its citizens from harm (Kim and Lee 1998). Disaster management in Korea is vested in a three-layered system—central government, metropolitan cities/provincial government, and local government levels. The central government formulates and implements basic national disaster and safety policies, declares major disasters, and provides human resources, technical and financial supports to local government affected by a disaster. The Central Safety Management Committee, Central Disaster and Safety Countermeasures Headquarters, Central Accident Settlement Headquarters, and Central Emergency Rescue Control Team are the main organizations to manage

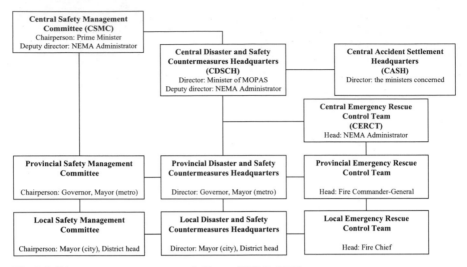

Fig. 9.3 Disaster management system in Korea. (NEMA 2012)

natural and man-made disasters at the central level. Provincial and metropolitan city governments have the autonomy to setup their own disaster management organizations to deal with disasters. Provincial and metropolitan city governments coordinate disaster operations with the central government in a large-scale disaster. Provincial and metropolitan city governments also support for disaster assistance to lower-level local governments. Local government has the primary responsibility and roles in taking the initiative to protect their citizens with regard to disasters (Col 2007).

Figure 9.3 shows the outline of the disaster management system of Korea. The Central Safety Management Committee (CSMC) is the government's lead organization for dealing with natural and man-made disasters. The CSMC is established at each level in accordance with the Disaster Safety Management Basic Act (enacted in 2004) and each committee is responsible for the management of natural and man-made disaster related issues under its authority. Chaired by the prime minister, the CSMC supervises and coordinates overall policy related to disaster and safety, and also promotes negotiations and coordination among the relevant departments. The Safety Management Committee of each metropolitan city, province, local city, county, and district is headed by the chairmen of the autonomous bodies in the local areas.

The Central Disaster and Safety Countermeasures Headquarters (CDSCH), directed by the minister of the Ministry of Security and Public Administration (MOSPA) and operated by NEMA, serves as an ad hoc organization for disaster management at the national level. The CDSCH takes charge of prevention and status control of natural disasters, immediate consequence management, longer-term recovery planning, and executing the necessary measures related to such disasters (Park 2005). The Disaster and Safety Countermeasures Headquarters (DSCH) also

Table 9.1 Disaster management organizations in the central and metropolitan city governments

Government	Disaster management organizations	
Central government	National Emergency Management Agency	
Seoul city	Fire & Disaster Headquarters	Urban Safety Office
Busan city	Fire Department	Urban Development Executive Office (Construction & Disaster Prevention Bureau)
Incheon city	Fire & Safety Management Department (Disaster Management Division)	
Deajeon city	Fire Fighting Head Office	Transportation and Construction Bureau (Disasters and Safety Division)
Gwangju city	Fire Safety Headquarters (Disaster Control Division)	
Deagu city	Fire Fighting Headquarters	Construction & Disaster Prevention Bureau (Disaster Management Division)
Ulsan city	Fire & Disaster Headquarters	Transportation and Construction Bureau (Civil Defense and Disaster Management Division)

operates in local autonomous jurisdictions such as provinces, metropolitan cities, cities, counties, and districts.

Meanwhile, the head of the competent ministries organizes a Central Accident Settlement Headquarters (CASH) to prevent or control events when disaster strikes or where there is a possibility for disaster to strike. Each competent ministry, including the Ministry of Public Administration and Security, the Ministry of Land, Transport, and Maritime Affairs, the Ministry of Environment, and the Ministry for Food, Agriculture, Forestry, and Fisheries, takes responsibility for the disaster which has occurred. In addition, the Central Emergency Rescue Control Team (CERCT), headed by the administrator of NEMA, supports CDSCH when a catastrophic disaster occurs and a massive search and rescue operation is needed (Park 2005). The CERCT operates the emergency rescue control team to supervise and control matters related to emergency rescue, command, and control at the disaster site. In metropolitan cities, provinces, counties, and districts, the Emergency Rescue Control Team is operated and headed by the chief of the firefighting headquarters.

The NEMA, under the MOSPA, is a single comprehensive agency to manage all types of disasters. The central government takes an integrated and comprehensive disaster management approach. However, disaster management organizations in the provincial, metropolitan, and local governments are administratively separated based on disaster types: natural disasters and man-made disasters (mainly fire). Among sixteen provincial and metropolitan governments, only four governments, including Incheon city, Gwangju city, Gyeonggi-Do, and Jeju-Do, have an integrated agency (department) to manage and control all types of disasters (Tables 9.1 and 9.2).

Table 9.2 Disaster management organizations in the provincial governments

Government (Province, Do)	Disaster management organizations	
Gyeonggi-Do	Firefighting & Disaster Headquarters	
Gangwon-Do	Fire Headquarters	Construction & Disaster Prevention Bureau (Disaster Prevention Officer)
Chungcheongbuk-Do	Headquarters of Fire Management	Bureau of Balanced Construction (Division of Flood Control and Disaster Prevention)
Chungcheongnam-Do	Fire Safety Office	Construction & Traffic Bureau (Division of Flood Control and Disaster Prevention)
Jeollanam-Do	Fire Department Headquarters	Construction & Disaster Prevention Bureau (Disaster Prevention Division)
Jeollabuk-Do	Fire Safety Headquarters	Construction & Transportation Bureau (Division of Flood Control and Disaster Prevention)
Gyeongsangbuk-Do	Fire Protection Headquarters	Dept. of Construction, Urban & Disaster Management (Division of Stream Management & Flow Control)
Gyeongsangnam- Do	Fire Service Headquarters	Urban & Disaster Prevention Bureau (Disaster Prevention Division)
Jeju-Do	Fire & Disaster Management Department	

Moreover, Tables 9.1 and 9.2 show that the names of administrative agencies for disaster management are different. The inconsistent and fragmented system for disaster management at the local level makes it hard to effectively coordinate and communicate between the central and local government in managing disasters. Local governments with two different disaster management organizations, Fire-related Headquarters and Disaster Management Division, have separate functions, responsibilities, and even different command lines in terms of mitigation, preparedness, response, and recovery. Provinces and metropolitan cities have their own separate publicly operating fire headquarters to provide fire services including fire protection, fire inspection and investigation, search and rescue, and emergency medical services. Fire departments play a critical role to protect lives from man-made and natural disasters, as well as emergencies. Governmental agencies regarding disaster management at the local government level are usually small divisions and not independent agencies.

Most local disaster management divisions fall under a construction and transportation related department in the local government administration. Because of this structural characteristics of local government for disaster management, most disaster mitigation projects at the local level are mainly related with structural mitigation measures regarding flood control and construction. While fire-related departments mainly focus on emergency response, including search and rescue, and protecting lives from disasters, disaster management departments have responsibilities for disaster mitigation, preparedness, and recovery projects. It would be very necessary

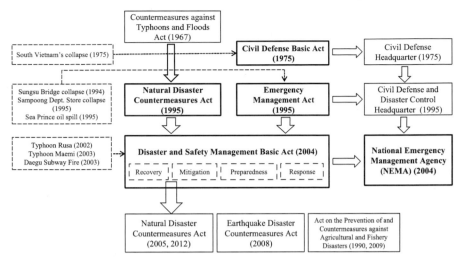

Fig. 9.4 Disaster policies in Korea

for local governments to have strong relationships and coordination between the fire department and disaster management division for enhancing performance and effectiveness in comprehensive disaster management.

9.4   Disaster Policies and Management in Korea

While the improving technologies and strengthening of facilities are important for disaster prevention, the enhancement and implementation of proactive disaster risk reduction policies for long-term mitigation and preparedness are increasingly emphasized for effective disaster prevention. Recently some policy changes regarding natural disasters in Korea have occurred, but the government still needs to take serious consideration to organizational structure and policy relating to disaster management. The government needs to take a more proactive approach to reduce disaster damages, developing long-term disaster policies from a consistent, long-term mitigation perspective, based on a nonstructural approach. Such an approach would help not only to reduce the potential for cumulative disaster damage in subsequent years but also to reduce the cost of disaster recovery.

This part provides an overview of the history of disaster policies and discusses their implementation's impact on disaster risk reduction. Figure 9.4 shows that the first national policy regarding disasters was the "Countermeasures against Typhoon and Flood Act" enacted in 1967, but it only focused on disaster relief and response against typhoons and floods, which are the main catastrophic disasters causing massive financial damages in Korea. After South Vietnam collapsed, the "Civil Defense Basic Act" was passed in July 1975 to protect human lives and reduce the economic damage resulting from enemy attacks, terrorism, or civil disturbance in local areas (Kim and Lee 1998). This act was enforced to deal with war and terrorism more severely than other acts. Since Korea has been divided into two countries, civil defense was a priority in the field of emergency management.

Several catastrophic man-made disasters which occurred in 1994 and 1995, including the collapse of Sungsu Bridge (1994), the collapse of Sampoong department store (1995), and the Sea Prince oil spill (1995), led to changes in disaster policies in Korea. The "Emergency Management Act" was passed in 1995 for the purpose of comprehensive management of man-made disasters.

Regarding policies for natural disaster, the "Natural Disaster Countermeasures Act" was established in December 1995 to manage all kinds of natural disasters in Korea. The Natural Disaster Countermeasures Act was based on the "Countermeasures against Typhoon and Flood Act" established in 1967 (Ha and Ahn 2008). This act includes methods for the prevention of natural disasters, the restoration of disaster-hit areas, investigation of relevant damage, rehabilitation costs, and other counter measures against natural disasters, such as torrential rain and typhoons.

The Natural Disaster Countermeasures Act was wholly amended by Act No. 7359 in January 27, 2005 and by Act No. 311 in August 23, 2012. The Act (enacted in 1995) lays down measures to be taken in order to prevent natural disasters which include: the development of a master plan to reduce damage caused by storm and floods; countermeasures against earthquakes and sea waves, snow-caused disasters, and droughts; development of a disaster prevention information system and a plan to cope with emergencies; and development and implementation of plans to restore disaster hit areas. The Act (amended 2005) further provides for: the establishment of the Disaster Information System and the Central and Regional Emergency Support System; formulation and implementation of Disaster Recovery Plans; promotion of research and development projects for preventing natural disasters; and penalties.

After the 1995 Kobe earthquake in Japan, the Korean government added "earthquake" in the revised National Disaster Countermeasures Act and established the 1st Earthquake Disaster Countermeasures Plan. Continued effort has been made for earthquake hazard mitigation with the 1st to 3rd Earthquake Disaster Countermeasures Plans and Earthquake Disaster Countermeasures Act enacted on March 28, 2008. The Korean government strengthened the earthquake resistant design code for building structures to cover a wide range of requirements for the seismic design of building structures (Higashino and Okamoto 2006). The national Korean Building Code are regulations that specify types of building structures, materials, and designs based on criteria such as strength, durability, flammability, and resistance to disasters including fire, water, wind, and earthquake. All homes, buildings, and structures must be inspected by a certified inspector based on the national Korean Building Code throughout the period of construction.

The Korean government established the NEMA, as an independent and coordinating agency on June first, 2004. NEMA under the MOSPA coordinates overall measures to counter natural and man-made disasters. NEMA launched the Disaster and Safety Management Basic Act (2004) to respond to emerging natural and man-made disasters. Both the Daegu underground fire and typhoon Maemi in 2003 were triggers to enforce this Act. The Disaster and Safety Management Basic Act (2004) was established to comprehensively manage all kinds of disaster in terms of disaster mitigation, preparedness, response, and recovery. Based on the previous experience from the above acts, such as the Civil Defense Basic Act of 1975, the Natural Disaster Countermeasures Act of 1995, and the Emergency Management Act of 1995, the government tried to manage social emergencies (including those relating to critical infrastructure and key resources) as well as natural disasters and man-made disasters through this Act.

The Disaster and Safety Management Basic Act (2004) stipulates that some ministries take charge of managing each type of disaster and enhancing safety from disasters. For example, the Ministry of Security and Public Administration takes charge of managing storm and flooding, and fire and explosion accidents, the Ministry of Land, Infrastructure and Transport takes charge of managing transportation safety and infrastructure incidents, and the Ministry of Environment is responsible for managing environmental pollution and chemical accidents. After the Disaster and Safety Management Basic Act (2004) was established, the Natural Disaster Countermeasures Act was wholly amended in 2005 to focus more on disaster mitigation and recovery because disaster response policies were duplicated both in the Disaster and Safety Management Basic Act (2004) and the Natural Disaster Countermeasures Act (1995). Table 9.3 shows the comparison of the contents of four natural disaster policies in terms of each chapter's title and the number of articles. Table 9.3 indicates that the Countermeasures against Typhoons and Floods Act (1967) consists of 52 articles arranged in six chapters. The Countermeasures against Typhoons and Floods Act (1967) included only four articles (Article 21 through 24) regarding disaster prevention and mitigation measures.

The Natural Disasters Countermeasures Act (1995) consists of 72 articles arranged in nine chapters. The Natural Disasters Countermeasures Act (1995) newly included Earthquake countermeasures (Chap. 5), Disaster recovery (Chap. 7), and Penal provision (Chap. 9), compared to the Countermeasures against Typhoons and Floods Act (1967). The Act (1995) had 11 articles about disaster prevention and mitigation measures (Chap. 4).

Two Natural Disasters Countermeasures Acts, which were wholly amended in 2005 and in 2012, had the same number of chapters and articles. But the Natural Disasters Countermeasures Act (2012) did not include the earthquake parts, which were Article 23, 24, and 25 in Chap. 2 of the Natural Disasters Countermeasures Act (2005). The central government enacted a comprehensive earthquake policy, the Earthquake Disaster Countermeasures Act (Act No. 9001), enacted on March 28, 2008. The Natural Disasters Countermeasures Act (2005, 2012) consists of 79 articles arranged in seven chapters. These two acts mainly focused on disaster prevention/mitigation and disaster recovery. This act included 30 articles (Article 4 through Article 33) in Chap. 2. But disaster response articles were removed from this act. This act also included Chap. 3 (Disaster information and emergency support) and Chap. 5 (Research and technology development).

Table 9.3 shows that articles regarding disaster mitigation measures were substantially increased through disaster policies over time. There were only four

Table 9.3 Comparison of the contents of four natural disaster policies

	Countermeasures against Typhoon and Flood Act (1967)	Natural Disasters Countermeasures Act (1995)	Natural Disasters Countermeasures Act (2005, (2012[a])
Chapter 1	General Provision (Article 1–2)	General Provision (Article 1–4)	General Provision (Article 1–3)
Chapter 2	Disaster Plan (Article 3–9)	Organization (Article 5–13)	Disaster Prevention of and mitigation (Article 4–33)
Chapter 3	Organization (Article 10–20)	Disaster Plan (Article 14–21)	Disaster information and emergency support (Article 34–45)
Chapter 4	Disaster Prevention and mitigation (Article 21–24)	Disaster Prevention and mitigation (Article 22–32)	Disaster recovery (Article 46–57)
Chapter 5	Disaster Response (Article 25–41)	Earthquake countermeasures (Article 33–35)	Research and technology development (Article 58–63)
Chapter 6	Supplementary Provisions (Article 42–47)	Disaster Response (Article 36–51)	Supplementary Provisions (Article 64–76)
Chapter 7	Penal Provisions (Article 48–52)	Disaster Recovery (Article 52–57)	Penal Provisions (Article 77–79)
Chapter 8	–	Supplementary Provisions (Article 58–67)	–
Chapter 9	–	Penal Provisions (Article 68–72)	–

[a] The Natural Disasters Countermeasures Act (2012) deleted Section 3 Earthquake, Article 23, 24, and 25 in Chapter 2 from the Natural Disasters Countermeasures Act (2005)

disaster mitigation articles in the Countermeasures against Typhoon and Flood Act (1967), but the Natural Disaster Countermeasures Acts (2012) included 30 articles regarding mitigation policies. Moreover, the Natural Disaster Countermeasures Acts (2012) included policies for disaster information and emergency support, which are critical information in making disaster mitigation strategies and projects.

The NEMA under the MOSPA has established the National Disaster Management System (NDMS), which is a nationwide information system, to prevent risk factors, to respond to emergency situations, and to support recovery and restoration from natural disasters (Shim and Lee 2012). This system consists of "the National Disaster Information Center," "Disaster Information Sharing System", "Disaster Management Information DB Center", "The Earthquake Response System," and "Disaster Information Dissemination System." The central government provides seven metropolitan cities, nine provinces, and related government agencies with natural and man-made disaster information using GPS (Global Positioning System), GIS (Geospatial Information System), GMS (Geostationary Meteorological Satellite), and GEOS (Geostationary Operational Environmental Satellite). Disseminating disaster information and data through interfacing with central and local governments on disaster situations and damage in real time plays a critical role in supporting decision making processes and emergency operating procedures to reduce disaster damages and to save lives.

Table 9.4 shows the comparison of the contents of disaster mitigation measures in four natural disaster acts. The Countermeasures against Typhoon and Flood Act (1967) and the Natural Disasters Countermeasures Act (1995) focused mainly on education and training, measures to manage supplies and materials, and measures to secure prevention equipment. These two acts focus more on structural mitigation measures rather than nonstructural mitigation measures. But, the Natural Disasters Countermeasures Act (2005) significantly amended its policy regarding disaster mitigation measures. This act included more nonstructural mitigation measures, such as Prior Consultations on Examination of Factors Influencing Disasters, Prohibition against Prior Permission for Development Project, Restrictions on Act of Constructing Buildings, and Formulation of Comprehensive Plans to Reduce Damage from Storm and Flood.

Prior Consultations on Examination of Factors Influencing Disasters predicts and analyzes disaster risk factors that can occur in various administrative plans and development projects. Prior consultations shall be requested by the heads of central and local administrative agencies when any development project before they finalize and permit the relevant administrative plan and development project. Administrative plans and development projects regarding land and city development, industry and distribution complex construction, energy development, transportation facility construction, river use and development, water resources and ocean development, and tourism complex and athletic facility construction should be given the preliminary assessment consultation of disaster impacts (Natural Disasters Countermeasures Act, 2012).

The local government is required by Article 16 of the Natural Disasters Countermeasures Act to develop and adopt its Comprehensive Plan to reduce damage from storm and flood, as prescribed by Presidential Decree. The local governments must have a NEMA-approved Local Comprehensive Plan in order to receive project grants, and resubmit it for approval every 5 years. NEMA subsidizes part of the cost of the project for reducing damage caused by storm and flood from the National Treasury after examining the project implementation plans. A comprehensive plan identifies and assesses the community's risks and vulnerabilities to natural hazardous events such as flooding, storms, typhoons, and landslides. The plan includes a set of goals related to the overall goal of hazard mitigation planning, including structural and non-structural mitigation measures, an assessment of existing mitigation measures, and a set of methods that collaborate with other city plans for disaster reduction. The plan also contains the plan implementation and maintenance strategies, including the plan adoption process, selection and prioritization of viable mitigation projects, financial strategies, and maintenance for compliance with central planning regulations.

The Natural Disasters Countermeasures Act (2005 and 2012) also included regulations to limit new development in vulnerable areas and established the Disaster Information System and Central Emergency Support System using various disaster maps as nonstructural mitigation measures. Moreover, the acts (2005 and 2012)

Table 9.4 Comparison of the contents of disaster mitigation measures in four natural disaster policies

	Countermeasures against Typhoon and Flood Act (1967)	Natural Disasters Counter-measures Act (1995)	Natural Disasters Countermeasures Act (2005, (2012[a])
Disasters	Typhoon, flood	Typhoon, flood, earthquake	Typhoon, flood, earthquake[a], snow, drought, tsunami
Disaster phase	Prevention, response, recovery	Prevention, response, recovery	Mitigation, recovery
Mitigation measures	The reform of organizations for the prevention	The reform of organizations for the prevention	Prior Consultations on Examination of Factors Influencing Disasters
	Education, training	Education, training	Prohibition against Prior Permission, etc. for Development Project
	Measures to manage supplies and materials	Measures to manage supplies and materials	Investigation and Analysis, etc. of Causes of Disasters
	Measure to secure prevention equipment	Measures to secure prevention equipment	Designation of Areas Vulnerable to Natural Disaster
	Designation of the vulnerable areas	Measures to secure disaster vulnerable facilities	Restrictions on Act of Constructing Buildings
		Designation of the vulnerable areas	Formulation of Comprehensive Plans to Reduce Damage from Storm and Flood
		Formulation of emergency plans for facilities subject to earthquake-proof designs	Establishment of Disaster Information System and Central Emergency Support System
			Design and Use of Various Disaster Maps Education and training disaster managers and officials

[a] The Natural Disasters Countermeasures Act (2012) deleted Section 3 Earthquake, Article 23, 24, and 25 in Chap. 2 from the Natural Disasters Countermeasures Act (2005)

provided guidelines for education and training for disaster managers and governmental officials. However, the acts (2005 and 2012) did not include detailed education and training programs for the public regarding disaster mitigation measures, compared with the previous 1995 act. The government needs to develop a systematic education program on disaster prevention for both the public and governmental officials. Education will help improve awareness of natural disasters, increase commitment to their resolution, promote a cooperative attitude, and increase self-reliance in disaster management. The acts need to improve mitigation measures for existing properties. Not only reinforcing or retrofitting existing buildings, but also an acquisition or buyout program will be necessary to resolve the fundamental vulnerability problems of existing buildings or houses in hazardous areas.

The following part examines the relationship between disasters and development, how development projects have the potential to aggravate or reduce vulnerability and disaster risk, and how disaster policies affect development projects to reduce disaster risk through the case example of the 2011 Seoul flood.

9.5 Disaster and Development: The 2011 Seoul Flood Case

Development processes have both positive and negative impacts on disaster risk. While disasters put human development at risk, human developments can contribute to disaster risk reduction (UNDP 2012). Over time, rapid urbanization increased the exposure of people and economic assets to natural disasters. Moreover, urban environmental degradation through the process of land development increases the disaster risk (Kreimer and Munasinghe 1992).

Korea has experienced rapid urbanization over the last 50 years. The rate of urbanization in Korea has increased from 35.8 % in the 1960s to 86.5 % in 2005 (Lee 2006; Jeon et al. 2010). Along with the rapid urbanization, the urban areas, especially Seoul Metropolitan city, have a very high percentage of impervious surface areas. The percentage of impervious surfaces in Seoul has significantly increased from 7.8 % in 1962 to 47.7 % in 2010. In particular, the coverage by impervious surfaces in the downtown and Gangnam areas (South of the Han River) of Seoul is more than 90 %. Increases in impervious surfaces due to urbanization result in decreased infiltration and increased surface runoff, peak flow, and the magnitude of flooding (White and Greer 2006).

In July 2011, the heaviest rains in 100 years, totaling over 530 mm, pounded Seoul and its surrounding areas, triggering flash floods and landslides that resulted in 60 deaths and 10 people missing. More than 4,500 people had been forced out of their homes and many houses were out of power. Because of this torrential rain, thousands of houses, main roads, and some subway stations in the downtown and Gangnam areas in Seoul were inundated and thousands of vehicles were left submerged on flooded roads. Seventeen residents were killed by multiple landslides in Mount Umyeon triggered by heavy rain in Southern Seoul. The excessive flood waters overwhelmed the city's drainage system, resulting in mass displacement of the residents, particularly those residing in low-lying areas. Increased torrential rains due to climate change and the lack of capacity of the drainage system causing frequent flooding in urban areas have created challenges for flood disaster risk management. Along with these challenges, the Korean government amended the Natural Disasters Countermeasures Act in 2012 to minimize potential causes of disasters inherent in various development projects by strengthening the preliminary consultation process on flood risk assessment and reduction for new development projects. Moreover, the Act requires all local and provincial governments to prepare a comprehensive storm and flood mitigation plan and to integrate hazard mitigation concepts and strategies into local comprehensive urban planning and management.

Under the regulation of the Natural Disasters Countermeasures Act, new development projects in urban areas with a targeted development area of 5,000 m^2 or less should be reviewed in consultation on the examination of factors influencing disasters and methods to reduce disaster risk. Local governments require new development projects to include disaster risk reduction plans and activities. Regarding flood risk, the stormwater runoff reduction design is one of the most used disaster risk reduction methods in urban development projects. In an attempt to reduce flood risk, the runoff reduction designs including permeable pavement, bioretention, green roof, and rooftop disconnection, are adopted in a development plan. These runoff reduction methods using Low Impact Development (LID) methods to reduce impervious cover and maximize retention of natural areas and undisturbed soil are currently adopted in urban development including new apartment complex projects and industrial complexes (Jeon et al. 2010).

The case of 2011 flooding in Seoul shows that there is a cause and effect relationship between disasters and economic development. This "focusing event" forced policy makers and disaster managers to evaluate development programs in the context of disasters (DMC 1997). Decision-makers and planners are increasingly aware that unplanned development problems often induce disasters. Through the disaster policies, such as Prior Consultations on Examination of Factors Influencing Disasters, Prohibition against Prior Permission for Development Project, Restrictions on Act of Constructing Buildings, and Formulation of Comprehensive Plans to Reduce Damage from Storm and Flood under the Natural Disaster Countermeasures Act (2012), the Korean government considers shifts and approaches in disaster risk management to reduce development-induced risks and vulnerability.

Conclusion

While human losses caused by natural disasters are decreasing, economic damages resulting from disasters have increased in Korea over the past five decades. The Korean government spends about $ 603 million annually for disaster recovery (NDIC 2012).

Several catastrophic events were the impetus to reform disaster policies over time in Korea. Moreover, the Korean government made significant changes to natural disaster organizations to protect lives and to mitigate economic damages from natural disasters. Since 2004, the Korean government established the NEMA as an independent agency to comprehensively manage natural and man-made disasters. However, Korea faces several challenges in disaster management at the local governmental level. Fragmented disaster management systems and organizations at the local government level and lack of coordination within and between disaster related organizations are responsible for ineffective and inefficient disaster management in Korea.

This chapter also demonstrated that there is a cause and effect relationship between economic development and disasters. While the process of urbanization generates and increases vulnerability and disaster risks, urban development integrated into sustainable development planning can decrease vulnerability and hazards.

While the Korean government has been making significant changes to disaster policies, focusing more on pre-disaster mitigation, communication and cooperation between relevant disaster institutions, including volunteer organizations, needs to be improved to minimize future disaster damages. Moreover, more disaster education and training for public officials and citizens should be emphasized in disaster policies to build a disaster resilient community.

References

Alexander, D. A. (2000). *Confronting catastrophe: New perspectives on natural disasters*. New York: Oxford University Press.

Birkland, T. A. (1997). *After disaster: Agenda setting, public policy, and focusing events*. Washington DC: Georgetown University Press.

Col, J. M. (2007). Managing disasters: The role of local government. *Public Administration Review, 67*(s1), 114–124.

Countermeasures against Typhoon and Flood Act. (1967). http://www.law.go.kr. Accessed 22 Sept 2012.

Disaster and Safety Management Basic Act. (2004). http://www.law.go.kr. Accessed 22 Sept 2012.

Disaster Management Center (DMC). (1997). Disasters and development: Study guide and course text for C280-DD02. Madison: University of Wisconsin-Madison.

Ha, K.-M., & Ahn, J.-Y. (2008). National emergency management systems: The United States and Korea. *Journal of Emergency Management, 6*(2), 31–44.

Higashino, M., & Okamoto, S. (Eds.). (2006). *Response control and seismic isolation of buildings*. New York: Taylor & Francis.

Jeon, J.-H., Lim, K. J., Choi, D., & Kim, T.-D. (2010). Modeling the effects of low impact development on runoff and pollutant loads from an apartment complex. *Environmental Engineering Research, 15*(3), 167–172.

Kapucu, N, & Ozerdem, A. (2013). *Managing emergencies and crises*. Boston: Jones & Bartlett.

Kim, P. S., & Lee, J. E. (1998). Emergency management in Korea and its future directions. *Journal of Contingencies and Crisis Management, 6*(4), 189–201.

Kreimer, A., & Munsinghe, M. (1992). *Environmental management and urban vulnerability*. Washington DC: The World Bank.

Lee, W. S. (2006). The trend and characteristics of urbanization of Korea by analysis of population census. *KRIHS Policy Brief, 106,* 1–4.

National Disaster Information Center (NDIC). (2012). http://www.safekorea.go.kr/dmtd/contents/room/newyearbk/New_sfdmg01.jsp?q_menuid=M_NST_SVC_01_03_05_03_01. Accessed 5 Sept 2012.

Natural Disasters Countermeasures Act. (1995). http://www.law.go.kr. Accessed 22 Sept 2012.

Natural Disasters Countermeasures Act. (2005). http://www.law.go.kr. Accessed 22 Sept 2012.

Natural Disasters Countermeasures Act. (2012). http://www.law.go.kr. Accessed 22 Sept 2012.

Nation Emergency Management Agency (NEMA). (2012). http://nema.korea.kr/gonews/main.do. Accessed 5 Sept 2012.

National Research Council (NRC). (1994). *Practical lessons from the Loma Prieta earthquake*. Washington DC: National Academy Press.

Park, D. (2005). Republic of Korea: National Reporting and Information on Disaster Reduction. Paper presented at the World Conference on Disaster Reduction, Kobe, Hyogo, Japan, 18–22 January 2005. http://www.unisdr.org/2005/wcdr/preparatory-process/national-reports/South-Korea-report.pdf. Accessed 24 Sept 2012.

Shim, J. H., & Lee, C. (2012). Strengthening disaster risk assessments to build resilience to natural disasters. In The World Bank (Eds.), *Improving the assessment of disaster risks to strengthen financial resilience: A special joint G20 publication by the government of Mexico and the World Bank* (pp. 207–209). Washington DC: The World Bank.

Sohn, K. T., Lee, J. H., Lee, S. H., & Ryu, C. S. (2005). Statistical prediction of heavy rain in South Korea. *Advances in Atmospheric Sciences, 22*(5), 703–710.

United Nations Development Programme (UNDP). (2012). Reducing disaster risk, a challenge for development. http://www.undp.org/content/undp/en/home/librarypage/crisis-prevention-and-recovery/reducing-disaster-risk-a-challenge-for-development.html. Accessed 5 Sept 2012.

White, M. D., & Greer, K. A. (2006). The effects of watershed urbanization on the stream hydrology and riparian vegetation of Los Peñasquitos Creek, California. *Landscape and Urban Planning, 74,* 125–138.

Part II
Disaster Resilience and Capacity Building

Disasters provide opportunities to change patterns of development, new resources to support those changes, and incentives to become more resilient and sustainable, reducing losses in future disasters. The whole community perspectives, networks and partnerships are highlighted in several government documents and scholarly work. Chapters in this part examine threats from different hazards, disaster preparedness and capacity building efforts in building resilient and sustainable communities. William Waugh and Cathy Yang Liu develop a model to include social vulnerability, the structure of the local economy, business diversity, nonprofit density, and government capacity. Engagement of the "whole community" and building social capital, as well as having effective community leadership, can mean the difference between success or failure in disaster recovery and sustainable community development. Steve Scheinert and Louise K. Comfort highlight the importance of governance structures in rapidly changing environments, the capacity to govern, as well as the resilience to respond to that changing environment for effective response and recovery. The chapter proposes a measure of resilience that is based in a complex adaptive systems approach and applies that measure to UN interventions in Bosnia-Herzegovina, which began in 1995, and in Haiti, which began in 2004 to validate its effectiveness. Samuel D. Brody examines the degree to which natural features of the landscape can support community-level resiliency with respect to flooding and flood impacts. The chapter provides a series of policy recommendations which can enhance a community's ability to cope with the persistent threat of flooding hazards. Melanie Gall and Brian J. Gerber consider the intersection of mitigation and resilience and examine how both are shaped by local and national economic development practices. The chapter traces developments in flood mitigation and flood risk management by considering three European countries with significant coastal and interior vulnerability to the flood hazard: The Netherlands, Great Britain, and Germany. Daniel Nohrstedt and Charles Parker examine learning and policy change as basis for building resilience to extreme events based on a comparative case-study of two storms in Sweden (2005 and 2007). Douglas Paton, David Johnston, Ljubica Mamula-Seadon, and Christine M. Kenney discuss the relationship between resilience, recovery, and development in relation to the 2009 Victoria, Australia wildfires and the 2011 Christchurch earthquake. Kiki Caruson, Osman Alhassan, Jesse

Sey Ayivor, and LegonRobin Ersing present the roles played by women in disaster resiliency efforts, the limits of their enfranchisement in the emergency management process, and the opportunities for the integration of a gender oriented approach to enhancing disaster resiliency among highly vulnerable populations using focus group data gathered from residents of several highly vulnerable migrant settlements in Ghana. Ralph S. Brower, Francisco A. Magno, and Janet Dilling illuminate recent struggles to implement the law and offer implications for disaster management theory and practice in the Philippines and recommend that Filipino leaders must foster democratic institutions that are as responsive to bottom-up problems as top-down interests; build cooperation across public, private, and voluntary sectors; and strengthen human development capabilities in parallel with economic development.

Chapter 10
Disasters, the Whole Community, and Development as Capacity Building

William L. Waugh and Cathy Yang Liu

10.1 Introduction

Disasters often provide opportunities to change patterns of development, new resources to support those changes, and incentives to become more resilient and sustainable, reducing losses in future disasters. In essence, policy inertia is broken, new voices are heard, constituencies are mobilized, champions emerge, and new development tools are brought to bear. However, some communities are able to use the opportunities created by disasters to address risks and to create momentum for redevelopment, while others are not able to do so. What distinguishes the successful communities in disaster recovery and resilience-building from the less successful? How are capacities built to manage environmental hazards and recovery processes? And how does local development planning help to build community capacities and increase resilience? Those are the questions to be addressed here. An integrated model will be built for assessing community characteristics and recovery-related capacities, and critical variables will be identified.

Community capacities are critical variables in long-term recovery and development and they can be nurtured to increase the likelihood of success. This analysis will examine the impact of disasters and subsequent development on local capacity building and the relationship between those capacities and community resilience and sustainability. It will also assess the Federal Emergency Management Agency's (FEMA's) new "Whole Community" approach to building resilience and sustainability. Examples will be drawn from recent natural disasters and post-disaster development efforts in the United States and in other nations. The analysis will also address the following questions in order to explicate the links between disaster

W. L. Waugh (✉) · C. Y. Liu
Georgia State University, Atlanta, GA, USA
e-mail: wwaugh@gsu.edu

C. Y. Liu
e-mail: cyliu@gsu.edu

N. Kapucu, K. T. Liou (eds.), *Disaster and Development,* Environmental Hazards,   167
DOI 10.1007/978-3-319-04468-2_10, © Springer International Publishing Switzerland 2014

recovery and community resilience. How can disasters and development be addressed in an integrated manner? How can the concept of resiliency be operationalized in a way that is useful as a framework to investigate the conditions that lead to stronger, safer, and more sustainable communities? What explains the resiliency of communities and regions? In other words, what factors account for the variation across jurisdictions and geographic units in the ability to respond and recover from a disaster? What are the various policy interventions and governance mechanisms that can be developed to improve the resiliency and sustainability of communities and reduce their vulnerability to natural disasters?

Capacity building may occur prior to the disaster, during the disaster response, and in the recovery process. For example, administrative capacity increases in the recovery process, social capacity is created through the linkage of community networks, and economic capacity grows as internal and external resources are leveraged to rebuild, redevelop, and re-imagine the community. The sense of community gets stronger and, at least initially, common goals tend to overcome social, political and economic divisions. In short, disasters can lead to development that reduces vulnerability and provides the capacity to better cope with future disasters. However, it is not always the case that disasters have positive long-term impacts, but it is usually the case that disasters encourage breaks with the past. In fact, disaster recovery seldom means a return to the pre-disaster condition. Indeed, following the 9/11 attacks there were proposals to expand the subway system in lower Manhattan to better link with airport and commuter line systems. Similarly, following the Katrina disaster there were proposals to rearrange business and residential districts and to limit development in the most vulnerable areas of New Orleans (Waugh and Smith 2006). And, following Hurricane Sandy, plans include updating the electrical grid and water system so that they will be less vulnerable to future storms and better able to support new development in coastal communities (Caldwell et al. 2012).

Of course, the stress and chaos of disasters are not ideal conditions for the consideration of new development strategies. Pressures to rebuild quickly can overwhelm land-use planning and building regulation processes. They can also strain local-construction resources such as skilled workers and building materials. For example, structural engineers have been in high demand in recent U.S. disasters. Recovery specialties typically top lists of employment opportunities. At best, new building codes are adopted in hopes of guiding reconstruction to reduce losses during future disasters. South Carolina strengthened its building codes after Hurricane Hugo in 1989, Louisiana strengthened its codes after Hurricane Katrina, and New York and New Jersey are planning to strengthen their codes since Hurricane Sandy in 2012. Florida communities already had strong building codes and they improved code enforcement following Hurricane Andrew in 1992. Despite those examples, nations and communities that do not have the political and administrative capacities to adopt and enforce appropriate and adequate land-use regulations and building codes and standards under normal circumstances, find it difficult to take proactive measures to reduce future losses. Disasters encourage pragmatic, quick action to house, feed, and employ residents, and provide little time to consider more sustainable alternatives (Barenstein 2013).

10.2 Sustainable Development as Capacity Building

If capacity building is the means to resilience for disaster relief, so is it increasingly the project of local economic development planning. Development planning in the twenty-first century aims at an increased standard of living in a community through a process of human and physical development based on equity and sustainability principles (Fitzgerald and Leigh 2002; Leigh and Blakeley 2013). Sustainability as a goal of economic development planning has been traditionally understood as *ecological* sustainability, but is increasingly associated with the broader concept of community resilience (Leigh and Blakeley 2013), or the ability to adjust to sudden or gradual shocks to the local economy. Similarly, Greenwood and Holt (2010) describe sustainable development as and orientation toward maintaining the various capital stocks that provide a quality of life for a community. These include not only traditional market capital, but also non-market good such as environmental, governmental, and social capital. This approach subsequently requires the ability to measure local capital stocks, creating the need for local indicators. Disasters—whether they are natural, man-made, or structural economic shifts—create threats to community capital stocks, and thus require resiliency planning. Additionally, attempts have been made to measure resilience capacity (Foster 2011, 2012) and sustainability practices based on the tripartite pillars of economic, equity, and environmental sustainability (Opp and Saunders 2013). Foster's "Resilience Capacity Index" uses 12 indicators of economic, socioeconomic, and community connectivity measures (Leigh and Blakeley 2013; Foster 2011).

10.3 Community Resilience

Community resilience has become the focus of the Federal Emergency Management Agency (Fugate 2010). It is a focus driven by the need to reduce reliance upon federal resources, especially federal funding, and to encourage local self-reliance in the face of a disaster. Federal assistance may be insufficient because financial resources are limited and the potential scale of catastrophic disasters, such as pandemics, will outstrip all federal resources, at least in the short term. Federal assistance may not just be slow in coming—it may not come for weeks or months and communities will have to fend for themselves. Even in circumstances like the recent Hurricane Sandy disaster, in which there is ample warning, adequate lead-time to implement emergency plans, and sufficient resources to evacuate most of the at-risk residents in coastal communities, assistance to some of the affected neighborhoods and communities may still be inadequate. The scale of the disaster can complicate the response. Damaged infrastructure and debris-filled roadways may also inhibit access to devastated communities. Therefore, residents have to be able to take care of themselves for a time period and local officials have to be able to make the best use of available resources until help arrives. The typical advice from emergency managers has been for residents to have 72-h kits with sufficient food, water, and other necessities to survive until help arrives. The expectation was that help would

(handwritten annotation: Global Warming = Worse)

arrive in 3 days. The advice now is to have 96-h kits or even 168-h (1 week) kits. Pandemic planners might suggest even larger emergency kits.

Surviving until help arrives is only the first step. Resilience is usually defined in terms of the capacity to recover quickly from catastrophic disasters. In Administrator Fugate's view, it is defined in terms of local capacity building, risk reduction, and preparedness (2010). Edwards identified four indicators of resilience from the literature: understanding community risks, advance mitigation planning and implementation, disaster preparation, and "whole community" engagement (Edwards 2013). Communities need the capacity to deal with disasters with minimal outside assistance; they need to invest in mitigation programs to reduce losses; and they need to plan for the disasters that they cannot prevent. Local capacity building is the key and having mechanisms to share critical resources can greatly expand local capabilities to cope with disasters, as well as with other challenges. Addressing known vulnerabilities can help close the gap between capabilities and demands. Community attributes also affect the capacities to recover. Sharing resources among the states through the Emergency Management Assistance Compact (EMAC), among cities and counties through statewide aid compacts, and between communities through other formal and informal arrangements can also bridge gaps, although neighboring jurisdictions may be hard pressed to share resources if they too are dealing with a disaster (Kapucu et al. 2009; Waugh 2007). The new "Whole Community" Approach adopted by FEMA recognizes that community resources are critical throughout the process of mitigating, preparing for, responding to, and recovering from disasters. Citizen engagement is the key. How to engage citizens is the challenge.

10.4 The "Whole Community" Approach

The importance of social capital for disaster mitigation and resilience cannot be underestimated. This principle is embedded in FEMA's post-Katrina "Whole Community" approach (Edwards 2013). In brief, the "Whole Community" approach enjoins officials to understand community complexity, recognize community capabilities and needs, foster relationships with community leaders, build and maintain partnerships, empower local action, and leverage and strengthen social infrastructure, networks, and other assets. The variables that affect capacities to recover quickly have to be understood in the context of particular communities. Officials need to understand the social, economic, political, and cultural dimensions of the community in order to understand its resources and vulnerabilities and how to build support for action. As recent disasters have demonstrated, social networks can be used for emergency warnings and to facilitate emergency operations. Pre-disaster, the community determines how it should address hazards and how it should prepare for disasters. The community should also determine response and recovery priorities. Community engagement is critical. Community resources ranging from individuals to large institutions can be essential in large disasters.

The challenge for officials is to open up planning processes to facilitate participation. Everyone should have a "seat at the table." Communities that have very

- Community councils;
- Volunteer organizations (e.g., local Voluntary Organizations Active in Disaster, Community Emergency Response Team programs, volunteer centers, State and County Animal Response Teams, etc.);
- Faith-based organizations,
- Individual citizens;
- Community leaders (e.g., representatives from specific segments of the community, including seniors, minority populations, and non-English speakers);
- Disability services;
- School boards;
- Higher education institutions;
- Local Cooperative Extension System offices;
- Animal control agencies and animal welfare organizations;
- Surplus stores;
- Hardware stores;
- Big-box stores;
- Small, local retailers;
- Supply chain components, such as manufacturers, distributors, suppliers, and logistics providers;
- Home care services;
- Medical facilities;
- Government agencies (all levels and disciplines);
- Embassies;
- Local Planning Councils (e.g., Citizen Corps Councils, Local Emergency Planning Committees);
- Chambers of commerce;
- Nonprofit organizations;
- Advocacy groups;
- Media outlets
- Airports;
- Public transportation systems;
- Utility providers; and
- Many others

(FEMA, 2011, p. 12).

Fig. 10.1 Partners to consider engaging

active community groups and participative processes for strategic planning, land-use planning, and other decision processes, can utilize the same mechanisms for engagement. A study of information sharing relative to risk at the community level, for example, concluded that citizen engagement and participative management processes greatly facilitated communication between and among residents, planning officials, and other officials about hazards and the need to address risks to life, property, and quality of life (Waugh 2009). In communities in which citizen engagement is limited, the challenge will be all the greater.

Community members should be asked to identify additional resources and capabilities (Fig. 10.1). Engagement engenders buy-in. The goal is to "(e)nable the public to lead, not follow, in identifying priorities, organizing support, implementing programs, and evaluating outcomes; empower them to draw on their full potential in developing collective actions and solutions" (FEMA 2011, p. 15). If possible, "(a)lign emergency management activities to support the institutions, assets, and networks that people turn to in order to solve problems on a daily basis" (FEMA 2011, p. 18).

Institutions are also a critical part of this new approach. Gazley discussed the importance of "Community Organizations Active in Disaster (COADs)" to the "whole community" approach, suggesting that developing and organizing community organizations around the goal of disaster relief can be a key component to future disaster relief, and can aid communities in, as she quoted from Manyena et al. (2011), "bounce forward" after a disaster by taking advantage of these new relationships and community engagement (Gazley 2013).

10.5   The Community and Disaster Recovery

Daniel Alesch et al. (2009) concluded that a number of factors influence a community's ability to recover from a major disaster. The severity of the disaster is one of those factors. Injuries and the loss of life, damage to the built environment, economic conditions prior to the disaster, are also factors. A large affluent community with many resources, a diverse economy, well-developed social networks, permanent anchor institutions that are likely to be rebuilt (such as universities, prisons, and military bases), and a quality of life that will encourage residents to stay and rebuild is clearly more self-reliant and resilient than a community lacking those attributes. Community leadership is critical. Haiti, for example had serious long-term economic problems, weak governance structures, and an extremely vulnerable built environment before it was struck by the earthquake on January 12, 2010. Persistent poverty and weak social institutions compounded the difficulty in recovering from the event. The prognosis for long-term development, even before the earthquake, was poor. International aid programs had not been able to overcome those obstacles and are still struggling to address social and economic ills. By contrast, Chile has managed its own recovery from a stronger earthquake and tsunami only months after the Haitian earthquake with minimal outside help because it does not suffer from the same social, economic, and political weaknesses that Haiti has (see Chap. 23 in this volume).

Daniel Aldrich expands upon the Alesch team's conclusions by identifying at least five dimensions in disaster recovery: social-psychological, organizational and institutional, economic, infrastructure, and governmental (2012, p. 7). His major variables are the quality of governance, the amount of external aid, the amount of damage, population density, demographics/socioeconomic conditions, and social capital. In some measure, the principal variable is social-psychological. He argues that "high levels of social capital—more than such commonly referenced factors as socioeconomic conditions, population density, amount of damage or aid—serve as the core engine of recovery" (2012, p. 15).

The secret to recovery, according to Aldrich (2012), is building social capital—maintaining and strengthening the social networks that hold the community together. Those social networks, in effect, amount to "social insurance." ' crease residents' willingness to help their neighbors, they help overcome rel to engage in collective action, and they strengthen residents' willingness to voice preferences and reduce the likelihood that they will simply leave the community. Residents will tend to work through their associations to help others and that participation in recovery will foster a sense of efficacy, control over their own fates that helps them recover psychologically.

However, Aldrich warns that social networks can dominate the recovery process, exclude those not within the network, and even encourage opposition to the preferences of the larger community. For example, some social groups made the siting of FEMA trailers difficult in Louisiana and Mississippi during the Katrina disaster and even precipitated violence against outsiders. In Japan, after the 1921 earthquake, violence was directed against Korean residents. Post-disaster divisions within the community largely mirror pre-disaster divisions.

Notwithstanding those caveats, Aldrich recommends that officials focus on bonding, bridging, and linking relationships—those that draw the community together, provide bridges to other social networks, and link community networks with external resources—and develop baseline data, i.e., profiles of communities, to facilitate work with the social networks. While some leading disaster scholars have found fault with Aldrich's methodology, in particular his categorization of social capital and his proxy variables, his general findings are consistent with the literature (see Aldrich forthcoming).

Barenstein and Leemann (2013) conclude that recovery is dependent upon the quality of local governance, rather than on central government control. The success of housing reconstruction projects is directly related to the participation of residents in the design and siting of new construction and the attention paid to local preferences by outside agencies, including central government and international NGOs. Local leaders should be accountable to their communities (Habertli 2013). If local property owners drive reconstruction, it is finished faster and cheaper and is more appropriate to the culture of the community. That was the experience in Gujarat, India, following the 2001 earthquake (Mantel 2013). Like Aldrich, Barenstein and Leemann conclude that local direction is essential but there is a danger that outside assistance will be captured by local elites and that their social networks will exclude nonmembers. Moreover, external assistance, international and domestic, tends to dissipate over time and aid agencies feel pressure to produce results quickly, often losing sight of their own development principles and often failing to be attentive to local preferences (Barenstein 2013).

Resilience is often defined in terms of capacity to recover from disasters quickly. Alesch and his team found some communities in the United States to be very resilient and some extremely vulnerable. Despite massive damage to infrastructure and loss of life, some communities recover quickly while others did not. The differences among communities were attributed to interventions by nongovernmental organizations, insured losses, and good community leadership. The resilience literature includes indices for social vulnerability based upon demographic criteria and hazard exposure. It also includes analyses based primarily on pre- and/or post-disaster economic conditions, such as methodologies developed to stimulate economic development. However, very little use has been made of measures of nongovernmental organization activity or governmental management capacity. The variables identified by Alesch et al. (2009) are remarkably similar. The importance of social networks and good leadership to disaster recovery and community resilience are very clear. Our objective is to bring together those variables and others common to the literature in order to develop a model to predict capacity to recover quickly and to assure that post-disaster redevelopment serves to reduce risks to life and property. An integrated approach to recovery necessarily has to address all or most of the dimensions. Therefore, we propose a model for increasing community resilience by adding dimensions to the current models and by encouraging collaborative relationships among public, private and nonprofit sectors.

Martin and Olgun (2011) describe the process as "enabling" communities rather than the federal government attempting to "coordinate" them. While Martin and Olgun focus primarily on enabling communities to manage the consequences of disasters, they suggest that the process of developing "multi-organizational alliances and partnerships" (MOAPs) can also increase the abilities of communities to prepare for, respond to, and recover from disasters. These MOAPs or networks may also increase the capacities of communities to address many other issues. They certainly have implications for community, as well as federal and state, leadership (Waugh 2013), as well as for program and policy design. The development of interorganizational coordination mechanisms in Turkey following the Marmara and Duzce earthquakes in 1999 is a good example of reform following poor disaster responses (Celik and Corbacioglu 2013).

10.6 The Integrated Model

We thus envision an enlarged and comprehensive framework of gauging the multiple dimensions of community resilience and capacity and in so doing suggest the metrics for local governments to evaluate their disaster preparedness (Fig. 10.2). These include:

1. The social vulnerability index which is composed of the demographic characteristics of the local population (age, race/ethnicity, immigrant status and language proficiency), their social and economic status (employment status, income, automobile availability, and household wealth), as well as housing structure characteristics (housing tenure, dwelling type, and lot size). All these statistics can be obtained from Census of Population and Housing as well as the American Community Survey on the county level, and have been summarized by some previous researchers (Cutter and Finch 2008; Flanagan et al. 2011).
2. Features of the local economy which includes the industrial composition among major North America Industry Classification System (NAICS) sectors as well as their relative size and growth as compared to elsewhere in the region or country. A more diversified economy, one that has more permanent institutions like public administration and education, healthcare services can be expected to be more disaster-resilient and recover at a faster rate. Detailed establishment, employ-

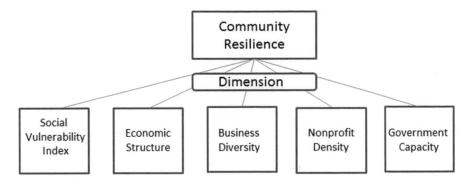

Fig. 10.2   An integrated model of community resilience

ment and payroll statistics on the county level are available from County Business Patterns database collected by the U.S. Census Bureau.

3. Size and characteristics of small businesses, especially minority-owned businesses in the area. Small businesses are viewed as engines of economic renewal in disaster-hit cities and regions, such as New Orleans, for their deep roots in the communities and capacity of rebuilding economic stability (U.S. Small Business Administration Office of Advocacy 2006). Minority-owned businesses play particular roles in anchoring community social life, nurturing social networks and contributing to social cohesion. Evidence suggests that immigrant businesses played catalyst roles in economic revitalization after September 11 in New York and the riots and earthquakes in Los Angeles in the 1990s (Bowles and Colton 2007). Survey of Business Owners offers detailed information on county-level women-, black-, Asian- and Hispanic- owned businesses and their associated characteristics.

4. Density of non-government organizations (NGO) in an area. NGOs can provide a wide range of services that satisfy the needs of local residents. Their flexibility and local knowledge position them especially well in economic recovery efforts and as potential partners of government agencies. Local for-profit and not-for-profit institutions together determine the density and diversity of the community's institutional completeness (Breton 1964) and thus disaster resilience. The National Center for Charitable Statistics (NCCS) is the clearinghouse of data on the nonprofit data in the United States and offers relevant information on the county level.

5. Government management capacity at the community, regional, and state levels. There are already metrics for government capacity, usually focusing on the federal and state levels, but adaptable to the community level. The major emergency management program standards include management capacities as a foundation for program elements, such as logistics, resource management, and emergency planning. Compliance with the standards, particularly the Emergency Management Accreditation Program (EMAP) standards, can serve as a measure of emergency management and administrative capacity. Mechanisms for resource sharing have also been identified as critical.

Empirical evidence demonstrates the variations in post-disaster recovery and economic development across communities with different endowments. Housing studies show that housing recovery trajectories depended on neighborhood demographic, socioeconomic, and housing characteristics with homes in low-income and minority neighborhoods recovering more slowly after Hurricane Andrew (Zhang and Peacock 2009). In examining the recovery efforts in response to Hurricane Katrina, Olshansky et al. (2008) confirmed the importance of previous plans, citizen involvement, information infrastructure, and external resources in post-disaster recovery. They also called for the need to establish a center for collecting and distributing data and news which would have better informed all parties.

Disasters also provide opportunities to review and evaluate the resilience of different development patterns and policy interventions. In examining 5 years of insured flood loss claims across 144 counties and parishes along the Gulf of Mexico, Brody et al. (2011) found that high-density development patterns significantly reduce amounts of reported property damage, while increasing percentages of low-density development greatly exacerbate flood losses. They suggest the type of community development design that can promote flood-resilient communities into the future. This and other studies offer valuable insights on post-disaster planning and capacity building in local communities and beyond.

The "Whole Community" approach fundamentally means taking advantage of regular, i.e., non-disaster, decision processes and relationships. To the extent that a community lacks a participative approach to local problem solving and policymaking, implementation may well encourage the development of new skills and capabilities. Also, the approach's emphasis on community determination of priorities should force external actors to defer to local prerogatives. Much the same approach is now guiding international disaster and development operations (Barenstein and Leeman 2013). Interestingly, the United Nation's cluster approach to humanitarian assistance has been criticized for failing to engage fully with local nongovernmental organizations and residents. It is not altogether clear how humanitarian assistance can be used to support local resilience, although there are strong arguments for "sustainable assistance" and more engagement with local residents (Levine et al. 2012). There is more agreement on the need to focus recovery/development efforts on rebuilding "communities" in the sense of social networks (Barenstein and Leemann 2013). Aldrich (2012), in fact, defines recovery as the building or rebuilding of social capital.

10.7 Building National Resilience—A New Vision

The new handbook published by the National Academies Press, *Disaster Resilience: A National Imperative* (COSEPUP 2012), urges the nation to take a new path in building disaster resilience: one that forges a culture of resilience on individual, household, community, and national level before instead of after they occur, which can potentially reduce disasters' costs afterwards. This new path departs from our present course in disaster relief and recovery with its huge financial burden as well as economic, social, and environmental losses. The ongoing demographic shifts and migration patterns of the country, which includes an aging population and migration towards coastal and southern regions imply greater challenges to disaster preparation in the years to come. This move from a reactive to a proactive approach can guide the nation forward under the new circumstances.

According to this handbook, resilience is defined as "the ability to prepare and plan for, absorb, recover from, and more successfully adapt to adverse events" (p. 16). A more disaster-resilient America is envisioned for 2030 in which information on risks and vulnerabilities are transparent and accessible; government entities, communities and businesses all have resilience strategies and plans; community coalitions are formed to provide essential services; and federal per capita recovery cost cease to rise. To achieve this vision does not only require policy decisions at various levels of governments, but also the collaborative efforts from all families and communities. Six specific recommendations are offered: (1) to create and maintain broad-based community resilience coalitions at local and regional levels; (2) the public and private sectors should invest in risk management strategies; (3) federal agencies should incorporate national resilience as a guiding principle in their mission and programs; (4) federal agencies need to coordinate resilience, through review, self-assessment and inter-agency communication; (5) to establish a national data center that document disaster-related data; and (6) the Department of Homeland Security should lead in developing a National Resilience Scorecard. Examples of this kind of measurement tool to track progress include the Coastal Resilience Index for coastal storms and the San Francisco Planning and Urban Research Association (SPUR) method for earthquakes (COSEPUP 2012). Other examples, such as the Baseline Resilience Indicator for Communities (BRIC) (Cutter et al. 2010) and the Resilience Capacity Index (RCI) (Foster 2011) use more quantitative measures to create a single indicator of resilience. Communities across the country can develop scorecards that are suitable to their specific contexts and needs.

U.S. Economic Development Administration (EDA) has recognized the need for economic adjustment planning or resilience planning and calls for design and implementation of strategies that facilitate localities' adjustment to external shocks on their economic bases. These shocks may consist of structural changes as a result of natural, economic, financial, as well as other causes. Thus, disasters and development can be better integrated through a more collaborative approach to local planning. The American Planning Association advocates linking hazard mitigation plans to local comprehensive plans. Risk information is more easily shared in communities in which the residents are engaged and involved in planning processes and the management of the jurisdiction is open and collaborative (Waugh 2010). Building local capacity necessarily requires the cultivation of linkages and the sharing of information. The building of social capital and, if Aldrich and Barenstein and Leemann are right, the building of governance mechanisms that assure participation by all segments of society in community decision making should be seen as critical.

References

Aldrich, D. P. (forthcoming). Response to my critics. *Risk, hazards, and crisis in public policy*.

Aldrich, D. P. (2012). *Building resilience: Social capital in post-disaster recovery*. Chicago: University of Chicago Press.

Alesch, D., Arendt, L., & Holly, J. (2009). *Managing for long-term community recovery in the aftermath of disaster*. Fairfax: Public Entity Risk Institute.

Barenstein, J. D. (2013). Communities' perspectives on housing reconstruction in Gujarat following the earthquake of 2001. In J. D. Barenstein & E. Leemann (Eds.), *Post-disaster reconstruction and change: Communities' perspectives* (pp. 71–100). Boca Raton: CRC Press.

Barenstein, J. E. D., & Leemann, E. (2013). Introduction. In J. E. D. Barenstein & E. Leemann (Eds.), *Post-disaster reconstruction and change: Communities' perspectives* (pp. xix–xxx). Boca Raton: CRC Press.

Bowles, J., & Colton, T. (2007). *A world of opportunity*. New York: Center for an Urban Future.

Breton, R. (1964). Institutional completeness of ethnic communities and the personal relations of immigrants. *American Journal of Sociology, 70*(2), 193–205.

Brody, S. D., Gunn, J., Peacock, W., & Highfield, W. E. (2011). Examining the influence of development patterns on flood damages along the gulf of Mexico. *Journal of Planning Education and Research, 31*(4), 438–448.

Caldwell, D., Wald, M. L., & Drew, C. (2012). Hurricane Sandy alters utilities' calculus on upgrades. *New York Times* (December 29). http://nytimes.com/2012/12/29/business/hurricane-sandy-alters-utilities-calculason-upgradesonJanuary9,2013. Accessed 29 Dec 2012.

Committee on Increasing National Resilience to Hazards and Disasters and Committee on Science Engineering and Public Policy (COSEPUP). (2012). *Disaster resilience: A national imperative*. Washington, D.C.: The National Academies Press.

Cutter, S. L., & Finch, C. (2008). Temporal and spatial changes in social vulnerability to natural hazards. *Proceedings of the National Academy of Sciences of the United States of America, 105*(7), 2301–2306.

Cutter, S. L., Burton, C. G., & Emrich, C. T. (2010). Disaster resilience indicators for benchmarking baseline conditions. *Journal of Homeland Security and Emergency Management, 7*(1), 1547–7355.

Edwards, F. (2013). All hazards, whole community: Creating resiliency. In N. Kapucu, C. V. Hawkins, & F. I. Rivera (Eds.), *Disaster resiliency: Interdisciplinary perspectives* (pp. 44–70). New York: Routledge.

Federal Emergency Management Agency. (2011). *A whole community approach to emergency management: Principles, themes, and pathways for action*. Washington, D.C.: FEMA (FDOC-008-1, December).

Fitzgerald, J., & Leigh, N. G. (2002). *Economic revitalization: Cases and strategies for city and suburb*. Thousand Oaks: Sage Publications, Inc.

Flanagan, B. E., Gregory, E. W., Hallisey, E. J., Heitgerd, J. L., & Lewis, B. (2011). A social vulnerability index for disaster management. *Journal of Homeland Security and Emergency Management, 8*(1), 1547–7355 (Article 3).

Foster, K. A. (2011). Resilience capacity index. http://brr.berkeley.edu/rci/. Accessed 21 May 2013.

Foster, K. A. (2012). In search of regional resilience. In M. Weir, N. Pindus, H. Wial, & H. Wolman (Eds.), *Urban and regional policy and its effects: Building resilient regions* (Vol. 4, pp. 24–59). Washington, D.C.: Brookings Institution Press.

Fugate, C. (2010). Opening plenary speech at the National Conference of the International Association of Emergency Managers, San Antonio, TX, October 31.

Gazley, B. (2013). Building collaborative capacity for disaster resiliency. In N. Kapucu, C. V. Hawkins, & F. I. Rivera (Eds.), *Disaster resiliency: Interdisciplinary perspectives* (pp. 107–121). New York: Routledge.

Greenwood, D. T., & Holt, R. P.F. (2010). *Local economic development in the 21st century: Quality of life and sustainability*. Armonk: M.E. Sharpe, Inc.

Habertli, I. (2013). Aid distribution after hurricane mitch and changes in social capital in two nicaraguan rural communities. In J. E. D. Barenstein & E. Leemann (Eds.), *Post-disaster reconstruction and change: Communities' perspectives* (pp. 31–54). Boca Raton: CRC Press.

Kapucu, N., Augustin, M. E., & Garayev, V. (2009). Interstate partnerships in emergency management: Emergency management assistance compact (EMAC) in response to catastrophic disasters. *Public Administration Review, 69*(2), 297–313.

Leemann, E. (2013). Communal leadership in post-mitch housing reconstruction in Nicaragua. In J. E. D. Barenstein & E. Leemann (Eds.), *Post-disaster reconstruction and change: Communities' perspectives* (pp. 3–29). Boca Raton: CRC Press.

Leigh, N. G., & Blakely, E. J. (2013). *Planning local economic development: Theory and practice* (5th ed.). Los Angeles: Sage.

Levine, S., Pain, A., Bailey, S., & Fan, L. (2012). The relevance of 'Resilience'? Overseas Development Institute, Humanitarian Policy Group, HPG Policy Brief 49, September.

Mantel, C. (2013). Ownership, control and accountability in post-tsunami housing reconstruction processes in Aceh, Indonesia. In J. E. D. Barenstein & E. Leemann (Eds.), *Post-disaster reconstruction and change: communities' perspectives* (pp. 55–70). Boca Raton: CRC Press.

Manyena, B., O'Brien, G., O'Keefe, P., & Rose, J. (2011). Disaster resilience: A bounce back or bounce forward ability? *Local Environment: The International Journal of Justice and Sustainability, 16*(5), 417–424.

Martin, J. R., & Olgun C. G. (2011). Emerging vulnerabilities and disaster risk management solutions for multi-hazard resilience. *Journal of Emergency Management, 9*(Special Issue 1), 13–26.

Olshansky, R. B., Johnson, L. A., Horne, J., & Nee, B. (2008). Longer view: Planning for the rebuilding of New Orleans. *Journal of the American Planning Association, 74*(3), 273–287.

Opp, S. M., & Saunders, K. L. (2013). Pillar talk: Local sustainability initiatives and policies in the United States-finding evidence of the "three E's": Economic development, environmental protection, and social equity. *Urban Affairs Review.* doi:10.1177/1078087412469344.

Suleyman, C., & Corbacioglu, S. (2013). Interorganizational coordination: Analysis of 2011 Van Earthquake. In N. Kapucu, C. V. Hawkins, & F. I. Rivera (Eds.), *Disaster resiliency: Interdisciplinary perspectives* (pp. 308–333). New York: Routledge.

U.S. Small Business Administration Office of Advocacy. (2006, April). *Entrepreneurship: The foundation for economic renewal in the Gulf Coast region.* Proceedings of Symposium in New Orleans, Louisiana.

Waugh, W. L. Jr. (2007). EMAC, Katrina, and the governors of Louisiana and Mississippi. *Public Administration Review, 67*(December 2007), 107–113.

Waugh, W. L. Jr. (2009). Where was the "FEMA Guy"…in Washington, in New Orleans, retired …? *Administration & Society, 41*(6), 748–751.

Waugh, W. L. Jr. (2010). Risk-based planning. Final Report for the Center for Natural Disasters (a DHS Center of Excellence), University of North Carolina at Chapel Hill.

Waugh, W. L. Jr. (2013). Management capacity and rural community resilience. In N. Kapucu, C. V. Hawkins, & F. I. Rivera (Eds.), *Disaster resiliency: Interdisciplinary perspectives* (pp. 291–307). New York: Routledge.

Waugh, W. L. Jr., & Smith, R. B. (2006). Economic development and reconstruction for the gulf after Katrina. *Economic Development Quarterly, 20*(3), 211–218.

Zhang, Y., & Peacock, W. G. (2009). Planning for housing recovery? Lessons learned from hurricane Andrew. *Journal of the American Planning Association, 76*(1), 5–24.

Chapter 11
Finding Resilient Networks: Measuring Resilience in Post-Extreme Event Reconstruction Missions

Steve Scheinert and Louise K. Comfort

11.1 The Challenges of Conceptualizing Resilience

In the aftermath of the 2010 Haitian Earthquake, a wide range of commentators noted the opportunity hidden in the disaster to redevelop Port-au-Prince, and Haiti generally. The disaster shattered existing structures of power and privilege and created an opening for innovative approaches to long-standing problems that would allow Haiti to emerge as a stronger society and government than had existed before the earthquake. As the level of progress that has been made since the earthquake has shown, this is a far easier opportunity to identify than to utilize. The question is how a community and society could seize such an opportunity. Recent scholarship on how communities prepare for and respond to extreme events, such as large-scale natural disasters or the aftermath of wars, focuses on the concept of resilience as the means to seize this opportunity (Comfort et al. 2010a). By presenting a policy direction that focuses on reconstruction and recovery, resilience offers an approach for reconstruction and recovery that could improve a community's ability to recognize the opportunity for improved development following a disaster, and act on it. Any community that effectively integrates resilient structures and behaviors with operations and an on-going development process will be better able to withstand disasters of all types, reduce their exposure to risk from disasters, incur less damage when extreme events occur, and regain continuity in operations faster.

Yet problems within this concept make this seemingly simple solution difficult to apply in specific events. The concept of resilience that scholars are currently discussing does not necessarily fit the goal of generating the most rapid return to pre-event conditions, the *status quo ante*. Resilience has multiple meanings, only one of which fits this policy approach. Wide and consistent application of resilience requires

S. Scheinert (✉)
University of Vermont, Burlington, VT, USA
e-mail: srsche@gmail.com

L. K. Comfort
University of Pittsburgh, Pittsburgh, PA, USA
e-mail: lkc@pitt.edu

N. Kapucu, K. T. Liou (eds.), *Disaster and Development,* Environmental Hazards,
DOI 10.1007/978-3-319-04468-2_11, © Springer International Publishing Switzerland 2014

a clear and consistent definition and measurements. The disagreement among scholars has led to a vague or poorly delimited concept that limits the ability of policy makers to apply a valid measure when designing policies and administrative systems. This situation has developed, because none of the conceptions of resilience are simple or clearly apply in all situations. More fully developing these conceptions can shed light on how each might apply in a given situation and so improve and clarify the application of resilience in planning for and responding to extreme events.

11.1.1   A Wide-Ranging Concept[1]

Researchers from multiple disciplines interpret the concept of resilience as best fits their discipline and apply it to their understanding of disaster recovery. Resilience currently takes on one of three primary conceptions. The first focuses on facilitating the most rapid return possible to *status quo ante* conditions following a shock (Homer-Dixon 2006). A second conception describes a more fluid system that returns less rapidly to a persistent initial state when faced with shocks (Scheffer et al. 2001; Walker et al. 2004). A third conception describes a flexible response that adapts to on-going changes in the situation (Arquilla and Ronfeldt 2001; Comfort et al. 2010b; Demchak 2010). Each conception emerges from a different body of literature, although all seek to establish the best way for a system to resume normal operations after a shock.

11.1.2   Resilience as Withstanding Damage

A resilient system or organization is one that can maintain operations after being stressed in some way (de Bruijne et al. 2010). For military planning, resilience means redundancy in materiel and personnel. When a unit or its materiel is destroyed in battle, the army can continue operations by deploying troops and materiel to replace those destroyed. Even in the context of the computer networks that maintain Wall Street's operations, resilience comes from having back-up facilities on which the network can rely when the primary hub of operations is destroyed (Homer-Dixon 2006). For a network of organizations, this means that the organizations involved are enabled to continue operating through stresses. These robust physical systems are resilient to the extent that they can absorb damage but maintain continuity of operations without disruptions in operations or changes in the manner of operation; replacement pieces are inserted for damaged pieces and the system resumes its prior operations using the new pieces.

In the context of armed struggle between networks, Arquilla and Ronfeldt (2001, p. x) argue that this capacity requires correctly operating on five key levels, "the

---

[1] Portions of this section of text, including Robust Physical Systems and Resilience as Adaptation, were first prepared for an internally-published doctoral dissertation (Scheinert 2012). Used with the author's permission.

technological, social, narrative, organizational, and doctrinal levels." Their explanation of this definition suggests preplanning as an approach to getting these levels correct. If it is not planned correctly, then the network will not achieve its goals. In contrast, a resilient network will treat errors in planning as a stressor to be overcome and will identify errors and adapt when it is operating incorrectly on any of those levels. Arquilla and Ronfeldt's theory does not account for adaptation to environmental changes in practice in the planned operation for each level during the mission, making it a static system. The 4R's of resilience that Bruneau et al. (2003) propose most closely match this type of resilience. The 4R's, robustness, rapidity, resourcefulness, and redundancy, focus on the physical operation of a system, on the system's ability to withstand a shock through the replacement of materiel and resources through its focus on reduced failure probability and reduced consequences from failures. The application of resourcefulness and rapidity, along with reduced time to recovery, does not fit the type of resilience needed to recover from disaster. The four R's concept focuses heavily on materiel and resources, bringing together aspects of what public administration researchers refer to as governing capacity, or having the material capability to perform governance tasks (Comfort 2005; Scheinert 2012). Organizations and systems certainly must have the capacity to perform, but resilience also includes how actors acquire resources and how resources are used, which this conception of resilience does not include.

11.1.3 Resilience as Return to a Persistent Status Quo Ante

Emerging from a range of literatures, the second conception presents resilience as a property of natural systems where the system will return to a persistent equilibrium after a shock. The range of post-shock conditions that support the return to that persistent equilibrium make up a basin of attraction (Scheffer et al. 2001). Resembling equilibria in economic theory, resilient systems will readily return to the central steady state of the basin (Holling 1973), with a larger basin corresponding to a more resilient system, since the system will be able to accommodate larger shocks while still returning to the same static equilibrium (Walker et al. 2004). This conception of resilience most nearly approaches the popular notion of seeking ways to hasten the inevitable return to the *status quo ante*.

The challenge for this approach comes when the basin of attraction and its equilibrium point in the system is or becomes a suboptimal or unsustainable space. Shocks to a system come in many sizes and dimensions. They include single events like the Haiti Earthquake of January 12, 2010, or the more recent landfall of Superstorm Sandy across one of the largest metropolitan areas in the United States. Shocks can also include rapidly-developing events that spread across a range of time, such as any form of armed conflict or more slowly-developing, but still large-scale dynamic processes like climate change. The consistent aspect of these shocks is the ability to fundamentally change the environmental conditions in which the system operates, as well as changing the set of actors that are present in the system (Comfort 1999; Comfort et al. 2010b). This fundamental change not only complicates the return to any *status quo ante*, but forces a reevaluation of whether or not such a return is advisable or even possible. This re-evaluation of prior conditions is illustrated by the debate that emerged in US newspaper articles following Superstorm Sandy about the wisdom of rebuilding along the beaches near New York City (Pilkey 2012). If, after such a shock, the old equilibrium is no longer tenable, then a new one must be found, but such efforts will be forced to fight existing mechanisms seeking to drive a return to the *status quo ante*.

11.1.4   Resilience as System Adaptation

When post-extreme event relief and reconstructions missions begin, whether after natural disasters or armed conflicts, a varying number of public, private, and nonprofit organizations rush into the area to begin delivery of various services which the local population needs. An *ad hoc* governance network forms among these organizations when they attempt to coordinate their efforts (Koliba et al. 2011). The network passes knowledge, information, materiel, and personnel among the organizations to support the response effort. A response emerges from the aggregate of the individual interactions among the organizations as they attempt to further their relief efforts (Prigogine and Stengers 1984; Holland 1995; Axelrod and Cohen 2000; Ostrom 2005; Comfort et al. 2010b). For this response to be effective, it must continue to meet the needs of those operating in the network and of those relying on its services (Provan and Milward 2001). For Provan and Milward (2001), this requirement means utilizing a network administrative organization that reviews the system and its network and guides the actors in strategies that best fit the situation.

The effectiveness of this approach relies heavily on the dynamics of the situation in which the response is operating. In a static situation, this approach can be used iteratively to bring a network into a closer match to the situation, with the network administrative organization that Provan and Milward claim a network requires (2001) directing the change by adjusting funding and system organization. In a rapidly changing situation, this approach fails since it would attempt to hit a moving target, as the needs of network members, its clients, and the best way to meet those needs continuously change. Networks are known for this ability to self-organize (Prigogine and Stengers 1984; Kauffman 1993; Axelrod and Cohen 2000; Arquilla and Ronfeldt 2001; Johnson 2001; Watts 2003), but, by requiring an administrative organization, this process jettisons this strength. Instead of asking whether a network meets the needs of those operating it and those relying on its services, questions about network resilience should focus on how networks can utilize self-organization among their component actors to identify changing needs and then adapt to meet those needs.

Self-organization in response to an extreme event response describes a complex adaptive system. In such a system, actors pursue their own separate interests, but they do not do so independently. They interact, forming interdependent networks of actors and resources that drive the system's operation, but also continually shift over time and across geographic space. Further, since each of these organizations are themselves complex adaptive systems, nesting within a larger complex adaptive system, their interactions transform extreme event response into the type of complex adaptive system of systems that Glass et al. (2011) identify for natural disaster response and political destabilization. In all levels of a complex adaptive system of systems, the behavior of the system is a product of how those actors behave, including the rules they follow, the goals they pursue, and how they adjust both rules and actions over time (Axelrod and Cohen 2000; Ostrom 2005). The ability of any actor in the system to make changes is contingent on gaining the information that can guide those actions (Hutchins 1995), at least as much as it is dependent on having the resources to implement those changes. This interdependence refocuses resilience from an emergent characteristic that can only be observed at the system level, as it occurs in basins of attraction, or as a function of resources in a robust physical system, into a product of the behavior of the actors. In this way, resilience is still an emergent characteristic, but one whose basis is known, allowing managers to recognize it as a process and generate resilience in their organizations. This means that defining resilience in an extreme event response requires a definition of resilience that focuses on the behaviors of the constituent actors rather than on the structure or behavior of the overall system. A system will become resilient when its constituent parts adapt their actions to changing conditions and are responsive to one another's changing needs and capacities.

Previous research has recognized this importance of the behaviors of an organization, community, and system to resilience. The 4R's of resilience from Bruneau et al. (2003) touches on this in calling for resourcefulness and rapidity. Adaptation-based resilience focuses on how actors in the system can be both resourceful and rapid. Davey et al. (2008) push resilience in this direction when they identify the requirements of a resilient response to a medical pandemic. For Davey et al. (2008), resilience exists in a networked response to a crisis and comes from the rapid implementation of a plan, high rates of plan compliance, a regional focus, and clear, verifiable criteria for terminating operations. This makes the transition observed in a system of networked organizations where resilience emerges from the actions of the organizations in the system.

This definition refocuses resilience on organizational actions but misses key operational confounders. Organizations sometimes fail to follow plans or outright ignore them (Kapucu and Demiroz 2011). Further, the reliance upon a pre-set plan assumes that the plan is and will remain effective during the response. Either of these occurring would undermine the system's ability to deliver services to its actors and clients, denying success even while maintaining resilience under the Davey et al. (2008) definition. When plans fail for either reason, organizations require information to adapt to a changing situation. Without timely, valid information, actors have no way of knowing if their actions might make a situation worse, in which case, rapid and resourceful action and adherence to improper plans could undermine system resilience by compounding threats to the system and its actors, and so raise rather than reduce the time to recovery. Resilience in dynamic conditions can be understood as a function of action, information, and information acquisition. Resilient behavior then includes acquiring information rapidly and applying it to guide adaptation, and engaging in adaptation to maintain and further the coherent operation of both actors and the system as a whole.

11.2   Measuring Adaptation—Based Resilience

The explosive growth in the public and scientific awareness of networks and network organization in social systems[2] is well documented. Popular scientific publications have now featured them (Johnson 2001; Barabasi 2002; Watts 2003) and the growth of social networking is now self-evident. With this awareness has come the recognition that these networks and their attendant complex systems are ubiquitous (Glass et al. 2011). For scientists, this realization of the ubiquity of networks and complex adaptive systems carries methodological implications for both understanding and researching the phenomena of networks as complex adaptive systems of systems. Resilience is one phenomenon that can emerge from networked organization of social systems (Comfort et al. 2010b), indicating that it is best understood through definitions that conceptualize resilience through networks. Of those listed above, only adaptation—based resilience grounds its definition in networks. Basins of attraction are closely related, since its definition is grounded in the closely related field of systems thinking. The Four R's concept focuses on physical systems, leaving little place for networks beyond ensuring robust computer networks. In those newly—recognized complex adaptive systems of systems, the most useful definition for understanding resilience is adaptation—based resilience.

Since adaptation requires information about the changing situation, circulating that information to those who need it, and then applying that information to direct effective changes is essential before finally making those changes (Comfort 1999; Comfort et al. 2010b). Information required for effective adaption is widely distributed geographically and temporally such that no one actor can obtain all of the relevant information directly. To meet this need, a smooth operation of the system requires communication and coordination for the information to reach the actors who seek it, after other actors have obtained it through direct observation (Hutchins 1995). This process defines the key components of adaptation—based resilience. A *presence on the ground* is required for actors to make direct observations of changing situations. They must then find ways to *facilitate interaction, communication and collaboration among actors and within organizations* that also constitute a *network of individuals* so that those seeking information can acquire it. Finally, actors must utilize that information and actually *adapt to those changing situations*. The greater the presence of these four constituent components in an actor's behavior, the greater that actor's resilience will be. The greater the resilience of the constituent actors, the greater will be the resilience of the system that emerges from the interacting individuals, organizations, and physical conditions.

---

[2] As opposed to computer networks, which, while not dissimilar, are fundamentally different.

11.2.1   Measuring Components

The challenge of measuring resilience is such that, even at this stage, with the concept separated into components, it still resists simple measures. The components are not simple variables that can be easily and directly measured. Following the example of such measures as the Polity Index (Marshall et al. 2010a), these components can be measured using ordinal scales. We propose the following scales for each component. Each scale starts with a base measurement, given the value '1,' used when an observed organizational action contains no resilient behaviors that fit that component and then increases with increasingly resilient behavior within its conceptual dimension. The rest of the scale is developed iteratively. Each component is built around a concept. The scales cover the range of activities that fit with that concept, as guided by the theories and literature discussed above and observed in the data, discussed below.

Presence on the Ground Presence on the ground represents an increasing variety in the types of actors present at events, with increasing scope for their actions by drawing from larger jurisdictional categories. The greater jurisdictional range of the actors present, the greater the presence on the ground will be. This scale records the possible mix of organization types that can be present on the ground.

Internal and External Coordination The communication and coordination component is properly addressed as two separate components, internal coordination and external coordination, which separately address how the actors coordinate within their own internal networks and with other actors in the system. In systems with unitary actors, such as individual persons, internal coordination can be largely assumed, but this is not the case for actors formed from the aggregation of individuals, such as any kind of organization, from a simple team of two people through to entire governments and intergovernmental organizations like the United Nations (UN) or the North Atlantic Treaty Organization (NATO). Measures of internal and external coordination represent increasing levels of communication and interaction between actors. Since communication is necessary for resilient action, a higher value is assigned to greater communication and coordination. *Adaptation*: Adaptation represents increasingly greater steps to apply the information gained through coordination and presence on the ground and to utilize the system beyond the individual actions that an actor takes in pursuing its goals. As organizations change their behaviors and their collaborators to account for changing needs in the system and account for their own shortcomings, the greater the measure of adaptation.

11.2.2   Combining Component Scores

The remaining challenge rests in how to combine these separate sets of ordinal scores into a single measure of resilience. Similar, existing indices can provide some guidance for this integration. Assigning positive and negative values to a single dimension and simply summing the scores to achieve a single measure of the component variables (Marshall et al. 2010b) is likely not an optimal, or even appropriate, approach for measuring resilience. Such a method evenly weights the transitions between each component measurement, as well as on transitions from one score to another across components, and places more weight on components with a greater number of possible states. The concept of resilience does not support a varying weight among the four components identified in this study, nor does it allow for the assumption that a transition between any two values in the same component or transitions between any two numerically 'equal' values of different components represent an equivalent increase or decrease in resilience. Indeed, the measures are defined by spreading values over the appropriate observed variable for each component individually, making any equality in size of transition or between the same observed numbers of different components, illusory. Rather, each has an independent scale that cannot be linearly combined with any of the others.

This would suggest that no combination is possible, but another index offers guidance on combining measurements of variables with widely divergent measurements methods and ranges. Cutter's Social Vulnerability Index (SoVI) (Cutter et al. 2003) builds an index of vulnerability from widely divergent variables, including the rates of poverty and presence of various minority groups, education levels, and per capita income, as well as median age and rental rates, to name just a few of the variables, that are present in a jurisdiction. The SoVI then combines them statistically, first normalizing each variable and then using a factor analysis to generate a single social vulnerability score that can be compared across all the jurisdictions used in the factor analysis. Being built from a set of variables with a wide variety of ranges and measures, even wider than those in the resilience components, provides useable guidance combining the above components into a single resilience score.

11.2.3   Resilience in Extreme-Event Response Networks

The operationalization of resilience is built around empirical observations. Theory guides the selection of the components so that their selection is grounded and meaningful, but the specific actions that qualify for each ordinal value, as well as the selection of the ranges and ordinal values, must emerge from empirical data. This ensures that the measurement of the components is properly scaled to provide a valid quantitative measure of a complex qualitative concept (King et al. 1994). Adaptation-based resilience emerges from studies based on response to extreme events, primarily natural disasters. With its application to complex adaptive systems of systems, it can also be applied to a wide range of events that have been identified as complex systems of systems, including armed conflicts (Glass 2011). A measure of resilience to be used in all types of complex adaptive systems of systems cannot rely on the idiosyncrasies of any one type. Building a measure using empirical data from more than one type will prevent that reliance.

Table 11.1 Case studies and data sources

Case	Haiti		Bosnia—Herzegovina
Study period	2004–Jan 11, 2010	Jan 12– December 31, 2010	1992–31 December 2002
Time of initial conditions	2004 Internal Ouster of Aristide		1992 Ceasefire Agreement 1995 Dayton Accords
Initiating documents	UNSC Res. 1542 (2004)		UNSCR Res. 743 (1992) UNSC Res. 1035 (1995)
Data sources	Formal Reports, MINUSTAH situation reports	UN OCHA sitreps, Consolidated Appeal documents	IFOR/SFOR CIMIC documents, formal reports, CAP documents, Content of newspapers
Data validation	Semi—structured interviews		Semi—structured interviews
Exploratory analysis	Factor analysis using principal factors method		
Combinatory analysis	Factor analysis using principal components analysis and summation of factor loadings		

11.3   Case Studies: Haiti and Bosnia—Herzegovina

To develop empirically grounded measures of the components of resilience we examine adaptive resilience in two cases of United Nations interventions, one that is primarily a response to a natural disaster and one that represented the stabilization and reconstruction effort that followed an armed conflict, though both cases include aspects of civil conflict and are chosen for their recent occurrence. Using both a civil conflict and a natural disaster that took place in the context of an unresolved civil conflict provides additional comparability between the cases, since both include conflict, while also providing a greater range to the situations from which the data are drawn so that the measure is not overly beholden to the idiosyncrasies of either armed conflict or disaster response interventions and allows application to the range of complex adaptive systems of systems that Glass et al. (2011) identify. Using recent events provides easy access to data that are widely available from public sources.

Table 11.1 lays out the cases that this study uses. Following the application of a complex adaptive systems approach, each case defines a set of initial conditions that serve as a baseline for that case (Comfort 1999; Axelrod and Cohen 2000). Variation and change in each case is measured against that case's baseline since comparability between cases of complex systems is limited. Comparison requires measuring internal changes and then comparing the amounts of change in each case. This functions as a comparison of systems and process rather than a comparison of practices and absolute levels of results (Comfort 1999). As both cases represent United Nations interventions, each uses a specific event and its associated UN resolutions to define the time frame of the initial conditions and initiation of the case study period.

11.4 Data

The actors involved in a response generate a compound unit of analysis that includes observed organizational actions, both unilateral and inter-organizational. These data map directly onto network datasets that record each organizational action as either a network dyad, when two organizations interact, or a monad, when an organization is observed acting unilaterally. The resulting dataset observes each organizational action and assigns a score to each of the components of resilience for that observation.

One of the challenges of analyzing any complex system is that each actor can only observe the events immediately around them, but is unable to observe and report on those events that occur outside its respective area of operations. A widely geographically distributed system adds another layer to this challenge since actors are limited in the number of locations that they can occupy, making it impossible for any one actor to directly observe the full functioning of the system (Hutchins 1995). As a result, any effort to observe and measure the full system must use multiple data sources, given that each provides a different piece of the whole. Acknowledging this limitation for any single source, we draw data on organizational actions from many sources for this analysis.

Chief among these sources is situation reports ("sitreps") produced by the major actors. In an extreme event, organizations publish reports about their actions in response to the event. Although this usually means that the organization records primarily its own actions, the reports constitute a detailed and rich source of data that can provide an extensive profile of the system when reports from multiple organizations contribute to the dataset. The network data for the Haiti case cover two periods in detail, 12 January 2010—1 February and June, 2010. They are drawn from 144 situation reports from 11 organizations that participated in the response to the 12 January 2010 Haitian Earthquake. Table 11.2 lists the breakdown of how many sitreps each organization provided for each timeslice in the analysis of the response to the Haitian Earthquake.

Situation reports did not provide comprehensive data for the stabilization and reconstruction effort in Bosnia-Herzegovina. Having occurred during the early- and mid-1990's, this UN intervention largely pre-dated the internet and the mass publication of situation reports that characterize disaster response operations today. As a result, data existed primarily in the form of newspaper reports, which the UN/ European Union's Office of the High Representative, which oversees Bosnian government, assembled for their relevance. Detailed analysis covered three separate months: December, 2000; July, 2001; and January, 2002. Data for these time periods were published in 12 situation reports and 37 sets of assembled, relevant news stories, for a total of 49 source documents. Table 11.3 lists the breakdown for which documents support the analysis of each timeslice of the analysis of the reconstruction efforts in Bosnia, including how many documents from that type support that analysis. Each data source is carefully reviewed to identify the organizational actions that it contains. Each action identified was coded to identify what organization

Table 11.2 Breakdown of network data sources for Haiti

Organizational Source	Static network	Timeslice 1: Jan. 12–Feb 1, 2010	Timeslice 2: June, 2010
Caribbean Disaster Emergency Management Agency (CDEMA)	11	11	0
UN Health Cluster	13	13	0
UN Logistics Cluster	24	21	3
United Nations Stabilization Mission in Haiti (MINUSTAH)	1	1	0
UN Office for the Coordination of Humanitarian Affairs (OCHA)	20	18	2
Pan-American Health Organization (PAHO)[a]	17	17	0
United Nations Environmental Program (UNEP)	1	1	0
United Nations Children's Fund (UNICEF)	7	6	1
Office of Foreign Disaster Assistance (OFDA), USAID	24	20	4
Water, Sanitation, Hygiene (WASH) Cluster	12	12	0
World Food Program (WFP)	13	13	1
Totals	144	133	11

PAHO is also the local division of the World Health Organization. Most actions performed by either PAHO or WHO were reported as having been done by PAHO/WHO or WHO/PAHO

Table 11.3 Breakdown of network data sources for Bosnia-Herzegovina (Table 11.3 was first produced for use in Scheinert (2012). They are reproduced here with the author's permission)

Data source	Static network analysis	Timeslice 1: Dec, 2000	Timeslice 2: July, 2001	Timeslice 3: Jan, 2002
OHR media round-ups	37	17	11	9
SFOR CIMIC reports	9	1	5	3
Disaster relief	1	1	0	0
Global IDP project	1	1	0	0
International Committee of the Red Cross	1	0	0	1
Total for each timeslice	49	20	16	13

took the action, what action was taken, when was it taken, what was its status at the time of the report, and with what other organizations the initiating organization coordinated, if applicable, in carrying out that action.

Data from news and situation reports require context for proper understanding and assessment. Field visits to both Haiti and Bosnia provide this context. The trip to Haiti took place in early May, 2010, and the trip to Bosnia took place in September, 2011. During each trip, researchers conducted interviews with managers involved in organizational decision making and operations in the aid system. These assessments of the system's structure and performance augmented and validated the formal data sources and provided guidance regarding how to operationalize the components of resilience in each context.

11.5 Component Measurements

The data source documents provide a wealth of organizational actions.[3] The data for
the Haiti case contain 3,344 observed organizational actions covering the two time
periods, January, 2004— pre earthquake, January 11, 2010, and Earthquake, January
12—December, 2010. The data for the Bosnia case contain 429 observed organizational actions covering the respective time frames listed above. After coding to
identify the organizations involved, actions taken, and the status of those actions,
researchers reviewed each observation and coded each for the four components of
resilience, assigning the value that most closely matched the values listed in the
scales, below. The scales are predetermined, drawn from the theories, literature, and
definitions discussed above, and review of the data prior to coding. The predetermined operationalizations identified specific tasks to be measured as certain values
for each component. As coding progressed, the process revealed discrete actions
that had not been previously observed and did not clearly fit any predetermined
measurement. When this occurred, the pre-assigned codes were reviewed and the
specific action was assigned a value if one could be identified. If not, a new value
for the relevant component was added where appropriate on that component's scale.
Finally, after finishing a complete coding of all identified actions, the assigned values were reviewed with the updated scales to ensure that the values fit both scale
and action appropriately. This method represents a qualitative effort to systematically measure the key variables involved in external humanitarian interventions following extreme events. In this way, an iterated process assigned measured values,
based on the actions that organizations took, and constituted an empirical review of
the proposed scales. The following operationalizations are the final result:

- *Presence on the Ground* (the mix of organizations by jurisdiction engaged in
  field activities)

    - 1: No Presence; organizations not engaged in field activities
    - 2: Only domestic organizations operating in the field (sub-national level)
    - 3: Only domestic, state-level (national, federal, state) organizations
    - 4: Only International organizations
    - 5: Combination of Local- and State-level organizations
    - 6: Combination of Local- and International-level organizations
    - 7: Combination of State- and International-level organizations
    - 8: Combination of Local-, State-, and International-level organizations

- Internal Coordination

    - 1: Action does not involve internal coordination
    - 2: Report includes communication between different parts of the same agency
        - example from Bosnia: Interaction and coordination between OHR offices
          in different cities

---

[3] These actions were sufficiently detailed to code for network analysis, although those analyses are
beyond the scope of this chapter. See Scheinert (2012), for these analyses.

- Haiti: large IGO's and governments deploying small teams, such as DMAT and S&R teams

- External Coordination

  - 1: Action does not involve external coordination
  - 2: Organization publishes report or makes requests for coordination
    - requests action in a task outside its own field of operation w/ no mention of reciprocation from the organization which receives the request
    - sitrep published to open system such as ReliefWeb
    - explaining/justifying your actions to another organization
  - 3: A meeting between two or more organizations for coordination. Note: The announcement of two or more organizations meeting is insufficient to qualify for external coordination; such a meeting requires a report of additional activity, including follow up actions, reports of agreements resulting from the meeting, or similar.
  - 4: Report includes cooperation between at least two organizations towards a goal

- Adaptation

  - 1: Action does not include indications of adaptation
  - 2: Content of interaction reports shows pattern of changes in actions or policies over time when compared against each other
  - 3: Report of interaction includes explicit change in an organization's actions or policies
  - 4: Interaction aimed at more closely tailoring organizations towards goal (example: Organizing separate Red Cross societies into unified society)
  - 5: Error Correction
    - When organization recognizes shortcomings in its knowledge and reaches out to a more knowledgeable organization in regard to the interaction
    - IO has low knowledge of the task it is engaged in while the RO has high knowledge
    - Knowledge score relationship is not sufficient in itself; interaction must show IO seeking cooperation (Ex.: OHR approving local administrative decision does not receive code 5)
    - Size Corrections: a large organization providing resources to a small organization.

While the development of the indices for each of these components sought to clearly differentiate them, interdependencies remain. Furthermore, internal and external coordination are very closely related practices that are prone to overlap. This potential overlap is an additional reason why direct, linear combination of the component measures is inappropriate. Application of Cutter et al.'s (2003) method of factor analysis and combination will therefore start with an exploratory factor analysis to identify any overlap. This includes normalizing the component measures before performing any analysis, which controls for component scales with differing

Table 11.4 Uniqueness scores from principal factors analysis

Component	Bosnia				Haiti		
	All data	1 Dec–31 Dec, 2000	1 July–31 July, 2001	1 Jan–31 Jan, 2002	All data	12 Jan–1 Feb, 2010	1 June–30 June, 2010
Presence on the ground	0.840	0.823	0.624	0.925	0.919	0.912	0.905
Internal coor-dination	0.809	0.591	N/A[a]	0.834	0.756	0.755	0.767
External coor-dination	0.798	0.900	0.604	0.792	0.679	0.683	0.636
Adaptation	0.760	0.552	0.960	0.791	0.858	0.861	0.769

This component saw no variation during this timeframe of this case and so no calculation could be made

ranges. The analysis uses the principal factors method of factor analysis, since this method is the best tailored for exploratory analysis (Shlens 2009). This method will define the number of factors[4] that the component variables should be included in the combinatory analysis, along with determining the level of independence between the factors. As with Cutter et al.'s method (2003), the combinatory analysis will use principal components analysis (PCA), since that is a proven method for limiting and combining data (Shlens 2009).[5] The results of this two-step analysis will indicate the appropriate combinations of each component variable as well as provide the factor loadings necessary for calculating a single, numeric measure of adaptation-based resilience in complex systems.

11.6   Results

The results split into two analytic steps. The principal factors analysis provides the results that measure the component overlap. The principal components analysis provides the calculations for combining the component measurements to acquire the final resilience score.

11.6.1   Component Overlap

Principal factors analysis shows the overlap between resilience components as measured in the data. For components to overlap, the analysis must show unique-ness scores for the overlapping component of 0.4 or less, and preferably less than

---

[4] These factors are produced internally to the factor analysis method and do not have a separate, conceptual definition. Such a definition can sometimes be proposed when the variables assigned to one or more factors show a pattern in what they measure, but this is not based in the method and was not possible in this analysis (see Table 11.4).

[5] All statistical analyses are run using Intercooled Stata 11.

Table 11.5 Descriptive statistics of calculated resilience scores (As a result of the normalization process, in all the cases the mean value for the resilience score is 0. Component scores are normalized prior to calculating the resilience score, making a non-normalized version of the resilience score impossible to calculate. Failing to normalize will generate a score that is weighted towards components with larger ranges)

Case and time period	Number of factors	Standard deviation	Minimum	Maximum
All Bosnia data	2	1.33	−0.80	4.77
Bosnia: December, 2000	2	1.58	−0.67	8.82
Bosnia: July, 2001	2	1.48	−1.17	4.09
Bosnia: January, 2002	3	1.78	−1.39	7.28
All Haiti data	3	1.66	−1.58	8.09
Haiti: post-earthquake	3	1.66	−1.59	8.10
Haiti: June, 2010	2	1.34	−1.54	4.75

0.3.[6] Table 11.4 contains the uniqueness scores for all four components in each timeframe for both cases. None of these scores are in the ranges for a measured overlap of the components. This indicates that there is minimal overlap between any of the components.

11.6.2   Resilience Scores

Factor loadings are summed to produce the resilience score. Cutter et al.'s (2003) method allows for various treatments within a linear combination of the loadings, based on how the variables are measured and their impact on social vulnerability. If a factor contains variables that increase vulnerability, then it is added to the overall score. When variables decrease vulnerability, they are subtracted, such that higher numbers represent more vulnerability. All of the components for this study are defined in such a way that higher values represent greater resilience, so that all factors can be added to create a measure where higher values represent greater resilience. Table 11.5 contains the number of factors that are used to calculate the resilience score in each time period under study and the basic descriptive statistics of the resilience score in each time period.

11.6.3   Measure Assessment

The results from the analyses described above provide evidence that the specification of resilience described here applies the conceptual definition of adaptation-based resilience, providing a strong grounding for its consideration as a measure for resilience in complex systems. The resilience score that the principal components analysis generates has the patterns of variance that indicate that the score can sup-

---

[6] Stata reports uniqueness (U) rather than communality (C) scores for its factor analysis results. Uniqueness and communalities are related measurements such that: $U = 1 - C$.

Table 11.6 Frequency count of adaptation component

Category	Haiti		Bosnia	
	Frequency	Percent (%)	Frequency	Percent (%)
No adaptation	2,459	74	324	76
Patterns of change	293	9	20	5
Explicit change in policy	71	2	25	6
Changing organizational structures	106	3	14	3
Error correction	415	12	46	11
Total	3,344	100	429	100

port conclusions about resilience both within and between cases, making a useful measure of adaptation-based resilience in complex systems.

11.7 Measure Appropriateness

An appropriate measure of adaptation-based resilience is both grounded in the literature and existing research on resilience and works well within the data (King et al. 1994). The above discussion grounds this measure. The analysis shows its fit to the data. For the resilience measure to be effective, the components must have clear lines of differentiation and include measures that vary across the range of the components, with no one component taking an undue portion of the dataset. Tables 11.4 and 11.6 present the findings relevant to establish the measure's appropriateness with the data. The data presented in Table 11.4 establishes the independence of the separate components and validate the distribution of concepts that describe resilience. These findings provide confirmation for conceptualizing resilience as the product of organizational presence on the ground, communication and collaboration, and adaptation.

The results in Table 11.6 present results that would appear more problematic for the proposed measure of resilience. Table 11.6 contains the frequency counts for one component of resilience, adaptation, in both cases under study. In this component the most frequent measurement was that the action did not include any resilient behaviors, and this pattern holds in both cases for all components of resilience. These low findings raise a possible question regarding the validity of the measurement or an over reliance upon one value of the scale. The field observations made in both Haiti and Bosnia show that, instead, this is a valid measurement. Both cases had very low levels of resilient behavior, suggesting that these findings reflect actual behavior in practice.

11.8 Measure Interpretation

Resilience is not a simple concept, and so any accurate measurement will not be simple to more or to interpret. Since the empirical measurements are embedded in the observed data and made at the dyad level rather than the system level, it is not immediately obvious how this measure indicates the level of resilience of a whole system. Understanding interpretation requires a more careful analysis of how the measurements and mathematical procedures interact. The coding method ensures that this anchor is not an artifact of the data used for this study. By coding each organizational action separately, it separates multiple resilient actions into different statistical observations, which would not be possible if the components overlap. Therefore, it is likely that, at most, only one resilient behavior will be observed at a time, leaving the rest of the components as 'not observed' on that action. This provides an anchor for the measure, which will remain present across all datasets, regardless of which case study supplies the action, where the minimum value can be expected to represent observations where all four components are measured as 'not observed'.

A mean value for multiple variables or for the same variable measured on different populations would move and not be comparable, but the anchor from the minimum value creates a floor that provides interpretable meaning to the minimum score values. A more resilient system will have more resilient behaviors, raising the mean value. Since the measurement floor is a fixed point, where no resilient behaviors are observed for any component in a given action, a minimum further below the mean indicates a greater prevalence of observed resilient behaviors, raising the normalized mean of the scale relative to the non-normalized measures, allowing the minimum to fall lower, and indicating a more resilient system. In this way, a lower minimum value indicates a more resilient system as average scores move further away from the measure's fixed floor that the floating minimum value represents. As with Cutter et al.'s method (2003), statistical combination through factor analysis allows for comparison within the system. The anchor provided by identifying only one behavior at time therefore allows for comparison across systems.

11.9 Implications for and Beyond Extreme—Event Response Systems

Understanding how communities, systems, societies, and organizations can behave resiliently will indicate how each can maintain operations and use extreme events to spur advancement and improvement. Each academic and operational field provides its own conceptual definition, though each concept remains useful across many fields. Resilience in a robust physical system relies on redundancy and interchangeable parts as characterized by the 4 R's components of robustness, rapidity, resourcefulness, and redundancy (Bruneau et al. 2003). Resilience in a basin of attraction has a fixed *status quo ante* to which the system will return with a varying rapidity. Resilience from adaptation requires understanding how systems change and adjust, as Davey et al. (2008) find by examining how a public health system should transition into and out of crisis operations with a rapidly-implemented plan to which actors adhere closely. This approach builds a system that is allowed to change rather than one that has some form of fixed operational structure or base, but focuses on only periods of transition.

What happens when the plan has flaws or when events move behind what the planners envisioned? This requires a conception of resilience that moves behind the key phase state change from normal to crisis operations and into how organizations behave during their crisis operations. Following from complex adaptive systems theory and distributed cognition, this conception of resilience requires organizations to acquire information and move it rapidly to the individuals within those and other organizations where it can be applied and behaviors adjusted to address changes in a rapidly changing environment. A measure of resilience must encode this information gathering, exchange, and adaptation so that when extreme events present their mix of challenge and opportunity, organizations can be ready to act and adapt in ways that will meet the challenge and seize the opportunity in conceptually similar but individually unique crises.

Acknowledgments This research was done with support from the University of Pittsburgh's Center for Disaster Management (CDM). This document was prepared with support from Vermont EPSCoR and funds from the National Science Foundation (NSF) Grant EPS-1101317. Any opinions, findings, and conclusions or recommendations expressed in this material are those the authors and do not necessarily represent the views of the NSF, Vermont EPSCoR, CDM, the University of Vermont or the University of Pittsburgh.

References

Arquilla, J., & Ronfeldt, D. (2001). *Networks and netwars: The future of terror, crime, and militancy*. Santa Monica: Rand.

Axelrod, R., & Cohen, M. D. (2000). *Harnessing complexity: organizational implications of a scientific frontier*. New York: Basic Books.

Barabasi, A.-L. (2002). Linked: How everything is connected to everything else and what it means for business, science, and everyday life. New York: Plume.

Bruneau, M., Chang, S. E., Eguchi, R. T., Lee, G. C., O'Rourke, T. D., Reinhorn, A. M., Shinozuka, M., Tierney, K., Wallace, W. A., & von Winterfeldt, D. (2003). A framework to quantitatively assess and enhance the seismic resilience of communities. *Earthquake Spectra, 19*(4), 733–752.

Comfort, L. K. (1999). *Shared risk: Complex systems in seismic risk*. New York: Pergamon.

Comfort, L. K. (2005). Learning from risk: A ten-year retrospective. Proceedings, International Forum on Constructing Safe and Secure Urban Society for the 21st Century, Kobe University Memorial Event for the 10th Anniversary of the Hanshin-Awaji Great Earthquake. Kobe, Japan: Kobe University, pp. 31–37.

Comfort, L. K., Boin, A., & Demchak, C. C. (Eds.). (2010a). *Designing resilience: Preparing for extreme events*. Pittsburgh: University of Pittsburgh Press.

Comfort, L. K., Oh, N., Ertan, G., & Scheinert, S. (2010b). Designing adaptive systems for disaster mitigation and response: The role of structure. In L. K. Comfort, A. Boin, & C. C. Demchak (Eds.), *Designing resilience: Preparing for extreme events* (pp. 33–61). Pittsburgh: University of Pittsburgh Press.

Cutter, S. L., Boruff, B. J., & Shirley, W. L. (2003). Social vulnerability to environmental hazards. *Social Science Quarterly, 84*(2), 242–261.

Davey, V. J., Glass, R. J., Min, H. J., Beyeler, W. E., & Glass, L. M. (2008). Effective, robust design of community mitigation for pandemic influenza: A systematic examination of proposed US guidance. *PLoS ONE, 3*(7), e2606. doi:10.1371/journal.pone.0002606.

de Bruijne, M., Boin, A., & van Eeten, M. (2010). Resilience: Exploring the concept and its meanings. In L. K. Comfort, A. Boin, & C. C. Demchak (Eds.), *Designing resilience: Preparing for extreme events* (pp. 13–32). Pittsburgh: University of Pittsburgh Press.

Demchak, C. C. (2010). Lessons from the military: Surprise, resilience, and the Atrium model. In L. K. Comfort, A. Boin, & C. C. Demchak (Eds.), *Designing resilience: Preparing for extreme events* (pp. 62–83). Pittsburgh: University of Pittsburgh Press.

Glass, R. J., Brown, T. J., Ames, A. L., Linebarger, J. M., Beyeler, W. E., Maffitt, S. L., Brodsky, N. S., & Finley, P. D. (2011). Phoenix: Complex Adaptive System of Systems (CASoS) engineering version 1.0. Sandia report. Albuquerque, NM: Sandia National Laboratories. SAND 2011-3446.

Holland, J. H. (1995). *Hidden order: How adaptation builds complexity*. New York: Basic Books.

Holling, C. S. (1973). Resilience and stability of ecological systems. *Annual Review of Ecology and Systematics, 4*(124), 1–23.

Homer-Dixon, T. (2006). *The upside of down: Catastrophe, creativity, and the renewal of civilization*. Washington DC: Island Press.

Hutchins, E. (1995). *Cognition in the wild*. Cambridge: MIT Press.

Johnson, S. (2001). Emergence: The connected lives of ants, brains, cities, and software. New York: Scribner.

Kapucu, N., & Demiroz, F. (2011). Measuring performance for collaborative public management using network tools and analysis. *Public Performance & Management Review, 34*(4), 549–579.

Kauffman, S. A. (1993). *The origins of order: Self-organization and selection in evolution*. New York: Oxford University Press, Inc

King, G., Keohane, R. O., & Verba, S. (1994). *Designing social inquiry: Scientific inference in qualitative research*. Princeton: Princeton University Press.

Koliba, C., Meek, J. W., & Zia, A. (2011). *Governance networks in public administration and public policy*. Boca Raton: Taylor & Francis/CRC Press.

Marshall, M. G., & Jaggers, K., & Principal, I. (2010a). Polity IV project: Political regime characteristics and transitions, 1800–2010. Political Instability Taskforce. http://www.systemicpeace.org/polity/polity4.htm. Accessed 31 July 2013.

Marshall, M. G., Gurr, T. R., & Jaggers, K. (2010b). *Polity IV project: Political regime characteristics and transitions, 1800–2009: Dataset users' manual*. Arlington, VA: Center for Systemic Peace, George Mason University.

Ostrom, E. (2005). *Understanding institutional diversity*. Princeton: Princeton University Press.

Pilkey, O. H. (2012). We need to retreat from the beach. *New York Times*. Op-Ed. http://www.nytimes.com/2012/11/15/opinion/a-beachfront-retreat.html. Accessed 14 Nov 2012.

Prigogine, I., & Stengers, I. (1984). *Order out of chaos: Man's new dialogue with nature*. New York: Bantam Books.

Proven, K. G., & Milward, H. B. (2001). Do networks really work? A Framework for evaluating public-sector organizational networks. *Public Administration Review, 61*(4), 414–423.

Scheffer, M., Carpenter, S., Foley, J. A., Folke, C., & Walker, B. (2001). Catastrophic shifts in ecosystems. *Nature, 413*, 591–596.

Scheinert, S. (2012). *International emergency response: Forming effective post-extreme event stabilization and reconstruction missions* (Doctoral dissertation). Proquest Dissertations and Theses (through D-Scholarship@Pitt: Institutional Repository at the University of Pittsburgh). UMI 3532916.

Shlens, J. (2009). A tutorial on principle component analysis: Derivation, discussion and singular value decomposition. La Jolla, CA: Systems Neurobiology Laboratory, Salk Institute for Biological Studies. http://www.snl.salk.edu/~shlens/pca.pdf. Accessed 29 Jan 2014.

Walker, B., Holling, C. S., Carpenter, S., & Kinzig, A. (2004). Resilience, adaptability and transformability in social-ecological systems. *Ecology and Society, 9*(2): article 5.

Watts, D. (2003). *Six degrees: The science of a connected age*. New York: Norton.

Chapter 12
The Role of Natural Functions in Shaping Community Resiliency to Floods

Samuel D. Brody

12.1 Introduction

Floods remain the costliest and most pervasive natural hazard in the United States and around the world especially in Europe. Property losses from flooding events have been steadily increasing since the mid-1900s and have now reached billions of dollars per year. For example, average annual property damage caused by floods has increased by a factor of 54 over the last five decades, from $ 51 million in the 1960s, to $ 2.77 billion per year in the 2000s (2000–2008, inflation adjusted to dollars in the year 2000) (Brody et al. 2011). From 1999 to 2009, insured flood losses under the National Flood Insurance Program (NFIP) totaled more than $ 35.5 billion nationwide, with the vast majority of this damage occurring in low-lying coastal areas. As a result of continued flood impacts, Federal Emergency Management Agency (FEMA) outlays to provide disaster aid and property insurance payments have also increased. As of 2011, for example, FEMA had borrowed almost $ 24 billion from the U.S. treasury to fulfill its financial obligations and is over $ 17.7 billion in debt (King 2011). Hurricane Sandy, which struck the eastern seaboard in 2012 will certainly exacerbate the economic situation and accentuate the need to mitigate rising costs in the future.

One understudied but essential component in developing more flood resilient communities is the way in which natural landscape functions work to reduce flooding and associated impacts. Specific features of the natural and physical environment serve to attenuate floods if left undisturbed by human activities. Protecting natural functions in the face of increasing development, particularly in coastal areas, may be critical as we move forward in adopting the next-generation of flood policies.

This chapter tackles the issue of the relationship between natural functions and flood mitigation head on by examining the degree to which natural features of the landscape can support community-level resiliency with respect to flooding and

S. D. Brody (✉)
Departments of Marine Sciences, Landscape Architecture and Urban Planning,
Texas A&M University, Galveston, Texas, USA
e-mail: sbrody@tamu.edu

N. Kapucu, K. T. Liou (eds.), *Disaster and Development,* Environmental Hazards,
DOI 10.1007/978-3-319-04468-2_12, © Springer International Publishing Switzerland 2014

flood impacts. First, specific features of the physical environment are evaluated for their role in mitigating the adverse impacts of floods. These include, among others, naturally-occurring wetlands, floodplains, soils, topography, and land cover. Empirical and statistical evidence from the author's research will be presented to support the concepts presented throughout the chapter. Based on the original evidence presented, a series of policy recommendations will be formulated that can enhance a community's ability to cope with the persistent threat of flooding over the long term.

12.2   Environmental Indicators of Flood Risk

The field of disaster reduction and hazard mitigation has long recognized community vulnerability as a place-based concept (Longhurst 1995). Susceptibility to the adverse effects of natural hazards, such as floods is predicated on the interaction between physical and socioeconomic conditions operating within a specific geographic domain (Cutter 1996). When characteristics associated with hazard risk (i.e. physical exposure to hazard events) coincide with human settlements, economic functions can be disrupted (Smith and Petley 2009). Recognizing that the social and built environmental components interact with hydrological and geophysical components provides the basis for a more systems-based approach to building hazard-resilient communities (Folke et al. 2005).

An integrated socio-ecological system framework provides insights into how to facilitate community development while minimizing risk (Lebel et al. 2006; Paton and Johnston 2006). Maintaining the critical functions of the natural environment while at the same time allowing for development help reduce exposure to hazards, such as flooding and storm surge (Godschalk 2003). Environmental landscape features, in this sense, essentially moderate the impacts of natural hazards on human populations and the built environment (Beatley 2009). The more in-tact these features, the greater the capacity of the natural environment to lessen the severity and impact of hazard events. Precisely where human communities persist within ecological and geophysical landscapes thus becomes a central aspect in determining the degree of risk and vulnerability. Communities that locate or build in risk-prone areas will be more likely to incur adverse impacts associated with natural hazards unless they can adopt appropriate coping strategies.

While scholars and practitioners seem to recognize the importance of location-based factors when seeking to develop communities in vulnerable areas, they often stop short when it comes to identifying and measuring specific features of the natural environment that provide a foundation for minimizing flood hazard impacts. In response to this lack of information, we propose below several baseline environmental indicators that serve as essential moderators between flood events and their degree of economic impact on local communities. Decision makers can use these measures to better understand their community's level of vulnerability (i.e. human structures and assets exposed to flood hazards) to flooding and help guide future development that is less damage-prone over the long term.

12.2.1   Floodplain Area

The 100-year floodplain (where there is a 1 % change of flooding every year) is a longstanding marker in the U.S. for determining the possibility of an area being inundated by a rainfall event. Structures within designated floodplain boundaries as determined by FEMA are generally considered at a greater risk of flood impacts unless mitigation measures are in place. In fact, over the course of a 30-year mortgage, there is a 26 % chance that a home in the FEMA-designated floodplain will experience flooding (NRC 2000). This spatial delineation guides local planning and development decisions, triggers insurance purchases and other household adjustments, and serves as the fundamental indicator for whether it is safe to build a structure on a particular site.

Based on the author's research, there is strong empirical evidence that floodplain area is a key indicator of risk. For example, Brody et al. (2007a) found that, on average, increasing areas of floodplain within Florida counties were correlated with greater dollar amounts of property damage from floods. A larger study examining 144 counties and parishes along the Gulf of Mexico coast showed a similar trend, where a percent increase in the area of land outside of the 100-year floodplain significantly reduced insured flood losses from 2001 to 2005 (Brody et al. 2012). In fact, this variable had the strongest effect on observed property damages from floods among all other indicators in the statistical model predicting flood damage. More specifically, a percent increase in these less vulnerable areas among Gulf coast counties/parishes was equivalent to, on average, $ 309,540,000 in savings from flood damage during the study period. This effect translates into an average $ 89.15 per-acre reduction in flood losses. Jurisdictions containing larger percentages of land outside the floodplain should have more opportunity to develop with less risk exposure and greater flexibility as to where development takes place. Development in these communities will also be, on average, less expensive because of lowered requirements for extensive drainage systems and associated infrastructure.

The importance of the floodplain as an environmental indicator of risk is further clarified when examining the precise location of a structure in relation to this boundary. In a parcel-level analysis examining over 9,700 flood damaged properties in the Clear Creek Watershed southeast of Houston, TX (see Fig. 12.1), we found that the position of a structure in relation to the 100-year floodplain has a major effect on the likelihood of increased flood damage. Those properties located outside the floodplain are less exposed to flood risk and therefore report significantly less property damage from flooding events. Although, it should be noted that research also shows that the floodplain boundary may not capture as much risk as originally thought (see Brody et al. 2013).

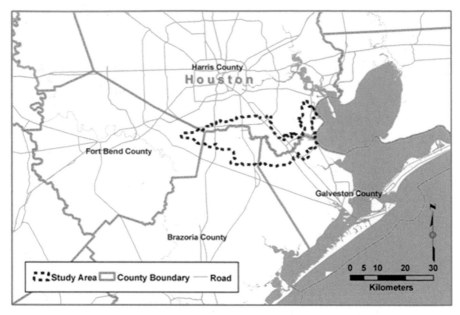

Fig. 12.1 Clear creek watershed study area

12.2.2 Soil Porosity

Soil porosity is another important landscape indicator of flood risk because it helps determine the rate of surface water infiltration (Tollan 2002; Chang and Frankzyk 2008). The amount of water that any given soil will infiltrate and retain depends primarily upon its texture and current moisture condition (Saxton and Shiau 1990). Porous soils, such as those with high sand content drain much more quickly than low porosity soils (e.g. clay), making them a potentially more suitable substrate for development. The potential for higher peak and mean annual flows from basins with low soil permeability is greater than that for basins with higher permeability soils, as higher permeability allows greater infiltration, more storage, and less run-off (Rasmussen and Perry 2000). Soil can be brought in from other areas and used to fill in a specific site for development, but at a larger scale this technique tends to not be effective for flood mitigation.

Localities containing soils with higher levels of porosity should incur significantly lower amounts of property damage from floods. Indeed, when considering flooding along the Gulf of Mexico coast, development on well-drained, sandy soils resulted in significantly lower amounts of reported damage. In fact, recent statistical results shows a percent increase in soil permeability translated into a $ 3,403,600 reduction in property damage from floods per year (Brody et al. 2012). These results suggest that jurisdictions on the west coast of Florida will have more opportunity to develop flood-resilient communities because of their sandy soils compared to the upper Texas coast, which is dominated by clay soils.

12.2.3   Naturally Occurring Wetlands

Naturally occurring wetlands are another critical natural feature when considering the risk to flood damage due to their ability to attenuate floods (Bullock and Acreman 2003). Wetlands naturally store, hold, and disseminate surface run-off in a way that can reduce peak riverine flows and mitigate the adverse impacts of flooding events. For these reasons, wetland systems are particularly effective in attenuating precipitation-based flooding. Maintaining the critical natural function of wetlands maximizes the water storage capacity of the landscape and can reduce flooding, even beyond the extent of the wetland area

The flood mitigation effectiveness of wetland systems is fairly well-documented (Mitch and Gosselink 2000; Lewis 2001; Bullock and Acreman 2003). Early research based on simulation models suggests that wetlands have the natural potential for reducing floods. For example, Ogawa and Male (1986) also analyzed a simulation model to evaluate the protection of wetlands as a flood mitigation strategy. Based on four scenarios of downstream wetland encroachment, ranging from 25 to 100 % alteration, these researchers found that increased encroachment resulted in statistically significant increases in stream peak flow. More recently, Massachusetts, federal and local governments collaborated to acquire 8,500 acres of wetlands along the Charles River for natural flood storage area. These wetlands were later estimated to have a per-acre present value for flood prevention of $ 33,370 (Fausold and Lilieholm 1996). Similarly, a study along the Des Plaines River in Illinois predicted that a marsh of only 5.7 acres could retain the natural runoff of a 410-acre watershed. Based on these results, it was estimated that only 13 million acres of wetlands (3 % of the upper Mississippi watershed) would have been needed to prevent the catastrophic flood of 1993 (Godschalk et al. 1999).

Recent observational research in Texas and Florida more precisely demonstrates the economic value of naturally occurring wetlands in reducing the adverse impacts of floods (Brody et al. 2011). Brody et al. (2007b) found that the development of wetlands significantly increased the number of exceedances in stream-flow across 85 watersheds in Texas and Florida. Also, using multiple regression models that controlled for socioeconomic and geophysical contextual characteristics, Brody et al. (2007c) showed that the loss of naturally occurring wetlands across 37 coastal counties in Texas from 1997 to 2001 significantly increased observed amount of observed property damage from floods. Based on the number of wetland permits granted over the study period, the authors found that, on average, wetland alteration added over $ 38,000 in property damage per flood. A parallel analysis for all counties in Florida showed even greater economic value of wetlands (Brody et al. 2007a). In this case, the alteration of wetlands increased the average property damage per flood at the county level by over $ 400,000. Based on this rate of change, wetland development costs the state over $ 30 million per year in flood losses. Wetlands are also a strong predictor of flood attenuation at the parcel level. Properties within the Clear Creek Watershed, for example, experienced significantly lower damage from floods when adjacent areas contained palustrine wetlands.

On an even larger scale, naturally occurring wetlands along the Gulf of Mexico coast appear to act as a key ecological indicator of resiliency when dealing with floods. An acre loss of naturally occurring wetlands from 2001 to 2005 increased property damage caused by flooding by an average of $ 7,457,549, which amounts to approximately $ 1.5 million per year across the study area (Brody et al. 2012). The role of wetlands in attenuating the adverse impacts of floods at the landscape scale should not be overlooked by planners and policy makers as they clearly provide an important protective and economic benefit to coastal communities. These natural features provide not only wildlife habitat and increased biological diversity in coastal areas, but also a critical buffer to storms which, as our models demonstrate, translates into significant dollar savings.

12.2.4 Pervious Surfaces

Another key landscape characteristic that can moderate flood impacts is the amount of area *not* covered by impervious surfaces, such as roads, rooftops, and parking lots (Arnold and Gibbons 1996). Pervious surfaces (e.g. riparian areas, green space, forest lands, grassland, etc.) serve important hydrological functions because they effectively absorb, store, and slowly release water (Tourbier and Westmacott 1981). Green infrastructure, such as swales, open space, retention ponds, etc. can reduce stormwater runoff, particularly when embedded into an urban context.

As described above, naturally occurring wetlands are able to perform flood attenuation functions when not disturbed or eliminated by development. But, other pervious land covers can also naturally mitigate flood impacts. Generally, "rougher" surface material over which storm water flows reduces velocity, peak flows, and potential flood impacts. Smooth surfaces like concrete have very low roughness values in comparison to heavy brush, which can more effectively capture and absorb flood waters. For example, forest lands surrounding development may play a role in determining the extent of damage experienced during a flood event. Forested areas lack impervious surfaces that reduce infiltration and expedite runoff into local streams. Trees and woody plants also help to slow and trap runoff, reducing flood peaks more so than other types of land cover (McCulloch and Robinson 1993).

Our empirical research supports this assumption. For example, in the Clear Creek Watershed parcel-level study, Brody et al. (under review) found that forested land cover surrounding a parcel in the watershed significantly reduced observed flood losses. It should be noted, however, that forest lands in coastal areas can be located in low-lying areas dominated by floodplains. For this reason, houses situated adjacent to or within forested bottom-lands can still be vulnerable to flooding.

Open space, such as parks and local preserves may reduce adverse impacts to houses located on adjacent parcels. Open space land use usually contains much less impervious surface compared to other development types, such as commercial or high-density residential. With the presence of ball-fields and playgrounds, drainage is usually well-maintained and runoff appropriately captured. Lastly, developed open space is most often integrated into the spatial fabric of local neighborhoods such that their flood-reducing effects will be more evident (Brody and Highfield

2013). In the same Clear Creek Watershed study cited above, the presence of actively-used open space within a 0.5 mile buffer surrounding each property, significantly reduced the amount of recorded flood damage from 1999 to 2009. Parks and recreation areas within this watershed consist mostly of green space and vegetation seems to provide a buffer against negative flood impacts.

Grassland, another pervious land cover commonly found in coastal areas in the southern U.S. also has potential to reduce the adverse impacts of floods. Herbaceous plants with non-woody stems may be enough to capture and absorb surface run-off without causing ponding. Grasslands were the most powerful predictor of reduced flood losses among all pervious land covers in the Clear Creek Watershed study. In fact, a percent increase in the area of grassland within a 0.5 mile buffer around a flood-damaged parcel translated, on average, into a flood loss savings of over $ 3,090 per structure.

12.2.5   Elevation

Another environmental landscape feature important to consider when developing flood-resilient communities is topography. The slope of a watershed affects both the temporal concentration and the amount of water storage. Generally, steeper slopes increase rainfall concentration, causing faster and higher stream peaks as well as mean annual flows (Stuckey 2006). Under these conditions, water bodies tend to overflow their banks more quickly and with less warning than do more gently-sloped watersheds. On the other hand, there is less depressional pooling of water on steep upper slopes where runoff sheds more quickly.

In low-lying coastal watersheds, the overall slope may be negligible, but very small changes in elevation may prove critical when predicting property damage from flood events. Coastal areas at or near sea-level are particularly vulnerable to riverine-based flooding and inundation from storm-surge. During Hurricane Katrina (2005), Ike (2008), and Sandy (2012), elevation proved a decisive environmental factor contributing to flood damage. The Clear Creek watershed exemplifies this issue. Based on LIDAR data, elevation within the watershed ranges from 0 to 19 m (the mean being just over 6 m), where the lowest values are generally located along tidally-influenced water bodies. A 1 m increase in elevation across this landscape was equivalent to, on average, a $ 1,788 decrease in insured flood losses from 1999 to 2009. Subtle changes in elevation within this and other coastal watersheds can be the determining factor in whether a home is flooded or not.

12.3   Policy Implications for Flood Risk Reduction

Based on the results of various studies presented above, it is evident that environmental characteristics and the overall functionality of watersheds are critical in reducing the adverse impacts of floods, particularly in coastal areas. However, the role of natural functions may be one of the most overlooked mitigation strategies for facilitating the development of flood-resilient communities. Planners and decision makers must better recognize the value of environmental landscape components as natural defenses against flood impacts and integrate these components into policies that seek to mitigate flood losses over the long term.

Perhaps the most effective approach to maintaining and leveraging the functionality of environmental characteristics of physical landscapes is to pursue an "avoidance" strategy of flood mitigation. This method guides and focuses human impacts away from critical natural areas not only to protect critical habitats, wildlife, water quality, etc., but to facilitate flood risk reduction. For example, keeping structures well outside the 100-year floodplain, which is considered the area most likely to flood, is a productive strategy for policy makers at the jurisdictional level. Land use tools, such as cluster development, density bonuses, and capital improvements programming have been shown to be effective in guiding development towards more resilient locations (for a more thorough discussion of development policies, see Beatley et al. 1994; Perlman and Miller 2005; Beatley 2009). Jurisdictions containing larger percentages of land outside the floodplain will have more opportunity and greater flexibility as to where development takes place. Development in these locations will also be, on average, lower in cost because of less stringent requirements for drainage systems, building code, and flood insurance purchases.

One of the more overlooked landscape characteristics in terms of reducing flood impacts is soil porosity. Soil structure may be considered at the site level, but rarely does this factor enter the minds of local and regional policy makers when they consider the location and extent of future development. However, evidence suggests that well-drained, porous soils effectively reduce storm-water run-off and associated flood losses. Coastal planners, policy makers, and developers should consider in their decisions soils alongside floodplains and any other indicator of vulnerability that could impact property and lives. This approach entails avoiding impermeable soils and keeping permeable soils unpaved as much as possible. Of course, regions consisting of permeable soils, such as the sandy coastal areas of west Florida are at a distinct advantage when pursuing a resilient approach to community development compared to the clay-based soils on the Texas coast. Some techniques that help reduce development pressure on impermeable soils include: conservation overlay zones, subdivision ordinances, site plan requirements, and public expenditure guidelines.

Ground elevation is another under-recognized landscape facet that can be an important driver in reducing flood impacts. Even in low-lying coastal areas where there is very little topographic change, a 3 ft change in elevation above sea level can make the difference in whether a property will flood. Raising buildings above Base Flood Elevation (BFE) is critical in these areas. On average, CRS communities that received BFE (also known as freeboard) credits reduced insured flood losses by over $ 960,000 per year (Brody and Highfield 2011). Instead, the most commonly-used solution to addressing elevation issues is to use fill material to raise a building site above the floodplain. This technique can alter the floodplain boundaries, increase flooding downstream and on neighboring properties, and in general exacerbate flooding with a larger area.

The role of wetlands in attenuating the adverse impacts of floods should also be a central component of flood reduction programs, as these natural features clearly provide an important protective and economic benefit to coastal communities. Naturally occurring wetlands provide not only wildlife habitat and increased biological diversity in coastal areas, but also a critical buffer to storms that translates into reduced flood risk and significant dollar savings. By enabling wetlands to persist, the water storage capacity of the landscape is maximized and flooding beyond the extent of the actual wetland can be minimized. In other words, protecting naturally occurring wetlands can prevent the inundation of a larger surrounding area. Empirical research at multiple scales, from region to parcel, shows that the absorbing properties of wetlands can reduce flood impacts, particularly for precipitation-based events. Land acquisition techniques, coastal buffers, and critical area setbacks are a few of the tools available to planners that, when implemented, can help protect wetland functions.

One of the most effective ways to maintain critical natural functions is the protection of open space. Open space land use designations can remove people and structures (aside from some recreational buildings) from the most flood-prone areas. Thus, the opportunity for property loss and economic disruption is eliminated. In particular, setbacks from or buffers around riparian areas make space for natural fluctuations of riverine systems and reduces adverse impacts to structures that would otherwise be placed in harm's way.

Even developed open space can reduce property damage from floods. Local parks, playing fields, and other recreational facilities consist primarily of green space that can act as a storm buffer to surrounding properties. Parks also often double as runoff detention areas to hold and slowly release water on to adjacent parcels. Open space land use can be strategically sited within neighborhoods not just to provide recreational opportunities, but also to reduce the adverse impacts of localized flooding. A national study of localities participating in the NFIP Community Rating System (CRS) demonstrated that they save, on average, approximately $ 200,000 per year in flood-related losses by protecting open space in the 100-year floodplain (Brody and Highfield 2013). In the Clear Creek Watershed parcel level analysis, a percent increase of adjacent open space land cover also significantly reduced observed flood losses.

Finally, the configuration of impervious surfaces and overall development pattern across landscapes should be considered important aspects of protecting natural functions and reducing the adverse impacts from floods. Jurisdictions along the Gulf of Mexico coast containing large amounts of clustered, high-intensity development patterns actually reported, on average, reduced amounts of property damage (Brody et al. 2011b). This finding supports the efficacy of well-defined urban cores associated with "smart growth" approaches to development. It appears that as long as dense urban development is situated away from vulnerable areas (such as the 100-year floodplain), this built-environment pattern can lead to more flood resilient local communities over the long term. Compact forms of development are better able to avoid vulnerable areas and focus development intensity on the most suitable land available (Stevens et al. 2009). Also, a more focused urban core may deter the release and subsequent development of flood-prone land elsewhere (White 2008). Finally, high density, well-connected urban areas may be more likely to have in place a flood mitigation infrastructure that can appropriately handle large amounts of runoff. Regional decision makers should generally encourage high-intensity, clustered development as compared to low-intensity, more sprawling development patterns. In general, patches of development should be connected and in close-proximity (Brody et al. 2013).

In contrast, low-intensity forms of growth can magnify the economic toll from floods by placing more people in harm's way. Low-density land conversion also generates larger total area of impervious surfaces and fragmentation of drainage networks, both of which can increase runoff and exacerbate flooding. Furthermore, unlike well-established urban centers, suburban and ex-urban communities are less likely to have adequate storm drainage systems and other infrastructure to accommodate increase increased surface runoff. Finally, sprawling residential development on the periphery of older urban areas and outside extraterritorial jurisdictions may be more likely to encroach on floodplains that were originally left as open space or low-impact uses.

Based on statistical models, every added square mile of low-intensity development increases flood damage along the Gulf coast by more than $ 5.7 million. In 2001, Harris County, TX, one of the most flood-damaged jurisdictions in the nation, contained over 227 square miles of low-intensity development. There are several policies local planners and decision makers can adopt to encourage the development of more concentrated urban cores. These tools include, among others, density bonuses, transfer of development rights, clustering, and conservation easements. Currently, these policies are only sporadically being used in coastal areas.

While the above-mentioned strategies to increase community resiliency to floods may seem challenging or even impractical, they actually fit quite well into existing mechanisms used manage development. Local comprehensive plans, development codes, and stand-alone hazard plans are already adopted in most coastal areas. These established policy instruments are thus opportunities to implement specific flood resilient policies, such as focusing development to less exposed areas, protecting open space, maintaining critical ecosystem services, facilitating specific development patterns, etc. Many of these policies also overlap with more traditional sustainability and environmental planning approaches. Decision makers can modify and build upon these existing plans and policies to increase the abilities of communities to cope with repetitive flooding.

12.4   Conclusion

This chapter has argued that natural functions play an important role in fostering more flood-resilient communities and should be integrated into flood mitigation initiatives at multiple scales. From the parcel level up to large coastal regions, empirical evidence suggests that environmental landscape features, such as wetlands, riparian areas, and soils can significantly reduce the magnitude of adverse impacts stemming from flooding events. Planners and decision makers must consider natural functions as flood mitigation tools along-side structural interventions, such dikes and levees if, as a nation, we wish to protect the property and safety of residents, particularly in vulnerable coastal areas. Future work should continue to evaluate the impact of development patterns and measure the effectiveness of specific land use planning strategies to inform communities on how they can reduce flood risk over time.

References

Arnold, C. L., & Gibbons, J. C. (1996). Impervious surface coverage: The emergence of a key environmental indicator. *Journal of the American Planning Association, 62*(2), 243–258.

Beatley, T. (2009). *Planning for coastal resilience: Best practices for calamitous times*. Washington, DC: Island Press.

Beatley, T., Brower, D., & Schwab, A. (1994). *An introduction to coastal zone management*. Washington, DC: Island Press.

Brody, S. D., & Highfield, W. E. (2011). Evaluating the effectiveness of the FEMA community rating system in reducing flood losses. Final report for FEMA mitigation division study, phase I. FEMA, Washington, DC.

Brody, S. D., & Highfield, W. (2013). Open space protection and flood losses: A national study. *Land Use Policy, 32,* 89–95.

Brody, S. D., Highfield, W. E., & Kang, J. E. (2011). *Rising Waters: Causes and consequences of flooding in the United States*. Cambridge, UK: Cambridge University Press.

Brody, S. D., Highfield, W. E., Ryu, H. C., & Spanel-Weber, L. (2007b). Examining the relationship between wetland alteration and watershed flooding in Texas and Florida. *Natural Hazards, 40*(2), 413–428.

Brody, S. D., Zahran, S., Highfield, W. E., Grover, H., & Vedlitz, A. (2007c). Identifying the impact of the built environment on flood damage in Texas. *Disasters, 32*(1), 1–18.

Brody, S. D., Zahran, S., Maghelal, P., Grover, H., & Highfield, W. (2007a). The rising costs of floods: Examining the impact of planning and development decisions on property damage in Florida. *Journal of the American Planning Association, 73*(3), 330–345.

Brody, S. D., Gunn, J., Highfield, W. E., & Peacock, W. G. (2011a). Examining the influence of development patterns on flood damages along the Gulf of Mexico. *Journal of Planning and Education Research, 31*(4), 438–448.

Brody, S. D., Highfield, W. E., & Kang, J. E. (2011b). *Rising waters: Causes and consequences of flooding in the United States*. Cambridge: Cambridge University Press.

Brody, S. D., Peacock, W., & Gunn, J. (2012). Ecological indicators of resiliency and flooding along the Gulf of Mexico. *Ecological Indicators, 18,* 493–500.

Brody, S. D., Kim, H. J., & Gunn, J. (2013). The effect of urban form on flood damage. *Urban Studies, 50*(4), 789–806.

Brody, S. D., Blessing, R., Sebastian, A., & Bedient, P. (2013). Examining the impact of land use/ land cover characteristics on flood losses. *Journal of Environmental Planning and Management*. DOI:10.1080/09640568.2013.802228.

Bullock, A., & Acreman, M. (2003). The role of wetlands in the hydrological cycle. *Hydrology and Earth System Sciences, 7*(3), 358–389.

Chang, H., & Franczyk, J. (2008). Climate change, land-use change, and floods: Toward an integrated assessment. *Geography Compass, 2*(5), 1549–1579.

Cutter, S. L. (1996). Vulnerability to natural hazards. *Progress in human geography, 20*(4), 529–539.

Fausold, C. J., & Lilieholm, R. J. (1996). *The economic value of open space*. Cambridge: Lincoln Land Institute Research Paper.

Folke, C., Hahn, T., Olsson, P., & Norberg, J. (2005). Adaptive governance of social-ecological systems. *Annual Review of Environment and Resources, 30,* 441–473.

Godschalk, D. R., Beatley, T., Berke, P., Brower, D., & Kaiser, E. J. (1999). *Natural hazard mitigation: Recasting disaster policy and planning.* Washington, DC: Island Press.

Godschalk, D. (2003). Urban hazard mitigation: Creating resilient cities. *Natural Hazards Review, 4*(3), 136–142.

Highfield, W. E., Norman, S., & Brody, S. D. (2012). Examining the 100-Year floodplain as a metric of risk, loss, and household adjustment. *Risk Analysis.* DOI:10.1111/j.1539-6924.2012.01840.x

King, Rawle, O. (2011) National flood insurance program: Background, challenges, and financial status. Washington, DC: Congressional Research Service.

Lebel, L., Anderies, J. M., Campbell, B., Folke, C., Hatfield-Dodds, S., Hughes, T. P., & Wilson, J. (2006). Governance and the capacity to manage resilience in regional social-ecological systems. *Ecology and Society, 11*(1), 1–15.

Lewis, W. M. (2001). *Wetlands explained: Wetland science, policy, and politics in america.* New York: Oxford University Press.

Longhurst, R. (1995). The assessment of community vulnerability in hazard prone areas. *Disasters, 19,* 269–270.

McCulloch, J. S. G., & Robinson, M. (1993). History of forest hydrology. *Journal of Hydrology, 150*(2–4), 189–216.

Mitch, W. J., & Gosselink, J. G. (2000). *Wetlands* (3rd ed.). New York: Wiley.

National Research Council (2000). *Risk Analysis and Uncertainty in Flood Damage Reduction Studies.* National Academy Press: Washington, DC.

Ogawa, H., & Male, J. W. (1986). Simulating the flood mitigation role of Wetlands. *Journal of Water Resources Planning and Management, 112*(1), 114–128.

Paton, D., & Johnston, D. (Eds.). (2006). *Disaster resilience: An integrated approach.* Springfield: Charles C. Thomas.

Perlman, D. L., & Miller, J. (2005). *Practical ecology for planners, developers, and citizens.* Washington, DC: Island Press.

Rasmussen, P. P., & Perry, C. A. (2000). *Estimation of peak streamflows for unregulated rural streams in kansas.* U. S. Geological Survey, Water-Resources Investigations, Report 00–4079.

Saxton, K. E., & Shiau, S. Y. (1990). Surface waters of North America; influence of land and vegetation on streamflow. In M. G. Wolman & H. C. Riggs (Eds.), *The geology of North America Vol. 0–1, surface water hydrology* (pp. 55–80). Washington, DC: The Geologic Society of America.

Smith, K., & Petley, D. N. (2009). *Environmental hazards: Assessing risk and reducing disaster.* New York: Routledge.

Stevens, M., Song, Y., & Berke, P. (2009). New Urbanist developments in flood-prone areas: Safe development, or safe development paradox? *Natural Hazards, 53*(3), 605–629.

Stuckey, M. H. (2006). *Low-flow, base-flow, and mean-flow regression equations for Pennsylvania stream,* U.S. Geological Survey, Scientific Investigations Report 2006–5130.

Tollan, A. (2002). Land-use change and floods: What do we need most, research or management? *Water Science and Technology, 45*(8), 183–190.

Tourbier, J. T., & Westmacott, R. (1981). *Water resources protection technology: A handbook of measures to protect water resources in land development.* Washington, DC: The Urban Land Institute.

White, I. (2008). The Absorbent City: Urban form and flood risk management. *Urban Design and Planning, 161*(DP4), 151–161.

Chapter 13
Hazard Mitigation, Economic Development and Resilience: A Comparative Analysis of Flood Control Policy and Practice in Germany, The Netherlands, and Great Britain

Melanie Gall and Brian J. Gerber

13.1 Introduction

Watersnood, Jahrhunderthochwasser, megaflood—those are terms used to describe floods that galvanized The Netherlands, Germany, and Great Britain. Coastal and riverine floods, along with wind storms, are the most significant and devastating natural hazards in Central Europe. Wind-driven, coastal flooding (i.e. storm surge) is largely brought on by winter storms whereas snowmelt, and stalling or slow-moving low-pressure systems trigger flooding along rivers such as the Rhine, Danube, or Thames rivers (see Appendix). Between 1998 and 2009, floods alone killed more than 1,100 individuals, displaced around half a million people, and caused more than EUR 52 billion in insured losses across Europe (EEA 2010). Central Europe, a region with significant economic development, is especially susceptible to coastal flooding and wind storms due to the shape and size of the North Sea basin and its attendant low-lying and subsiding coastlines.

As Europeans engineered river systems, built levees, straightened rivers, and dredged drainage canals, they also changed the impacts of both riverine and coastal floods. Hundreds of years ago floods, including storm surge, resulted in large losses of human lives. Structural mitigation practices have reduced direct loss of human life from flood incidents, though not eliminating that particular effect entirely. Today's impacts have shifted to being primarily economic in nature. Both over-development in floodplains and growing populations in

M. Gall (✉)
Department of Geography, University of South Carolina,
Columbia, SC, USA
e-mail: melanie.gall@sc.edu

B. J. Gerber
School of Public Affairs, University of Colorado Denver,
Denver, CO, USA
e-mail: brian.gerber@ucdenver.edu

N. Kapucu, K. T. Liou (eds.), *Disaster and Development,* Environmental Hazards,
DOI 10.1007/978-3-319-04468-2_13, © Springer International Publishing Switzerland 2014

hazard-vulnerable areas are key driving factors behind the acceleration of hazard losses from flooding. In short, more effective early warning systems enable timely evacuation and save lives but the built environment—economic assets such as buildings and critical infrastructure—are, of course, not easily removed from harm's way. And even as more traditional flood mitigation practices (i.e. structural engineering solutions such as barriers and drainage approaches) are supplemented with newer, non-structural approach to flood risk management (which refers to an emphasis on both effective water management practices and sustainable development practices) the risk-increasing effects of economic development and population growth in flood-prone areas have caused hazard losses to skyrocket (as evidenced by the costs of the 2013 and 2002 floods in Europe; see next section).

This context of increasing adverse economic effects is exacerbated by the threat of climate change (see also Chap. 12 in this volume). Sea-level rise challenges the protection level of structural flood barriers such as flood gates, storm surge barriers, and levees (dikes), which are common flood mitigation measures in European countries such as Great Britain, Germany, and The Netherlands. Furthermore, the damage from riverine flooding is also predicted to rise due to increases in peak flow and extreme precipitation events combined with both the continued development and existing concentration of assets in floodplains (Mechler and Kundzewicz 2010). Indeed, the 2013 flooding along the Danube and Elbe rivers hinted at the economic effects of future floods: failure of dams and levees, isolated communities, interruption of urban life, devastating blows to local economies, and closures of major transportation routes (interstates, rivers, railways). According to German Rail, its 2013 first half year profits shrunk by 30 % largely due to the spring floods and an unusually long and cold winter (Kaiser 2013). Thus, severe weather associated with climate change, and especially the flood hazard, threatens the economic well-being of both coastal and interior European regions. Rivers and coasts offer essential economic resources (e.g., transportation, fresh water supply, and irrigation) but the concentration of people and property in floodplains and coastal areas has now created an ideal environment for both economic growth and escalating losses from natural hazards—the flood hazard in particular.

In this chapter we trace developments in flood mitigation and flood risk management by considering three European countries with significant coastal and interior vulnerability to the flood hazard: The Netherlands, Great Britain, and Germany. By doing so we seek to consider the intersection of flood mitigation and flood risk management with economic development practices. That is, we provide an overview of the degree to which the incentives associated with, and the socio-political effects of economic development imperatives affect hazard mitigation policy choices and community resilience. Variations in the physical characteristics of the flood hazard across the three countries permit us to consider how underlying management demands posed by the hazard affect policy and practice across individual national settings as well as the European Union in general.

13.2   Living with Floods: Comparative National Context

Addressing the flood hazard, related vulnerabilities and consequences is a complex process largely dependent on the ability to identify the specifics of the hazard in a defined spatial context, identify assets at risk, and understand those interdependencies. The traditional way to address the flood hazard was through structural mitigation actions (e.g., construction of dykes, water diversion, etc.), which led to an alteration of the hazard itself by prohibiting the encroachment of water onto land. A more comprehensive way to address the flood hazard is to conceptualize the challenge as one of flood risk management. Flood risk management is itself a somewhat ambiguous term, but it essentially refers to the notion of dealing with both acute disturbances (i.e. short term extreme incidents from rainfall or storm surge) and chronic hazard exposure (both longer-term trends) from more of a systems perspective (de Bruijn et al. 2007). This means linking policy and practices of both broader water management and sustainable development and giving a stronger role to non-structural mitigation actions (e.g., education, land use planning, elevation of homes, etc.), which attempt to minimize the adverse impacts of flooding. That is, national flood risk management strategies combine and coordinate structural and non-structural actions in order to reduce the risk, lessen both the environmental and socioeconomic impacts of a disturbance, and this creates a flood-resilient environment where a community, region or nation can either operate as close to normal during an extreme incident or recover as quickly as possible from said incident. But finding the right mix between these practices and policies and how to pay for them are contentious issues. Influenced by the type of flooding (e.g., riverine, storm surge, flash floods), risk tolerance, and choice of flood risk mitigation strategies, each of the countries discussed in this chapter follows different pathways to living with floods. When it comes to effective flood risk management, it is the interaction between flood risk and society that matters—not just protection standards and flood risk.

Hazard mitigation, in general, represents a critical link to community resilience in the face of disasters. The concept of mitigation refers to the idea of identifying potential hazards and vulnerabilities, including assessing the likelihood of occurrence, and identifying proactive measures to limit adverse effects. More formally defined, mitigation is an "advance action taken to reduce or eliminate the long-term risk to human life and property" (Godschalk et al. 1999, p. 5). In European and international contexts, the term *mitigation* is much less utilized than in the United States and instead frequently replaced with *risk management*. This chapter uses both terms interchangeably.

13.2.1   The Netherlands

The Netherlands can be thought of as the "poster child" for structural flood mitigation; as the saying goes: "*God created the earth, but the Dutch created The Netherlands.*" The country is essentially a delta formed by three rivers: the Rhine (headwaters: Switzerland), the Scheldt (headwaters: Belgium), and the Meuse

Table 13.1 Socioeconomic and environmental statistics for The Netherlands, Germany, United Kingdom and for comparative purposes also the United States. (Source: CIA World Factbook 2013)

	The Netherlands	Germany	United Kingdom	United States
Size	41,543 km^2	357,022 km^2	243,310 km^2	9,826,675 km^2
Population in million (2013 est.)	16.8	81.1	63.4	316.7
Population density (2010)	492.2 km^2	229.0 km^2	256.8 km^2	34.01 km^2
GDP per capita (2012 est.)	$ 42,900	$ 39,700	$ 37,500	$ 50,700
Waterways	6,214 km	7,467 km	3,200 km	41,009 km
Coastline	451 km	2,389 km	12,429 km	19,924 km
Highest elevation	322 m (Vaalserberg,)	2,963 m (Zugspitze)	1,343 m (Ben Nevis)	6,194 m (Mt McKinley)

(headwaters: France). Until the tenth century much of the agricultural land and population was above sea level (Tol and Langen 2000). However, similar to the processes at the mouth of the Mississippi River, drainage of peatlands and marshes for pasture, cultivation and settlement allowed organic soils to dry out, oxidize, and compact which subsequently led to subsidence below sea level. While a cycle of draining, compacting, and subsiding ensued to provide farmland and sustain the country's growing population, not surprisingly catastrophic storm surges (between the ninth and thirteenth century, the country lost vast stretches of land to the sea) and riverine flooding soon followed (see Appendix, Rijkswaterstaat 2013).

Starting in the thirteenth century, newly formed levee boards (so-called *water-ships*) began managing water, levees (dikes) and constructing comprehensive levee systems. By 1350, levees protected all major rivers in The Netherlands (Tol and Langen 2000). However, the Dutch' flood problems continued to persist: on the leeward side, neighboring Germany pursued large scale deforestation; and on the windward side, severe storm surge events altered not only coastlines but also river discharge behavior; and the little Ice Age (AD 1550–1850) produced new flood hazards (so-called ice jams) along with increased sedimentation and silting. Subsequently, land was reclaimed (polders), levees grew higher and wider and new types of flood control structures were invented (e.g., cross-dikes, dike rings, flood gates/sluices, diversions, river channeling). Starting in the nineteenth century, taxation and public dike maintenance were introduced to alleviate landowners from flood control duties and enabling centralized and coordinated flood risk management (Tol and Langen 2000). With the onset of the industrial revolution, engineers made waterways navigable to respond to navigation and transportation demands (Table 13.1).

The year 1926 marked the last year of a major riverine flood–though it was not the last natural disaster in The Netherlands. On January 31, 1953, a rapidly moving, low pressure system (966 mb) travelled across the North Sea producing intense winds (gusts up to 125 mph) and storm surge that hit The Netherlands and Great

Table 13.2 Strategies to flood risk management in The Netherlands, Great Britain and Germany

	The Netherlands	Great Britain	Germany
Importance of Structural Flood Mitigation	High	Moderate	Moderate
Importance of non-structural flood mitigation	Low	Moderate	Moderate
Flood protection standards	Regulatory standards vary between 4,000 and 10,000-year return periods	Planning standards distinguish between three flood zones: high (100-year return period for riverine, 200-year for coastal), medium (1,000-year), and low (beyond 1,000-year)	Planning standards vary between 30- and 100-year return periods
Flood insurance	Only available for flash floods as add-on policy	Included in homeowners insurance	Available as add-on policy

Britain during high tide. More than 1,800 people perished in The Netherlands and more than 300 residents died in Great Britain including a ferry accident with 133 passengers (Lamb 1991). On the British island, more than 1,200 flood protection structures were breached and 24,000 homes damaged. In The Netherlands, 45 % of dikes and 47,300 buildings received damage—despite the Dutch Surge Warning Service established in 1916, which accurately predicted the event. In general, home insurance did not cover this event (RMS 2003a) and most financial aid came through government and non-governmental sources. For both countries, this event remains the costliest disaster since World War II (RMS 2003a).

The event was a turning point for The Netherlands. The Dutch Government initiated the Deltaworks—a massive flood defense project consisting of a network of dikes, locks, and dams with protection levels ranging between 4,000 and 10,000 years. The final storm surge gate of the Deltaworks was completed in 1997 (RMS 2003a). The main objective of the Deltaworks—costing in total more than U.S. $ 7 billion—was to shorten the coastline and thereby reduce dike upkeep efforts in the Dutch interior. To solidify its commitment to flood protection and public safety, the national government guaranteed these protection levels in the Water Embankment Act of 1995 with sliding safety standards based on the number of people and assets at risk—more assets and people translating into more stringent standards (Aerts 2009). Given the structural protection offered and governmental investments in flood defense over past decades, no flood insurance programs are offered in The Netherlands. It should be noted that the Dutch insurance industry distinguishes between flooding from rivers, storm surge, and flooding from extreme precipitation events (flash floods), the latter of which is covered by some insurance companies, including commercial and agricultural losses (Botzen et al. 2010) (Table 13.2).

13.2.2   Great Britain

Across the North Sea, the British government reacted differently to flooding disasters in the mid-twentieth century. Not only did the island experience catastrophic storm surge during the 1953 disaster, but it had also experienced the worst riverine flooding since the mid-eighteenth century in 1947. This was also true for Germany and other European countries at the same time. The 1947 flooding was caused by rapid snowmelt and persistent heavy rainfall after an extremely harsh winter. The disaster affected 30 out of 40 English counties (RMS 2007). The combination of severe risk from riverine as well as coastal flooding let the British government to fortify flood defenses in large, urban areas and to leave smaller towns and rural areas less protected. In contrast to The Netherlands, flood coverage is now a standard component of residential and commercial insurance policies since the 1960s—not as an add-on or additional rider, but as comprehensive coverage. Residential insurance is not mandatory (except for mortgage holders) and poorer households tend to forego it if possible. Only 30 % of low-income households possess comprehensive coverage (Crichton 2005).

Despite these efforts, the level of flood risk has increased in Great Britain. Since World War II, half of all new housing units have been place in flood-prone areas according to the Association of British Insurers (RMS 2007). As a result, Great Britain has seen some of its worst flooding in recent years such as 2000, 2007, and 2012. In 2007, much of the loss came from flash flooding—heavy rainfall within a short period that overwhelms drainage systems—and the failure of flood walls, levees, etc. Aside from buildings, the event badly damaged or interrupted utility supplies (e.g., water and power), agriculture, railways and even major roadways (e.g., M5). The 2007 flooding set a U.K. record as the costliest flood in terms of insured losses (EUR 3.5 billion) as well as the wettest summer since record-keeping started in 1766 (Stuart-Menteth 2007). It was also the country's largest peacetime emergency since World War II (Pitt 2008). The after-action report, or the so-called Pitt Review (2008), highlighted a broad failure in flood management across all sectors of preparedness, response, and mitigation and compiled a list of nearly 100 recommendations.

Given the increase in exposure in Great Britain, insurers are threatening to leave the market and/or drop flood coverage claiming inadequate land use planning and insufficient investments into flood protection. This issue was reiterated by the Pitt Review and the Foresight programme managed by the Office of Science and Technology (Foresight 2004), which had previously concluded that significant investments and "hard choices need to be taken" (p. 2) to avoid unacceptable levels of risk. The British government responded and promised to increase flood defense spending to GBP 800 million by 2010/2011 (Stuart-Menteth 2007). This promise evaporated and instead a 30 % cut to the Department for Environment, Food and Rural Affairs (DEFRA) was implemented in 2010 slashing DEFRA's budget to GBP 400 million per year for 2011 to 2015—a drastic reduction to capital spending, mostly flood defense spending according to AIR Worldwide (Hughes and Gambrill 2012).

13.2.3   Germany

Depending on the track, many storm systems that affect The Netherlands and Great Britain also impact Germany. The 1962 storm surge was such an event, which claimed 315 lives in Germany alone and destroyed more than 6,000 buildings in Hamburg (Appendix, Lamb 1991). Similar to Great Britain, Germany suffers from coastal and riverine flooding. Earthen levees run alongside many major rivers in Germany, especially as they wind through flat and low-lying areas. Many of these levees are over 100 years old and frequently experience failures during flood stages lasting several days/weeks. Along the Elbe river, nearly 40 % of the 1,200 km of levees are structural insufficient (IKSE 2004). Floodplain development, straightening of river channels, and reliance on temporary flood barriers are main drivers behind accelerating flood damage in Germany.

The years of 2013 and 2002 were particularly devastating along the Elbe and Danube rivers. Some residents along the Elbe river were still not recovered from 2002 and without proper flood protection when the 2013 flood hit (e.g., Gohlis neighborhood in Dresden; Lose 2013). The 2002 record flood cost more than EUR 11 billion and heavily impacted the city of Dresden (Munich Re 2003). The flood interrupted the railway line between Dresden and Prague for more than 4 months (RMS 2003b). Europe-wide, the 2002 floods also affected Austria, the Czech Republic and Slovakia, making it the costliest European natural disaster with a loss total of more than EUR 20 billion (EEA 2010)—at least until 2013.

Only around 15 % of the 2002 loss was insured (EUR 1.8 billion; Thieken et al. 2006). According to Thieken et al. (2006), the small share of insured loss can be explained by the unusually high portion of damaged infrastructure and commercial buildings rather than residential properties, which incurred less than half of the damage (Kron 2004). Only 4 % of residences and 10 % of home contents carry flood insurance (Thieken et al. 2006) with even fewer businesses being covered by commercial flood insurance. As a result of the low market penetration, the German government granted EUR 7.1 billion in reconstruction aid and EUR 500 million individual aid along with funds from the ad-hoc established EU Solidarity Fund (EUR 444 million) and private donations (EUR 350 million) (Thieken et al. 2006). Thus, the government shouldered more than 60 % of the loss. Such a financial and economic burden is not sustainable for a country in the long run and when faced with repetitive events.

In Germany, homeowners can add riverine flood coverage to their insurance plan but it is not part of the standard policy; storm surge is not insurable (Botzen and van den Bergh 2008; Thieken et al. 2006). Premiums are risk-based and prohibitively expensive in some high risk areas. In contrast to The Netherlands and Great Britain, flood insurance availability has varied within Germany. Comprehensive policies that included flooding were available in former East Germany, of which 30 % are still in effect in the Eastern states of Germany (Kron 2004). Additionally, the State of Baden-Württemberg mandated comprehensive homeowners insurance until 1994 (Thieken et al. 2006) and about 80 % of property owners still have flood coverage (Kron 2004).

Overall, flood risk management strategies and flood experiences vary between
The Netherlands, Great Britain, and Germany although they are united in their high
levels of flood risk and exposure. Each country possesses a unique vulnerability
profile: The Netherlands relies extensively on structural measures compared to
more moderate investments in flood defense by Germany and Great Britain, where
residents have some access to flood insurance. All three countries though grapple
with mounting costs and/or losses of floods and the added threat of climate change
and sea-level rise—not just in regard to the protection of their residents but also in
regard to their ability to maintain stable and growing economies.

13.3 Linkages between Economic Development and Flood Risk Management

13.3.1 Economic and Regional Benefits

Water provides a range of functional purposes: it provides critical resources
(e.g., drinking and irrigation water, food supply), ecosystem services, transportation
of individuals and goods, recreation, energy generation, and other related inputs
relevant to economic activities and development (e.g., as a component of large-
scale industrial processes). Not surprisingly many of these factors attracted settlers
allowing large population centers to flourish and grow along rivers and coastlines
(e.g., Paris, Amsterdam, London, Hamburg, Cologne, Vienna, Venice, etc.). Rivers
such as the Rhine and Danube as well as ports such as Rotterdam and Hamburg represent economic lifelines for Europe. For instance, the port of Rotterdam, the largest
port in Europe and the 5th largest port in the world, connects the industrial heart-
land of The Netherland and Germany to the world, and about 434.6 million tons
of goods were cleared in 2011 (RIA 2013). About 300 million tons of goods are
shipped along the Rhine river, which makes it Europe's most important waterway
(Bundeszentrale fuer politische Bildung 2013). Any transportation interruption
ripples through Europe and the world economy: crop prices rise; the availability
of goods as disparate as cars and car parts and cut flowers declines and results in
reduced economic output.

13.3.2 Economic Burden from Flooding

In order to sustain life and economies in its vicinity, water is managed by balancing the previously mentioned benefits with certain drawbacks (e.g., floods, drought
cycles, water-borne diseases, contamination). In terms of flooding as a hazard, due to
the improvements in record keeping (see Table 13.3), forecasting and early warning
has significantly improved in the past decades and has subsequently reduced the loss
of human life from such incidents. However, at the same time, the economic impacts

Table 13.3 Climate record keeping in Central Europe. (Adopted from Pfister 2001)

From	Frequency	Availability	Source
ca. AD 700	At best yearly	Sporadic	Documents
ca. 1250	seasonal	Almost uninterrupted	Documents
ca. 1500	Daily to monthly	Almost uninterrupted	Documents
ca. 1680	Daily	Sporadic	Instruments, anecdotal accounts
ca. 1860	Daily	Uninterrupted	Instruments, official accounts

of floods and natural hazards in general are not as well understood. Despite the fact that floods along with wind storms are major threats to the European economy, there is no comprehensive, public record/database of losses (direct, indirect, insured, or un-insured)—neither at the EU level nor by any individual member state. Existing loss databases operate either at the international level (e.g., EM-DAT) or are proprietarily held by reinsurance companies and provide only an incomplete picture (Gall et al. 2009). As a result, the cumulative socioeconomic effect of floods is largely elusive and is based on fairly crude estimates of economic impact in the short term; seldom are long-term effects on development estimated with much precision (if at all). This lack of information makes an evaluation of the costs and benefits (avoided losses) challenging and conceals the true burden of the flood hazard on all European economies.

13.3.3   Managing the Burden and Benefits in an Uncertain Environment

This also translates to a fundamental challenge in the flood risk management and economic development linkage. As noted above, it is known that there are major economic losses associated with flooding—even with imperfect and imprecise information about specific losses for a given incident. It is also known that vulnerability to the flood hazard in Europe, including the three countries profiled here, is increasing. However, estimating the effects that such increasing vulnerability has on future economic development is problematic. As noted, floodplain development has not been significantly curtailed in many parts of Europe, which will predictably result in greater hazard losses from flood incidents as the major floods of 2002 and 2013 have demonstrated.

13.3.4   Flood Risk Management Beyond National Boundaries

On top of accurate economic effect measurements which hampers planning and land use management practices, Europe grapples with a related set of challenges: political obstacles to comprehensive flood risk management beyond national boundaries. For most of the European Union's history, planning occurred at the national level and the management of water and water bodies was compartmentalized and conducted by a multitude of entities with different, often conflicting, interests and objectives (Begum et al. 2007). Treaties and basin-wide, transboundary consortia such as International Commissions for the Protection of the Rhine or Danube rivers related to flood risk management only tangentially or implicitly. The result was intensive development in floodplains without any harmonization or integration of objectives of economic growth, land-use planning, and sustainable water management. The problem was furthermore aggravated by a complex network of local, regional, national, and supranational authorities with varying responsibilities and policies in regard to water management.

An illustration of the political challenges can be seen in the United Kingdom (UK) Parliament's response to the major flooding incident in the summer of 2007 as described above. Again, the Pitt Review offered nearly 100 recommendations related to improving administrative systems for prediction and early warning, prevention, emergency management practices and new approaches to promote more effective resilience and recovery (Pitt 2008). However, it is useful to note that while the UK Parliament did approve—for the first time—a statutory flood and coastal erosion risk management strategy for the whole of the UK, certain gaps still exist, such as building regulations designed to promote greater flood resilience and making more explicit flood risk information to potential home buyers (Van Alphen et al. 2011).

Further, the UK Environment Agency's approach taken toward addressing climate change is to recommend a "managed adaptive approach" coinciding with DEFRA's flood and coastal risk management strategy (UKEA 2011). This managed adaptive approach can be seen to stand in contrast with a precautionary management approach though (Van Alphen et al. 2011). Simply put, a precautionary approach requires that even in the absence of complete information on a phenomena such as climate change and sea level rise, it is the obligation of governing authorities to minimize the potential risk exposure to a community—especially when the hazard might be of a low probability category (though that might not be entirely clear) but is also of high consequence (including irreversible harms to a community and its resources or assets). This principle requires taking actions such as preventing development in the floodplain or making changes to existing development in hazard-prone areas. What can be regarded as regulatory hesitancy in Great Britain is therefore a function of misaligned incentives in policy making that tend to occur regardless of national setting. In effect, development imperatives tend to outweigh more effective risk management strategies (Burby 2006; 1998).

13.3.5   The 2007 European Flood Directive

European flood risk and water management has evolved in recent years. Tol et al. (2003) identified four trends: internationalization, integration, democratization, and ecologicalization. Existing treaties and national strategies have been supplemented

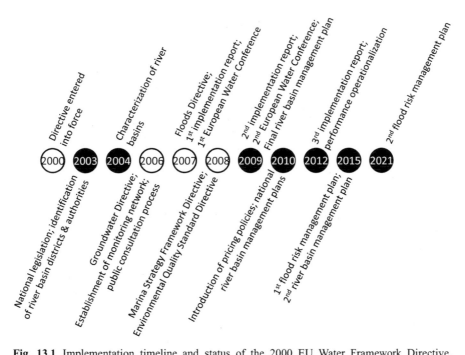

Fig. 13.1 Implementation timeline and status of the 2000 EU Water Framework Directive. (Adapted from European Commission DG Environment 2010)

with the groundbreaking *2000 European Union (EU) Water Framework Directive* and the *2002 Integrated Strategy on Prevention, Preparedness and Response to Natural, Man-made and Other Risks*. The directive consists of several legally binding regulations geared towards integrated and international river basin management and water protection to safeguard clean groundwater, river, lakes, and coastlines (Fig. 13.1). The *2007 EU Floods Directive* (2007/60/EC) is part of this framework and requires member states to assess areas at risk from riverine and coastal flooding, map the flood extent, determine people and assets at risk, and to devise mitigation actions. Preliminary assessments were due in 2013 and initial flood risk management plans are slated for publication in 2015. For the first time in EU history—and in the history of many of its member nations—flood risk must be mapped. To support this effort, the EU Directorate General of the Joint Research Centre as well as the Directorate General on Research Policy have made significant investments in hazard research such as the development of a large-scale flood modeling system (LISFLOOD) or FLOODsite (www.floodsite.net), which was a EUR 14 million, 5-year, project funded under Framework Programme 6 (FP6). (For a more detailed discussion of actors, legislative procedures and governing bodies of the EU see Arellano et al. 2007).

This legislative overhaul of risk management strategies could position the EU to better control its flood losses but also to improve long-term climate change adaptation. Indeed, such efforts represent a possible template for mitigating the risks associated with other climate-related hazards.

13.4   Future Challenges to Flood Risk Management

It is unclear if the recent catastrophic floods in Germany and Great Britain are indications of a new "normal" ushering in an "era of catastrophes" (Kunreuther and Michel-Kerjan 2009a). Attributing specific events to climate change is difficult. But what is already happening and what is likely or very likely to occur in the future is: the rise in sea-level of the North Sea, increase of warm days with more heat waves and drought periods, slight decrease of cold days, and increase in extreme precipitation events along with more frequent severe flooding (IPCC 2012; Tank and Lenderink 2009).

This will affect Germany, Great Britain, and The Netherlands in a multitude of ways. For the Rhine river, some predict an increase of winter precipitation by 40%, a decrease in summer precipitation by 30% (Lenderink et al. 2007) and an increase in annual damage between 50 and 230% (te Linde et al. 2011). How much of this precipitation will fall as snow, though, will have significant implications for the probability of spring floods in Germany, which tend to be a product of combined snowmelt in the Alps and slow-moving/stalling weather systems with heavy rainfall. In The Netherlands, observed yearly precipitation has already gone up by 18% since 1906 and the 10-day precipitation total has risen by 29% (Tank et al. 2006; Tank and Lenderink 2009). According to te Linde et al. (2011), the Lower Rhine area in Germany has the highest flood risk due to its lower flood protection standards.

Aside from Germany, The Netherlands exhibits potential for exceptional flood damage in the future. More than half of the Dutch population, about 9 million people, lives already below sea level (Aerts 2009). Given that law sets safety standards, The Netherlands will have to keep up with sea-level rise and higher flood peaks. Based on climate change predictions, this translates into higher discharge values (measured in m^3/s) for riverine flood return periods. It means that flood defenses must be improved in order to keep up with higher discharges and to maintain protection levels. Aerts et al. (2008) predict an order of magnitude increase in flood risk for sea-level rise between 50 and 80 cm. A sea-level rise of 70 cm would drop the current protection level for the cities of Rotterdam, The Hague, and Amsterdam from 10,000 years to 100 years.

Combined with a growing population living below sea level, many are promoting flood adaptation measures in addition to structural engineering. These possible structural solutions include reinforcement of storm surge barriers, widening of river beds (e.g., *Give Room to the River* program), dike reinforcement, and beach nourishment, which according to the 2008 Delta Commission report (Aerts et al. 2008) would require more than EUR 100 billion in new spending through the year 2100. These are massive spending and construction projects aimed at over-dimensioning flood projects and creating multiple lines of defense to maintain or improve protection levels and to reduce the likelihood of failure.

But there is a flip side to structural flood mitigation. Flood defenses tend to provide a "false sense of security" (Tobin 1995) and homeowners are either unaware or underestimate the residual risk. In return, such seemingly "safe" areas attract additional development (Burby 2006) leading to an even greater disaster should flooding occur. As the tragic 2005 events in the City of New Orleans showed, a false sense of security combined with other factors (e.g., poverty, lack of transportation, uncertainty over sheltering guidance) led many residents to stay in the city despite clear and urgent warning messages. Sudden and unexpected levee failures triggered swift rising and deadly flood levels within the levee-enclosed city. The drainage of a city largely below sea-level proved difficult, time-consuming, and costly but above all provided a significant obstacle to recovery (Comfort et al. 2010). Quick recovery—not flood mitigation became the mantra. Legitimate questions such as—What areas will be safe in the future? What are appropriate building codes? Which areas should be rebuilt? What is the long-term recovery plan?—received only ad-hoc answers based on inadequate empirical support and limited prospects for sustainability (Olshansky and Johnson 2010). Long-term planning that mitigates future impacts played only a secondary role. Thus, exclusive reliance on structural mitigation has the potential to create a disaster of unimaginable proportions in a location such as The Netherlands.

Consequently, some researchers advocate for supplementary, non-structural solutions to mitigate the impacts of failing flood defenses or lower levels of protection (Wesselink 2007; Meyer et al. 2012). Examples of non-structural measures that might be effective in this domain include risk education, improvements in flood mapping, hardening of buildings and infrastructure, and increased provision of insurance to property owners. The British approach of flood risk management embraces both structural and non-structural mitigation actions. A recommended spending increase from GBP 570 million to GBP 1 billion by 2035 (UKEA 2009) is aimed at maintaining protection levels, which tend to be at the 100-year level, and combined with non-structural efforts. The U.K. Environment Agency clearly states (2009): "It is neither technically feasible nor economically affordable to prevent all properties from flooding. Therefore, a risk-based approach is taken to achieve the best results possible using the budget and resources available" (p. 13). The report also underscores the importance of limiting development in the floodplain and implementing better building standards. To promote a more sustainable development approach, the Association of British Insurers no longer insures properties built against the recommendations of the Environment Agency after January 2009. Nevertheless, it is still possible to build in the floodplain, which may lead to an increase in flood levels and a widening of the floodplain for others. While this policy protects the insurance industry it does not necessarily mitigate actual losses.

Demands on the insurance industry for more involvement and increased market penetration reverberate through all three countries. The insurance industry could affect short-term adjustments and directly influence the behavior of the insured. Policies are generally renewed or cancelled annually and insurance companies can decide to adjust premiums along with deductibles, maximum pay-out and/or modify policy exceptions, restrictions, and so forth. Thus, incentivizing flood mitigation could motivate homeowners by means of lower premiums and deductibles.

However, improvements in certain insurance practices are not a panacea. In fact, insurance programs for the sake of transferring losses rather than reducing them are unsustainable. The unaffordability or unavailability of insurance to residents may lead politician to "demand" a way out for residents. In the United States, for instance, flood insurance is provided by the federal government and many states now serve as wind insurers of last resort. Unfortunately, the U.S. National Flood Insurance Program is heavily in debt due to actuarially inaccurate premiums and decades of uncurbed floodplain development. A similar fate could affect the State of Florida, which made wind insurance affordable—not just available—and thereby priced private insurers out of the market. The State of Florida might in fact be overly leveraged and incapable of paying-out future claims in the case of a large-scale hurricane(s) (Kunreuther and Michel-Kerjan 2009a).

Relatively little is known in regard to European flood insurance, consumer behavior and decision-making. Unlike the widely studied and heavily criticized National Flood Insurance Program in the United States, research on hazards, risk communication, and particularly socioeconomic aspects are in its infancy in Europe; even less is transferred from theory into practice (Höppner et al. 2012). Despite centuries of living with floods and recurring devastating floods, European countries have only recently discovered the need for better flood risk management as indicated by the 2007 Flood Directive. Many European countries have yet to realize the full potential of non-structural flood mitigation actions (see Fig. 13.2), social capital (Wood et al. 2013), and capacity building (Kuhlicke et al. 2010)—i.e. managing flood risk beyond probabilistic protection levels.

After decades (even centuries) of severe flooding and escalating losses across Europe, the need for concerted flood mitigation actions has ultimately led to legislative action. This is a promising beginning. Based on existing research, it appears though that Europe's ability to recover from a disaster might be as important for its economic development as its ability to avoid losses in the first place. Hallegatte et al. (2007) found that slow recovery hampers economic growth over time—not just in the immediate aftermath of the disaster. In fact, the level and duration of economic decline depends largely on the economic system's ability to adjust through redistribution, expansion, and production substitution (Bockarjova et al. 2007).

Current EU policies focus largely on loss avoidance and loss sharing rather than coping and adaptation capacity as this chapter illustrates. Perhaps such capacity could be developed when hazard risk management is folded into the larger context of climate change adaptation. Although the EU has long been committed to climate change mitigation and the reduction of greenhouse gases, the European Commission has only recently initiated the development of an EU climate adaptation strategy (Aakre et al. 2010).

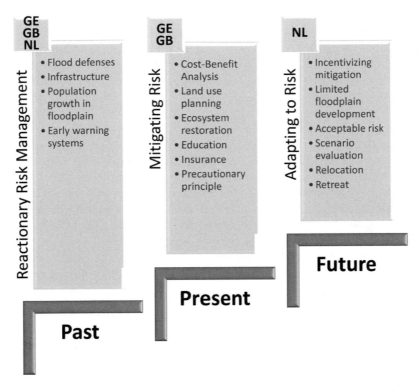

Fig. 13.2 Flood risk management over time as it applies to The Netherlands, Germany, and Great Britain

13.5 Conclusion

Living with floods is a long-standing practice in Europe—although more explicit in some countries (e.g. The Netherlands) than others. While structural flood mitigation (e.g., dikes, polders) permitted economic growth, the continued concentration of assets in flood-prone areas has created both an economic and even an ethical problem: more resources, people, and infrastructure are now protected by flood mitigation efforts that are inadequate to cope with the flood hazard within the context of a new climate. Sea-level rise and more intense (and perhaps more frequent) floods threaten economic hubs like London and entire countries such as The Netherlands; as such, climatological changes in hazard profiles will necessitate new approaches to flood mitigation. Maintaining economic growth and preparing society for future conditions requires re-thinking and re-considering the acceptability and feasibility of adaptation costs to floods. In the face of climate change, flood-prone countries like The Netherlands, the United Kingdom and Germany are in the midst of developing new flood risk management strategies and visions for the future. Many of these new visions are shaped by the countries' history and experience with both floods and flood mitigation. Some promote more integrated approaches such as flood risk managements whereas others rely on technological solutions.

Changing behavior and planning decisions for the benefit of hazard mitigation and climate change adaptation requires leadership, properly designed incentives across public and private sectors, and a long-term commitment at national, sub-national, and international levels. There will be many difficult decisions in terms of financial investments, land-use restrictions, building code enforcement, and so on—none of which will come easily. What, however, will change swiftly are insurance policies and insurance premiums. Thus, making insurance available as a risk-sharing tool is reasonable—even effective when stipulating flood mitigation and risk education (Kunreuther and Michel-Kerjan 2009b).

In terms of economic impacts from future floods, it appears that election cycles and myopic planning horizons have detrimental effects on long-term risk management strategies—as the British budgetary cuts to key agencies after the 2007 floods illustrated. Hallegatte et al. (2007) pointed out that the economic costs of extreme events may be much larger than their initial direct losses since the impacts of disasters are amplified and accumulate over time without adequate prevention and mitigation. Thus, unsustainable short-term recovery decisions may have significant long-term repercussions that increase the disaster burden in the future and slow-down future economic growth.

Appendix

Significant historic flood events that caused major damage in The Netherlands, Germany, and Great Britain. Events listed in cursive affected all three countries. Note that Germany and Great Britain experience flooding on an annual basis but only major events are included below.

Netherlands			Germany			Great Britain		
Year	Type	Fatalities	Year	Type	Fatalities	Year	Type	Fatalities
	120 BC[a]			Storm surge				
	AD 800[a]			Storm surge				
			1012[b]	Riverine				
			1051[b]	Riverine				
	1164[a]			Storm surge		< 20,000 fatalities		
			1179[b]	Riverine				
			1209[b]	Riverine	> 1,000			
	1219[a]			Storm surge		< 36,000 fatalities		
			1235/36[b]	Riverine				
	1287[a]			Storm surge		50,000–80,000 fatalities		
			1304	Storm surge	271			
			1306[b]	Riverine				
1313[c]	Riverine							

Netherlands			Germany			Great Britain		
Year	Type	Fatalities	Year	Type	Fatalities	Year	Type	Fatalities
1315[c]	Riverine							
			1342[b]	Riverine	< 6,000			
1362[a]	Storm surge							11,000–30,000 fatalities
			1413[b]	Riverine				
1421[d]	Storm surge	<10,000						
			1432[b]	Riverine				
			1436[a]	Storm surge	>180			
			1451[b]	Riverine				
			1501[b]	Riverine				
1509[a]	Storm surge							
1530[a]	Storm surge					1530[a]	Storm surge	
1532[a]	Storm surge							
			1546[b]	Riverine				
			1551[b]	Riverine				
1552 or 1553[a]	Storm surge							
			1558[b]	Ice jam				
			1561[b]	Ice jam				
1570[a]	Storm surge (All saints flood)							> 20,000 fatalities
			1595[b]	Riverine				
			1613[*]	Riverine	2,261			
1626[c]	Ice surge							
			1633[b]	Riverine				
			1634[a]	Storm surge	6,000			
			1655[b]	Riverine				
			1682[b]	Ice jam				
						1694[a]	Storm surge (Culbin sands disasters)	
1703[a]	Storm surge							8,000 fatalities
			1709[b]	Ice jam				
1717[a]	Storm surge							11,000 fatalities
			1732[b]	Riverine				
			1756[a]	Storm surge				
						1774[f]	Riverine	
			1778[b]	Riverine	63			
			1784[b]	Riverine	> 1,000			
1792[a]	Storm surge							
1825[a]	Storm surge	825	1825[a]	Storm surge				
						1829[a]	Riverine	
1855[c]	Riverine							
1861[c]	Riverine							
						1864	Dam failure	270
			1909	Riverine				
						1912[e]	Riverine	
			1916[a]	Storm surge				

Netherlands			Germany			Great Britain		
Year	Type	Fatalities	Year	Type	Fatalities	Year	Type	Fatalities
1926[c]	Riverine							
			1927	Riverine	160			
						1928[a]	Riverine	14
			1938[a]	Storm surge				
			1947[f]	Riverine		1947[f]	Riverine	
						1952[e]	Riverine	34
1953[g]	Storm surge	1,800				1953[g]	Storm surge	307
1962[a]				Storm surge		315 fatalities in Hamburg		
1976[a]			Storm surge (Capella storm)			52 fatalities		
			1970[b]	Riverine				
						1978[a]	Storm surge	1
						1979[h]	Riverine	
						1987[h]	Riverine	
						1990[j]	Storm surge	
			1993[i]	Riverine				
1995[i]	Riverine							
1998[i]	Flash flood		1998[b]	Riverine		1998[h]	Riverine	
			1999[b]	Riverine				
			2000[b]	Riverine		2000[i]	Riverine	
			2002[b]	Riverine	21			
						2004[h]	Riverine	
			2005[b]	Riverine		2005[e]	Riverine	
2007[i]			Storm surge (Cyclone Tilo)					
						2007[i]	Riverine and flash flooding	13
						2009[i]	Riverine	1
						2012[i]	Riverine and flash flood	10
			2013[b]	Riverine				

[a] Lamb (1991)
[b] Hochwassernachrichtendienst Bayern
[c] Tol and Langen (2000)
[d] Encyclopaedia Britannica
[e] Stuart-Menteth (2007)
[f] RMS (2007)
[g] RMS (2003a)
[h] Safetynet Systems, Ltd.
[i] EM-DAT
[j] Zong and Tooley (2003)
[k] Grünewald (Grünewald 2006)

References

Aakre, S., Banaszak, I., Mechler, R., Rübbelke, D., Wreford, A., & Kalirai, H. (2010). Financial adaptation to disaster risk in the European Union. *Mitigation and Adaptation Strategies for Global Change, 15*(7), 721–736.

Aerts, J. C. J. H. (2009). Adaptation cost in The Netherlands: Climate change and flood risk management. National climate research The Netherlands. http://www. climateresearchnetherlands.nl/highlights/10354094/Adaptation-cost-in-the-Netherlands-Climate-Change-and-flood-risk-management. Accessed 10 Aug 2013.

Aerts, J. C. J. H., Sprong, T., & Bannink, B. (Eds.). (2008). *Aandacht Voor Veiligheid.* Den Haag: Deltacommissie.

Arellano, Ana Lisa Vetere, De Roo, A., & Nordvik, J.-P. (2007). Reflections on the challenges of EU policy-making with view to flood risk management. In S. Begum, M. Stive, & J. Hall (Eds.), *Flood risk management in Europe* (pp. 433–468). Dordrecht: Springer.

Begum, S., Stive, M. J. F., & Hall, J. W. (Ed.). (2007). *Flood risk management in Europe—innovation in policy and practice.* Dordrecht: Springer.

Bockarjova, M., Steenge, A. E., & Van der Veen, A. (2007). Structural economic effects of large-scale inundation: A simulation of the Krimpen dike breakage. In S. Begum, M. Stive, & J. Hall (Eds.), *Flood risk management in Europe* (pp. 131–154). Dordrecht: Springer.

Botzen, W. J. W., & van den Bergh, J. C. J. M. (2008). Insurance against climate change and flooding in The Netherlands: Present, future, and comparison with other countries. *Risk Analysis, 28*(2), 413–426.

Botzen, W. J. W., van den Bergh, J. C. J. M., & Bouwer, L. M. (2010). Climate change and increased risk for the insurance sector: A global perspective and an assessment for The Netherlands. *Natural Hazards, 52*(3), 577–598.

Bundeszentrale fuer politische Bildung. (2013). Der Rhein. Bonn: Bundeszentrale fuer politische Bilding. http://www.bpb.de/geschichte/zeitgeschichte/geschichte-im-fluss/135612/der-rhein. Accessed 10 Aug 2013.

Burby, R. J. (1998). *Cooperating with nature: Confronting natural hazards with land-use planning for sustainable communities.* Washington D.C: Joseph Henry Press.

Burby, R. J. (2006). Hurricane Katrina and the paradoxes of government disaster policy: Bringing about wise governmental decisions for hazardous areas. In W. L. Waugh (Ed.), *Shelter from the storm: Repairing the national emergency management system after Hurricane Katrina* (pp. 171–191). Philadelphia: *Annals of the American Academy of Political and Social Science.*

U.S. Central Intelligence Agency (CIA). (2013). The world factbook. https://www.cia.gov/library/publications/the-world-factbook/index.html. Accessed 10 Aug 2013.

Comfort, L. K., Birkland, T. A., Cigler, B. A., & Nance, E. (2010). Retrospectives and prospectives on Hurricane Katrina: Five years and counting. *Public Administration Review, 70*(5), 669–678.

Crichton, D. (2005). *Flood risk & insurance in England and Wales: Are there lessons to be learned from Scotland?* London: Benfield Hazard Research Centre.

De Bruijn, K. M., Green, C., & Johnson, C. (2007). Evolving concepts in flood risk management: Searching for a common language. In S. Begum, M. Stive, & J. Hall (Eds.), *Flood risk management in Europe* (pp. 61–75). Dordrecht: Springer.

European Commission DG Environment. (2010). *Water Is for life: How the water framework directive helps safeguard Europe's resources.* Luxembourg: Publications Office of the European Union. doi:10.2779/83017.

European Environment Agency (EEA). (2010). *Mapping the impacts of natural hazards and technological accidents in Europe: An overview of the last decade.* Technical Report No. 13/2010. Copenhagen: Publications Office of the European Union. doi:10.2800/62638.

Foresight. 2004. Future Flooding. London: Department of Business Innovation & Skills. http://www.bis.gov.uk/foresight/our-work/projects/published-projects/flood-and-coastal-defence/project-outputs. Accessed on August 10th, 2013.

Gall, M., Borden, K. A., & Cutter, S. L. (2009). When do losses count? Six fallacies of natural hazards loss data. *Bulletin of the American Meteorological Society, 90*(6), 799–809.

Godschalk, D. R., Beatley, T., Berke, P., Brower, D. J., & Kaiser, E. J. (1999). *Natural hazard mitigation: Recasting disaster policy and planning.* Washington D.C: Island Press.

Grünewald, U. (2006). "Extreme Hydro(meteoro-)logische Ereignisse Im Elbegebiet. *Österreichische Wasser- Und Abfallwirtschaft, 58*(3–4), 27–34.

Hallegatte, S., Hourcade, J.-C., & Dumas, P. (2007). Why economic dynamics matter in assessing climate change damages: Illustration on extreme events. *Ecological Economics, 62*(2), 330–340.

Höppner, C., Whittle, R., Bründl, M., & Buchecker, M. (2012). Linking social capacities and risk communication in Europe: A gap between theory and practice? *Natural Hazards, 64*(2), 1753–1778.

Hughes, B. Y. Thomas, & Gambrill, S. (2012). Summer floods in the UK: Comparing 2012 and 2007. Boston: Air Worldwide. http://www.air-worldwide.com/Publications/AIR-Currents/2012/Summer-Floods-in-the-UK-Comparing-2012-and-2007/. Accessed 10 Aug 2013.

International Commission for the Protection of the Elbe River (IKSE). (2004). Dokumentation Des Hochwassers Vom August 2002 Im Einzugsgebiet Der Elbe. Magdeburg: http://www.ikse-mkol.org/fileadmin/download/gescannte_Publikationen/DE/IKSE-Dokumentation_Hochwasser_2002.pdf. Accessed 10 Aug 2013

IPCC. (2012). *Managing the risks of extreme events and disasters to advance climate change adaptation.* In C. B. Field, V. Barros, T. F. Stocker, & Q. Dahe (Eds.), Cambridge: Cambridge University Press. doi:10.1017/CBO9781139177245.

Kaiser, S. (2013). Gewinneinbruch: Deutsche Bahn Verdient Fast 30 Prozent Weniger—SPIEGEL ONLINE. *Der Spiegel Online.* http://www.spiegel.de/wirtschaft/unternehmen/gewinnein-bruch-deutsche-bahn-verdient-fast-30-prozent-weniger-a-913140.html. Accessed 10 Aug 2013.

Kron, W. (2004). Zunehmende Ueberschwemmungsschaeden: Eine Gefahr Fuer Die Versicher-ungswirtschaft? In ATV-DVWK (Ed.), *Proceedings of the Bundestagung, September 15–16* (pp. 47–63). Hennef.

Kuhlicke, C., & Steinführer, A. (2010). Social capacity building for natural hazards: A conceptual frame. Leipzig: CapHaz-Net Consortium. http://caphaz-net.org/outcomes-results/CapHaz-Net_WP1_Social-Capacity-Building2.pdf. Accessed 10 Aug 2013.

Kunreuther, H. C., & Michel-Kerjan, E. O. (2009a). *At war with the weather: Managing large-scale risks in a new era of catastrophes.* Cambridge: The MIT Press.

Kunreuther, H. C., & Michel-Kerjan, E. O. (2009b). Encouraging adaptation to climate change: Long-term flood insurance. Washington D.C.: Resources for the Future. http://opim.wharton.upenn.edu/risk/library/RFF-IB-09-13.pdf. Accessed 10 Aug 2013.

Lamb, H. H. (1991). *Historic storms of the North Sea, British Isles and Northwest Europe.* Cambridge: Cambridge University Press.

Lenderink, G., Buishand, A., & Deursen, van W (2007). Estimates of future discharges of the river rhine using two scenario methodologies: Direct versus delta approach. *Hydrology and Earth System Sciences, 11*(3), 1145–1159.

Lose, S. (2013). Hochwasser in Dresden: Elbepegel Steigt Weiter—DNN-Online. *DNN-Online,* June. http://www.dnn-online.de/dresden/web/dresden-nachrichten/detail/-/specific/Hochwasser-in-Dresden-Elbe-naehert-sich-7-50-Meter-erste-Stadtteile-evakuiert-1850832252. Accessed 10 Aug 2013.

Mechler, R., & Kundzewicz, Z. W. (2010). Assessing adaptation to extreme weather events in Europe—editorial. *Mitigation and Adaptation Strategies for Global Change, 15*(7), 611–620.

Meyer, V., Priest, S., & Kuhlicke, C. (2012). Economic evaluation of structural and non-structural flood risk management measures: Examples from the Mulde River. *Natural Hazards, 62*(2), 301–324.

Munich, Re. (2003). *Annual review: Natural catastrophes 2002.* Munich: Munich Re.

Olshansky, R. B., & Johnson, L. (2010). *Clear as mud: Planning for the rebuilding of new Orleans.* Chicago: APA Planners Press.

Pfister, C. (2001). Klimawandel in der Geschichte Europas: Zur Entwicklung und zum Poten-zial der historischen Klimatologie. *Österreichische Zeitschrift fuer Geschichtswissenschaften, 12*(2), 7–43.

Pitt, M. (2008). The pitt review: Learning lessons from the 2007 Floods. London: The National Archives. http://archive.cabinetoffice.gov.uk/pittreview/thepittreview/final_report.html. Accessed 10 Aug 2013.

Rijkswaterstaat. (2013). History of water management. Utrecht: Rijkswaterstaat. http://www.ruimtevoorderivier.nl/meta-navigatie/english/history-of-watermanagement/. Accessed 10 Aug 2013.

Risk Management Solutions (RMS). (2003a). 1953 U.K. Floods: A 50-year retrospective. Newark: RMS. https://support.rms.com/publications/1953_Floods_Retrospective.pdf. Accessed 10 Aug 2013

Risk Management Solutions (RMS). (2003b). Central Europe flooding, August 2002. Newark: RMS. https://support.rms.com/publications/Central%20Europe%20Floods%20Whitepaper_final.pdf. Accessed 10 Aug 2013.

Risk Management Solutions (RMS). (2007). 1947 U.K. river floods: 60-Year retrospective. Newark: RMS. https://support.rms.com/publications/1947_UKRiverFloods.pdf. Accessed 10 Aug 2013.

Rotterdam Investment Agency (RIA). (2013). Rotterdam is Europes most important port. Rotterdam Investment Agency. http://www.rotterdaminvestmentagency.com/page/Rotterdam+is+Europes+most+important+port/2023/en/. Accessed 10 Aug 2013.

Stuart-Menteth, A. (2007). U.K. summer 2007 floods. Newark: Risk Management Solutions, Inc. (RMS). https://support.rms.com/publications/UK_Summer_2007_Floods.pdf. Accessed 10 Aug 2013.

Tank, A. K., & Geert, L. (Ed.). (2009). *Climate change in The Netherlands*. De Bilt: Royal Netherlands Meteorological Institute (KNMI).

Tank, A. K., Geert, L., Ulden, Aad Van, Katsman, C., Keller, F., Bessembinder, J., Burgers, G., Komen, G., Hazeleger, W., & Drijfhout, S. (2006). KNMI climate change scenarios 2006 for The Netherlands. De Bilt, The Netherlands. http://www.knmi.nl/klimaatscenarios/knmi06/WR23mei2006.pdf. Accessed 10 Aug 2013.

Te Linde, A. H., Bubeck, P., Dekkers, J. E. C., de Moel, H., & Jeroen, C. J. H. Aerts (2011). Future flood risk estimates along the river Rhine. *Natural Hazards and Earth System Science, 11*(2), 459–473.

Thieken, A. H., Petrow, T., Kreibich, H., & Merz, B. (2006). Insurability and mitigation of flood losses in private households in Germany. *Risk Analysis, 26*(2), 383–395.

Tobin, G. A. (1995). The levee love affair: A stormy relationship? *Water Resources Bulletin, 31*(3), 359–367.

Tol, Richard. S. J., & Langen, A. (2000). A concise history of Dutch river floods. *Climatic Change, 46*(3), 357–369.

Tol, Richard. S. J., Grijp, van der N., Olsthoorn, A. A., & van der Werff, P. E. (2003). Adapting to climate: A case study on riverine flood risks in The Netherlands. *Risk Analysis, 23*(3), 575–583.

United Kingdom Environment Agency (UKEA). (2009). *Investing for the future: Flood and coastal risk management in England*. Bristol: http://www.environment-agency.gov.uk/research/library/publications/108673.aspx. Accessed 10 Aug 2013.

United Kingdom Environment Agency (UKEA). (2011). *Adapting to climate change: advice for flood and coastal erosion risk management authorities*. Bristol: http://publications.environment-agency.gov.uk. Accessed 10 Aug 2013.

Van Alphen, J., Bourget, L., Craig, Elliott, Fujita, K., Riedstra, D., Rooke, D., & Tachi, K. (2011). *Flood risk management approaches*. Arlington: Institute for Water Resources, U.S. Army Corps of Engineers.

Wesselink, A. J. (2007). Flood safety in The Netherlands: The Dutch response to Hurricane Katrina. *Technology in Society, 29*(2), 239–247.

Wood, L. J., Boruff, B. J., & Smith, H. M. (2013). When disaster strikes… How communities cope and adapt: A social capital perspective. In C. D. Johnson (Ed.), *Social capital: Theory, measurement and outcomes* (pp. 143–169). ebook: NOVA Publishers.

Zong, Y., & Tooley, M. J. (2003). A historical record of coastal floods in Britain: Frequencies and associated storm tracks. *Natural Hazards, 29*(1), 13–36.

Chapter 14
The Public Policy Dimension of Resilience in Natural Disaster Management: Sweden's Gudrun and Per Storms

Daniel Nohrstedt and Charles Parker

14.1 Introduction

Resilience research as well as crisis and disaster management research underscores the importance of crisis-induced learning as a prerequisite for increased societal resilience. These literatures view extreme events as benchmarks for appraising the viability of policies that were designed to ensure adaptation and robustness to hazard as well as potential catalysts for change. Extreme events may sometimes expose performance gaps and highlight the need for change and reforms to develop better policies and practices to heighten preparedness for future contingencies. From a theoretical perspective, crisis-induced learning and change may be depicted as a linear process of problem identification, a search for better practices, and implementation of revised policies and structures. Reality is more complex, however, and research reports many barriers to constructive learning and change in the wake of extreme events.

Past research has documented the variability in crisis-induced policy outcomes; whereas some crises fuel creativity and constructive learning and policy change, other events are followed by more conservative policy reactions or stasis (Nice and Grosse 2001; Boin et al. 2009). The type of change can vary as well; in some cases crises only feed minor organizational development and change, while other cases result in more fundamental processes of institutional renewal, organizational redesign, and policy change. Understanding these variations is a legitimate concern for researchers and practitioners alike. The contribution of this chapter is to conduct a comparison between two major storms in Sweden, 'Gudrun' in December 2005 and 'Per' in January 2007, in an effort to unveil plausible factors and processes that can help shed light on the dynamics of crisis-induced policy change.

D. Nohrstedt (✉) · C. Parker
Department of Government, Uppsala University, Uppsala, Sweden
e-mail: Daniel.Nohrstedt@statsvet.uu.se

C. Parker
e-mail: Charles.Parker@statsvet.uu.se

N. Kapucu, K. T. Liou (eds.), *Disaster and Development,* Environmental Hazards,
DOI 10.1007/978-3-319-04468-2_14, © Springer International Publishing Switzerland 2014

On 8 January 2005 super storm Gudrun slammed in to the west coast of southern Sweden. Gudrun was the most devastating storm in modern Swedish history. Its hurricane force winds (33 m/s with gusts of 42 m/s) wreaked an extensive path of destruction across southern Sweden. Millions of trees were uprooted and broken and the combination of strong winds, falling trees and broken branches blocked roads, stopped rail traffic, and damaged electricity and telecommunications infrastructure, which caused phone service to stop working and 730,000 to lose power, some for up to 45 days. Sweden's crisis management system and the robustness of its critical infrastructure were exposed as inadequate. Two years after Gudrun had revealed Sweden's surprising societal vulnerabilities to a disruptive storm, on 14 January 2007, Sweden was hit by another powerful storm. Storm Per, while not as violent as Gudrun, actually affected a larger geographical area and it battered southern Sweden with maximum force winds of 29 m/s. While Per caused extensive forest damage, the damage was only 20–25 % of the destruction caused by Gudrun. Once again many people lost power, but far fewer customers, 440,000, were affected and power was restored much faster in most cases than 2 years previously. What, if anything, was learned from Gudrun to Per and what lessons this case may hold for small states, and by extension developing states, attempting to increase societal resilience to extreme disruptions is the subject of this chapter.

We proceed in four steps. First, we conduct a descriptive analysis of the aftermath of the 2005 storm Gudrun in an effort to identify what beliefs (ideas and motives) were elevated as key lessons from this event. Second, by focusing on legislative action, structural and organizational change we investigate if those beliefs were transferred into policy change (see Chap 3 in this volume). Third, we analyze the response to the storm Per in 2007 in an effort to see if the revised policies that were implemented after Gudrun had intended effects. In the final step, we engage in an explorative analysis to generate a number of candidate explanations for variations in crisis-induced policy change following disruptive crises.

14.2   A Policy Subsystem Approach to Learning, Policy Change, and Resilience

This chapter examines the dynamics of resilience through the prism of policy process research. Our point of departure is the insight that the capacity of societies to respond to extreme events is regulated by formal and informal policies and institutions (Handmer and Dovers 2007). Therefore, to better understand the development of crisis response capacities we need more detailed knowledge about the processes by which actors learn from previous events and how these lessons may feed policy and institutional change.

In line with Comfort et al. (2011, p. 273), we agree that resilience is elusive and adhere to the perception of resilience as an important if not critical feature of a society or organization, which in turn "is the outcome of a long-term process." Clearly, understanding related challenges of prevention, mitigation, and crisis management requires many different analytical approaches, frameworks, and theories. As one precondition for making societies more sustainable and robust, resilience research emphasizes the importance of adaptive capacity. The argument holds that resilience depends on the ability of social systems to adjust to external events by building organizational capacity (networks) and mobilizing resources (natural and social) (Thompkins and Adger 2004; Wise 2006). Meanwhile, recent research emphasizes the limitations of adaptive capacity and social learning as pathways to resilience. For example, Adger et al (2009) conclude that "adaptation to climate change is limited by the values, perceptions, processes, and power structures within society" (p. 349). Likewise, crisis and disaster scholars observe that processes of learning and preparedness for extreme events are constrained by a variety of social and political factors (Boin et al. 2009; Nice and Grosse 2001). Given these insights, there is a need for more systematic research assessing the prospects for resilience-building through the lens of public policymaking (Boin 2009; Gerber 2007; Kapucu and Van Vart 2006).

As a public policy problem, resilience is a legitimate concern for policy process research. Resilience poses critical policy design challenges (Clark et al. 1979), which resonates with the need for theoretical tools to better understand processes of problem formulation, influence, and policy choice. Policymaking in this area generally includes a vast number of stakeholders and policy experts whose interactions are a classical focus in policy process studies. Nevertheless, we note that relatively little work has been done to incorporate insights from policy process research into the study of resilience. Some researchers (Anderies et al. 2004; Fiksel 2006; Lebel et al. 2006; Ludwig et al. 2001; Vogel et al. 2007) have taken steps in this direction but more work clearly remains to clarify the public policy dimensions of resilience particularly in the context of extreme events.

A policy process perspective on resilience-building puts the spotlight on specific questions about learning, change, and stability. Generally speaking, policy process research refers to "the systematic study of the interactions among people in the development of public policy over time" (deLeon and Weible 2010, p. 23). In order to simplify the study of these interactions one common approach involves taking the policy subsystem as the unit of analysis, which provides direction for describing and simplifying the policymaking environment and for explaining change and stability in policymaking in relation to extreme events (Weible and Nohrstedt 2012).

14.2.1 Subsystems, Policy-Oriented Learning and Policy Change

To advance the understanding of the relationship between crisis, learning, and policy change researchers should explore and compare multiple frameworks and theories. In this chapter, we rely on a policy subsystem approach to simplify and structure our analysis of learning and policymaking between storm Gudrun and storm Per. This approach is by no means exhaustive, but it provides basic concepts (subsystem, learning, and policy change) and explanations for variations in crisis-induced policy changes.

A policy subsystem is a theoretical construct to simplify the analysis of policy-making, including processes of learning and policy change. The starting-point is the observation that policymaking is complex and, therefore, actors need to special-ize in order to be influential. Specialization takes place within policy subsystems, which hence consist of participants that regularly seek to influence any given area. Policy subsystems are composed by a functional or substantive component (emer-gency management policy, energy policy, water policy, etc.) and one geographical component (from the local level to regional to national and international levels). Subsystems are vertically nested (where policymaking at one level is to some extent influenced by actions at the other levels) and horizontally interdependent (influ-enced by actions and developments in adjacent substantive areas) (Sabatier and Jenkins-Smith 1999).

The key definitional properties of a policy subsystem—the substantive area and the territorial area—provide direction to simplify the study of policy learning and change by empirical analysis. In this chapter, we study processes of learn-ing and change in two partially overlapping national subsystems: critical societal infrastructure (in this case electricity supply and telecommunications) and emer-gency management.

The notion of policy subsystems provides direction in simplifying the analysis of policy learning and change. We adopt a simple conceptual distinction between policy-oriented learning and policy change. Influenced by May (1992), we start by distinguishing between social learning, political learning, and policy learning. While social learning captures revised problem perceptions in society at large and political learning refers to revised strategies for advocating policy ideas, policy learning focuses explicitly on "the viability of policy interventions and implementa-tion designs" (May 1992, p. 335). In practice, actors' understanding of interventions or implementation is founded on underlying normative and instrumental policy be-liefs. This can include beliefs about the underlying causes of problems, the overall seriousness of problems, the priority that should be accorded to various policy in-struments, the involvement of different actors in any given policy subsystem, and the proper distribution of authority among the relevant actors. Thus, learning refers to relatively enduring alterations of these belief system components, which may change gradually over time in the face of new information or as the result of some external shock. Learning may be constrained to one or a few actors or within the broader policy subsystem as a whole (Sabatier and Jenkins-Smith 1993, p. 42).

Policy beliefs are also crucial elements when it comes to defining policy change. The theoretical literature addressing the ontology and epistemology of policy change is vast (Capano 2012). Our conceptualization of policy change traces back to the idea of the policy subsystem as the unit of analysis. The substantive component of any given policy subsystem in practice translates into single or multiple policy programs that seek to operationalize the goals and instruments designed to address societal problems. In other words, policy or governmental programs—sanctioned by authoritative policy decisions—can be seen as containing normative as well as instrumental belief system components. Policy change is understood as alterations in these program components (Sabatier and Jenkins-Smith 1999).

14.2.1.1   Explaining Crisis-Induced Policy Change and Implementation

Variations in policy change across subsystems provide opportunities to search for underlying explanations. Although we proceed exploratively, the existing literature offers some guidance for this part of the analysis as well. Overall, subsystem theorists recognize the importance of pre-existing structures and interactions in explaining the role and impact of external shocks in policymaking. The point is that crisis events do not occur in a political or administrative vacuum; rather, they interfere with ongoing processes and debates in any given policy area. It matters a great deal, for example, if a subsystem exposed to some shock is characterized by ongoing policy controversies between competing policy coalitions or if the system is predominantly consensual (Birkland 2006). In return, the variables and mechanisms that come into play after a crisis are likely to be determined by preexisting patterns of interaction in a subsystem (Nohrstedt and Weible 2010).

Following the notion of punctuated equilibria, policy impacts are also likely to be affected by the intensity and type of attention. Decades ago, Anthony Downs (1972) depicted policymaking as an issue-attention cycle, which modeled the policy process according to "a systematic cycle of heightening public interest and then increasing boredom with major issues" (p. 39). Accordingly, heightened public attention increases the likelihood for public action. Similarly, May (1991) distinguishes between policies with publics and policies without publics as a means of recognizing more enduring patterns of attention in some subsystems. Following the same logic, the general punctuation hypothesis recognizes the importance of "situations in which information flows into a policymaking system, and the system, acting from these signals from the environment, attends to the problem and acts to alleviate it, if necessary" (True et al. 2007, p. 176). However, the link between attention and policy action is not as straightforward. Policy change also depends on the longevity and type of attention as well as the ability of policy actors to overcome cognitive and institutional transaction costs in order to mobilize for change. Whether the understanding or 'image' of events (including its underlying causes, severity, and proper policy implications) is widely accepted or subject to disagreement among policy actors is important as well (Baumgartner and Jones 1991).

In some cases, people may agree on the nature and role of extreme events while in other cases framing contests may erupt regarding the 'true' nature, underlying causes and implications of events (Boin et al. 2009). Finally, crises may change the terms for access to and usage of policymaking venues. That is, dramatic events may move the locus of decision-making into new settings and also change the conditions for accessing these settings by including new participants while excluding others (Nohrstedt and Weible 2010; Pralle 2003; Sabatier and Weible 2007).

In summary, while the literature has found that crises are often necessary for policy change, they are not sufficient (Sabatier and Jenkins-Smith 1999). For example, in some cases crises serve as the necessary driving force to overcome institutional inertia, entrenched interests, perceptual blinders, and active skepticism towards change. However, the occurrence of a disruptive crisis is often insufficient to spark policy change due to the interaction between subsystem actors engaging in mobilization of political resources either in support or opposition of the status quo (Parker and Dekker 2008; Sabatier and Jenkins-Smith 1999).

A legitimate question from the perspective of resilience is: do policy changes actually lead to better performance and preferred outcomes over time? There is plenty of evidence to suggest that policy reforms—particularly those initiated in the wake of disruptive crises—are oftentimes symbolic political actions that are never fully implemented into new or altered structures, instruments, or practices ('t Hart 1993; Boin et al. 2005). Likewise, decades of implementation research bear witness to the many political, bureaucratic, organizational, cultural, and psychological pitfalls that may hinder transformation of policy lessons into new and better practices (Hjern 1982; Mazmanian and Sabatier 1983; O'Toole 2000; Parker and Stern 2002; Schofield 2001). Hence, the capacity of societies to build resilience to extreme events, in parts, depends on their ability to overcome these barriers to implementation (Boin and Otten 2006; Gunderson 2003). In this chapter, we seek to trace plausible effects of policy-oriented learning on crisis response actions.

Against this theoretical background, we structure the analysis by addressing three empirical questions: What policy beliefs changed as the result of the storm Gudrun and did those changes result in any revision of (critical infrastructure and emergency management) policy programs? Did policy changes have intended effects during the response to storm Per? What factors may explain processes of policy change and implementation in this case?

14.3 Empirical Analysis: Learning and Policy Change After Gudrun

14.3.1 Critical Infrastructure: Electricity Supply and Telecommunications

Storm Gudrun paved the way for several lessons regarding critical infrastructure and some of these lessons were eventually transformed into new policies. One of the early insights drawn from Gudrun was related to the vulnerability of the electricity grid. Although Gudrun caused extended outages in electricity and telecommunication systems, most attention was initially directed at electricity supply. Government representatives early on attributed responsibility to the electricity companies and stressed the need for stricter requirements regarding the replacement of overhead power lines with underground cables and creation of compensation following extended power outages. The problem with telecommunication disruptions was primarily framed in terms of a challenge to the crisis response system.

This was partly due to the fact that the storm demonstrated that an effective crisis response depended on well-functioning mobile tele-communications, which led to calls for 'roaming' capabilities (automatic extension of connectivity across tele-communication providers).

Several inquiries were carried out to review the Gudrun experience in order to draw lessons for the future. One of these inquiries, by the Swedish Civil Contingencies Agency (KBM) (2005), presented a number of recommendations that sought to improve the crisis response capacity. In its report, KBM identified several areas where changes were needed, including the crisis management capacity of Swedish tele-communication companies, identification of minimal levels of consumer access to electricity and telecommunications, the need for insular power and roaming capabilities in emergency situations.

In addition, the Government assigned the Swedish Energy Agency to review the terms for legislative change regarding electricity supply. The Swedish Energy Agency acknowledged that many of the needs Gudrun had revealed were, in fact, already on the policy agenda. Moreover, two prior reports had investigated the terms for increasing the security of the electricity supply and compensation for electricity outages. Whereas one of these inquiries resulted in a voluntary commitment on behalf of Swedish electricity companies to take action in order to increase security, the other inquiry presented its findings just 2 months prior to Gudrun. In retrospect, the Energy Agency concluded that if the actions suggested by the inquiry (particularly regarding improved public-private collaboration) had been implemented in a timely manner they would have reduced the impact of Gudrun on the electricity system (Narby 2009).

Regarding telecommunications, sources suggest that Gudrun provided an opportunity to reinvigorate and add urgency to ideas that were already on the policy agenda (Birkland 2006). For example, the Post and Telecom Authority (PTS) presented two reports (2005a, b) assessing the consequences of Gudrun for telecommunication systems, which both underscored the need for several changes that had been introduced prior to Gudrun. These changes, including possibilities of roaming and legislative change to clarify the obligations of cell-phone companies, were all part of the PTS strategy between 2003 and 2005.

In addition to these public inquiries, some private sector actors initiated evaluations as well. Some electricity companies focused their attention on technical aspects of the electricity grid. They concluded, for instance, that the new insulated cables (which had previously replaced uninsulated cables) were vulnerable as well and could in fact worsen the damage since they could tear down utility poles. Based on this insight, the electricity companies questioned the utility of underground cable methods and started to assess alternative technical solutions (Swedish Energy Agency 2007, p. 46, 53). The electricity companies also initiated stakeholder dialogues targeting municipalities, regional administrative boards and other authorities through which they offered a number of promises regarding crisis preparedness and various measures to increase the robustness of the electricity grid. Eventually, a few months after Gudrun and despite the initial skepticism towards underground cables, some electricity companies invested additional resources in the burial of underground cables. By the end of 2006, some 3,000 km out of a total of 17,000 km had been completed.

Some lessons from Gudrun regarding electricity supply and telecommunications were eventually formalized through legislative change. First, in October 2005, the Swedish Government (bill 2005/2006:27) presented a number of proposals regarding changes in electricity legislation some of which were directly linked to the Gudrun experience. This included specification of a 24-hour limit for acceptable power outages (a law which came in effect in 2011) and establishment of compensation levels for extended power outages, which the Government hoped would provide an incentive for the electricity companies to reduce the vulnerability of the electricity grid. In addition, following the need for insular power electricity, legislation was changed to enable temporary power production during power outages. Second, concerning tele-communication legislation, the Swedish Parliament (2005) decided that the demands for delivery of mobile tele-communications in times of crisis should match the demands set for fixed-line communication. In fact, this proposal was introduced in 2003 but at that time it was seen to conflict with EU-legislation. This view had changed in 2005 when policymakers agreed that the interpretation made in 2003 had been too narrow (Narby 2009).

14.3.2   Crisis Management

Shifting focus to the crisis response system, it can be noted that Gudrun coincided with an ongoing review scrutinizing the Swedish crisis management system following the devastating 2004 Boxing Day Tsunami just 2 weeks before the storm. As a result, when Gudrun struck, the issue of crisis management was already at the top of the public agenda because of the magnitude of the tsunami disaster and the massive criticism the government received for its slow response. Interestingly, one of the main critiques in the wake of the tsunami was the government's inability to learn from prior crises as a basis for improving crisis planning and preparation.

Similar to the electricity supply issue, a number of evaluations and inquiries were launched in order to analyze what lessons could be drawn from Gudrun for the Swedish crisis management system. The overall conclusion by the Swedish Civil Contingencies Agency (2005, p. 4, authors' translation) was that "for this type of event the basic principles for the crisis management system work well at the local and regional level." This narrative illustrates a general view at the time that the response to the storm was effective. Nevertheless, evaluations identified a number of lessons from Gudrun although they focused primarily on the need for change as a means to streamline the existing system. In summary, the evaluations of Gudrun identified two major systemic flaws: information and collaboration.

A key insight regarding information management following the storm was that the flow of information between responsible authorities and to the public did not work properly. For example, in some cases the electricity companies did not pass on reliable forecasts concerning power outages to local and regional authorities.

Communication failures were mainly attributed to the lack of proper structures to ensure shared situation awareness, which was another problem that had been on the policy agenda for a number of years before Gudrun. By the time of the storm in early 2005, several projects were already exploring various solutions to achieve coordination of crisis information with the purpose of establishing a template for shared situation awareness (Hansén 2009). Thus, in retrospect, Gudrun was a focusing event which provided additional momentum to calls for action that were introduced in the wake of the tsunami crisis to address practices for situation awareness in a more coherent fashion. In 2006, this process resulted in the establishment of a unit for preparedness and analysis within the Prime Minister's Office.

Additional efforts were undertaken to further investigate the merits of merging the two major civilian agencies in Sweden working on crisis and emergency management and to establish an agency to lead the public response in crisis situations. Thus, efforts to create better structures for shared situation awareness coincided with ongoing processes to reform the crisis management system. A year later, however, in early 2007, the newly elected Alliance government (center-right) altered these proposals. The new government went ahead with plans to merge the Emergency Management Agency and the Swedish Rescue Services Agency, however they scraped the idea for a new crisis agency and decided instead to establish a crisis management coordination secretariat inside the Prime Minister's Office (Narby 2009, p.25).

The lesson regarding the need for improved practices for shared situation awareness also led to some legislative change regarding the responsibilities of local and regional authorities. Following a bill in 2006 (2005/2006:133), changes were made to clarify the flow of information from the municipalities (local level) to the county administrative boards (regional level). Accordingly, the municipalities were prescribed to pass on situation reports to the county administrative boards, which were assigned to coordinate these reports.

Gudrun also exposed gaps in the capacity to achieve collaborative crisis responses. While the information management issue was primarily framed as a problem that could be curbed by structural reforms at the national level, collaboration problems were essentially attributed to the local and regional levels. Once again, the need for improvement in this area had been recognized prior to Gudrun. Actors at local, regional, and national levels were already working on a number of measures to improve the conditions for collaborative crisis management when the storm struck and Gudrun provided another crucial event that fed into this process. In essence, the storm provided a test to evaluate these recently implemented practices for improved collaboration and resulted in a number of lessons—primarily concerning coordination between local and regional authorities and the private sector (Swedish Civil Contingencies Agency 2005). Some of these lessons led to immediate organizational changes, mainly concerning the composition of local and regional crisis management networks. Based on the insight that communication with private sector actors (representatives of electricity and tele-communication companies) was too slow, some counties decided to include these actors in their crisis management networks. These solutions were partly the result of lesson-drawing and the sharing of best practices across counties. Even if Gudrun did not lead to any legislative change

Table 14.1 Comparison between Gudrun and Per. (Source: Swedish Energy Agency 2007, p. 11)

	Gudrun (2005)	Per (2007)
Maximum wind-force over mainland (m/s)	33	29
Quantity of destroyed forest (millions of cubic meters)	70	16
Number of customers with power cuts	730.000	440.000
Longest power cuts (days)	Appr 45	Appr 10
Maximum time to restore regional grids (days)	7	1
Reallocated capital as a result of power cuts (billion SEK)	4–5	1,8–3,4
Restoration costs for network operators (million SEK)	2,400	650
Total compensation from insurance companies to affected customers (million SEK)	4,000	550

related to collaborative management, the government pointed to the need to speed up the process of establishing regional crisis management councils to facilitate collaboration and coordination between local and regional actors. Several counties had established such councils before Gudrun and hence the storm highlighted the need to accelerate this process further (Narby 2009).

14.4   Policy Effects During Per

In this section, we turn attention to our second research question: did policy changes introduced after Gudrun have the intended effects during the response to the storm Per? To reiterate, the Gudrun storm was followed by a number of policy changes related to electricity supply, tele-communications, and the Swedish crisis management system. Specifically, these changes aimed at improvements in the following areas: compensation requirements in cases of extended power outages, insular power production, increased replacement of overhead power lines with underground cables, measures to safeguard regional networks, shared situation awareness, and collaborative management.

When considering if these changes had any meaningful effects in relation to the response to storm Per, one must keep in mind several caveats. Primarily, there is a risk that positive effects associated with post-Gudrun reforms are exaggerated due to differences in event magnitude. Nonetheless, as the data presented in Table 14.1 demonstrate, Gudrun had more serious consequences than Per.

To start with, it appears that the crisis response by the electricity production companies was more effective in relation to Per compared to Gudrun. In the wake of Gudrun, the electricity companies were criticized for their relatively slow response, but following Per, most stakeholders agreed that they had taken their responsibilities more seriously and performed better. The Swedish Energy Agency (2007, p. 15) found that restoration work after Per was significantly faster than after Gudrun, with power being restored for most customers within 1–2 days. There were also fewer power losses caused by failures in regional electricity networks. While this was partially due to the fact that Storm Per was weaker than Gudrun, the Energy Agency also credited the improved performance to measures taken to better protect regional networks. These measures included a tree securing program, which by 1 January 2007 had safeguarded some 80% of the networks from trees, that was carried out by the regional network operators, the increased reliance on underground power cables, and the investments made in the network, which increased dramatically after Gudrun (Swedish Energy Agency 2007, p. 15, pp. 24–25).

One of the lessons that the government drew from Gudrun was that private actors such as the electricity companies needed to be held responsible for securing the power grid to be able to better withstand eventualities such as storms of Gudrun's magnitude. To that end, legislation was passed that increased compensation for periods of extended power outage, required the electrical grid to be secured within 2 years, and set a 2011 deadline for an all-encompassing underground grid structure. As always implementation can be difficult and a number of power companies attempted to resist these reforms and suggested an alternative model that they felt would be more feasible to implement (Dagens Nyheter 2006). The government did not back down and the Energy Agency, in their report on Per, noted that after Gudrun the electrical companies had intensified their work to safeguard regional networks from falling trees and by 1 January 2007, 80% of the network had been safeguarded from trees. Investments in the network doubled or tripled after Gudrun and by 2007 one of the large power companies had increased the percentage of underground cables to 30% and estimated the number would be 60% by 2010 (Swedish Energy Agency 2007, p. 25, pp. 48–49). Smaller companies lagged in implementation and would require much more long term work. Thus, while operators plan to maintain large parts of the overhead line network, after Gudrun and the new legislation, they had increasingly concentrated on burying power lines in forest areas, which made a difference in reducing power cuts when Per struck.

A number of improvements were made between Gudrun and Per regarding structures and procedures for crisis management. To begin with, the system of shared situation awareness appears to have functioned more effectively during the response to Per. New routines and shared understandings about the content of situation reports among county and municipality representatives helped improving the process of establishing regional situation reports after Per. Likewise, national and cross-sector situation reports produced by the Swedish Contingencies Agency paved the way for an improved response to Per. Nonetheless, some unresolved problems with the situation reports still remained, particularly regarding the balance between description and analysis. Some of these efforts resulted from the Gudrun experience but it can also be noted that some routines—for example the use of telephone-conferences—were unrelated to Gudrun and were rather based on experiences from other disruptions in 2006.

Additional improvements had also been made with respect to organizational collaboration. Again, part of the explanation can be attributed to the fact that Per was a less severe storm compared to Gudrun, but the evidence suggests that improvements may also be explained by lessons drawn after Gudrun. Many actors bear witness to the fact that the crisis response to Per was fairly easy to coordinate; there was relatively little uncertainty about the situation and the crisis response proceeded as planned. Even if some problems remained during Per, collaboration across public and private sectors had improved since Gudrun. One explanation was the work initiated by some electricity companies after Gudrun to develop their organizations to facilitate collaboration with other organizations. Another explanation for these improvements is experiential learning (March and Olsen 1975). Many people involved with the response to Per had also worked together during Gudrun, which facilitated swift problem diagnosis and collaboration within pre-established response networks. On this basis, Gudrun provided an analogy and a practical experience, which helped reduce uncertainty in the wake of Per. In addition, several actors came closer together during exercises and education efforts organized after Gudrun, which created stronger inter-personal ties among actors in between the two events. By contrast, it should be noted that some public organizations still had problems collaborating in other areas with private actors. For example, some County Administrative Boards experienced difficulties collaborating with TeliaSonera to restore regular telecommunications in the wake of Per (Swedish Energy Agency 2007, p. 211).

Another notable improvement in relation to the response to Per was the better communication of weather warnings. During Gudrun, many actors had difficulties interpreting the warnings issued by the Swedish Meteorological and Hydrological Institute (SMHI). After Gudrun, SMHI revised its two-level weather threat scale by adding a third level (indicating extreme weather conditions). Per was eventually classified as a level-three contingency and most actors appear to have taken the warnings more seriously compared to Gudrun.

To conclude, we have documented a number of changes that appear to have contributed to improved performance in the response to Per when compared to Gudrun. However, we must be cautious in our assessments and the question remains to what extent these improvements can be attributed to the structural, legislative, and policy changes we observed rather than to the fact that Gudrun rather provided a recent vivid analogy which helped increase organizational vigilance and heightened the readiness to respond. The response to Per is perhaps more likely attributed to a combination of organizational reform and experiential learning.

14.5   Discussion

In this chapter, we conducted a comparative policy subsystem analysis to contrast policy developments across critical infrastructure and crisis management subsystems in the wake of crisis. The objective is to search for underlying explanations for policy learning and change, which ultimately seeks to strengthen resilience to natural disasters. This case-study demonstrates the utility of a policy subsystem approach and points to several noteworthy patterns that may be of interest for policy change analysis more generally.

By the time the storm Gudrun struck Sweden in January 2005, crisis management already had been prominently placed high on the public agenda following the 2004 Boxing Day tsunami disaster just 2 weeks before. Thus, at the time, the crisis management capacity of Swedish authorities and the Government was being closely scrutinized (Brändström et al. 2008). Following the impact of Gudrun on Swedish society—particularly the extended power outages—the storm added another negative experience to an already existing critical public narrative questioning the Swedish system's ability to respond to extreme events. Most of the attention after Gudrun was directed at the electricity supply issue, whereas organizational aspects of the crisis response received less public attention. This offers a candidate explanation for the pace of the policy response addressing legislation related to electricity and tele-communications. The Government was being criticized for forcing legislative changes related to electricity and tele-communications shortly after Gudrun at the expense of taking the time to carry out a careful review and proper preparation. By contrast, changes in the crisis management system proceeded slowly and without any major public interest. In summary, the amount and type of attention is a key factor explaining the pace of policy action in the wake of crisis.

Many of the changes and reforms related to mitigation and preparedness initiated after Gudrun had in fact been on the agenda before the storm happened. This confirms the expectation that crises rarely elevate novel policy solutions; instead they are more likely to renew or accelerate on-going reform initiatives and provide momentum for pre-existing solutions (Birkland 2006). Developments in Sweden also suggest that implementation of revised practices and structures is slow in the absence of disruptive crises.

This analysis points to a number of issues where policymakers took action to correct systemic flaws and these efforts appear to have resulted in improved performance and results when Per hit. Two examples include efforts to improve shared situation awareness and the creation of incentives that spurred network operators to begin burying power lines and to reduce lines on poles in forest dense areas thereby lowering the risk of power cuts as a result of storms and snow (Swedish Energy Agency 2007, p. 49). In both these cases, policymakers initiated several changes and measures which resulted in a more robust infrastructure, a swifter and more competent response by involved parties, and an enhanced ability to respond to and bounce back from the problems that did occur.

Nonetheless, because there had been limited time to implement all measures, difficulties in getting all the involved actors to buy into and carry out the preferred reforms, and due to unclear roles and expectations, Per revealed that Sweden was still quite vulnerable to a major storm, even one not as powerful as Gudrun. For example, 440,000 customers were affected by power cuts and the event demonstrated that improvements in a number of areas were still needed. Cooperation with external parties still needs further development, more training is needed for support resources in crisis organizations, such as switchboard operators and for information provision, better training and routines are needed so network operators, telecom operators, and municipalities can better cooperate, and there is still a need to ensure networks are in place and maintained with network operators, municipalities, police and rescues services prior to large hazard events (Swedish Energy Agency 2007, pp. 37–38).

This case also tells us that resilience to natural hazards depends on a combination of policy intervention, organizational change, and experiential learning. Although it is hard to objectively evaluate the effects of policy changes carried out in the wake of Gudrun, the evidence reviewed here confirms that this event was followed by a number of legislative actions and structural changes that had already begun to increase the robustness of the electricity grid and the effectiveness of the crisis management system. In addition to these policy changes, it also seems that organizational change and experiential learning were important sources for improvement and something that had a direct impact on the crisis response in relation to the storm Per 2 years later. This was one of the reasons that the Energy Agency (2007) concluded that for the most part the crisis management for Per was more effective than it had been in relation to Gudrun. Since many of the organizations and individuals that responded to Per had been involved with Gudrun, they knew what to do and performed better. The risk, however, is that much of this knowledge that was gained from experiential learning remains in the minds of individuals rather than in their organizations and will be lost as attention fades, people forget, or when people leave. One practical lesson is to establish organizational practices or mechanisms to ensure that the lessons learned by individuals are being properly disseminated within and across organizations. Spreading lessons across organizations is particularly pressing as a basis for improving the capacity to collaborate in the pursuit of shared objectives (Comfort et al.2011).

Understanding the link between disaster, resilience, and development requires consideration of formal processes of legislative change and organizational redesign as well as informal processes of experiential learning. This chapter has illustrated that policy reforms based on prior experience can improve performance over time by creating new incentives for action, by adding resources for mitigation, and by clarifying roles and responsibilities. But when it comes to heightened vigilance and readiness to take swift action in the face of acute events, factors such as pre-established personal relationships and historical analogies are equally important to reduce uncertainty and increase crisis preparedness.

Finally, our comparison across subsystems suggests that policy action to improve crisis preparedness is mediated by problem tractability and bureau-political dynamics. Looking at the aftermath of Gudrun, the protection of critical infrastructure (electricity supply and telecommunication) and the organizational issues regarding the crisis management system raised complex but quite different challenges for policymaking. While electricity supply and telecommunication mainly involved technical issues and matters of political regulation, the re-organization of the crisis management system raised questions about the distribution of power and political responsibility in times of crisis. The extent to which these areas were exposed to political pressures appears to have varied as well; while electricity supply and telecommunication policies were heavily criticized in public, there are no signs that these areas were subject to any major political-bureaucratic controversy. By contrast, crisis management policies received very limited public attention but the political stakes were higher particularly following the Tsunami debacle. These factors—along with the change in Government after the 2006 election—may account for the level and pace of policy change after Gudrun. Policy changes related to electricity and telecommunication were relatively uncontroversial and could therefore be implemented quite swiftly, whereas changes related to crisis management, the fusion of two civilian crisis management agencies and the creation of a crisis unit within the Prime Minister's office, were more drawn out partly due to the politicized nature of crisis management in the wake of the tsunami-disaster.

14.6   Conclusion

The human capacity to learn from experience is one critical aspect of resilience (Comfort et al. 2011). Systems' ability to withstand and bounce back from extreme events requires capacities of adaptation, which involves continuous monitoring and evaluation of feedback from multiple sources as a basis for systemic improvement. Meanwhile, as recognized in the introduction to this volume, further development of resilience research in the context of extreme events hinges on answering more specific questions about the nature of resilience and factors that explain variations in adaptation over time and across systems. These questions require theoretical approaches that simplify the complex systems and processes that govern learning, adaptation, and change. The theoretical contribution of this chapter is to demonstrate how policy process theory may inform this effort.

Given that policy interventions are needed to address resilience, one challenge is to better understand how policy-oriented learning and policy change may be transformed into more effective practices for crisis management. Here loom a range of psychological, organizational, and political hurdles that need to be overcome to improve preparedness and response to extreme events (Parker et al. 2009). Policy process theory is a suitable basis to more fully understand policy learning and change as the result of interactions between events, institutions, and actors. Specifically, policy process analysis sheds light on developments and factors that can help explain why certain problems receive attention by policymakers, why they take action to address those problems, and under what conditions actions can be successfully implemented. In this chapter, we have studied the policy process in two overlapping subsystems over time between two major crisis events. By comparing the response to two storms in Sweden, we have analyzed the process from learning, to policy change and implementation, and finally to a subsequent critical test of revised policies and practices. We can draw a number of conclusions that add to our understanding of why efforts to strengthen community resilience succeed and why they fail.

While disruptive crises oftentimes impede careful analysis and identification of lessons for the future (Boin et al. 2009; Stern 1997), our study illustrates the ability of policy actors to overcome such barriers to the improvement of systems and practices. It is clearly not easy to go from lessons observed to lessons internalized and implemented. A key challenge is how to capture, institutionalize, and implement experiential learning before experience fades. One practical lesson in this regard is to avoid the fallacy of "fix-it-and-forget-it" approaches to crisis learning and change. This is a balancing act between analyzing, storing and institutionalizing information about successes and failures on the one hand, and the ability to be vigilant and flexible in order to better respond to unknown future events on the other (Comfort et al. 2011).

A policy process perspective helps us understand how systemic improvements can be achieved by changes in statues, formal rules, and organizational structures. Yet, while policy interventions may fix some problems, resilience also depends on the 'soft' aspects of crisis governance. Above, we have cited evidence that the response to the second storm (Per) was facilitated by pre-established personal relationships and fresh memories from storm Gudrun. These relationships and experiences reduced uncertainty and provided a template for action when Per hit 2 years later. These observations underscore the importance of using processes of communication and organizational infrastructure to make sure that experience is properly stored and shared among managers and decision makers. Part of this challenge is to exploit organizational structures and practices to ensure the transfer of experience within and across organizations while maintaining awareness that the next crisis may pose different challenges and require different responses.

This study has demonstrated that many problems can be, in fact, swiftly and successfully addressed by policymakers and organizations in the wake of crisis, which increases society's ability to withstand extreme events. But we have also documented examples of solutions that have been on the policymaking agenda for some time without resulting in any concrete action or systemic improvement. This insight may not come as a surprise given the incrementalist model of policymaking and notions of organizational and bureaucratic inertia, which predict that development of policy solutions often proceeds slowly and without any major shifts in the status quo (Lindblom 1979). These examples illustrate the difficulties associated with a proactive approach to resilience and the importance of crises as potential triggers for reform.

Finally, this chapter demonstrates how new insights regarding the dynamics of resilience can be gained by alternative methodological approaches to crisis analysis. Specifically, by a comparative process analysis of two subsystems over time across two similar hazard events we have documented how lessons are learned, how these lessons may feed changes in policy, and how revised practices are put to the test. Yet, from a methodological perspective this approach is a double-edged sword. While this approach has clear merits as it enables assessment of the impact of policy learning and change over time, it also brings analytical challenges—particularly regarding the ability to assess the effectiveness of policy interventions vis-à-vis differences in magnitude between events. However, this difficulty reflects the reality of resilience and crisis preparedness in which policymakers continuously have to prepare for unanticipated, unstructured and unknown events.

References

Adger, W. N., Dessai, S., Goulden, M., Hulme, M., Lorenzoni, I., Nelson, D. R., Naess, L. O., & Wolf, J., & Wreford, A. (2009). Are there social limits to adaptation to climate change? *Climatic Change, 93,* 335–354.

Anderies, J., Ostrom, E., & Janssen, M. (2004). A Framework to analyze the robustness of social-ecological systems from an institutional perspective. *Ecology and Society, 9*(1), 18.

Baumgartner, F., & Jones, B. (1991). Agenda dynamics and policy subsystems. *The Journal of Politics, 53*(4), 1044–1074.

Birkland, T. (2006). *Lessons of disaster: Policy change after catastrophic events.* Washington, D.C: Georgetown University Press.

Boin, A. (2009). The new world of crises and crisis management: Implications for policymaking and research. *Review of Policy Research, 26*(4), 367–377.

Boin, A., & Otten, M. (2006). Beyond the crisis window for reform: Some ramifications for implementation. *Journal of Contingencies and Crisis Management, 4*(3), 149–161.

Boin, A., 't Hart, P., Stern, E., & Sundelius, B. (2005). *The politics of crisis management: Public leadership under pressure.* Cambridge: Cambridge University Press.

Boin, A., 't Hart, P., & McConnell, A. (2009). Crisis exploitation: Political and policy impacts of framing contests. *Journal of European Public Policy, 16*(1), 81–106.

Brändström, A., Kuipers, S., & Daléus, P. (2008). The politics of tsunami responses: Comparing patterns of blame management in Scandinavia. In A. Boin, A. McConnell, & P. 't Hart (Eds.), *Governing After Crisis* (pp. 114-147). Cambridge: Cambridge University Press.

Capano, G. (2012). Policy dynamics and change: The never-ending puzzle. In E. Araral, S. Fritzen, M. Howlett, M. Ramesh, & X. Wu (Eds.), *Routledge handbook of public policy* (pp 451-461). London: Routledge.

Clark, W., Jones, D., & Holling, C. (1979). Lessons for ecological policy design: A case-study of ecosystem management. *Ecological Modelling, 7*(1), 1–53.

Comfort, L., Boin, A., & Demchak, C. (2011). *Designing resilience: Preparing for extreme events.* Pittsburgh: University of Pittsburgh Press.

Dagens, N. (2006). Eon går emot regeringen om elpriser, published 2 January, 2006, http://www.dn.se/ekonomi/eon-gar-emot-regeringen-om-elpriser/. Accessed 20 June 2013.

deLeon, P., & Weible, C. (2010). Policy process research for democracy: A commentary on Lasswell's vision. *International Journal of Policy Studies, 1*(2), 23–34.

Downs, A. (1972). Up and down with ecology: The issue attention cycle. *Public Interest, 28,* 38–50.

Fiksel, J. (2006). Sustainability and resilience: Towards a systems approach. *Sustainability: Science, Practice, and Policy, 2*(2), 14–21.

Gerber, B. (2007). Disaster management in the United States: Examining key political and policy changes. *Policy Studies Journal, 35*(2), 227–238.

Gunderson, L. (2003). Adaptive dancing: Interactions between social resilience and ecological crises. In F. Berkes, J. Colding, & C. Folke (Eds.), *Navigating socio-ecological systems: Building resilience for complexity and change* (pp. 33–52). Oxford: Oxford University Press.

Handmer, J., & Dovers, S. (2007). *Handbook of disaster and emergency policies and institutions.* London: Earthscan.

Hansén, D. (2009). Effects of buzzwords on experiential learning: The Swedish case of 'Shared Situation Awareness'. *Journal of Contingencies and Crisis Management, 17*(3), 169–178.

Hjern, B. (1982). Implementation research: The link gone missing. *Journal of Public Policy, 2*(3), 301–308.

Kapucu, N., & Van Wart, M. (2006). The evolving role of the public sector in managing catastrophic disasters: Lessons learned. *Administration and Society, 38*(3), 279–308.

Lebel, L., Anderies, J., Campbell, B., Folke, C., Hatfield-Dodds, S., Hughes, T., & Wilson, J. (2006). Governance and the capacity to manage resilience in regional social-ecological systems. *Ecology and Society, 11*(1), 19.

Lindblom, C. (1979). Still muddling, not yet through. *Public Administration Review, 39*(6), 517–526.

Ludwig, D., Mangel, M., & Haddad, B. (2001). Ecology, conservation and public policy. *Annual Review of Ecology and Systematics, 32,* 481–517.

May, P. (1991). Reconsidering policy design: Policies and publics. *Journal of Public Policy, 11*(2), 187–206.

May, P. (1992). Policy learning and failure. *Journal of Public Policy, 12*(4), 331–354.

March, J., & Olsen, J. (1975). The uncertainty of the past: Organizational learning under ambiguity. *European Journal of Political Research, 3*(2), 147–171.

Mazmanian, D., & Sabatier, P. 1983. *Implementation and public policy.* Glenville: Scott Foresman.

Narby, P. (2009). *From experience to implementation: Learning within the Swedish crisis management system between storm Gudrun and Per.* Stockholm: CRISMART (unpublished manuscript, in Swedish)

Nice, D., & Grosse, A. (2001). Crisis policymaking: Some implications for program management. In A. Farazmand (Ed.), *Handbook of crisis and emergency management* (pp. 55–68). New York: Marcel Dekker

Nohrstedt, D., & Weible, C. (2010). The logic of policy change after crisis: Proximity and subsystem interaction. *Risks, Hazards, and Crisis in Public Policy, 1*(2), 1–32.

O'Toole, L. (2000). Research on policy implementation: Assessment and prospects. *Journal of Public Administration Research and Theory, 10*(2), 263–288.

Parker, C. F., & Stern, E. (2002). Blindsided? September 11 and the origins of strategic surprise. *Political Psychology, 23*(3), 601–630.

Parker, C., & Dekker, S. (2008). September 11 and post-crisis investigation: Exploring the role and impact of the 9/11 commission. In A. Boin, A. McConnell, & P. 't Hart (Eds.), *Governing after crisis: The politics of investigation, accountability and learning* (pp. 285–282). Cambridge: Cambridge University Press.

Parker, C., Stern, E., Paglia, E., & Brown, C. (2009). Preventable catastrophe? The hurricane Katrina disaster revisited. *Journal of Contingencies and Crisis Management, 17*(4), 206–220.

Post and Telecom Authority. (2005a). Electronic communication and the storm January 8-9 2005: How to establish robust communications? (in Swedish). http://www.pts.se/upload/Documents/SE/Rapport_stormen_8_9_jan05_2005_9%20doc.pdf.

Post and Telecom Authority. (2005b). *PTS report to KBM after the storm January 8–9 2005.* (in Swedish)

Pralle, S. (2003). Venue shopping, political strategy and policy change: The internationalization of Canadian forest advocacy. *Journal of Public Policy, 23*(3), 233–360.

Sabatier, P., & Jenkins-Smith, H. (1993). *Policy change and learning: An advocacy coalition approach.* Boulder: Westview Press.

Sabatier, P., & Jenkins-Smith, H. (1999). The advocacy coalition framework: An assessment. In P. Sabatier (Ed.), *Theories of the policy process* (pp. 117–168). Boulder: Westview Press.

Sabatier, P., & Weible, C. (2007). The advocacy coalition: Innovations and clarifications. In P. Sabatier (Ed.), *Theories of the policy process* (2nd ed., pp. 189–220). Boulder: Westview Press.

Schofield, J. (2001). Time for a revival? Public policy implementation: A review of the literature and an agenda for the future. *International Journal of Management Reviews, 3*(3), 245–263.

Stern, E. (1997). Crisis and learning: A conceptual balance sheet. *Journal of Contingencies and Crisis Management, 5,* 69–86.

Swedish Civil Contingencies Agency. (2005). Crisis management in the wake of the storm (in Swedish). http://ndb.msb.se/Document/Report/633275274240143750.pdf.

Swedish Energy Agency. (2007). *Storm Per: Lessons for a more secure energy supply after the second severe storm in the 21st century.* Stockholm: The Swedish Energy Agency. http://webbshop.cm.se/System/TemplateView.aspx?p=Energimyndigheten&view=default&cat=/Broschyrer&id=87944afba6814e48808ea05f5c641e06'.

Swedish Government. (bill 2005/2006:27). Secure electricity grids. http://www.regeringen.se/content/1/c6/05/13/45/a13a3cfb.pdf.

Swedish Government. (bill 2005/2006:133). Collaboration in crisis: Towards a safer community. http://www.regeringen.se/content/1/c6/06/05/71/fb61f905.pdf.

Swedish Parliament. (2005). Measures in communications following the 2005 storm in southern Sweden. http://www.riksdagen.se/sv/Dokument-Lagar/Kammaren/Protokoll/_GS09104/.

't Hart, P. (1993). Symbols, rituals and power: The lost dimension of crisis management. *Journal of Contingencies and Crisis Management, 1*(1), 36–50.

Thompkins, E., & Adger, N. (2004). Does adaptive management of natural resources enhance resilience to climate change? *Ecology and Society, 9*(2), 1–14.

True, J., Jones, B., & Baumgartner, F. (2007). *Punctuated equilibrium theory: Explaining change and stability in public policymaking.* In P. Sabatier (Ed.), *Theories of the policy process* (2nd ed., pp. 155–188). Boulder: Westview Press.

Vogel, C., Moser, S., Kasperson, R., & Dabelko, D. (2007). Linking vulnerability, adaptation and resilience science to practice: Pathways, players and partnerships. *Global Environmental Change, 17*(3–4), 349–364.

Weible, C., & Nohrstedt, D. (2012). The advocacy coalition framework: Coalitions, learning and policy change. In E. Araral, S. Fritzen, M. Howlett, M. Ramesh, & X. Wu (Eds.), *Routledge handbook of public policy* (pp. 125–137). London: Routledge.

Wise, C. (2006). Organizing for Homeland security after Katrina: Is adaptive management what's missing? *Public Administration Review, 66*(2), 302–318.

Chapter 15
Recovery and Development: Perspectives from New Zealand and Australia

Douglas Paton, David Johnston, Ljubica Mamula-Seadon and
Christine M. Kenney

15.1 Introduction

Despite the loss and disruption disasters create for affected communities, it has become increasingly apparent that people and communities can surmount the challenges they encounter by drawing upon and using their own resources in ways that positively influence their recovery. This highlights the potential for people to be resilient. The need to develop this resilient capacity was identified by the Hyogo Framework for Action 2005–2015 (ISDR 2005) as the key element of a disaster risk reduction (DRR) strategy. To pursue this goal, it is necessary to define resilience.

Resilience is variously defined in terms of "bouncing back" (e.g., Klein et al. 2003) or adapting (e.g., Norris et al. 2008; Paton and Johnston 2006; Pelling and High 2005). Definitions of resilience have also encompassed the ability of a system to overcome hazard impacts (Adger et al. 2002; Bruijn 2004; Gaillard 2007) and sustain people's livelihood in the face significant environmental disruption (Adger et al. 2002; Chambers and Convey 1991). Other authors have introduced a need for resilient systems to be self-organizing and capable of learning, innovation and creativity (Carpenter et al. 2001; Berkes et al. 2003, Gaillard 2007; Marschke and Berkes 2006). Marshall and Marshall (2007) expanded on the latter to highlight a need include people's beliefs in their capabilities and defined resilience as comprising perception of risk associated with change, perception of ability to learn, plan

D. Paton (✉)
School of Psychology, University of Tasmania, Tasmania, Australia
e-mail: Douglas.Paton@utas.edu.au

D. Paton · D. Johnston · C. M. Kenney
Joint Centre for Disaster Research, Massey University, Palmerston North, New Zealand

D. Johnston
GNS Science, Lower Hutt, New Zealand
e-mail: david.johnston@gns.cri.nz

L. Mamula-Seadon
University of Auckland, Auckland, New Zealand

N. Kapucu, K. T. Liou (eds.), *Disaster and Development,* Environmental Hazards,
DOI 10.1007/978-3-319-04468-2_15, © Springer International Publishing Switzerland 2014

and self-organize, perception of the ability to cope, and level of interest in change. These capabilities are realized through the mobilization of the community capacities required to cope with and recover from shocks and maintain and enhance capacity over time (Chambers and Convey 1991). The notion of mobilizing capacities has been reflected in defining resilience as a process.

Paton and Johnston (2006) defined resilience as a process described by the interdependent capability of people, communities and societies to use their resources and abilities to anticipate, cope with, adapt to, recover from, and learn from the demands, challenges, and changes encountered before, during, and after hazard events. Similarly, Norris et al. (2008) identify adaptive capacity as a process enacted through linking social capital, economic development, information and communication, and community competence.

Taken together these conceptualizations point to a need to identify the capacities that exist at personal, community, and societal levels and how they are integrated to facilitate individual and collective ability to cope, adapt, recover and learn from environmental hazard consequences. It is also important to operationalize resilience to develop an empirical understanding. Norris et al. (2008) and Paton and Johnston (2006) identified not only potential sources of adaptive capacity (e.g., social capital) but also discussed mechanisms (e.g., community competence, empowerment, trust) that serve to integrate different levels of analysis (e.g., link community and civic agency levels) to develop societal adaptive capacity. They also discussed the importance of identifying what people have to adapt to.

To empirically examine resilience and provide the evidence base required to support resilience assessment and intervention planning, it is necessary to assess resilience when people are experiencing significant levels of disruption from hazard events (Bruijn 2004; Carpenter et al. 2001; Gaillard 2007; Klein et al. 2003; Paton and Johnston 2006). Furthermore, it is important to appreciate that the demands and challenges people will encounter and need to adapt to change over the course of what can be a long recovery and rebuilding period (Gaillard 2007; Norris et al. 2008; Paton and Johnston 2006).

The objective of this chapter is to contribute to understanding natural hazard resilience by examining people's experience of recovery from the consequences of significant natural hazard events. The content is based on research on the 2009 Victoria, Australia bushfires (McAllan et al. 2011; McLennan and Elliot 2012; Whittaker et al. 2009) and work by the authors on the 2011 Christchurch, New Zealand earthquake. In Christchurch, focus group interviews were conducted in each of five affected suburbs in July 2011. These data were complemented with that from twenty individual in-depth interviews (Mamula-Seadon et al. 2012; Paton 2012). The timing of data collection allowed assessment of recovery experience from both the February earthquake and a significant aftershock in June 2011.

These analyses were also used to test the validity of a community engagement-based theory of resilience developed empirically in New Zealand (Paton 2008). This theory, based on the concept of empowerment (Dalton et al. 2001), developed a multi-level approach to operationalizing resilience expressed in terms of the complementary relationships that exist between empowered people and communities (assessed by, for example, self-efficacy, community participation, collective efficacy, and place attachment) and social/agency settings that empower people (assessed using empowerment and trust). While empirical support for this theory was forthcoming (Paton 2008), examining people's recovery experience can be used to validate the theory and identify how it can be developed. If people's accounts of their experience corroborate theoretical conceptualizations of resilience, this would increase the utility of theory as a guide for resilience assessment and planning for future events (Norris et al. 2008; Paton 2008; Pelling and High 2005).

This chapter examines three issues. The first discusses people's accounts of what they had to contend with and how they responded to the disruption and challenges encountered. This discussion identified differences between community groups with regard to how well they responded to these challenges. Consequently, the second issue explored concerns identifying how individual and community variability in adaptive capacities (Norris et al. 2008; Paton and Johnston 2006) contribute to explaining differences in the reported effectiveness of community recovery activities. The third issue addressed concerns the relationship between recovery and development and discusses how lessons learned from these events are facilitating development in communities that can expect to face hazard events in the future. Discussion of these issues commences with an overview of events that have changed Australian and New Zealand approaches to DRR and post-disaster development.

15.1.1 The Events

On 7 February 2009 Victoria experienced Australia's worst day of bushfires in recorded history. Extreme weather conditions—high temperatures (over 44°C), low relative humidity (less than 10%), and strong winds (over 100 km/hr)—resulted in fires that caused 173 fatalities, destroyed 2,000 homes and created economic, social, and environmental costs in excess of US $4 billion (Victorian Bushfires Royal Commission 2010).

On 22 February 2011, an M_W 6.3 earthquake struck Christchurch, causing 185 deaths, over 7,000 injuries and in excess of US $12 billion in damage (Bannister and Gledhill 2012). Some 100,000 homes (approximately half the housing stock) were damaged, and about 7,000 homes were rendered uninhabitable. Some 3,000 of the 5,000 businesses in the Christchurch CBD were displaced and 1,200 CBD building had to be demolished. More than half of the road network had to be replaced. The Canterbury earthquake sequence was characterised by a high level of seismicity over an extended period within close proximity to Christchurch throughout 2011 and into 2013.

The scale, complexity, national significance and duration of these events provide opportunities to build understanding of how the actions of people, communities and governments can, individually and collectively, facilitate or hinder recovery and development in the aftermath of a major disaster. The first issue addressed concerns what people had to contend with and how these demands changed over the course of recovery.

15.2   Changes in Recovery Demands over Time

The Christchurch earthquakes resulted in respondents recognizing inadequacies in their structural readiness. Structural readiness refers to adjustments that increase the survivability of and level of physical protection offered by the home and property during the period of experiencing hazard effects. For earthquakes it involves, for example, securing the house to its foundations and strapping chimneys to prevent their collapse. For wildfires, it entails, for example, creating a defensible space around the home to protect against ember attack. Such actions enhance the capacity of the home to protect its inhabitants and increase the likelihood of people having a habitable dwelling during the recovery period. Being able to remain in a habitable home increases people's availability within a neighborhood to offer mutual assistance and social support to others and to be available to contribute to economic recovery (as customers and employees). The lack of structural preparedness was identified as increasing the adaptive demands people faced (e.g., having to move out of the home).

The second area of concern identified by people was their having inadequate essential supplies (e.g., stored food and water for all those in the home for 3-5 days) and a lack of household emergency planning. Respondents reported how their lack of such readiness and planning made adjusting to the loss of water, power and sewerage services more challenging than need be. Not only did this reduce their capacity for self-reliance, it meant diverting attention to accessing essentials like water and temporary toilet facilities rather than attending to family needs and planning how best to respond to their circumstances (e.g., dealing with livelihood issues). Respondents also identified their lack of psychological preparedness as an issue affecting their resilience, particularly with regard to coping with the impact of repeat aftershocks and adapting to changes in living conditions, loss of social relationships and livelihood disruptions throughout the several months over which recovery took place.

As the situation stabilized, people moved into the recovery phase of their disaster experience. During the first few weeks (although aftershocks required people to run through response-recovery cycles several times), people discussed how being able to work with neighbors and other community members to develop (self-help) groups to confront local demands (e.g., removing rubble, proving mutual support, setting up community meeting places, taking care of those with special needs) contributed to their resilience. Subsequently, respondents described how re-establishing community groups or forming new ones played significant roles in resilience (e.g., organizing local efforts to repair homes, identifying and meeting local needs). Recovery was also influenced by the ability of community groups to represent diverse (needs differed from community to community depending on, for example, cultural status, socio-economic status, impact of loss of CBD employment etc.) community needs to agencies

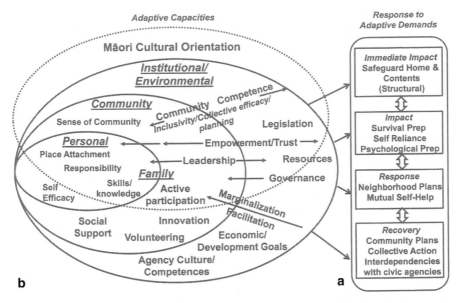

Fig. 15.1 Summary of **a** responses to adaptive demands (over time), and **b** the adaptive capacities and interdependencies at personal, community, cultural and Institutional/environmental levels identified in the Australian and New Zealand research

and to secure the resources each community needed to take responsibility for their recovery. Thus, it was evident that what people had to contend with changed over time (e.g., from loss of utilities to making neighborhoods safe to dealing with government agencies) as they negotiated physical, social and institutional demands that changed over the course of a prolonged period of recovery (summarized in Fig. 15.1).

While all focus groups acknowledged experiencing similar adaptive demands, they differed with regard to how the communities they represented dealt with recovery demands. The existence of such differences provided an opportunity to explore how their respective community capacities and characteristics (Norris et al. 2008; Paton and Johnston 2006) influenced recovery (e.g., their ability to develop neighbourhood groups, deal with external agencies etc.). The in-depth interviews identified several factors that influenced people's vulnerability and resilience to hazard consequences and thus the quality of their recovery.

15.3   Individual Level Influences on Vulnerability and Resilience

In Victoria and Christchurch alike (Mamula-Seadon et al. 2012; McLennan and Elliot 2012; Paton 2012; Whittaker et al. 2009), factors contributing to vulnerability were isolation and loss of social networks (e.g., people moving away, particularly following aftershocks); loss of social support, and progressively dwindling psychological resources. In contrast, being resilient was linked to confidence in self-help and self-sufficiency; flexibility and adaptability to respond to changing demands; a 'mind-set' of getting the job done (cf., self-and collective efficacy); engaging with others and maintaining relationships (cf., community participation); willingness to care, share and listen to others' stories; being knowledgeable about and sensitive to community dynamics and cultures; and having a sense of belonging and identity.

Developing a group identity had several benefits. It created a sense of connectedness to people and place and generated a sense of commitment to the recovery and well-being of the community and its members. It also facilitated mutual understanding and support; increased innovation; and facilitated the development of trust and credibility (Winstanely et al. 2011). These findings validated the definitions of resilience introduced above, particularly regarding the important role trust plays in conceptualizing resilience, particularly in circumstances characterized by uncertainty that increases people's dependence on others (Paton 2008). The valuable role social connectedness played in resilience was reinforced by data from focus groups that addressed social context issues in more detail.

15.4   Community Recovery

Analyses of the Christchurch experience identified that just because recovery took place in a community context it could not be assumed that the quality of engagement within a community or between a community and civic agencies was sufficient to facilitate recovery (Appley 2012; Mamula-Seadon et al. 2012; McBrearty 2012; Paton 2012; Winstanley et al. 2011). Several community developmental trajectories were identified in the recovery context. The degree to which community groups transitioned into functional recovery resources for their members was influenced by several factors. One such factor was people's early experiences, particularly with regard to when they received external assistance.

For example, focus group respondents who described their community recovery experience as poor reported how their receiving external aid in the period immediately following the 2011 earthquake acted to constrain their developing functional relationships. They raised the possibility that the early provision of external assistance deprived the group of a context in which to develop the community competencies (e.g., inclusivity, forming a sense of community, problem solving (collective efficacy)) required to function more effectively in recovery settings in which they had to take responsibility for their own recovery. This is consistent with theoretical arguments that community competencies such as active community participation and collective efficacy influence resilience (Paton 2008). Support for this contention came from community groups whose members stated that it was collectively developing strategies to meet their initial survival needs that laid the foundation for their developing both a sense of community and the skills required to deal with recovery demands (Mamula-Seadon et al. 2012). Thus while the early receipt of food, water and clothing and so on did expedite initial recovery, the fact that it limited opportunities for people to collectively confront issues of shared significance may have prevented the development of the core community competencies (Norris et al. 2008; Paton and Johnston 2006) required to deal with the unique demands each community had to contend with (more so when having to deal with repeated aftershocks). Nor could it be assumed that pre-existing groups (e.g., Residents' Associations that had a strong locational presence) would provide a viable social recovery resource.

Focus group respondents from two suburbs stated how their belief that their membership of pre-existing residents' groups would help them deal with their recovery issues was not realized (Paton 2012). The common denominator in their accounts was the presence of passive community leadership. That is, community leaders who decided to wait for government directives about what they should do. Thus a lack of willingness to take responsibility for actions in the absence of such directives meant that the group did not provide a forum for their engagement in their own recovery, with the ensuing sense of being in limbo undermining their recovery. However, those groups who did evolve into more effective recovery resources identified several reasons for this. These reasons addressed the following areas: leadership; community processes, planning and prioritizing; volunteering; and working with business and government. It is to a discussion of these characteristics that this chapter now turns. It starts with the pivotal issue of community leadership.

15.4.1 Community Leadership

In both Australia and New Zealand, a key predictor of resilience was the emergence of leaders from within community groups. Emergent leaders not only helped bring people together, they facilitated a coherent community response and linked the community with external agencies and specialists to secure the resources and help required to meet local needs. Thus leadership acted both as a community resource and as an example of a mechanism linking community with government and other agencies to empower community recovery (Norris et al. 2008; Paton 2008). Effective leaders were defined as possessing:

- A strong sense of commitment to helping others within the community
- Knowledge of how the community works/how to get the community working
- Good local knowledge (people and place)
- Leadership experience and willing to take action in novel, challenging circumstances.
- Good connections or knowing who/how to call on government and business

A need for multiple leaders was recognized (e.g., some management skills to take enduring roles in community recovery, others with specialist knowledge, such as builders, to lead repair work) as a characteristic of a resilient community. In Victoria, community leaders not only volunteered, they were also elected. Having elected leaders increased the perceived legitimacy of community groups, not only in the eyes of other community members, but also by government agencies (McAllan et al. 2011). This made it easier to secure government funding for community initiatives. As leaders in both locations were volunteers who had to function under exceptional circumstances, a need for leadership training and support to be included in DRR and recovery planning for future disasters was identified. As a recovery resource, the effectiveness of community groups was not just due to the presence of leaders, it was also a function of the capabilities of the communities in which they lived.

15.4.2   Community Processes, Planning and Priorities

In both Victoria and Christchurch (Mamula-Seadon et al. 2012; McLennan and Elliot 2012), the characteristics of community groups identified as contributing to resilience were:

- Being proactive and willing to take action to deal with emergent needs;
- Bottom-up planning and community inclusivity
- Being able to operate on scarce or unreliable information
- Creating a register of skills and resources within a community
- Identifying vulnerable community members and how to meet their needs
- Identifying a specific contact to coordinate outreach and donations
- Empowering, encouraging and supporting community initiative

These accounts reiterate the value of the community competencies (e.g., inclusivity, empowerment, and innovation) identified as pivotal to resilient recovery (Norris et al. 2008; Paton 2008; Paton and Johnston 2006). A need for new, community-specific ways of communication that addressed specific and diverse community needs and that was available when people need it was also identified as playing a significant role in expediting community recovery (Winstanley et al. 2011).

In both Christchurch and Victoria a pivotal communication device was including a broad range of people in planning activities. In Victoria, this process was expedited by establishing Community Recovery Committees (CRC) that took responsibility for eliciting and communicating the diverse views that emerged from community meetings (McAllan et al. 2011). The CRCs also took responsibility for managing the inevitable conflict that emerged as a result of having to deal with resource limitations and constraints, dealing with competing needs and priorities and managing diversity. The importance of the latter derived from the fact that while considerable effort was put into eliciting ideas for recovery and development planning, not all ideas could be actioned, at least in the short term. Consequently, it was very important to explain the process of prioritising and how and why particular ideas were being implemented within given timeframes in order to sustain community integrity and prevent conflict and dissent from reducing the functional and socially supportive aspects of community life. In Christchurch, this role was predominantly the responsibility of community leaders. The importance of ensuring continuity of group processes was especially evident in Christchurch in the context of experiencing several after-shocks that meant that community groups had to work through the response-recovery cycle several times.

In Christchurch, an additional social recovery dynamic arose from the unique cultural dimension Māori bring to the social landscape. The discussion of the Māori experience also highlights the need for cultural diversity to be considered in recovery planning and development initiatives.

15.4.3   Māori Perspectives on Recovery

The Māori demographic (25,725 individuals), constituted 4.1 % of the urban population (Statistics New Zealand 2012a). Although Māori resided in all suburbs, most lived in low socioeconomic areas, particularly the Eastern suburbs (Statistics New Zealand 2012b). Because these areas were most significantly impacted by the earthquakes, Māori were disproportionately affected (Canterbury Earthquake Royal Commission 2012).

New Zealand legislation imposes a statutory obligation on local and national governance to address the needs and aspirations of Māori communities by honouring the principles of partnership, protection, and participation as an affirmation of the Treaty of Waitangi (Royal Commission on Social Policy 1987). Paton (2007) suggests that indigenous community recovery strategies are relevant across the continuum of hazard mitigation preparedness, response and recovery, and to wider themes of sustainable development. Māori communities have culturally determined infrastructure and localised systems for ensuring community resilience based on knowledge gained in response to social, natural/environmental and political (colonisation) change. Exemplars include the collaborative Māori response to the disaster and the cross-cultural approach to community recovery and resilience which regional authorities have developed in partnership with the local Māori tribe. Consequently, disaster preparedness and management practices are embedded in tribal, organisational, sub-tribe and family behaviours and actions. Thus community competencies, such as active participation and inclusivity, which underpin resilience, are implicit elements of Māori culture.

The Māori response in Christchurch demonstrates how interrelated cultural knowledge, values and practices constitute technologies which act to ensure the resilience of Māori, the wider urban community and surrounding environment. Such technologies have potential for innovating local and national disaster preparedness planning and risk management strategies through facilitating the development of contextually relevant initiatives which foster and sustain community resilience (Paton and Jang 2011).

Following the earthquake the local Māori response was immediate and shaped by the cultural value 'aroha ki te tangata—love to all people' (Marae TVNZ 2011). Cultural values and practices which facilitate community well-being and resilience have been noted in Māori communities following disaster (Boulton and Gifford 2011; Hudson and Hughes 2007; Proctor 2010), but rarely documented in full. In

Christchurch, Te Rūnanga o Ngāi Tahu have tribal responsibility for ensuring the wellbeing of the Canterbury community and environment.

Within 24 h of the February earthquake, Te Rūnanga o Ngāi Tahu relisted the tribal 0800 Kai Tahu telephone number as a 24 h emergency contact centre for accessing support information and assistance (Sharples 2011). It was agreed that the Māori response to the earthquake would be led by Te Rūnanga o Ngāi Tahu, and that intertribal networks would be operationalized to access social and material resources. Ngāi Tahu subsequently mediated communication and collaborative decision making (cf., collective efficacy, empowerment) with Government ministries, local authorities relevant NGOs and various tribes to facilitate a coordinated emergency response (Te Puni Kōkiri 2011a).

Community needs and concerns as well as information regarding resources were communicated to stakeholders through daily debrief meetings at Rehua marae, emergency support networks, tribal websites and facebook links (Te Puni Kōkiri 2011b). Ngāi Tahu community responders, Māori wardens, students and community workers from the Red Cross, Salvation Army, and other local, groups facilitated rapid delivery of food, water and other necessities to approximately 15,000 households (Te Rūnanga o Ngāi Tahu 2012a). The 12 Ngāi Tahu marae, (tribal community centres) also opened as shelter and support centres. Within a week of the February 22, 2011 quake all tribal marae in the South Island were hosting evacuees, followed by several marae in the North Island (Te Puni Kōkiri 2011b). Māori wardens were deployed to provide security services and conduct needs assessments (Te Rūnanga o Ngāi Tahu 2012a). This last example introduces the role volunteering played as a recovery resource.

15.4.4   Volunteering

Volunteering took many forms (e.g., local people assisting welfare services, members of social and sporting clubs and churches organizing social events) and contributed to resilience in several ways. These included it being therapeutic and it giving people a sense of control in otherwise uncontrollable circumstances. Volunteering also enhanced people's capacity to cope, and enhanced their sense of community and collective efficacy (Paton 2012). Volunteering also increased opportunities for people to be actively involved in community life (cf., community participation) in meaningful ways (e.g., doing something of value for others) and helping rebuild the social fabric of the community. These examples lend support to the need for the inclusion of constructs such community participation and collective efficacy in theoretical conceptualizations of resilience (Paton 2008).

However, volunteering also had certain negative consequences (e.g., over-commitment to volunteer work, having too few people to fill roles and burnout). Thus, if it is to fulfil its potential as a recovery resource, and make an enduring contribution to social capital, attention needs to be directed to, for example, the selection and training of volunteers and developing procedures covering issues such as termination processes, time-out periods, and succession planning (over possibly long periods of time) (McAllan et al. 2011).

The leadership, community processes and volunteer activities discussed above represent resources that contribute to creating empowered communities. However, in complex disaster recovery environments, communities also need information and resources that cannot be met from within the community but only from the wider societal context; from NGOs, government agencies and businesses. The quality of the relationships between communities and NGOs, government and businesses that evolve in the recovery environment will thus influence the degree to which communities are empowered to enact their own recovery initiatives. This introduces a need to consider how relationships with NGOs, government and businesses influence recovery.

15.4.5 Relationships with Government, NGOS and Businesses

The important role relationships with government and businesses played in meeting community needs was endorsed in Victoria and Christchurch (Mamula-Seadon et al. 2012; McAllan et al. 2011). This reinforces the theoretical contention that being able to define and represent needs to external sources plays an important role in resilience (Norris et al. 2008; Paton 2008). It is, however, important to appreciate that the relationship between community recovery and business and economic recovery is one characterized by interdependence (e.g., community members as employees and consumers) and not just as one defined in terms of agencies and businesses being unidirectional providers of resources to communities (Paton and McClure 2013). Notwithstanding, the latter is important. The relationship between government agencies and businesses and communities can facilitate or marginalize community recovery. Communities can influence this relationship by being proactive.

For example, in Victoria, it emerged that while many businesses wanted to support communities, they often did not know who to contact or what different communities needed (McAllan et al. 2011). Consequently, a need to proactively approach businesses was seen as an important issue in future disaster recovery planning. In Victoria, the CRCs played a highly instrument role in developing and sustaining the functional relationships between community and government and local businesses that facilitated the ability of civic agencies to empower community recovery. Similar examples emerged in Christchurch.

An example of an effective proactive response in Christchurch reflected the fact that Ngāi Tahu had effective network linkages with several NGOs, local authorities, and Government departments (as an Iwi but also through Ngāi Tahu in Senior institutional positions) prior to the earthquake. Positive community-agency relationships in Christchurch were facilitated by government and relief agencies using community leaders as the conduit into communities. Community recovery was also facilitated by groups being able to work through dedicated recovery organizations (e.g., CERA, CanCERN) to meet their needs.

However, evidence of agency marginalizing community response was also evident. Some Christchurch respondents identified frustration at being reassigned to a new government agency representative each time they contacted a specific agency. This identifies a need for all agencies that will interact with the public to develop specific crisis management procedures and processes, agency culture change, and the selection and training (management and staff training issues such as cultural competence, listening and communication skills etc.) of staff for crisis response roles (especially where staff are operating at some distance from the event and those directly and indirectly affected). Issues of marginalization also emerged from the Māori experience.

With regard to Māori recovery, marginalization was evident in the fact that coordinating an integrated response with Emergency Management was delayed (by 8 days) and eventually had to be negotiated through external mediators (Solomon 2012). The delayed coordination of the Māori response with the formal disaster and emergency management infrastructure contributed to duplication or the absence of services (e.g., fresh water delivery, portable toilets, and primary health services) in some regions (Batt 2012; Potangaroa 2011; Solomon 2012). Marginalization was evident from focus group data (Paton 2012) describing the government emergency response as often poorly coordinated, lacking local contacts, and not knowing of local distribution locations. Agencies were perceived to be working in isolation with little coordination of functions, reducing their capacity to empower community initiatives.

These examples illustrate how agency activity within a recovery environment can empower or disempower community recovery. This makes it important to consider how a capacity for creating empowering social settings in recovery contexts can be facilitated. At a macro level, an important influence on empowering community recovery is governance.

15.5 Governance

Governance is an important area when conceptualizing the consequences of large scale disasters that simultaneously deplete capital stock and services, and requires the performance of a host of complex societal activities within the compressed time period of post-disaster recovery (Bach 2012). Bureaucracies often do not adapt well to this circumstance and new governmental organizations typically emerge to provide more resources. Significant post-disaster adaptations to pre-disaster governance structures also inform future disaster recovery management policy and governance approaches that will likely be taken. This section discusses the recovery related institutional transformation following the September 4, 2010 and February 22, 2011 Canterbury earthquakes and the implications this has for future post-disaster recovery governance frameworks in New Zealand and around the world.

Prior to September 2010, New Zealand had a well-established all-hazards approach to emergency management supported by an institutional framework of decentralized decision making in a tiered government system with emphasis on the bottom-up escalation of responsibilities. Within this framework, the coordinated involvement of all stakeholders was underpinned by a comprehensive array of regulatory instruments and policies, implemented by central and local government agencies. Recovery, reconstruction and development were to be led by local government and supported by the central government. Thus, at the time of the September 2010 earthquake, the responsibility for recovery rested with local government in Canterbury and with the Ministry of Civil Defence & Emergency Management nationally.

In the first days following the September 4, 2010 earthquake, elements of local and central government organized and acted within that pre-existing institutional framework. Then, institutional adaptations began to occur. The first occurred on September 14, 2010 with central government's passage of the Canterbury Earthquake Response and Recovery Act and formation of the Canterbury Earthquake Recovery Commission (CERC). This commission unified local governance in Canterbury to advise central government and channel information between central and local government. The Act also enabled some critical regulatory and policy flexibility for all levels of government to execute in managing response and recovery. The second adaptation occurred in April 2011, with the central government's passage of the Canterbury Earthquake Recovery Act 2011 and the creation of a new central government department, the Canterbury Earthquake Recovery Authority (CERA). When the state of emergency ended in May 2011, CERA effectively assumed primary responsibility for recovery in the region, superseding previous arrangements.

Experience has demonstrated that, notwithstanding the policies and frameworks that promote decentralisation and integrated, holistic frameworks and processes, much more needs to be learned about how local communities succeed in normal, pre-disaster conditions, and how they respond and recover afterwards. Thus the lessons from social analyses need to be incorporated within the policy domain. This provides another argument for incorporating empowerment in conceptualizations of resilience (Paton 2008) as it provides a social framework capable of facilitating this kind of integration. Governance represents an overarching approach to conceptualizing the creation of *empowering* settings (Dalton et al. 2001) that can complement the development of *empowered* communities (Paton 2008). One aspect of this that is directly relevant concerns the integration of recovery and development.

15.6 Linking Recovery and Development

In Victoria and particularly in Christchurch, development is still underway. This section discusses how development is being facilitated in two general ways. The first reflects the adoption of an integrated emergency management philosophy that conceptualizes hazard readiness and recovery in an holistic way that affords opportunities to learn from experience.

Some recovery initiatives were established to facilitate social development. For example, a Psychosocial Recovery Advisory Group (PRAG) was developed to provide expert, timely, quality evidence-based advice on psychosocial issues

(Mooney et al. 2011). The PRAG adopted a strengths-based approach for conceptualizing recovery and facilitating longer-term social development. The strengths-based approach mobilizes cultural, spiritual and social resources with an emphasis on development through community participation and engagement. It is thus an example of a strategy that contributes to creating empowered communities. This development initiative was complemented by the government establishing an empowering setting in the form of a Community Forum, established by the Canterbury Earthquake Recovery Minister that comprises 34 representative of a broad cross-section of Canterbury: business, ethnic interests, residents associations, community groups.

Development will be further facilitated by its emerging from the broader Canterbury Recovery Strategy that provides a unifying higher level vehicle for integrative development plans such as those covering built heritage, education, Iwi Maori, and building community resilience in ways that recognizes the need for planning to be based in community engagement principles and to encompass physical, social and cultural recovery. Planning is thus attempting to accommodate the role of spiritual and cultural values and practices in facilitating the development of social capital. Planning is also being driven by the goal of ensuring that cultural values and practices are maintained by ensuring that cultural identity is used as a community development resource. In regards to Ngāi Tahu and Māori kaupapa (Values), identities are mutual constitutive and collectively shape behaviours and cultural practices.

Development planning is responding to the community's desire for a more compact central city. The current conceptualization of how recovery and development can be linked is through the Transitional City concept. This describes Christchurch as being in a transitional phase—from recovery through to the return of a functioning central city—and is a critical platform for future sustainable development. This transition time provides opportunities to test new ideas, explore new concepts, and look at new ways to bring people, business and investment back to the central city.

The design concept for the development blueprint is to develop a greener, more accessible city with a stronger built and social identity. It foresees a city for all peoples and cultures, recognising, in particular, Ngāi Tahu heritage and places of significance. Following the 'Have your say day!' The broader community as a whole requested that within the central Core they would like a strong cultural hub showcasing Māori culture. Te Rūnanga o Ngāi Tahu tribal development initiatives are shaping the longer term recovery and resilience of Māori in Christchurch. Māori workforce development is being fostered through tribal investment in He Toki ki te Rika (the Māori Trade Training Scheme) in which the existing knowledge, experience and expertise of partner organisations is leveraged to up skill Māori for the recovery of Canterbury (Te Rūnanga o Ngāi Tahu 2012b).

In keeping with statutory obligations the active participation of Māori stakeholders in strategizing for future community resilience and urban sustainability is also facilitated. Prior to the Christchurch Earthquakes, Māori resources, practices and emergency management strategies had not been integrated into pre-disaster planning or emergency response polices in any meaningful way. Te Rūnanga o Ngāi Tahu has used its legislated authority as guardians of the land and natural resources within Canterbury to secure a statutory governance role in the Christchurch rebuild stipulated in the Canterbury Earthquake Recovery Authority Act (2011) (CERA 2012). Māori concerns and recommendations relating to Christchurch development planning have been communicated directly to the Crown via established links between Te Rūnanga o Ngāi Tahu, the New Zealand cabinet, relevant government agencies and local authorities. At the regional level, Ngāi Tahu cultural and historical knowledge, values and processes are informing integrated civil/disaster preparedness (DRR) and risk management strategies for the Canterbury region (CERA, Canterbury Earthquake Recovery Authority 2012). Te Rūnanga o Ngāi Tahu is also engaged in a collaborative partnership with national and local authorities, to facilitate the rebuilding of urban Christchurch (CERA Canterbury Earthquake Recovery Authority, Christchurch City Council & Te Rūnanga o Ngāi Tahu 2012). Ngāi Tahu intergenerational knowledge pertaining to environmental history, previous land use and composition is informing urban planning, contributing to environmental sustainability, and facilitating the resilience of the wider community.

15.7   Conclusion

Conducting interviews with people as they navigated their way through a long and complex recovery made it possible to assess the challenges that people had to negotiate in disaster recovery. Interviewees all identified the important role structural and survival readiness would play in increasing resilience for future events. Examining how physical, community and societal recovery challenges were dealt with by different groups identified how the adaptive capacities of people (e.g., self-efficacy, place attachment), communities (e.g., inclusivity, sense of community, collective efficacy) and government agencies and businesses (e.g., empowerment, trust) play interdependent roles in facilitating recovery and development in the aftermath of major disasters.

The adaptive capacities that influenced the effectiveness of people's responses to recovery demands are summarized in Fig. 15.1 and provide some level of empirical corroboration of theoretical conceptualizations of resilience (e.g., Klein et al. 2003; Norris et al. 2008; Paton 2008; Paton and Johnston 2006; Pelling and High 2005). The contents of Fig. 15.1 depicts both the resources at person, community and societal levels identified as facilitating recovery, and constructs that play pivotal roles in linking or integrating levels of analysis to create a more holistic model of societal resilience. For example, it depicts "community competence" and "leadership" as constructs that help link communities to government agencies and businesses (Norris et al. 2008; Paton 2008). The bi-directional arrows next to these constructs are used to indicate the influence of these constructs in both mobilizing community activity and linking community to societal-level resources. The constructs of "empowerment" and "trust" are also linking concepts, but ones that act in ways that link societal agencies with communities. All four of these constructs thus play complementary roles in resilience, with the former contributing to empowered communities and the latter to creating empowering settings. The concepts of facilitation and marginalization are depicted with a uni-directional arrow, signifying that their current mode of operation is top-down. Governance is depicted in the same way. It was suggested, however that both concepts could enhance recovery by conceptualizing governance and agency and business relationships as occurring within an empowerment framework. This conceptualization remains tentative until more work is conducted. The contents of Fig. 15.1 could be used to formulate research questions and hypotheses to further empirical research into resilience.

Analysis highlighted the need to include a cultural dimension in conceptualizations of community resilience (particularly in multi-cultural countries such as Australia and New Zealand). This appears in Fig. 15.1 as a level of analysis intersecting capacities reflecting those emanating from a predominantly more European (in New Zealand, Pakeha) perspective. Depicting the Māori perspective in this way highlights how several adaptive capacities are implicit aspects of Māori cultural life and society. It also highlights the need for development planning to be based on community engagement programs that encompass shared cultural learning.

In Christchurch, development is continuing. The Development blueprint seeks to develop a city with a stronger built and social identity. From a social perspective, the adaptive capacities people identified as being influential in their recovery reflect community competencies and characteristics that are developed in and are sustained by people's everyday life experiences. This opens the way for development to occur through integrating risk management with community development activities in ways that are more likely to be perceived, by community members and civic authorities alike, as building sustained social capital in ways that can contribute to increasing adaptive capacities over time (Norris et al. 2008; Paton and Jang 2011).

References

Adger, W. N., Kelly, P. M., Winkels, A., Huy, L. Q., & Locke, C. (2002). Migration, remittances, livelihood trajectories, and social resilience. *Ambio, 31,* 358–366.

Appley, R. (2012). The new Brighton community response following the Canterbury earthquake. *Tephra, 23,* 12–17.

Bach, R. L. (2012). Mobilising for resilience: From government to governance. *Tephra, 23,* 36–39.

Bannister, S., & Gledhill, K. (2012). Evolution of the 2010–2012 Canterbury earthquake sequence. *New Zealand Journal of Geology and Geophysics, 55*:3, 295–304.

Batt, J., Atherfold, C., & Grant, N. (2011). Te Arawa responds. *Kai Tiaki Nursing New Zealand, 17*(9), 38.

Berkes, F., Colding, J., & Folke, C. (2003). *Navigating social-ecological systems: Building resilience for complexity and change.* Cambridge: Cambridge University Press.

Boulton, A., & Gifford, H. (2011). Resilience as a conceptual framework for understanding the Māori experience: Positions, challenges and risks. In T. McIntosh, M. Mulholland (Eds.), *Māori and social issues* (Vol. 1). Wellington: Huia Publications.

Bruijn, K. M. D. (2004). Resilience indicators for flood risk management systems of lowland rivers. *International Journal of River Basin Management, 2,* 199–210.

Canterbury Earthquakes Royal Commission. (2012). *Interim report.* Christchurch: Canterbury Earthquakes Royal Commission.

Carpenter, S., Walker, B., Andries, J. M., & Abel, N. (2001). From metaphor to measurement: Resilience of what to what? *Ecosystems, 8,* 941–944.

CERA Canterbury Earthquake Recovery Authority, Christchurch City Council, & Te Rūnanga Ngāi Tahu. (2012). *Christchurch central recovery plan Te mahere 'Maraka Ōtautahi'*. Christchurch: Authors.

Chambers, R., & Convey, G. R. (1991). *Sustainable rural livelihoods: Practical concepts for the 21st century*. Brighton: Institute of Development Studies.

Dalton, J. H., Elias, M. J., & Wandersman, A. (2001). *Community psychology*. Belmont: Wadsworth.

Gaillard, J.-C. (2007). Resilience of traditional societies in facing natural hazards. *Disaster Prevention and Management, 16*, 522–544.

Hudson, J., & Hughes, E. (2007). *The role of marae and Māori communities in post disaster recovery: A Case study*, GNS Science report, 2007/15, GNS Science, Wellington, New Zealand.

International Strategy for Disaster Risk Reduction (ISDR). (2005). *Hyogo framework for action 2005–2015: Building the resilience of nations and communities to disasters*. Kobe: International Strategy for Disaster Reduction.

Klein, R., Nicholls, R., & Thomalla, F. (2003). Resilience to natural hazards: How useful is this concept? *Environmental Hazards, 5*, 35–45.

Mamula-Seadon, L., Selway, K., & Paton, D. (2012). Exploring resilience: Learning from Christchurch communities. *Tephra, 23*, 5–7.

Marae Investigates, T. V. N. Z. (2011). Interview with Mark Solomon the Kaiwhakahaere of Te Rūnanga o Ngāi Tahu. http://www.youtube.com/watch?v=4vVmM99VUqg. Accessed 18 Dec 2012.

Marschke, M. J., & Berkes, F. (2006). Exploring strategies that build livelihood resilience. *Ecology and Society, 11*(4). http://www.ecologyandsociety.org/issues/article.php?id=1730.Accessed 19 Dec 2012.

Marshall, N. A., & Marshall, P. A. (2007). Conceptualizing and operationalizing social resilience within commercial fisheries in Northern Australia. *Ecology and Society, 12*(1). http://www.ecologyandsociety.org/vol112/iss1/art1/. Accessed 19 Dec 2012.

McAllan, C., McAllan, V., McEntee, K., Gale, W., Taylor, D., Wood, J., et al. (2011). *Lessons learned by community* recovery committees of the 2009 Victorian bushfires. Victoria. Australia: Cube Management Solutions.

McBrearty, T. (2012). Challenging disaster management through community engagement. *Tephra, 23*, 32–35.

McLennan, J., & Elliot, G. (2012). Community bushfire safety issues: Findings from interviews with residents affected by the 2009 Victorian bushfires. Bushfire Cooperative Research Centre Fire Note, 98, October 2012.

Mooney, M. F., Paton, D., de Terte, I., Johal, S., Karanci, A. N., Gardner, D., Collins, S., Glavovic, B., Huggins, T. J., Johnston, L., Chambers, R., & Johnston, D. M. (2011). Psychosocial recovery from disasters: A framework informed by evidence. *New Zealand Journal of Psychology, 40*, 26–39.

Norris, F. H., Stevens, S. P., Pfefferbaum, B., Wyche, K. F., & Pfefferbaum, R. L. (2008). Community resilience as a metaphor, theory, set of capacities, and strategies for disaster readiness. *American Journal of Community Psychology, 41*, 127–150.

Paton, D. (2007). *Measuring and monitoring resilience*. GNS Science report 2007/18, GNS Science, Wellington, New Zealand.

Paton, D. (2008). Risk communication and natural hazard mitigation: How trust influences its effectiveness. *International Journal of Global Environmental Issues, 8*, 2–16.

Paton, D. (2012). *MCDEM Christchurch community resilience project report*. Wellington. New Zealand: Ministry of Civil Defence and Emergency Management.

Paton, D., & Jang, L. (2011). Disaster resilience: Exploring all-hazards and cross cultural perspectives. In D. Miller & J. Rivera (Eds.), *Community disaster recovery and resiliency: Exploring global opportunities and challenges*. London: Taylor & Francis.

Paton, D., & Johnston, D. (2006). *Disaster resilience: An integrated approach*. Springfield, Ill.: Charles C. Thomas.

Paton, D., & McClure, J. (2013). *Preparing for disaster: Building household and community capacity*. Springfield, Ill.: Charles C. Thomas.

Pelling, M., & High, C. (2005). Understanding adaptation: What can social capital offer assessment of adaptive capacity? *Global Environmental Change, 15,* 308–319.

Potangaroa, R., Wilkinson, S., Zare, M., & Steinfort, P. (2011). The management of portable toilets in the Eastern suburbs of Christchurch after the February 22, 2011 earthquake. *Australasian Journal of Disaster and Trauma Studies, 2011-2,* 35–48.

Proctor, E. (2010). *Toi tu te whenua, toi tu te tangata: A holistic Māori approach to flood management in Pawarenga*. Masters thesis, Waikato University, Hamilton, New Zealand.

Royal Commission on Social Policy (1987). The princples of the Treaty of Waitangi. In *The royal commission on social policy: Te Kōmihana A Te Karauna Mo Ngā Āhuatanga-A-Iwi. Discussion Booklet No. 1* (pp. 4–5 and pp. 14–23). Wellington: Author.

Sharples, P. (2011). Media release from the Māori Party 28 February, 2011-*Te Puni Kōkiri supporting Ngāi Tahu response to Canterbury earthquake*. http://www.maoriparty.org/index.php?pag=nw&id=1592&p=tepuni-kokiri-supporting-ngai-tahu-response-to-canterbury-earthquake/. Accessed 11 Dec 2012.

Solomon, M. (2012). Opening address. *Recover reconnect rebuild. MASS (Māori Academy of Social Science) conference*. 28–30 November, Canterbury University, Christchurch.

Statistics New Zealand. (2012a). *Census of population and dwellings: Table builder*. http://www.stats.govt.nz/tools_and_services/tools/TableBuilder/2006-census-pop-dwellings-tables/culture-and-identity/ethnic-group.aspx. Accessed 19 Dec 2012.

Statistics New Zealand. (2012b). *Interactive map boundary*. http://apps.nowwhere.com.au/StatsNZ/Maps/default.aspx. Accessed 19 Dec 2012.

Te Puni Kokiri. (2011a). *Earthquake bulletin 1*. http://.tpk.govt.nz/en/newsevents/new/archive/2011/2/25/earthquake-bulletin1/. Accessed 20 Dec 2012.

Te Puni Kokiri. (2011b). *Earthquake bulletin 2*. http://.tpk.govt.nz/en/newsevents/new/archive/2011/2/26/earthquake-bulletin1/. Accessed 20 Dec 2012

Te Rūnanga o Ngāi Tahu. (2012a). *Aoraki Matatū annual report*. Christchurch: Author.

Te Rūnanga o Ngāi Tahu. (2012b). Media release 19 December 2012-*Canterbury Māori trades training programme receives a $1 m boost from government*. http://www.ngaitahu.iwi.nz/News/Media/Media-Releases/2012/trades-programme-receives-boost.php. Accessed 20 Dec 2012.

Victorian Bushfires Royal Commission. (2010). Final report. Available online at http://www.royalcommission.vic.gov.au/Commission-Reports. Accessed 10 Jan 2013.

Whittaker, J., McLennan, J., Elliott, G., Gilbert, J. Handmer, J., Haynes, K., & Cowlishaw, S. (2009). *Human behaviour and community safety. Victorian 2009 bushfire research response final report October 2009*, East Melbourne: Bushfire CRC. www.bushfirecrc.com/managed/resource/chapter-2-human-behaviour.pdf. Accessed 19 Dec 2012.

Winstanely, A., Cronin, K., & Daly, M. (2011). *Supporting communication around the Canterbury earthquakes and other risks*. GNS Science miscellaneous report 2011/37. p. 39.

Chapter 16
Disaster and Development in Ghana: Improving Disaster Resiliency at the Local Level

Kiki Caruson, Osman Alhassan, Jesse Sey Ayivor and Robin Ersing

16.1 Introduction

Economic losses from disasters since 2000 are in the range of $ 2.5 trillion, a figure at least 50 % higher than previous estimates by the United Nations (UNISDR 2013). The year 2012 marked the third consecutive year where annual, world-wide, economic losses exceeded $ 100 (UNISDR 2012). Among highly vulnerable localities smaller scale disasters can cause widespread suffering. Most women and children living in seriously deprived neighborhoods suffer numerous environmental burdens in domains such as water, sanitation, waste disposal, housing, drainage, food contamination, inadequate hygiene, pest control, and indoor and outdoor air pollution (Cutter et al. 2010; Songsore et al. 2009). Women have every incentive to contribute to mitigation and preparedness activities as they often have the most to lose when disaster strikes—not only economically, but physically and emotionally as well. As a result, women are key stakeholders and natural candidates for community leadership in promoting disaster resiliency at the local level.

In 2012, the International Day for Disaster Reduction (IDDR), an event created and sponsored by the United Nations, was dedicated to the central role women and

K. Caruson (✉)
Department of Government and International Affairs, University of South Florida,
4202 East Fowler Avenue, SOC 107 Tampa, Florida 33620, USA
e-mail: kcaruson@usf.edu

O. Alhassan
Institute of African Studies, University of Ghana, Legon, Accra, Ghana
e-mail: aosman@ug.edu.gh

J. S. Ayivor
Institute for Environmental and Sanitation Studies, University of Ghana, Legon, Accra, Ghana
e-mail: jsayivor@ug.edu.gh

R. Ersing
School of Public Affairs, University of South Florida, 4202 East Fowler Avenue,
SOC 107 Tampa, Florida 33620, USA
e-mail: rersing@usf.edu

N. Kapucu, K. T. Liou (eds.), *Disaster and Development,* Environmental Hazards,
DOI 10.1007/978-3-319-04468-2_16, © Springer International Publishing Switzerland 2014

girls play in effective disaster risk reduction and in ensuring resilient communities at the local level.[1] The theme of the event: "Women and Girls: The Invisible Force for Resilience" not only recognizes the importance of acknowledging the contribution of women, it also concedes that women often operate at the margins of national and sub-national policy decision-making. In the words of a senior program coordinator at Ghana's National Disaster Management Organization (NADMO), "women and girls are essential to enhancing disaster resiliency for their contributions at the local level to activities such as vulnerability assessment, community organization, and risk reduction, but their opinions and voices are not always accorded priority attention" (NADMO), Senior Program Coordinator 2012).

This goal of the chapter is to assess the potential for Ghanaian women inhabiting highly vulnerable localities to advance disaster resiliency within their communities. Women in deprived localities are essential to disaster risk reduction efforts, but their knowledge and ideas are not systematically leveraged as an asset to community efforts at establishing resilient neighbourhoods. Using focus group data gathered from residents of several highly vulnerable areas in Ghana, the authors present a portrait of the current roles played by women in disaster resiliency efforts, the limits of their enfranchisement in the emergency management process, and the opportunities for the integration of a gender oriented approach to enhancing disaster resiliency among highly vulnerable populations. Research concerning the strategies employed by women that enhance resiliency through mitigation and preparedness is especially critical to our understanding of best practices in risk reduction as well as designing more inclusive strategies for implementing these practices (Kapucu et al. 2013).

16.2 Local Relevancy and Capacity Building

Local relevancy is a fundamental component of capacity-building among populations with limited resources (Shaw and Krishnamurthy 2009). Of critical importance is the formulation of low cost policies that build upon existing assets within the community and its residents (UN-Habitat 2008). For example, in 2008, emergency management professionals identified the "traditional extended family system of caring for its members" as fundamental to the resiliency of communities affected by widespread flooding in Northern Ghana (Antwi 2008). Record flooding was reported in the Volta Basin region in Ghana in 2007, 2009, 2010 and 2011 leaving numerous port communities devastated (Armah et al. 2010; Ministry of the Interior, Republic of Ghana 2011). The support network created by familial ties enhanced the ability of flood victims to cope and rebound from the flood events. Robust

---

[1] International Day for Disaster Reduction (IDDR) was established by the United Nations in 1989, and as of 2009, October 13th of each year is recognized across the international community as a day to promote a global culture of disaster reduction. United Nations Resolution 44/236 signed December 22, 1989 and United Nations Resolution 64/220 signed December 21, 2009. See http://www.un.org/en/documents/ for access to copies of U.N. Resolutions.

evidence supports the approach of integrating local knowledge with available scientific and technical approaches (IPCC 2011).

Self-generated knowledge by local populations "can uncover existing capacity" within the community and important shortcomings….local participation supports community-based adaptation [and transformation] to benefit management of disaster risk and climate extremes (IPCC 2011). Empowering communities to be resilient is vital. If emergency management activities are controlled by external organizations, it is less likely that the community will return to normalcy in the event of a disaster (Shaw and Krishnamurthy 2009). The involvement of local residents, including men, women, children and seniors, in disaster risk reduction activities, including promoting education and information dissemination and cooperation across entities and organizations, is the key to effective disaster management (Caruson and MacManus 2012; Comfort et al. 2010; Waugh and Tierney 2007). In Ghana, highly vulnerable communities have begun to introduce disaster resiliency topics into the curriculum of elementary schools—many of these children will not advance in their schooling creating a limited window of opportunity for exposing young citizens to how they can contribute to improving their living conditions by reducing their risk to disaster events—often exacerbated by poor sanitation, unsafe building structures, and lack of potable water.

The application of the theme of "positive deviance" to local situations of disaster management is also useful in so far as individuals within a single community demonstrate varying levels of resiliency in the face of an extreme event. Examining the process of adaptation among successfully resilient community members assists in the identification of approaches, techniques, processes, or principles that might be applied to a larger segment of the affected community. The term "positive deviance" was popularized by Marian Zeitlin in the context of child nutrition. From the perspective of child nutrition, positive deviants are children who grow and develop adequately in low-income families living in impoverished environments, where a majority of children in the same environment suffer from growth retardation and malnutrition (Zeitlin et al. 1990; Zeitlin 1991). Where resources (financial, supplies, equipment) are often limited but communities continue to experience increasingly adverse effects from natural events such as flooding, drought, hurricanes, and fire, identifying the locally-based factors that contribute to resiliency among some members or segments of a community is critical.

The concepts of local relevancy and the identification of situations of positive deviance define how the data that inform this study are evaluated. Among deeply impoverished communities, locally derived strategies for improving disaster resiliency are often the only option. Simple strategies and a creative approach to identifying assets can set the stage for community development. For example, plastic containers (bottles and bags) litter many Ghanaian communities. In slums, plastic rubbish clogs roadways and drainage areas. The practice of recycling plastic has become something of a cottage industry in some settlement neighborhoods. Community-led programs designed to educate residents of the benefits of recycling led to the creation of recycling "centers" located outside residences managed by the women of the area.

16.3   Literature Review and Background: Women, Disaster Resiliency and Development

Across the globe, women play a critical role in the ability of a community to mitigate, prepare for, and recover from a disaster because of their central function as primary care providers for both the young and the aging (Poole 2005; Poole and Issacs 1997). Women are not only caretakers of vulnerable populations, such as children and the elderly, their own vulnerability increases dramatically in the context of disaster events. Hazard researchers have linked gender inequality to a number of negative effects for women as a result of their increased vulnerability to extreme events (Morrow 1998). In the last decade, natural catastrophes have left their mark around the globe, and in each case, women were disproportionately more likely to suffer the consequences.

In a recent study of more than 140 natural disaster events occurring over a period of two decades (1981–2002), Neumayer and Plümper (2007) found no significant differences in death rates relevant to natural disasters when socioeconomic status between men and women were equal. However, when women experience inequality, they are significantly more likely to die as a consequence of the disaster. Notable tragedies include the 1991 Bangladesh floods where five times more women than men lost their lives; the 1995 Kobe, Japan earthquake where the death rate for women was one and one-half times that of men; and the 2004 Southeast Asia tsunami where nearly four times more women than men perished from the flood waters (Chew and Ramdas 2005). The aftermath of a disaster also paints a bleak outlook for women survivors as they encounter harsh and sometimes violent conditions such as increased exposure to domestic and sexual violence, lack of access to relief services, inadequate compensation for losses, sustained economic hardship, long-term displacement, limited health care, and a lack of voice in recovery and rebuilding efforts (UNISDR 2009).

The role of women in disaster-related community development continues to receive increased attention both in terms of efforts to promote hazard risk reduction and rebuilding sustainable livelihoods post-disaster. In this context, successful mitigation efforts require attention to the specific needs of women and support for a gender-equal approach to development (UNISDR 2009). Although women are often recognized for participating in local efforts to mobilize relief efforts and provide for basic needs, rarely is the collective voice of women acknowledged and integrated into the policy-making realm (Chew and Ramdas 2005). Indeed, at the core of community development are social and economic structural forces that leave women among the most vulnerable in a hazard event. Global development efforts to reduce disaster risk and strengthen resiliency advocate greater access to education for women and girls thus helping to lessen the impact of poverty (UNISDR 2010). Also, development planners emphasize the value of building upon the social capital of women, particularly their use of social networks to communicate, organize, and broker for resources (UNISDR 2009). Several case studies suggest the need to recognize women as an integral asset in rebuilding post-disaster, particularly if given the opportunity to take on more traditional male roles (Yonder et al. 2005).

In Africa, although women play a vital role in household and local level natural disaster management, policies that address the impact of disasters and resiliency efforts often favor the livelihoods of men. In many cases, such a policy assumption remains despite evidence that women's needs and capacities can be quite different from their male counterparts (WEDO 2008). Both the Hyogo Framework for Action (HFA), a 10 year strategic plan to build disaster resilience across nations and within communities, and the Millennium Development Goals (MDG) promulgated by the United Nations set a priority to use knowledge, innovation and education to build a culture of resilience locally and globally. The Post-2015 United Nations Development Agenda integrates gender equality into all of its goals, but also identifies empowering girls and women as a stand-alone goal and one that can "catalyse progress" toward sustainable development (United Nations, Monrovia Communiqué 2013). One emphasis has been to integrate a gender perspective in all disaster management policies and decision making as a way to value the knowledge and experiences of women, and incorporate their unique needs into pre and post disaster risk assessments.

16.4 The Ghanaian Context

Ghana is one of West Africa's more successful nations in terms of per capita income, human development, and conflict mitigation. Ghana, however, is especially vulnerable to risks from multiple weather-related hazards. The country is prone to coastal flooding, flash flooding in urban areas, as well as flooding from water management such as periodic spills of excess water from both the Volta and Densu river basins. The low mitigation and adaptation capacities of many deprived communities in Ghana create an environment where people are vulnerable to modest changes in the weather. Climate change has resulted in an increase in the instances of flooding and the magnitude of the events across the globe (Mercer 2010; Pelling and Wisner 2009). Disasters are on the rise not only from changes in the weather but also because more people today find themselves at risk. Coastal development in Florida, North Carolina and New Jersey has exposed more people to hurricanes and storms. Likewise, rapid growth in cities in developing nations in Africa and Asia "has made millions more vulnerable to heat waves and floods" (Miller 2012, p. 52).

Ghana's national emergency management agency, the National Association of Disaster Management Organization (NADMO), is modeled after the Federal Emergency Management Agency in the U.S. with a national unit headquartered in the capital city of Accra and regional units dispersed in central and northern Ghana. Created by an act of parliament in 1996 with a mandate to coordinate disaster management in Ghana, NADMO became operational in January of 1997. The agency's primary goal is to empower self-sufficiency at the local level. The need to work with migrant settlement populations is critical for organizations like NADMO as this type of community is likely to increase as the world's population grows and access to land decreases. Sustainability in this context becomes essential.

The leadership of NADMO recognizes the critical importance of enfranchising local residents in disaster risk reduction activities. Underfunded and lacking sophisticated equipment, NADMO's capacity to provide comprehensive service is limited, especially to areas far from Accra (where all heavy equipment is housed). Its strength is the agency's multidisciplinary approach (public health, geological hazards, nuclear safety, pest control…) and its role as a coordinating body managed by local representatives who serve ten districts and numerous rural zones.

In Ghana access to land (on which to build in urban areas and farm in rural areas) is a major challenge for immigrants, including female workers and other wage earners in the informal business sector. Individuals often inhabit areas that are flood prone or they reside in unplanned settlements characterized by a high degree of risk and vulnerability. Such "recalcitrant" populations pose specific challenges to disaster management efforts. For example, lakeshore flooding in 2010 along the Volta basin displaced more than 25,000 people in a single region. Those most affected acknowledged that they had ignored a buffer zone created by the Volta River Authority which prohibits human habitation adjacent to the shore line (University of Ghana 2012). Those living along the shores of the Volta Lake are mostly migrants without land ownership rights who choose to settle along the prohibited areas of the lake because it offers a livelihood from commercial fishing, processing, and retail sale.

An example of the heightened vulnerability of deprived communities in Ghana is the cholera outbreak that occurred between March and September of 2010 and resulted in more than 4,500 cases of cholera and more than 60 deaths. The cholera epidemic highlights the multi-hazard nature of vulnerability in low income communities in both rural and urban Ghana. Perennial flooding, poor sanitary conditions, erratic power supplies, inadequate access to potable water, and poor drainage that encourages the breeding of mosquitoes and houseflies, all contribute to the magnitude of a crisis event such as the cholera outbreak. In 2010, health facilities and workers found their resources stretched to the breaking point.

Climate change resulting in more severe weather events coupled with the particular socio-economic conditions of women in Ghana mean that any disaster is likely to have a large impact on the female population. Women are not sufficiently represented in high decision-making levels and organizations. Ghana is one of fifteen countries in Sub-Saharan Africa with less than 10 % of the national legislature represented by women (8.3 % in Ghana) (UNECA 2012). Female participation in the scientific disciplines and in the structures in place for environmental and climate change issues is also limited. This depresses the ability of women to articulate their specific concerns to affect mitigation and adaptation policy measures.

In addition, women and men occupy distinct positions in the economy largely as a result of a gender division of labor within households and society at large. This division allocates the bulk of reproductive activities to women, leaving men time to pursue more market-valued activities and resulting in extensive gender segregation across sectors of the economy (UNECA 2004). Women's unpaid labor is critical for livelihoods and the security of households and family members. It involves repetitive and time-consuming tasks often described as household chores, such as collection of firewood, water fetching, childcare, sweeping and home cleaning, garbage disposal and cooking. Ghanaian women spend more than two times as much time on domestic work as men. In Ghana, women's unequal land rights affect their access to other resources and their economic, social and political status. Women's access to land is affected by tenancy arrangements, inheritance and land use patterns. Thus, although women may have land usage rights, their access to this resource depends on its availability and the goodwill of the men who control it (UNECA 2004, 2012). For women in Ghana, land ownership is largely considered a "secondary" or "derived" right granted by brothers, fathers, or husbands (Alhassan and Manuh 2005; Alhassan 2009).

16.5 Field Research and Focus Groups

Ghana's customary and traditional gender roles for men and women, coupled with the existence of highly vulnerable communities routinely affected by flooding and fire events, offers an excellent venue for assessing the unique ways women can contribute to the safety and sustainability of their communities and allows for an evaluation of the factors that enhance disaster resiliency. The field research conducted for this analysis represents a collaborative project between the University of South Florida and the University of Ghana to engage in the development of a cross-national community-based disaster resilience initiative to empower and equip women as local-level leaders in disaster risk reduction through active participation in mitigation, preparedness, response, and recovery roles. The Global Hazard Resilience through Opportunities for Women (GHROW) initiative is designed to build capacity around disaster risk reduction knowledge, skills, and advocacy so women can use their social and communication networks to inform, educate and mobilize others within their communities. The project is also designed to strengthen connections between women and other stakeholders in the disaster management community (e.g. NGOs, faith groups, community organizations, private enterprise, and government agencies) to ensure a sustainable exchange of information and access to resources.

Data were collected through rapid assessments of the vulnerabilities and resources of the areas under evaluation coupled with focus group conversations with community residents. Three sites were selected for inclusion in the study including: Sabon Zongo, an unplanned urban settlement or "slum" located in the western part of metropolitan Accra, the capital city of Ghana, and two Volta Lakeshore communities located in Yeji (Brong Ahafo Region) and Buipe (Northern Region). Flooding is a perpetual event and causes serious sanitation and health risks during rainy season in all three localities. Fire is also a hazard among the communities. The Sabon Zongo is a settlement that is home to an immigrant community of primarily Muslim residents (from Mali, Burkina Faso and Ghana's northern regions) nearing 25,000. In many respects, Sabon Zongo is like a village within a city; the roads are unpaved and difficult to navigate, sanitation services are not widespread, and cattle and goat farming exist within a densely populated, urban environment. Sabon Zongo is one of twelve communities within Greater Metropolitan Accra that ranks lowest in terms of environmental health and disaster monitoring and is characterized as "absolutely deprived" of basic services such as potable water, waste disposal, sanitation, drainage, pest control and pollution abatement. These communities have generally been excluded from any meaningful planning in terms of the built environment, environmental conditions and health services. Even where services are provided, a lack of financial and human capacity at the metropolitan government level limits service provision. In these settlement communities, it is women and children who suffer the most from inadequate service provisioning (Songsore et al. 2009).

The communities of Yeji and Buipe are fishing villages located at the shoreline of the Volta Lake. Commercial fishing and its supporting businesses dominate the economies of both localities. The specific areas where focus groups were held are situated in designated flood zones deemed unsafe by the Volta Basin Authority.[2] These communities resemble the urban slums: unpaved roads, limited sanitation, congested living conditions, and a lack of potable drinking water. A cholera outbreak was in progress at the Yeji site. The residents of the three communities are poor, illiterate, undereducated, and largely marginalized from the policy-making process.

Focus groups are especially relevant in Ghana, a country with a strong oral tradition. All community participants had experienced an extreme hazard event— flooding and/or fire. Questions focused on information about the gender specific challenges in dealing with a disaster event; availability, accessibility, and adequacy of community resources; and the identification of strategies for improving resiliency including leveraging existing social ties, enhancing communication networks, peer-to-peer education and outreach, leadership development, and change management within the locality. During focus group sessions, residents often claimed: "We're on our own. We cannot depend on help from the outside." Their resiliency is home grown.

Sabon Zongo[3] Among its residents, the most critical risks to the community are perceived to be flooding, exacerbated by the lack of a drainage system or inadequate waste management where drainage culverts exist, and fire. The outbreak of fire requires community involvement as the municipal fire services cannot access much of the area because of the narrow (and clogged) streets and due to the lack of directional coordinates for locations with the slum. Seasonal rains cause repeat

---

[2] The Volta Basin Authority (VBA) was established in 2007 to coordinate water management policies for the six countries of the Volta River Basin (Ghana, Burkina Faso, Togo, Mali, Benin and Côte d'Ivoire). The Economic Community of West African States (ECOWAS), believing that there might be future conflicts in the region over the sharing of water resources, initiated the formation of the VBA (Ghanaweb 2007). For years, the 400 000 km^2 Volta Basin had been one of the few trans-boundary basins in Africa which had no formal agreement in place for cross-border cooperation and management. Ghana and Burkina Faso together use 85 % of the basin's water.

[3] Focus group conversations with key female and male stakeholders concerning disaster resiliency were facilitated by Ms. Jane Amerley Oku, a well-respected Sabon Zongo community leader (and resident) and former elected member of the Accra District Assembly.

flooding. Refuse is not stored in a central location but is often left in the streets leading to clogged drains and flooding of homes. Ghana lacks a cohesive building code and where laws do exist they are not consistently enforced. The result in the Sabon Zongo is a jumble of buildings held together with all matter of materials. Buildings often encroach on roadways and drainage systems.

Given the gender dynamics of Ghanaian society, the focus groups included both female and male participants. At the Sabon Zongo site, twelve women of various ages and backgrounds were joined by five men—two of whom are respected elders within the Muslim community. Women participants articulated their more vulnerable position by discussing barriers to education, their roles as caretakers for the sick—exposing them to illness and disease, their lack of a voice in community –wide decision-making, and the fact that they, as women, are responsible for the bulk of preparedness and recovery activities (clean-up). And yet, the women of the Sabon Zongo are an entrepreneurial group. They are the primary caretakers of the home and of children and the elderly, but they also are very active in establishing micro-businesses and identifying ways of augmenting their income through the sale of imported items, food preparation, rent from buildings/land acquired through ancestry, and in one case, the creation of a public toilet in a private dwelling—the family then charged a small fee for use of the toilet by community residents. The women of the community have organized various groups including a Ladies Association that pools its resources to help members with celebrations associated with a birth or wedding. The groups meet regularly and are an important source of support for members. Ironically, despite the fact that most women cannot qualify for a bank loan on their own—their husband or wage earner must sign the loan document— banks often prefer to make micro-loans to women as they are perceived to be more reliable and more likely to repay the loan than are men. One female focus group participant was the beneficiary of such a practice and had used the funds to begin a small business venture.

In comparison to the lakeshore communities of Yeji and Buipe, Sabon Zongo has received more attention from NGOs and government entities. A United Nations Habitat WaterAid project built drainage canals in a large section of the settlement. The Accra Metropolitan Authority championed a policy of "a toilet in every home" and assisted with the installation of community and residential toilet facilities. In addition, an agreement among the government of Ghana, UN Habitat and the Sabon Zongo settlement produced a community management model designed to build capacity in the context of operations, maintenance and sustainability of several projects including: drainage routes, a twenty seat public toilet, toilet facilities for two schools, and two solid waste bins and assorted sanitation equipment Although it is an unplanned community, Sabon Zongo does not occupy an area deemed unsafe by the Ghanaian government. This is not the case for the lakeshore settlements.

The Volta Basin The Volta Lake was created as a result of the construction of the Akosombo Dam between 1961 and 1965 designed for hydro-electric power generation. The lake flooded parts of the low-lying Volta River Basin, resulting in the formation of one of the world's largest man-made lakes, covering 3,283 square miles (8,502 square kilometres) or 3.6% of Ghana's total land area. The creation of the Volta Lake and the attractions it offered including fishing and farming activities resulted in the migration of people from other parts of the country (and outside of Ghana) to the basin. Following serious anthropogenic pressures resulting from the migration of people into the Volta basin, the Volta River Authority (VRA) was created by an Act of parliament in 1961 to protect the basin from excessive degradation. The VRA prohibits human activities and settlements below the 280 foot contour line and within a 50 m buffer zone. Limited action to curb encroachment has taken place, however, and monitoring of human activities along the basin is seemingly nonexistent. For instance, the absence of any visible landmarks, such as boundary pillars, defining the area of the buffer zone has encouraged residents to establish unlawful settlements under the guise of unawareness.

Despite extensive flooding and devastation in 2010, settlements continue to appear along the shores of Lake Volta. The residents of these shoreline communities make a living from fishing and despite the danger find it easier to sell their products directly from the shore. Following extensive flooding in 2010, residents were warned not to resettle along the shore, but whole communities have repopulated in flood zones out of economic necessity.

The Volta basin is susceptible to climate change and weather variability due to the nature of water intake and discharge. Small changes in precipitation in the basin cause proportionately large changes in runoff. The Volta basin runoff exhibits higher temporal variability than does the basin rainfall. At the point where precipitation reaches 85 % of average annual rainfall, roughly half of subsequent precipitation is discharged from the basin (van Zwieten et al. 2011). The result is habitual flooding along the river banks where many vulnerable human settlements and economic activities are situated.

Yeji represents an important port community of approximately 25,000. The vast majority of participants in the Yeji focus group had experienced the devastation of the 2010 flooding. The need for basic survival caused many to return to the barrier zone—disaster mitigation and preparedness are "luxuries" for a people living hand to mouth. The greatest perceived risks are seasonal flooding and a lack of potable water (causing disease), electricity and basic sanitation services. Without electricity, emergency response after sundown is complicated and time consuming. In 2010, the flooding of the region began in darkness. Should the residents choose to move from the shore to a nearby slum where living conditions are markedly better, rent would be required for housing whereas the flood zone is "free" but devoid of any level of government/public services. Women residents articulated many of the same sentiments regarding gender inequality as were expressed in urban Accra. As a group, they were also enterprising like the women of Sabon Zongo and microbusinesses proliferated, but there exists a lack of any economies of scale. Like their urban counterparts, most of the women possessed cell phones, several carried more than one. The cell phone is ubiquitous in Ghana. Women in this community can travel to the market where vendors charge cell phone batteries for a small fee. The women use the cell phones to communicate with husbands or other male family members as they fish on the lake. Fishermen can "call-in" the results of their work and expedite the sale of fish to local merchants. Women also use their cell phones as a means of communicating with each other and have developed extensive phone trees for sharing information across the settlement.

The Buipe lakeshore settlements also experienced extensive flooding in 2010. The comments made by focus group participants—ten women of various ages and eight men representing the commercial interests in the community—made similar statements as the residents in Yeji. The focus group was conducted in what is commonly known as the "Bridge Community." The approximately 5,000 residents outnumber the indigenous population of the area, but are considered migrants because of their ethnic minority status. According to the settlement residents, the greatest disaster risk is flooding; more immediate hazards include a lack of potable water, nonfunctioning public toilets, a lack of basic sanitation services, and many expressed that the quality of life has deteriorated since the 2010 flooding. Like their counterparts in other parts of the country, despite their poverty and illiteracy, most women own cell phones.

A common theme among the focus group discussions across Sabon Zongo, Yeji and Buipe was the extensive social networks women have developed within their communities. Both the male and female participants readily agreed that women, in general, are more adept at organization and planning activities. The communities rely on the women's social networks for information, communication, and economic support. In Buipe, the women's group calls itself "Miwuenenyo"—local language for "Let's Do It Better." The group meets weekly and consists of approximately 50 members—all women. Their mission is to support one another and the community at large. An inability to secure small loans from external organizations led the group to pool its resources to establish a small micro-loan program. Innovation is at work in this community, as it is in Yeji and Sabon Zongo, but resources are extremely limited and the challenge is to better utilize existing assets with local relevancy rather than imprint the community with unsustainable goals developed externally.

16.6 Discussion

In Ghana, as in many parts of the world, national and regional disaster resources (equipment in particular) are insufficient and government resources have not kept pace with the level of disaster events. Local communities must often "go it alone" in the context of all phases of emergency management—not just immediate recovery efforts. Locally relevant and simple strategies are of critical importance. The settlements of Sabon Zongo, Yeji, and Buipe represent "recalcitrant populations" in so far as they lack the resources to engage in comprehensive disaster mitigation efforts and have little alternative for relocation from hazardous areas. Nevertheless, the residents of these communities have assets that can be leveraged to improve their resiliency.

The communities' greatest asset is their residents. In particular, familial ties and social organizations offer a particularly robust support network within poor and underserved communities. Among the female population residing in vulnerable settlements, organizational infrastructure exists that could be better leveraged for greater disaster resiliency. Women belong to social groups that meet regularly and provide opportunities for information sharing and the pooling of resources. Women have formal information communication structures often in the form of phone trees that offer an avenue for rapid information dissemination.

Significant potential exists for leveraging wide-spread cell phone ownership—among all segments of the population. Cell phones are easily accessible and prevalent. Most women in the settlements own at least one. Settlement communities, however, have yet to fully utilize this technology as a disaster resiliency tool. FM radio, "town criers" or "drum beats" remain the primary sources for information about emergency events. Despite wide-spread illiteracy, a cell phone emergency notification system could be constructed to alert residents to hazards or emergencies via voice messages or text messages utilizing symbols. Most women know little more than how to place a call using their cell phone. Education concerning how to better utilize the technology they already possess is important in the context of disaster resiliency efforts.

In addition, women are on the front-line of disaster preparedness and recovery and thus represent an important resource of information about what works and what doesn't. Too few government officials, NGOs, or researchers ask these questions post-disaster. The women themselves are not fully aware of what agencies and organizations are responsible for emergency management. The social and information networks established by women's groups augmented by their familiarity with the impact of disasters make women ideal candidates for developing protective measures. In this context "protective measures" refers to simple actions that can help to save lives, contribute to public health and safety, and protect the built environment. Examples from the communities evaluated include identification of a designated location for sheltering children in the event of a natural disaster; in the case of Sabon Zongo, the use of garbage containers to keep refuse from clogging drainage routes and the promotion of centralized garbage collection and recycling; in the case of the Volta Lake communities, placing cement blocks at the base of housing or utilizing concrete (or sandcrete) instead of a mixture of clay, mud and wood sticks to construct homes and buildings.

Across communities (Sabon Zongo, Yeji, and Buipe) women feel they do not have enough of a voice in disaster planning and recovery—their input is not commensurate with their responsibility in the event of a disaster. There is a critical need for greater identification and mentoring of local leaders—both female and male—especially male leaders willing to include women in the disaster management decision-making process. Two such individuals stand-out within the settlement communities as examples of the impact a single person can make toward improving the human condition. In Sabon Zongo, Ms. Jane Amerley Oku, a former elected member of the Accra District Assembly and founder of the Janok Foundation, is a settlement resident and widely respected community leader. Her efforts have led to the greater enfranchisement of women in disaster preparedness and emergency management activities, improvements to sanitation in the settlement, and myriad other activities that benefit the residents of Sabon Zongo. As a Christian woman in a predominantly Muslim community, her leadership is a prime example of "positive deviance." One that illustrates the importance of utilizing her experience to empower others and establish a pipe-line of future leaders. In Yeji, Mr. Yahaya Ibrahim is an example of a male community leader committed to mobilizing young people and women for self help initiatives, including disaster resiliency. Like Ms. Oku, Ibrahim is a former elected official from his community—serving in the Pru District Assembly His activism has empowered the Yeji lake settlements to work together to make decisions that improve resiliency including moving "critical" infrastructure such as toilets, schools and other community buildings further from the lakeshore to areas of higher ground, including the rental of an apartment in central Yeji to enhance the safety of the settlement's children who travel on foot to school and are at risk when crossing commercial roadways.

16.7   Conclusion

Resiliency can thrive in the worst of conditions. Employing local knowledge and a creative approach to identifying "community assets" to build capacity translates into the possibility for resiliency and sustainable community development even among the poorest localities. Impoverished communities have the ability to be innovative and proactive. The communities included in this study are evidence of this phenomenon.

An important catalyst is leadership at the local level that recognizes and values the unique position of women within the community. The lessons learned from the settlement populations in Ghana include the recognition that local residents perceive great value in their own capacity to be resilient. In each community, focus group respondents emphasized that "the people" represent the greatest asset in the context of disaster. This recognition allows for personal networks to be leveraged, in locally relevant ways, for the purpose of enhancing communication, the dissemination of information, knowledge sharing, leadership building, for the purpose of enhancing resiliency. Women represent the cornerstone of any such effort. As such, the social ties and networks of women are important factors in development efforts to promote sustainable change and give voice to this vulnerable population in the decision-making realm of global disaster politics and policies.

References

Alhassan, O. (2009). Customary land documentation in the Wasa Amenfi District, Western Ghana. *Ghana Journal of Geography, 1,* 95–114.

Alhassan, O., & Manuh, T. (2005). *Land registration in Eastern and Western regions of Ghana (Securing land rights in Africa research report 5).* Nottingham: IIED/Russell Press.

Antwi, V. S. (28 February 2008). ActionAid Ghana helps reduce disaster risk. The Statesman. http://www.thestatesmanonline.com/pages/news_detail.php?newsid=5874&section=1. Accessed 28 Feb 2008.

Armah, F. A., Yawson, D. O., Yengo, G. T., Odoi, J. O., & Afrifa, E. K. A. (2010). Impact of floods on livelihoods and vulnerability of natural resource dependent communities in Northern Ghana. *Water, 2*(2), 120–139. doi:10.3390/w2020120.

Caruson, K., & MacManus, S. A. (2012). Interlocal emergency management collaboration: vertical and horizontal roadblocks. *Publius: The Journal of Federalism*, (42)1, 162–187.

Chew, L., & Ramdas, K. N. (2005). Caught in the storm: The impact of natural disaster on women, December. Retrieved from The global fund for women website: http://www.globalfundforwomen.org/storage/images/stories/downloads/disaster-report.pdf.

Comfort, L. K., Boin, A., & Demchak, C. C. (Eds.). (2010). *Designing resilience: Preparing for extreme events*. Pittsburg: University of Pittsburg Press.

Cutter, S. L., Burton, C. G., & Emrich, C. T. (2010). Disaster resilience indicators for benchmarking baseline conditions. *Journal of Homeland Security and Emergency Management, 7*(1), 1547–7355. doi:10.2202/1547-7355.1732.

Intergovernmental Panel on Climate Change (IPCC). (2011). IPCC special report on managing the risks of extreme events and disasters to advance climate change adaptation (SREX), November. Retrieved from the Intergovernmental Panel on Climate Change website: http://www.ipcc-wg2.gov/SREX/report/.

Kapucu, N., Hawkins, C., & Riveria, F. (2013). *Disaster resiliency: Interdisciplinary perspectives*. New York: Routledge.

Mercer, J. (2010). Disaster risk reduction or climate change adaptation: Are we reinventing the wheel? *Journal of International Development, 22*(2), 247–264.

Miller, P. (2012). Weather gone wild. *National Geographic*, (September), 30–55.

Ministry of the Interior, Republic of Ghana, National Disaster Management Organization (NADMO). (2012, 2011). National platform report and 2011 report on flooding. http://www.nadmo.gov.gh/index.php/ component/search/?searchword=volta%20 basin & searchphrase=all & Itemid=241.

Morrow, B. H. (1998). *The gendered terrain of disaster*. Westport: Praeger Publishers.

National Disaster Management Organization (NADMO). (2012). Senior program coordinator (October 12). Interview by Kiki Caruson, University of South Florida (In person.) Accra Ministries, Accra Ghana.

Neumayer, E., & Plümper, T. (2007). The gendered nature of natural disasters: The impact of catastrophic events on the gender gap in life expectancy, 1981–2002. *Annals of the Association of American Geographers, 97*(3), 551–566.

Pelling, M., & Wisner, B. (Eds.). (2009). *Disaster risk reduction: Cases from urban Africa*. London: Earthscan.

Poole, M. (2005). *Family: Changing families, changing times*. Crows Nest: Allen and Unwin.

Poole, M., & Issacs, D. (1997). Caring: A gendered concept. *Women Studies International Forum, 20*(4), 529–536.

Shaw, R., & Krishnamurthy, R. R. (Eds.). (2009). *Disaster management: Global challenges and local solutions*. Boca Raton: CRC Press.

Songsore, J., Nabilia, J. S., Avle, S., Bosque-Hamilton, E. K., Amponsah. P., & Alhassan, O. (2009). Integrated disaster risk and environmental health monitoring: Greater Accra metropolitan area, Ghana. In M. Pelling & B. Wisner (Eds.), *Disaster risk reduction: Cases from urban Africa*. London: Earthscan.

United Nations Economic Commission for Africa (UNECA). (2004). Economic report on Africa. http://www.new.uneca.org.era.era2004.aspx.

United Nations Economic Commission for Africa (UNECA). (2012). Assessing progress in Africa toward the millennium development goals. MDG Report 2012. http://www.new.uneca.org/mdgreports/mdgreport2012.aspx.

United Nations International Strategy for Disaster Reduction (UNISDR). (2010). *Disaster risk reduction: An instrument for achieving the millennium development goals*. New York: United Nations.

United Nations. Monrovia Communique. (2013). A new global partnership: Eradicate poverty and transform economies through sustainable development—The report of the high-level panel of eminent persons on the post-2015 development agenda, May 30, 2013. http://www.post2015hlp.org/wp-content/uploads/2013/02/Monrovia-Communique-1-February-2013.pdf.

United Nations Office for Disaster Risk Reduction (UNISDR). (2009). Global assessment report on disaster risk reduction. http://www.unisdr.org/we/inform/ publications/9413.

United Nations Office for Disaster Risk Reduction (UNISDR). (2012). Economic losses from disasters set new record in 2012, March 14, 2013. http://www.unisdr.org/archive/31685.

United Nations Office for Disaster Risk Reduction (UNISDR). (2013). Global assessment report, 2013. http://www.un.org/apps/news/story.asp?NewsID=44911&Cr=disaster&Cr1=risk.

University of Ghana Institute for Environmental and Sanitation Studies. (2012). The effects of Volta floods on Volta Lake shore communities (Information Note: 015, May 2012). http://www.ug.edu.gh/iess/post/briefs/The%20Effects%20of%20Volta%20Floods%20on%20Lake%20Volta%20Shore%20Communities%28Jesse%20Ayivor%29.pdf.

UN-Habitat. (2008). *The state of African cities report 2008: A framework for analysis.* Nairobi: UN-Habitat.

van Zwieten, P. A. M., Béné, C., Kolding, J., Brummett, R., & Valbo-Jørgensen, J. (2011). Review of tropical reservoirs and their fisheries: The cases of Lake Nasser, Lake Volta and Indo-Gangetic basin reservoirs. (FAO Fisheries and Aquaculture Technical Paper No. 557). Rome: Food and Agriculture Organization of the United Nations.

Waugh, W., & Tierney, K. (Eds.). (2007). *Emergency management: Principles and practices for local government* (2nd ed.). Washington, DC: ICMA.

Women's Environment and Development Organization (WEDO). (2008). Case study: Gender, human security and climate change: Lessons from Bangladesh, Ghana and Senegal. New York: WEDO. http://www/wedo.org/wp-content/uploads/hsn-study-final-may-20-2008.pdf.

Yonder, A., Akcar, S., & Gopalan, P. (2005). *Women's participation in disaster relief and recovery.* New York: Population Council.

Zeitlin, M. (1991). Nutritional resilience in a hostile environment: Positive deviance in child nutrition. *Nutrition Reviews, 49*(9), 259–268.

Zeitlin, M., Ghassemi, H., Mansour, M., Levine, R. A., Dillanneva, M., Carballo, M., & Sockalingam, S. (1990). *Positive deviance in child nutrition: With emphasis on psychosocial and behavioral aspects and implications for development.* Tokyo: United Nations University Press.

Chapter 17
Evolving and Implementing a New Disaster Management Paradigm: The Case of the Philippines

Ralph S. Brower, Francisco A. Magno and Janet Dilling

> *It has often been said that in cultural terms we were in the*
> *convent for 300 years and in Hollywood for fifty. We prayed the*
> *rosary in the morning and dreamed of Marilyn Monroe at night.*
> Antonio Meloto, Builder of Dreams (2009), p. 64

17.1 Introduction

In the 1950s the Republic of the Philippines was seen as a shining economic light for its Southeast and East Asian neighbors. In the decades since most of them have overtaken the Philippines, leaving many Filipinos to wonder "what happened to us"? Tony Meloto introduces one prevalent account: three hundred and fifty years of colonial diktat, first under Spain and the Catholic Church, and then under the United States, sapped a sense of self-determination from the collective Filipino psyche, even though many contemporary individuals and groups exhibit a "can do" attitude. While Meloto alludes to this imposition of culture, we submit that the Philippines' stunted development also reflects a failure to foster institutions that promote inclusive property ownership and democratic and economic participation (e.g., Acemoglu & Robinson 2012; North et al. 2009). In our subsequent conclusions we return to this interplay between local context and institutions in the development process of the Philippines and illustrate its importance for building disaster resilient systems.

For more than three decades disaster management in the Philippines, as in many developing countries, reflected this development malaise; what prevailed was a focus on disaster response and recovery with the military and national police as central actors (Alexander 2002; Gaillard 2011). In the early 1990s Nongovernmental Organizations

R. S. Brower (✉) · J. Dilling
Florida State University, Florida, USA
e-mail: rbrower@fsu.edu

J. Dilling
e-mail: jdilling@fsu.edu

F. A. Magno
De La Salle University, Manila, Philippines
e-mail: magnofra@gmail.com

N. Kapucu, K. T. Liou (eds.), *Disaster and Development,* Environmental Hazards,   289
DOI 10.1007/978-3-319-04468-2_17, © Springer International Publishing Switzerland 2014

(NGOs) and progressive political leaders began to recognize how the response and recovery emphasis aligned poorly with development goals. Armed with nationalistic pride borne of the non-violent overthrow of Ferdinand Marcos' dictatorship a few years earlier, they began to push in the 1990s for a realignment of participants and responsibilities. After several failed legislative attempts a consortium of civil society groups, business leaders, and university experts enlisted the help of legislative champions and succeeded in establishing the 2010 Disaster Risk Reduction and Management Act.

The process was noteworthy for the ways it brought civil society leaders to the table with military leaders who, 25 years earlier, had been tasked by Marcos with stamping out civil society groups. The new law introduced the language of community-based disaster risk reduction and community resiliency while injecting good governance principles and an understanding about root causes of vulnerability— poverty, landlessness and land use policies. The law also ensured the participation of civil society groups in councils charged with overseeing disaster risk reduction and management at each level of government.

We begin by laying out the social, political, cultural, and historical context in the Philippines on which past and present disaster management is premised. We offer a description of natural and human-created disaster patterns of the Philippines, a brief history of recent disasters, a characterization of the system of actors in Philippine disaster management, and reports of novel innovations in various local communities. We also provide a discussion of the ongoing challenges that practitioners and advocates still encounter in their efforts to mainstream disaster management into the country's development goals. We conclude by illuminating through the Philippine case the interwoven dynamics of development, vulnerable populations, economic and political elite interests, and disaster management.[1]

17.2 Context of the Study

With an estimated 2012 population of more than 92 million, the Republic of the Philippines is the 12th most populous country in the world (NSO 2012). It is a nation of 7,107 islands distinguished by three main geographical divisions: Luzon (largest island, in the North), Mindanao (second largest island, in the South), and the Visayas (central region of numerous islands).

As of December 2012 the three divisions were apportioned into 17 administrative regions, 80 provinces, 140 cities, 1,494 municipalities, and 42,026 baran-

---

[1] While this manuscript was being completed, on November 8, 2013, Typhoon Haiyan—known locally as Yolanda—made landfall in the central Philippines, bringing strong winds and heavy rains that resulted in flooding, landslides, and widespread damage. As of January 2014 Typhoon Haiyan was reported to have caused 6,201 deaths, affected an estimated 16 million people, and destroyed or damaged more than one million homes, as well as public infrastructure and agricultural land, across 44 provinces (NDRRMC 2014).

gays[2] (NSCB 2012). Unlike states and local governments in the U.S., these several levels of local government units (LGU) take primary authority from the national government. Police and numerous other public services in the Philippines are directed from the national government.

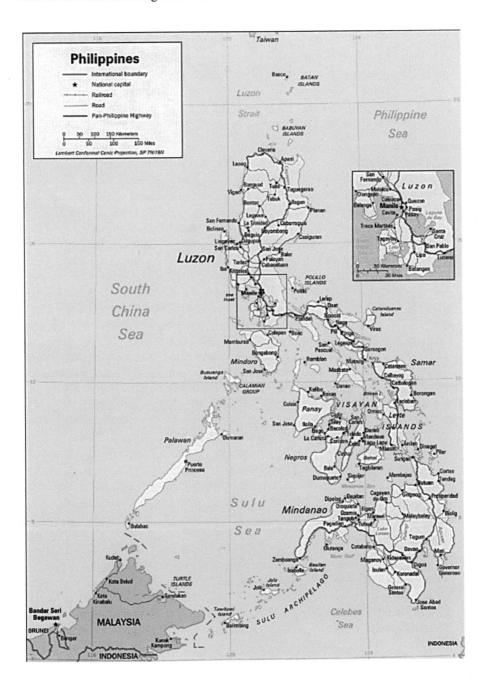

The Republic of the Philippines has a representative democracy modeled in many ways on the U.S. system. In 1987 a new constitution was adopted shortly after Corazon Aquino's accession to the presidency following non-violent protests that ousted the Marcos dictatorship. The constitution reestablished a separation of powers with a president, bicameral national legislature, and an ostensibly independent judiciary. It is governed as a unitary state with the exception of autonomous regions in Muslim areas of Mindanao and in the Cordillera region of northern Luzon (U.S. Department of State 2013).

17.2.1   Disasters in the Philippines

In the recent World Risk Report 2012 (Alliance Development Works 2012) the Philippines ranked third out of 173 countries in terms of susceptibility to disasters, behind only the small South Pacific island nations of Vanatu and Tonga. The Philippines has held this position for two years in a row. Because of their proximity to the ocean these island countries are significantly exposed to natural hazards such as storms, flooding, and sea level rise. In the case of the Philippines overall risk is further exacerbated by vulnerability due to under development.

The Philippines is located on the Pacific Rim feature known as the "ring of fire"; it is subject to major seismic fault lines and has approximately 220 volcanoes, of which 22 are considered active (UNFAO n.d.). Philippine volcanoes are regarded as the most deadly and costly in the world because of the frequency and damage of eruptions. Mudflows, frequently exacerbated by seasonal heavy rains, often occur in conjunction with eruptions.

In recent years, however, weather events have caused the most serious disasters in the Philippines. Weather patterns in the Philippines are generally dictated by prevailing winds. These consist of monsoons from the southwest (*habagat*) from approximately May to October and the cooler, drier northeast monsoon (*amihan*) from November to early May. Typhoons, known as Bagyo *and arriving in an easterly direction from the Philippine Sea on the Pacific side of the country,* frequently hit Luzon and the Eastern Visayas from June to November. The typical habagat is a slow-moving storm that draws moisture off the South China Sea and often causes flooding on land from continuous rainfall. In August of 2012, for example, a "hanging" habagat deluged Manila for nearly two weeks, flooding up to a third of the metro area (Whaley 2012), causing more than a 100 deaths from flooding and landslides, and driving 440,000 to evacuation centers.

Typhoons, on the other hand, are faster moving storms packing substantial winds that may do millions of dollars of damage. Since their seasons roughly coincide, typhoons' effects are often enhanced by the habagat, producing heavy rains that linger for days over wide areas of the country. The country experiences an average of 20 typhoons per year; about half of these make landfall and inflict destruction (PAGASA 2009). In the fall of 2009 typhoons Ondoy (international name Ketsana) and Pepeng (Parma) left nearly 1,000 dead, displaced millions of others, damaged thousands of homes and other infrastructure, and destroyed crops. The cost of damage was equivalent to 2.7% of GDP and constituted a major setback to Philippine development (AusAID 2011). Recent tropical storm patterns have been less predictable. During the past two Decembers severe tropical storms have struck the southern island of Mindanao, which is well south of the usual path of typhoons.

In December 2011 tropical storm Sendong (Washi) struck eastern Mindanao, dropped torrential rain at higher elevations, exited into the Sulu Sea, and eventually crossed the westernmost island of Palawan. The downpour raised river levels precipitously, killing hundreds of people in flooding in the northern Mindanao cities of Cagayan de Oro and Iligan, and causing boating and other fatalities in its wake. At least 1,268 were killed and property damage was estimated at approximately $ 50 million (NDRRMC 2012). Typhoon Pablo (Bopha) swept ashore on eastern Mindanao in early December 2012. It devastated high elevation mining and farming communities, then proceeded toward the west following a path slightly to the south of Sendong. The United Nations Office for the Coordination of Humanitarian Affairs has reported (UNOCHA 2013) casualties of 1,067 dead and 834 missing and damage at $ 1.04B. It was by far the most destructive storm ever to hit the Philippines and ranked as the most deadly catastrophe in the world in 2012 (Domingo 2013).

Apart from seismic activity and typhoons, heavy rains also prove destructive when mudslides engulf rural settlements built near steeply sloped land that has been denuded by forestry or farming. An unfortunate example was a February 2006 mudslide in Southern Leyte which buried the entire population of nearly 1200 in Guinsaugon Village (Stone 2006).

17.2.2 Social Vulnerability

Of the 92 million inhabitants of the Philippines more than one third reside in Metro Manila and other provinces of Central Luzon (NSCB 2012). A national population growth rate between 1995 and 2000 of 3.21% decreased to an estimated 1.95% for the period from 2005 to 2010, but still remains among the highest in the world. The country's median age is 22.7 years, which reflects the high birth rate over the last several decades (NSO 2008).

Many observers regard poverty as the Philippines' most critical social problem. More than one-quarter (26.5%) of the population fell below the poverty line in 2009 (National Statistical Coordination Board 2011). Although poverty levels have declined over the last two decades, the gains have been disappointing by comparison to other Southeast Asian nations, such as Indonesia, Thailand, and Vietnam (NSCB 2011).

The United Nations estimates that the Philippine population is 59% urbanized and this percentage continues to increase. Between 1960 and 1995 the Philippines' urban population grew at an annual rate near 5% but then slowed slightly to approximately 3%. As in other developing countries, urban-rural migration has driven much of this increase, and at a rate faster than employment growth. This results in significant levels of urban poverty and slums (The World Bank n.d.).

Poverty reduction in the Philippines has not kept up with growth in GDP and has been uneven over time (NSCB 2011). Observers blame high population growth, high unemployment and inflation, and large income disparities (ADB 2009; NEDA 2011). One obvious implication of poverty is the proliferation of informal settlers.

Informal settlements Estimates of the numbers of informal settlers vary wildly. The Philippine National Census Office defines informal settlers as "households occupying a lot rent-free without the consent of the owner" (quoted in Cruz 2010, p. 2) and estimated that 550,771 households in the Philippines as of August 1, 2007 were living as informal settlers. Quezon City, largest of the NCR cities, had more than 90 thousand such households, and Rizal Province and Davao City each had more than 20 thousand households (Cruz 2010). A separate accounting undertaken in 2007 by the National Housing Authority used a more expansive definition to include "homeless and underprivileged citizens"; this other study identified 544,609 informal settler households in the NCR alone (Cruz 2010). Contrast this with an accounting from the United Nations Human Settlements Programme, which estimated in 2003 that 2.5 million slum dwellers lived in 526 different communities in Manila alone. This included both informal settlers and more established households (UNHSP 2003). The World Bank has estimated that half of the area's 12 million residents are slum dwellers (The Hindu 2010). These vastly differing numbers should raise concerns; if the methods of the UNHSP and World Bank are to be trusted, the much lower numbers from in-country agencies suggest faulty institutionalization of reliable data collection and analysis methods. By any accounting informal settlers represent a significant population of families in Metro Manila and more broadly in the Philippines.

The United Nations estimated in 2003 that the majority of those living in Manila's urban slums had been there at least two decades (UNHSP 2003), suggesting that the rural-urban migration pattern is not new. They settle "on vacant private or public lands, usually along rivers, near garbage dumps, along railroad tracks, under bridges and beside industrial establishments" (UNHSP 2003, p. 215). These locations are otherwise vacant because they are vulnerable to flooding, landslides, traffic accidents, and toxins. Many informal settlers are characteristically unemployed, although some work as domestic helpers, tricycle and jeepney drivers, part-time construction laborers, street-side vendors, or recycling foragers. The location of settlements is thus influenced by both the availability of vacant property and nearby employment.

Governments at all levels have been largely reactive rather than proactive in solving the problems of these vulnerable populations. As one of our informants noted:

> The flooding that occurs every year in various levels of calamity comes about primarily because local governments have granted building permits for developers to build homes or industries in waterways and squatters put up shanties in the slum areas nobody wants.

Land use and water management practices in many parts of the country exacerbate the potential threats to informal settlers in vulnerable locations. During typhoon Ondoy in October 2009, supervisors released excess water from dams on the Agno

River in Pangasinan Province about 100 miles north of Manila out of fear the dams would breach. Unfortunately, because the decision was taken quickly and without adequate warning systems the surge of water led to scores of drownings among informal settlers in the downriver floodplain, and thousands of other residents were rescued from rooftops after 30 out of 46 towns along the river were inundated (The Guardian 2009).

Another informant offered this observation from the August 2012 flooding in Manila:

> The situation is made worse because of denuding of tree cover in high elevations to the East in Rizal and Bulacan Provinces, which contributes to the rapid runoff that then deluges the Marikina and Pasig River watersheds and Laguna de Bay. In addition to low lying areas within flood plains, squatters often perch their shanties on steeply sloped land, which then gives way when the soil becomes saturated.

Informal settlers were conspicuous in media coverage of the August 2012 floods in Metro Manila. Some drowned and others died when mudslides buried homes built on vulnerable ground. But the scene also provoked government leaders to get tough with informal settlers. One public official drew special attention to those "living in creeks and along Pasig River and San Juan River as they serve as obstructions to the easy flow of water" (Villas 2012). In this account informal settlers are no longer vulnerable casualties of the flooding: they are its culprits. Thus orders came from the Secretary of the Department of Interior and Local Government to NCR mayors to begin clearing 125,000 informal settlers and relocating them to other government designated sites (Villas 2012). Another source reported that the President had ordered Public Works Secretary Rogelio Singson "to clear water channels of all 'obstructions,' to the point of 'blowing up houses if they (residents) won't leave within a certain period'" (Elona 2012).

A few days later the plans were put on hold when the Vice President acknowledged that 979 informal settler families who had been relocated from Manila to a site in Rizal province before the floods had to be further evacuated to higher ground when the relocation site flooded. Another proposed relocation site, the Vice President admitted, also ended up under water (Reyes 2012). But before these problems came to light a settler clearing effort that went forward in another NCR city, Makati, resulted in a violent confrontation between police and a group of informal settlers. At least 22 people were injured as demolition workers from the Makati City government evicted 86 families who refused to leave, and the city subsequently arrested nine of the settlers and pledged to sue them. In a statement to the media the mayor said:

> We are determined to hold accountable those who have instigated the violence and lawlessness in the area and bring them to justice. From the outset, we have followed the law to the letter and fully complied with due process in handling the matter. The affected residents have been granted many concessions and ample time to leave the area voluntarily, so they had no reason to resort to violence. (quoted in Reyes 2012)

A similar incident in 2010 in Quezon City led to a rock-throwing conflict between demolition workers and informal settlers who resisted an eviction notice that gave the settlers seven days to move to a government-designated relocation site. They were being moved to accommodate a private development (The Hindu 2010).

Despite various efforts over the years by NCR and national government leaders to remove informal settlers from ostensibly vulnerable areas along rivers, railroads, and roadways, relocation efforts have rarely succeeded. Various informants from our field study related stories about government efforts to relocate NCR slum dwellers to rural areas in nearby provinces in which most of those relocated soon returned to the urban area when they discovered the resettlement sites provided neither employment opportunities nor public transportation. The underlying conflict in interests and values illuminated in these events reinforces a need for better assessment tools to identify underlying dynamics that can be leveraged to reduce disaster risk for these vulnerable populations as well as comprehensive planning that goes beyond short-term solutions aimed at merely moving the settlers out of harm's way.

Terrorism and conflict The Philippines also has a significant history of human-made disasters, including periodic violent strife between the government and both communist and radical Muslim elements. Recent examples include the massacre of 58 people in November 2009 in the town of Ampatuan in Mindanao, an August 2010 hostage crisis in Manila in which 8 Chinese tourists and the gunman were killed, and a January 2011 bus bomb that killed 5 and injured another 14 in the NCR City of Makati. In the Muslim areas of Mindanao the government is engaged in ongoing political negotiations with the largest separatist organization, the Moro National Liberation Front. Other militant groups remain, although with diminishing numbers, in some rural areas, including the Moro Islamic Liberation Front, the communist New People's Army, and the Abu Sayyaf (BBC News 2007; Schiavo-Campo & Judd 2005).

17.3   Disaster Management in the Philippines

Presidential Decree No. 1566 (PD 1566), issued by President Marcos in 1978, provided the foundation for disaster management in the Philippines. It created Disaster Coordinating Councils at all levels from the national government down to the barangay level, and local executives were designated as chairs at every level. At the top was the National Disaster Coordinating Council (NDCC), a policy making body headed by the Secretary of National Defense. Other members, totaling 18 in all, included the secretaries of public works, transportation and communications, social welfare and development, agriculture, education, finance, labor, justice, trade and industry, local government, health, and natural resources; the Armed Forces Chief of Staff; and the President's Executive Secretary (Pedroso 2010). As a policy making body, the NDCC did not engage in disaster management operations, but delegated those responsibilities to the Office of Civil Defense. Each of the Philippines 17 administrative regions also had its own Regional Disaster Coordinating Council (RDCC). Each local government unit (LGU)—province,

---

Table 17.1: National Disaster Risk Reduction and Management Council

Chair: Secretary, Department of National Defense

Vice Chairpersons:

Secretary, Department of Interior and Local Government: Disaster Preparedness
Secretary, Department of Social Welfare and Development: Disaster Response
Secretary, Department of Science and Technology: Disaster Prevention and Mitigation
Director-General, Natonal Economic Development Authority: Disaster Rehabilitation and
Recovery

Members:

All other department secretaries
Chief, Philippine National Police
Executive Secretary to the President
Secretary-General, Philippine Red Cross
Presidents of Government Service Insurance System and Social Security System
President, Union of Local Authorities of the Philippines
Presidents of League of Provinces, League of Cities, League of Municipalities, Liga ng Mga
Barangay
Three representatives of Civil Society Organizations
One representative from the private sector

Executive Director: Administrator, Office of Civil Defense
Secretariat: Office of Civil Defense

---

city, municipality, and barangay—similarly had a disaster coordinating council (Agsaoay-Saño 2010).

The Disaster Risk Reduction and Management Act was signed into law by President Gloria Macapagal Arroyo in May, 2010. This law was the culmination of several failed legislative efforts to reform PD 1566, dating back to 1997. In the sections that follow we describe the structures created by the new law and the advocacy activities that brought the spirit of community-based approaches and disaster risk reduction.

National Disaster Risk Reduction and Management Council The 2010 law substantially reorganized the oversight structure. The NDCC was replaced by the National Disaster Risk Reduction and Management Council (NDRRMC), which spread responsibility to a number of additional actors. As before, the Secretary of the Department of National Defense chairs the NDRRMC, but four Vice Chairpersons were also established by the new law. The Vice Chairpersons' responsibilities align with the four phases of disaster management as defined in international disaster management circles (see Table 17.1).

The new law expanded the governing body to more than 40 members, including all the major government agency heads and presidents of the local government leagues. The law also assigns memberships to the Philippine Red Cross, four CSOs, and one representative from the private sector. The Administrator of the Office of

Civil Defense acts as executive director of the National Council and the OCD serves as its secretariat.

The Disaster Risk Reduction and Management Act was a product of nearly two decades of advocacy. As early as 1992 the National Land Use Committee had acknowledged that disaster preparedness was not generally integrated into overall development planning (cited in Agsaoay-Saño 2010). The traditional approach regarded natural disasters as unforeseeable events, but overlooked the source of casualties and suffering originating from vulnerable populations and insufficient development capacity. Vulnerability and under development, in turn, arose from social inequalities and government policies that catered to status quo interests (Heijmans & Victoria 2001).

A review of Presidential Decree 1566 in 1997 acknowledged the need to use the calamity fund for preparedness, to focus on pre-disaster risk management, amend public school curricula and development planning to integrate disaster management, and institutionalize civil society organizations' participation in disaster management (Agsaoay-Saño 2010). Despite this growing awareness, at least nine separate bills were introduced in Congress between 1998 and 2004, all seeking to integrate development goals into disaster management, and all failed. Many were re-filed during the 13th Congress (2004–2007) and again failed.

As a signatory to the 2005 Hyogo Framework for Action (HFA) the Philippines was required to: (a) produce a baseline assessment of its disaster risk reduction capabilities and needs, (b) create an implementing mechanism for the HFA, (c) publish a summary of its HFA programs and update them periodically, (d) develop review procedures to analyze and assess ongoing vulnerability and risk, especially for hydrometeorological and seismic hazards, (e) report progress in existing international agreements concerning sustainable development, (f) consider, as appropriate, approving or ratifying international legal instruments relating to disaster reduction, and (g) promote integration of risk reduction for present and future climate change into disaster risk strategies (ISDR 2005, pp. 14–15). Subsequently, several bills were filed in both houses of the 14th Congress (2007–2010) with the objective of shifting from a reactive to a proactive approach to disaster risk management (Agsaoay-Saño 2010). At this point many in the civil society community stepped forward to inspire legislators to underscore community-based disaster risk reduction and community resiliency and address the root causes of vulnerability.

Disaster Risk Reduction Network, Philippines At the center of the advocacy was a group of civil society, local government, academic, and other actors who came together as the Disaster Risk Reduction Network of the Philippines (DRRNet-Phils). Two key actors in the formation of the DRRNet were the Philippine International NGO Network (PINGON) and the international NGO, Christian Aid (Agsaoay-Saño 2010). PINGON was an association of international NGOs active in the Philippines; among them were Oxfam, Christian Aid, Plan International, and World Vision Development Foundation. Other central actors in the DRRNet included groups from three leading Manila universities, University of the Philippines, Ateneo de Manila, and De La Salle University; business interests represented by the Corporate Network for Disaster Response, and Philippine NGOs and community based organizations from throughout the archipelago.

The civil society sector of the Philippines was well-suited to the task. Where the voluntary sector in the U.S. and other Western countries embodies the instrumental professionalization of third party government (Salamon 1987), civil society organizations (CSO) in the Philippines are among the most politicized in the world (Clarke 1998; Hilhorst 2003), a reflection of centuries of struggle against colonial powers and, more recently, the martial law dictatorship of Ferdinand Marcos (Cariño & Fernan 2002).

In 2007 and 2008 Christian Aid initiated advocacy for replacing the Philippines' traditional response and recovery disaster management approach with a disaster risk reduction emphasis. They persuaded PINGON to enlist a broader set of advocacy partners. In June 2008, PINGON arranged a meeting of 31 civil society organizations and community leaders as potential advocates for a community-based approach to disaster risk management. From this meeting sprang the DRRNet as an umbrella formation to advocate for a disaster risk reduction emphasis in pending Congressional legislation. The DRRNet eventually boasted a membership of some "300 CSOs, communities, practitioners and advocates" for the Hyogo Framework and community-based disaster risk management (Agsaoay-Saño 2010, p. 133).

The DRRNet lobbied both houses of Congress. Their actions included drafting passages of the legislation, pushing for specific Implementing Rules and Regulations for the legislation, participating in public hearings, conducting research, providing evidence-based presentations and community education, briefing the media, and networking among member organizations and with the various local government leagues (Agsaoay-Saño 2010). Representatives Biazon and Guingona were eventually brought on board as legislative champions of the bill.

The advocates consistently pushed to replace the traditional reactive paradigm with one that emphasized disaster preparedness, recognition of the connection between disaster risk reduction and sustainable development, participation from community-based groups in governance and oversight, attention to vulnerable populations, and strengthening community capacity. DRRNet member organizations also promoted local advocacy and worked simultaneously with LGUs to establish local enabling ordinances.

The new law reflects much of the language and substance that the DRRNet had sought. The law gives legitimacy to community-based disaster risk management, and community resiliency becomes an important goal. The law addresses root causes of disaster vulnerabilities explicitly—poverty, landlessness and land use policies. Good governance principles are also addressed in the law's wording, as were community-level action and responsibility.

The law also created disaster risk reduction and management offices at lower levels of government—region, province, city, and municipality—and designated that four seats on each DRRM council at each level be assigned to civil society organizations. Coordinating Councils in the barangays were eliminated and their responsibilities shifted to existing Barangay Development Councils. This assignment of civil society seats in all the councils enhances the likelihood that community level advocates will have a voice in shaping future local ordinances and gives community based organizations a role in communications between their communities and their local government units. Institutionalizing collaborative roles for local government and civil society groups in this way signified a major paradigm shift toward a proactive disaster risk reduction approach that the DRRNet advocates had sought.

Calamity Funds At the national level Philippines disaster funding had been traditionally tied to the National Calamity Fund (NCF). In the previous structure the NDCC recommended and the President approved disaster funding for departments and agencies with implementation responsibilities, as well as to LGUs. LGUs were required to allocate 5% of estimated revenues to a Local Calamity Fund (LCF). Historically the majority of calamity funds were set aside to provide immediate response and relief in the wake of disasters; Local Calamity Funds were restricted to use only for circumstances when the LGU's legislative body officially declared a state of calamity. In 2003 a joint circular from the Department of Budget and Management (DBM) and the Department of Interior and Local Government (DILG) permitted LGUs to use their LCFs for preparedness and other pre-disaster activities as well (Agsaoay-Saño 2010).

The 2010 Disaster Risk Reduction and Management Act substantially altered the use of local calamity funds. Of the 5% appropriation, now called the Local Disaster Risk Reduction and Management Fund (LDRRMF), only 30% could be allocated to a Quick Response Fund (QRF) in standby for immediate disaster response and relief. The balance of the LDRRMF accrues to a special trust fund for supporting disaster risk reduction and management activities over the next 5 years, and unexpended funds were to be returned to the general fund after 5 years.

DRRNet members expressed disappointment about two aspects of the bill: failure to commit new national budget resources to disaster risk reduction and failure to replace the Office of Civil Defense, supervised by the military as operational center for disaster management. Network members had advocated for a new central disaster management entity in the Office of the President. Some also suggested that because of new budget restrictions imposed on local governments and the absence of additional funding, local governments will perceive the new law as an unfunded mandate and thus resist implementation.

Implementing the Disaster Risk Reduction and Management Act. Project NOAH. In July, 2012, the Philippine government unrolled the Nationwide Operational Assessment of Hazard (NOAH) project.[3] Project NOAH establishes eight technological and management components to serve disaster risk reduction. Those components include: (a) distribution of 600 automated rain gauges and 400 water level monitoring stations for 18 major river basins, to be completed by December 2013; (b) a light detecting system to produce flood inundation and hazard maps in 3D for flood-prone and major river systems, to be completed by December 2013; (c) a LIDAR and computer-assisted system to identify landslide-prone areas, to be

---

[3] All information presented about NOAH is from Official Gazette (2012).

completed by December 2014; (d) a coastal hazard system to measure wave surge, wave refraction, and coastal circulation to help solve coastal erosion problems, to be completed by December 2014; (e) a flood early warning system, to be completed by December 2013; (f) local Doppler Radar systems for sensing sea surface characteristics; (g) landslide early monitoring and warning systems in 50 or more sites by some time in 2013; and (h) a weather hazard information system that employs television, a web portal, and information and education activities to equip LGUs and communities with real-time information on potential impending hazards. In the years before and after the 2010 Disaster Risk Reduction and Management Act various informants repeatedly emphasized to us the need for adequate early warning systems, so these technological systems offer a distinctive step in the right direction. We discuss some of their ongoing challenges below.

Progress for some LGUs Various informants from NGOs and from the local government leagues acknowledge that progress in implementation among local governments is quite uneven, but a few stand out as exemplars; San Mateo in Rizal Province is illustrative. On June 24, 2010, leaders of 15 San Mateo barangays, municipal officials, the regional director for the Office of Civil Defense of the National Capital Region, the Center for Disaster Preparedness (a Philippine CSO), and community based organizations (CBO), met to learn about new requirements from the DRRM law, barely a month old. By August 13, 2010, the Municipality of San Mateo had enacted its own DRR ordinance (Ordinance No. 2010–2015), which created coordinating councils and assigned roles as designated in the DRRM law. Despite tensions among LGU leaders and some community-based organizations, we presume that San Mateo was one of the first in the country to implement its new responsibilities (Banaba Disaster Risk Reduction Project 2010).

Other local governments that have been cited for their advances in disaster risk reduction include: Makati City, NCR; Municipality of San Francisco, Camotes Islands, Cebu Province; Albay Province; Barangay Cunsad, Municipality of Alimodian, Iloilo Province; Montalban, Rizal Province; and Hinatuan, Surigao del Sur Province (Legarda 2012).

Community-based initiatives A handful of CBOs, having been active in community-based disaster management (CBDM) earlier, are now role models for others trying to find their way with the new law. Victoria noted (2002) that one such organization, Buklod Tao, a people's organization in San Mateo, had been helping other communities build CBDM capacity since 1997. Most of Buklod Tao's members lived in vulnerable informal settlements at the junction of two flood-prone rivers, so that disaster preparedness for them was about simple survival.

> After a one-day… seminar in June 1997… three disaster management teams were organized and emergency rescue and evacuation plans were detailed (including fabrication of 3 fiberglass boats using local expertise and labor and practice rescue maneuvers in the river).... Two months (later)… a typhoon hit the community again. Although several houses were swept away by the waters, no one was killed and many people were able to save their belongings. Since then, when typhoons hit the area everybody can be brought to safety because of flood-level monitoring, early warning, evacuation, rescue operations, and relief assistance activities of the organization. (Victoria 2002, p. 5)

Word of their activities spread, and for over a decade they have been training other communities in disaster risk reduction and preparedness. They helped organize the June 2010 meeting in San Mateo described above. Two weeks later they organized a stakeholders' consultation at a nearby resort. Sixty two participants came, representing various community-based and religious associations, Philippine-based CSOs, international NGOs, and barangay, municipal, and provincial disaster management officials. Buklod Tao's leader, Noli Abinales described a Participatory Capacity and Vulnerability Assessment that Buklod Tao had recently completed, using both children and adults as community assessors. Invited professionals explained various aspects of the new law and disaster risk reduction more generally. Finally, participants generated lists of perceived local disaster vulnerabilities, possible remedies, and agencies that could potentially help with the remedies (BDRRP 2010).

Buklod Tao continues to push for evacuation facilities and mitigation against rising river water. They created their own early warning systems and maintain rescue teams for themselves and neighboring communities. Pineda (2012) described the "Pandora" system that she, her students, and Buklod Tao created using computer communications and text messaging to both upload and download vital information between community members and available warning systems. Their red-orange rescue boats, by then numbering half a dozen, were prominently featured in photojournalism covering the Manila floods of August 2012. Despite many of their members' homes becoming inundated, not one soul was lost in the floods.

We also observe various indicators of systems and public servants performing disaster functions as intended by the new law. During our field work in July 2012 NGO leaders observed that a vulnerable hillside in Rizal Province near Manila had recently collapsed under heavy rain. The local DRRDM manager, they noted, promptly barred the property from future housing development, using new authority provided under the 2010 law. Similarly, as water subsided from August 2012 flooding, 441,000 people were still reportedly in crowded evacuation camps. The Chief of the Office of Civil Defense (OCD) observed that these occupants needed food and care for at least another seven days, but there was little concern of running out of food or supplies. "The government has a month's worth of relief goods if necessary," noted the Chief. "We have substantial emergency supplies" (Agence French-Presse 2012a). Disaster relief resources are manifestations of improved preparation and planning and may reflect the fruits of more flexible uses of calamity funds.

Gaillard (2011) argues that the Philippines, like other developing countries, suffers from a flawed view that disasters are merely unpredictable and uncontrollable events, and that people's inaccurate perceptions and inadequate preparations for the risks compound the casualties. He illustrates this position with an account about flooding in 2011 in Pampanga Province, approximately 50 miles north of Manila. In an interview with a newspaper columnist the provincial Governor reflected that: "I sent some boats to rescue (some residents) and one councilman even died while trying to rescue them. But the residents still refused to leave, so I have decided to withhold relief goods from them." In a later column the journalist praised the Governor for "teaching recalcitrant residents lessons in civic and personal responsibility and refusing to abet their stubbornness which, moreover, puts the lives and safety of rescuers and officials at risk" (Gaillard 2011, p. 30). Apparently both the official and the journalist have little understanding about the ways that disasters' consequences are directly related to development capacity, public education, and comprehensive planning that anticipates and mitigates potential hazards. It is against this mindset among government officials and many others in Philippine society that the advocates for the new law are toiling.

Resistance to the law from above and below Advocates argue that the new law gave governors and mayors greater discretion over their local calamity funds since they now have greater flexibility to expend part of the money on mitigation and preparedness. But in many instances these elected leaders find the new provisions more restrictive than before. One informant suggested that:

> Many mayors considered the calamity funds to be discretionary money and don't like that the new 2010 Disaster Management law requires larger percentages to be directed to preparation and mitigation. Besides, the politicians know that photo ops in which they hand out money and relief supplies when folks are sensitized to a current calamity carry greater salience than ribbon cuttings for new evacuation centers or instances in which building and zoning regulators merely do their jobs.

Several informants reported that political jockeying near the top of the system has also created frustration. Some NGO leaders reported grumblings about the new Vice Chairmen flexing their newfound system muscles, often using their new roles in ways that stalled LGUs' attempts to make mitigation and preparedness purchases. Until these undercurrents are resolved, LGU leaders are left in the dark as to the limits of their financial options and their capacity to plan comprehensively for their communities' disaster vulnerabilities.

Limitations of new technologies and public confidence about them Despite the advent of Project NOAH, many LGUs are well behind the curve in integrating hazard mapping with comprehensive land use plans. The Office of Civil Defense, NDRRMC (2011) reported that:

> Most LGUs, particularly third and fifth class municipalities and less capable cities do not have sufficient capacities to prepare a comprehensive land use plan that integrates risk factors as bases for planning. Moreover, basic information on risk, such as hazard maps are most often not available or have not been prepared for lack of capacity, expertise, resources, or data.... Technical assistance to LGUs is of utmost importance (p.26).

In Compostela Valley Province in eastern Mindanao in December 2012 a deadly incident from typhoon Pablo graphically illustrated this reality. Nearly 80 villagers and soldiers died in the New Bataan village of Andap when a wall of water engulfed two emergency shelters and a military camp.[4] The Bureau of Mines and Geosciences had identified 80 % of the area's Mayo River valley as a hazard zone; but because local governments either ignored or were incapable of responding to this

---

[4] Sources suggested that up to 20 villagers and soldiers may have drowned when a wall of water overcame an army truck transporting them to an emergency shelter (Mangosing 2012).

hazard information many villagers fled to a designated shelter that was itself located in a hazard zone, with deadly results (Marquez 2012).

Moreover, local leaders have other pressing issues; practices that stave off hunger often take precedence over sustainable development and disaster risk reduction. Manifesting denial of the underlying hazards, Compostela Valley's governor insisted that mining at higher elevations had not caused the flooding. He rejected the claims of officials who felt his devastated communities should relocate. "It's not possible to have no houses there because even the town center was hit. You mean to say the whole town will be abandoned?" he asked the media. He challenged the hazard classification and urged national leaders to review the hazard maps. He insisted that earlier floods had been less threatening and had not endangered the fateful emergency centers located in a village hall and health center (Marquez 2012).

New technologies may also sometimes breed overconfidence. A few days into typhoon Pablo, United Nations disaster officials were praising the effects of NOAH's early warning systems. When initial reports listed 274 casualties the UNISDR, referencing the larger casualty numbers from the previous year's typhoon Sendong, asserted that Pablo's toll could have been much worse. "This time big improvements in the early warning systems have saved many lives," said the head of the agency's Asia office (Agence French-Presse 2012b). This assertion proved both premature and inaccurate. Even President Aquino drew on early reports to suggest the country was learning to improve its natural disaster performance. He pointed out the "big difference" in casualties compared to other storms, noting that the more than 500 dead or missing from Pablo was still less than the 1200 deaths from Sendong a year earlier (Alibe 2012). And as the death toll rose he expressed annoyance that so many residents remained in harm's way despite his orders to the military and LGUs to evacuate them before the storm (Marquez 2012).

A few days later a Manila journalist reported:

> The true extent of the devastation from typhoon "Pablo" is just starting to emerge, and it's much worse than previously reported.... Pablo destroyed not only nipa huts and other structures made of light materials but also concrete buildings, including schools and gymnasiums used as evacuation centers. The communities look like they were hit by a tsunami. (Pamintuan 2012)

The subsequent toll of Pablo's dead and missing gives little cause for rejoicing about early warning systems. Philippine government meteorologists had accurately predicted the severe storm. Communicating its potential threat to rural communities at high elevations in Compostela Valley, connecting its likely impact to inadequate hazard maps and plans, and getting local leaders and citizens to heed the warning, are quite different challenges.

In the case of Sendong in 2011 American meteorologists warned the Philippines that this storm carried more rain than Ondoy had delivered in 2009. A leading Philippine online weather forecast site also predicted the storm would pass Cagayan de Oro and Iligan City. But the ball got dropped somewhere between the Office of Civil Defense and local government officials. Raging river waters were at their highest during the night, and this compounded the failure of communication warning systems.

Others note that wrinkles in existing systems and costs remain as serious barriers to local groups and average Filipinos' using the early warning systems. In the case of the Pandora project with Buklod Tao (see above), for example, the cost of cell phone service from major service providers makes the project cost prohibitive in the longer term (Pineda 2012). One of the principals involved in the project also observed that information available from NOAH is aggregated in large numbers that are not fine-grained enough to provide information that local communities can use for their specific needs (personal communication, Mavic Pineda, January 30, 2013). These observations suggest a need for fine-tuning and community input.

17.4 Lessons Learned

In this section we describe theoretical and practical implications that arise from the case of Philippine disaster management. We assume that these two sets of propositions are intertwined, and we address them simultaneously rather than separately. In particular, we punctuate our depiction of practical implications from our field study with conjectural applications from institutional economic development theories (e.g., Acemoglu & Robinson 2012; North et al. 2009).

The limits of rational information systems We note first an inherent tension between technological solutions and on-the-ground constraints. In the Philippines and other developing countries, we often find technology being imposed from the top of the authoritative system, presumably to enhance control and serve leaders' demands for accurate and timely information for decisions. But among soldiers charged with rounding up vulnerable citizens and transporting them to an emergency shelter in Compostela Valley, the assumption of linear rationality designed into information and warning systems may not align with unpredictable weather behavior nor with nonlinear human actions guided by urgency, panic, and fear.

In societies with mature institutional development—that is, with economic and political institutions that afford equality of access—we would expect to find frontline responders who are capable and empowered to provide timely information when local conditions change or directives do not match local circumstances. But the Compostela Valley situation suggests a social order characterized, not by rule-of-law and impersonal roles, but by diktat from elite individuals or coalitions. North et al. (2009) characterize the former situation as an open-access order and the latter as a limited-access or "natural" state. This system failure is thus institutional in origin: technology imposed from elites "who know best" onto non-elites who may have reason to trust neither the elites' intentions nor their own capabilities to offer timely corrections to practice. For such situations, governments, businesses, and communities may be able to minimize casualties by continuously training people and keeping them mindful of disaster systems and dynamics, but a more foundational intervention is to level the institutional playing field in order to empower non-elites to provide bottom-up information and to facilitate elites' acknowledging and using such vital input.

Regarding ostensibly rational information systems, we suggest two concerns based on our analysis. First, given evidence from events surrounding typhoon Pablo, we submit that executives in developing countries are prone to over-estimate the ability of warning and disaster systems to protect their vulnerable populations. Their over confidence is due in part to misunderstanding the disjuncture between rational information systems at their disposal and the worldviews and perceptions of vulnerable people. Here again the institutionalized distance between dominant elites and vulnerable Filipinos with little influence over political and economic systems is illuminated. This distance will not shrink until broader political and economic institutions provide relatively open access to all Filipinos.

We are also concerned that the various components of monitoring and warning ought not be tightly coupled, as they may be susceptible to the "normal accidents" phenomenon identified by Perrow (1984). Perrow argued that complex systems typically have many nonlinear properties such that unanticipated interaction arises from multiple failures, and the effects on the system can be neither anticipated nor understood by humans managing the systems. In order to avoid these effects system designers should build modular rather than tightly coupled components into the system, since modules permit human intervention more readily among the components.

We suggest that the normal accident dynamic may be exacerbated under limited-access institutional systems in three ways: a) elite hubris, as clearly evidenced in the early stages of Pablo, renders decision makers unresponsive to warning signals, especially when they originate from non-elites; b) often non-elites are not likely to understand their potential contributions to report changing conditions nor to feel empowered to act on the insights they do possess; and c) tightly coupled designs are more likely—and modular designs less likely—because elites do not trust underlings with human judgments in the system's coupling points.

Disaster and development, a false dichotomy During the 1990s development scholars and practitioners advanced the idea that NGOs must transition from relief work to "real" development (see Korten 1990; Macrae & Zwi 1994). For some observers disasters were interruptions in the longer-term work of development, and development actors were presumed to shift from relief to rehabilitation and then to development work as required by conditions. Some experts believed that such transitions are problematic, since short-term international relief undermines local institution-building. They believed, nonetheless, that efforts to link relief and development are important (Buchanan-Smith & Maxwell 1994; Eade & Williams 1995). Others worried that humanitarian aid undermines homegrown actors in developing countries because it creates dependency on donors and international NGOs and binds local groups to managerialism and "development thinking" from the North (Wallace 2000).

During the past decade, fortunately, various voices have promoted the mainstreaming of disaster risk reduction as a governance process for all aspects of sustainable development. As the UNDP has noted (2010):

building resilient communities in disaster-prone countries requires that: a) underlying risk factors are continuously considered in all relevant sectors; and b) risk reduction standards and measures are an integral part of the planning and delivery of core development services and processes, including education, environment, and health (p. 1).

In hindsight, since the four-phase model has existed since the early1980s, and activities of the United Nations and World Bank in the 1970s and 1980s referred distinctly to preparation and mitigation as important components of disaster management,[5] it seems surprising that the obvious link between development and disaster mitigation and preparation had not been recognized in this earlier tortured debate among development scholars. Moreover, as we noted above in the account about the Governor of Pampanga (Gaillard 2011), the mainstreaming prescription has a long way to go before it is a conventional mindset of Philippine officials.

We suspect that two practical dynamics reinforce the false dichotomy and inhibit the adoption of mainstreaming prescriptions. First, the intense media focus on disaster crises typically creates disproportionate recognition of immediate response activities and a resultant distortion in resources directed toward crisis relief. Second, the prominent influence of and spending for military and police in developing countries is inversely related to spending for eradicating poverty and reducing risks for vulnerable populations. And since the military is good at delivering rubber rafts, helicopters, and rescue vehicles, this is where emphasis falls in the disaster activities of many developing countries.[6]

Institutional theorists assert that in limited-access social orders elite coalitions need military and police leaders to enforce control over non-elites and the extraction of societal resources for elites' rent-taking (Acemoglu & Robinson 2012; North et al. 2009). Recall that the Philippines 2010 Disaster Risk Reduction and Management Act retained system leadership with the Defense Secretary rather than move it to the Office of the President, as local government and civil society elements had advocated. We see in this the persistence of countries' extractive, limited-access institutional structures that protect elite interests and exclude access for the masses.

We submit that these distorted emphases partially blinded development practitioners to the inherent connection between development and disaster risk reduction. It took growing awareness of global climate change to make the connection clear. We submit that disaster management, when seen through the lens of the four- phase

---

[5] In the U. S. a model encapsulating the activities of emergency management in four phases—mitigation, response, recovery, and preparedness—emerged in conjunction with President Carter's creation of the Federal Emergency Management Agency (FEMA) in 1979 (Giuffrida 1985). Recognition of the four phases spread incrementally to other countries and was reflected in numerous development grants of the World Bank in the early 1980s. Moreover, creation of the United Nations Disaster Relief Office in 1971 "to promote the study, prevention, control and prediction of natural disasters" (UNISDR 2013) reveals early attention to the preventative and mitigation components of international disaster management.

[6] See Brower and Magno (2011) for an expanded explanation of these two dynamics.

model, is synonymous with development, not an unfortunate interruption of it. Unfortunately, mainstreaming of disaster preparation and mitigation activities into development planning is unlikely to succeed until countries such as the Philippines transition in their broader governance structures from limited-access to open-access institutional systems.

Poverty vs. risk In the view from 50,000 feet, disaster risk reduction and development that alleviates poverty go hand-in-hand, but in the daily life of a Compostela Valley banana farmer massive flooding is a once in a millennium event, whereas hunger is a daily struggle. Established traditions built around one's livelihood become routines that are hard to change. In poor countries like the Philippines, those without other employment risk life and limb to feed families, and the government can do little more than warn them of danger. "It's not only an environmental issue, it's also a poverty issue," noted Environment Secretary Ramon Paje. "The people would say, 'We are better off here. At least we have food to eat or money to buy food, even if it is risky'" (Marquez 2012).

As noted earlier, government officials had declared a danger zone for nearly 80 % of the valley due to mountain slopes, high gradient rivers, and logging that stripped hills of trees that can minimize landslides and absorb the rain. Logging had been banned after Sendong a year earlier, but local officials turn a blind eye as it continues illegally. "But somehow we would like to protect their lives and if possible give them other sources of livelihood so that we can take them out of these permanent danger zones," said Secretary Paje (Marquez 2012).

Even as officials in situ were contending with hundreds of cadavers, grieving survivors, and the devastated farming economy of Compostela Valley, the chairwoman of the Climate Change Committee held forth from the Senate chambers in Pasay City in Metro Manila:

> Typhoon 'Pablo' has unveiled the vulnerability of our Mindanao communities to typhoons, landslides and flashfloods. Our local leaders, therefore need to appreciate how disasters are linked inextricably to the vicious cycle of poverty, socioeconomic inequality and environmental degradation. (Yamsuan 2012)

She suggested that geohazard maps can assist community planners, for example, in identifying areas that should be restricted from further housing development.[7] Notwithstanding the wisdom and good intentions that may prompt such advice, one might regard her remarks as somewhat detached and poorly timed. We see a similarity, moreover, in the urgent actions in the immediate aftermath of flooding in Manila in August 2012 of government officials who were intent on clearing informal settlers from near waterways where their homes were "obstructions to the easy flow of water." Unfortunately, their plans to relocate the informal settlers, like earlier resettlement programs, would merely have shuttled them to other flood prone areas or to settings far removed from jobs and public transportation. As observers we are struck by the disjuncture between elite attitudes and strategies, on the one hand, and

---

[7] The Senator has an online infomercial promoting her ideas: http://www.youtube.com/watch?v=r-a7wR4YicI

vulnerable people's everyday realities, on the other. We suggest that this disconnect reflects again the significant economic and political divide between elites and the masses that is inherent in limited-access institutional systems.

In fairness to political leaders we acknowledge that it is better to do something than nothing, and drastic actions and enlightening the public are often accomplished more easily when crisis has sensitized the public's attention. What will not work is to assume that disasters can be mitigated by simply moving poor people out of harm's way and keeping people from building in hazardous locations. Planning needs to be more holistic; it needs to engage aspects of the cultural, social, political, economic, and environmental. If, as the Senator suggested, "disasters are linked inextricably to the vicious cycle of poverty, socio-economic inequality and environmental degradation," then the remedies need also to address the institutional causes of poverty that continue to impede Philippine development. Secretary Paje's desire to provide other sources of employment and move people out of hazardous areas is a step in the right direction, but the necessary structural changes go much deeper.

In addition, the failure of industrialized nations to take stronger steps to cut carbon emissions gives them substantial culpability in the disaster vulnerability of the Philippines and other countries affected by global climate change. The structural factors that exacerbate the Philippines' disaster vulnerability do not stop at the shores of the Sulu and Philippine seas. In the shorter term, substantial risk reduction for vulnerable populations is still a distant goal for developing countries like the Philippines.

Our analysis has illuminated numerous indicators in disaster management practice that reflect extractive or limited-access systems in the Philippines' fundamental institutional structures (Acemoglu & Robinson 2012; North et al. 2009). More conspicuously, widespread poverty and large vulnerable populations are obvious outcomes of economic and political institutional structures characterized by unequal access. We submit that effective disaster management systems occur where quality government is exercised, that is, where impartiality of treatment thrives (Rothstein 2011). And although it may be on the output side of political systems that impartiality is judged, we believe it is on the input side, in the political activities that lead to open-access, equality-serving institutions, that impartiality is determined.

Philippine disaster management does not lack for best practice advice from its international partners; the language of community-based disaster risk reduction, community resiliency, good governance, and root causes of vulnerability are featured prominently in the 2010 law. But even the system's leaders acknowledge that performance by local governments and commitment to the new practices by local government leaders are uneven. Shortcomings in implementing the new law originate, in our view, in the fundamental institutions of Philippine society, and working out the wrinkles and stumbles rests ultimately in the nation's movement toward open-access, equality-serving political and economic institutions. What is required for lasting success includes: strengthening democratic institutions to be as responsive to bottom-up problems as top-down interests; significant cooperation across public, private, and voluntary sectors; and, recognition that human development capabilities must be strengthened in parallel with economic development.

References

Acemoglu, D., & Robinson, J. A. (2012). *Why nations fail: The origins of power, prosperity, and poverty*. New York: Crown Business.

ADB. (2009). *Poverty in the Philippines: Causes, constraints and opportunities*. Pasig City: Asian Development Bank.

Agence French-Presse. (2012a). Flood deaths climb to 66. Inquirer global nation online, August 11, 2012. http://newsinfo.inquirer.net/247755/flood-deaths-climb-to-66. Accessed 11 Aug 2012.

Agence French-Presse. (2012b). UN hails Philippines early typhoon warning systems. Inquirer global nation online, December 6, 2012. http://globalnation.inquirer.net/58871/un-hails-philippines-early-typhoon-warning-systems. Accessed 12 Dec 2012.

Agsaoay-Saño, E. (2010). Advocacy and support work for the disaster risk reduction and management (DRRM) law in the Philippines. In L. Polotan-dela Cruz, E. M. Ferrer, & M. C. Pagaduan (Eds.), *Building disaster resilient communities: Stories and lessons from the Philippines* (pp. 129–143). Quezon City: CSWCD–UP.

Alexander, D. (2002). From civil defense to civil protection and back again. *Disaster Prevention and Management, 11,* 209–213.

Alibe, T. (2012). 322 dead, hundreds missing due to 'Pablo.' ABS-CBN News. December 5, 2012. http://www.abs-cbnnews.com/nation/regions/12/05/12/238-dead-hundreds-missing-due-pablo. Accessed 2 Feb 2013

Alliance Development Works. (2012). World risk report 2012. Berlin: Alliance Development Works. http://www.nature.org/ourinitiatives/habitats/oceanscoasts/howwework/world-risk-report-2012-pdf.pdf. Accessed 27 Jan 2013.

AusAID. (2011). Philippines. http://www.ausaid.gov.au/country/country.cfm?CountryID=31&Region=EastAsia. Accessed 20 Feb 2011.

Banaba Disaster Risk Reduction Project (BDRRP). (2010). *Buklod Tao, partners launch Banaba Disaster Risk Reduction Project*. Bukluran: Official Publication of the BDRRP, July-August.

BBC News. (2007). Guide to the Philippines conflict. http://news.bbc.co.uk/2/hi/asia-pacific/1695576.stm. Accessed 29 Sept 2012.

Brower, R. S., & Magno, F. A. (2011). A 'third way' in the Philippines: Voluntary organizing for a new disaster management paradigm. *International Review of Public Administration, 16,* 31–50.

Buchanan-Smith, M., & Maxwell, S. (1994). Linking relief and development: An introduction and overview. *IDS Bulletin, 25*(4), 2–16.

Cariño, L. V., & Fernan, R. L. III. (2002). Social origins of the sector. In L. V. Cariño (Ed.), *Between the state and the market: The nonprofit sector and civil society in the Philippines* (pp. 27–60). Quezon City: Center for Leadership, Citizenship and Democracy, University of the Philippines, Diliman.

Clarke, G. (1998). *The politics of NGOs in South-East Asia: Participation and protest in the Philippines*. London: Routledge.

Cruz, J. E. (2010). *Estimating informal settlers in the Philippines*. Paper presented at the 11th National Convention on Statistics, Mandaluyong, Philippines. October, 4–5, 2010.

Domingo, R. W. (2013). "Pablo" was deadliest catastrophe in the world in 2012. Inquirer global nation, January 11, 2013. http://globalnationinquirernet/61367/pablo-was-deadliest-catastrophe-in-the-world-in-2012. Accessed 11 Jan 2013.

Eade, D., & Williams, S. (1995). *The Oxfam handbook of development and relief*. Oxford: Oxfam Publications.

Elona, J. (2012). Lim: Clearing of Manila's informal settlers in danger zones long overdue. Philippine daily inquirer online. August 14, 2012. http://newsinfo.inquirer.net/249893/lim-clearing-of-manilas-informal-settlers-in-danger-zones-long-overdue. Accessed 29 Sept 2012.

Gaillard, J. D. (2011). *People's response to disasters: Vulnerability, capacities and resilience in Philippine context*. Angeles City: Center for Kapampangan Studies.

Giuffrida, L. O. (1985). FEMA: Its mission, its partners. *Public Administration Review, 45*, (Special Issue), 2.

Heijmans, A., & Victoria, L. (2001). *CBDO-DR: Experiences and practices in disaster management of the Citizens' disaster response network in the Philippines*. Quezon City: Center for Disaster Preparedness.

Hilhorst, D. (2003). *The real world of NGOs: Discourses, diversity and development*. London: Zed.

International Strategy for Disaster Reduction. (2005). *Hyogo framework for action 2005-2015: Building the resilience of nations*. Geneva: United Nations Office for Disaster Risk Reduction.

Korten, D. C. (1990). *Getting to the 21st Century: Voluntary action and the global agenda*. West Hartford: Kumarian Press.

Legarda, L. (2012). Disaster risk reduction can be done. Philippine daily inquirer, December 31, 2012. http://newsinfo.inquirer.net/333069/disaster-risk-reduction-can-be-done. Accessed 3 Feb 2013.

Macrae, J., & Zwi, A. (Eds.). (1994). *War and hunger: Rethinking international responses to complex emergencies*. London: Zed Books.

Mangosing, F. (2012). 43 dead, 25 injured in Compostela Valley as 'Pablo' sweeps Mindanao. Inquirer news online, December 4, 2012. http://newsinfo.inquirer.net/318357/35-drown-in-flash-floods-in-compostela-valley. Accessed 2 Feb 2013.

Marquez, B. (2012). Safety, need compete in typhoon-hit Philippines. Philippine daily inquirer, December 7, 2012. http://newsinfo.inquirer.net/320221/safety-need-compete-in-typhoon-hit-philippines. Accessed 7 Dec 2012.

Meloto, A. (2009). *Builder of dreams*. Madaluyong City: Gawad Kalinga Community Development Foundation, Inc.

National Statistical Coordination Board. (2011). 2009 official poverty statistics. Makati, Philippines: National Statistical Coordination Board. http://www.nscb.gov.ph/poverty/2009/Presentation_RAVirola.pdf. Accessed 29 Sept 2012.

National Statistical Coordination Board (NSCB). (2012). List of regions. Makati, Philippines: National Statistical Coordination Board. http://www.nscb.gov.ph/activestats/psgc/listreg.asp. Accessed 29 Sept 2012.

National Statistics Office (NSO). (2008). Official population count reveals. Manila, Philippines: National Statistics Office (NSA). http://www.census.gov.ph/data/pressrelease/2008/pr0830tx.html. Accessed 29 Sept 2012.

National Statistics Office. (2012). Census of the Republic of the Philippines. http://www.census.gov.ph. Accessed 26 Jan 2013.

NDRRMC. (2012.) Final report on the effects and emergency management re tropical storm Sendong. National disaster risk reduction and management council. February 10, 2012. Philippines, Quezon City.

NEDA. (2011). Philippines development plan 2011–2016. Pasig City, Philippines: National Economic Development Authority. http://www.neda.gov.ph/PDP/2011-2016/default.asp. Accessed 29 Sept 2012.

North, D. C., Wallis, J. J., & Weingast, B. R. (2009). *Violence and social orders: A conceptual framework for interpreting recorded human history*. Cambridge: Cambridge University Press.

Office of Civil Defense, NDRRMC. (2011). National progress report on the implementation of the Hyogo Framework for Action (2009–2011). Manila, Philippines: Office of Civil Defense, NDRRMC. March 27, 2011. http://www.preventionweb.net/english/countries/asia/phl/. Accessed 27 Nov 2012.

Official Gazette. (2012). Project NOAH (Nationwide Operational Assessment of Hazards). Office of the President of the Philippines. July 6, 2012. http://www.gov.ph/about-project-noah/. Accessed 1 Feb 2013.

PAGASA. (2009). Member Report to the ESCAP/WMO Typhoon Committee, 41st Session. Philippine Atmospheric, Geophysical and Astronomical Services Administration. http://www.typhooncommittee.org/41st/docs/TC2_MemberReport2008_PHILIPPINES1.pdf. Accessed 20 Feb 2011.

Pamintuan, A. M. (2012). Worse than sendong. The Philippine star online. December 14, 2012. http://www.philstar.com/opinion/2012-12-14/885754/worse-sendong. Accessed 13 Dec 2012.

Pedroso, K. (2010). From NDCC to NDRRMC. The Philippine daily inquirer, October 20, 2010. http://newsinfo.inquirer.net/inquirerheadlines/nation/view/20101020-298696/From-NDCC-to-NDRRMC. Accessed 29 Jan 2013.

Perrow, C. B. (1984). *Normal accidents: Living with high risk technologies*. New York: Basic Books.

Pineda, M. V. G. (2012). Exploring the potentials of a community-based disaster risk management system (CBDRMS), the Philippine experience. *International Journal of Innovation, Management and Technology, 3,* 708–712.

Reyes, F. (2012). Binay says relocation sites for informal settlers to be revisited, investigated. Philippine daily inquirer online, August 17, 2012. http://newsinfo.inquirer.net/251900/binay-says-relocation-sites-for-informal-settlers-to-be-revisited-investigated. Accessed 29 Sept 2012.

Rothstein, B. (2011). *The quality of government: Corruption, social trust and inequality in international perspective*. Chicago: University of Chicago Press.

Salamon, L. M. (1987). Of market failure, voluntary failure, and third-party government: Toward a theory of government-nonprofit relations in the welfare state. *Nonprofit and Voluntary Sector Quarterly, 16,* 29–49.

Schiavo-Campo, S., & Judd, M. (2005). *The Mindanao conflict in the Philippines: Roots, costs, and potential peace dividend. Social development papers on conflict prevention and reconstruction*. Washington, D.C: The World Bank.

Stone, R. (2006). Too late, Earth scans reveal the power of a killer landslide. *Science* (Online), March 31, 2006. http://www.sciencemag.org/content/311/5769/1844.summary?maxtoshow=&HITS=10&hits=10&RESULTFORMAT=&fulltext=guinsaugon&searchid=1&FIRSTINDEX=0&resourcetype=HWCIT. Accessed 20 Feb 2011.

The Guardian. (2009). Philippines landslides kill 160 after fresh floods. October 9, 2009. http://www.guardian.co.uk/world/2009/oct/09/philippines-landslides-storms-parma. Accessed 28 Jan 2013.

The Hindu. (2010). Thousands of slum dwellers resist eviction in Philippines. The Hindu. September 23, 2010. http://www.thehindu.com/news/international/article786444.ece. Accessed 28 Jan 2013.

The World Bank. (n.d.). Issues and dynamics: Urban systems in developing East Asia, Philippines. Urbanization dynamics and policy frameworks in developing East Asia. Washington, DC: The World Bank. http://siteresources.worldbank.org/INTEAPREGTOPURBDEV/Resources/Philippines-Urbanisation.pdf. Accessed 27 Jan 2013.

United Nations Development Programme. (2010). Disaster risk reduction, governance and mainstreaming. New York: United Nations Development Programme. http://www.undp.org/content/dam/undp/library/crisis%20prevention/disaster/4Disaster%20Risk%20Reduction%20-%20Governance.pdf. Accessed 12 Feb 2013.

United Nations Food and Agriculture Organization UNFAO. (n.d.). *On solid ground: Natural disasters of all kinds rank high in the Philippines*. Rome: United Nations Food and Agriculture Organization.

United Nations Human Settlements Programme. (2003). *The challenge of slums: Global report on human settlements*. London: Earthscan.

United Nations International Strategy for Disaster Risk Reduction. (2013). Who we are: History. http://www.unisdr.org/who-we-are/history#assistance. Accessed 31 Aug 2013.

United Nations Office for the Coordination of Humanitarian Affairs (UNOCHA). (2013). Philippines: Typhoon Bopha situation report no. 15. January 15, 2013. http://reliefweb.int/sites/reliefweb.int/files/resources/full%20report_186.pdf. Accessed 30 Jan 2013.

U.S. Department of State. (2013). Background note: Philippines. http://www.state.gov/r/pa/ei/bgn/2794.htm#gov. Accessed 26 Jan 2013.

Victoria, L. P. (2002). Community based disaster management in the Philippines: Making a difference in people's lives. Quezon City, Philippines: Center for Disaster Preparedness. http://unpan1.un.org/intradoc/groups/public/documents/APCITY/UNPAN025912.pdf. Accessed 30 Jan 2013.

Villas, A. L. T. (2012). Robredo: Act on informal settlers. Tempo.com. August 17, 2012. http://www.tempo.com.ph/2012/robredo-act-on-informal-settlers/. Accessed 29 Sept 2012.

Wallace, T. (2000). Development management and the aid chain: The curse of NGOs. In D. Eade (Ed.), *Development and Management* (pp. 18–39). London: Oxfam GB.

Whaley, F. (2012). Rain floods a third of Manila area displacing thousands. The New York Times, August 7, 2012. http://www.nytimes.com/2012/08/08/world/asia/flooding-in-philippines-grows-worse-as-thousands-flee-manila-and-desperate-residents-are-trapped-on-roofs.html?_r=0. Accessed 27 Jan 2013.

Yamsuan, C. (2012). Planners urged to study maps of geohazards. Philippine Daily Inquirer Online. December 13, 2012. http://newsinfo.inquirer.net/323763/planners-urged-to-study-maps-of-geohazards. Accessed 13 Dec 2012.

Part III
Recovery, Development, and Collaborative Emergency Management

This part of the book focuses on long term disaster recovery, sustainable development, and networks and partnerships for effective disaster governance. Fatih Demiroz and Qian Hu examine the role of nonprofit organizations and civil society in disaster recovery and development from a collaborative emergency management perspective. Frances L. Edwards and Dan Goodrich evaluate the contributions of multi-disciplinary collaboration to the development of resiliency in California as a model for other states and nations confronted with intractable natural hazards. Simon A. Andrew and Sudha Arlikatti examine housing policy changes in India through the lens of housing recovery programs initiated in the aftermath of the Killari earthquake in the Latur district of Maharashtra in 1993, the Bhuj earthquake in Gujarat in 2001, and the Great Tsunami of 2004 in the Nagapattinam district of Tamil Nadu. Sana Khosa outlines the existing disaster management system in Pakistan along with the recent developments and improvements in plans and policies for disaster response and recovery. The chapter also addresses the response and recovery efforts that have been underway since the 2010 floods which have focused on reconstructing homes, community infrastructure such as schools, roads and bridges, addressing massive internal displacement, and the loss of livelihoods especially for farmers affected by the destruction of agricultural land. Rajib Shaw chapter provides insight on community based recovery in three cities: Kamaishi in Iwate prefecture (with focus on school based recovery program), Kesennuma in Miyagi prefecture (with focus on community development in temporary housing), and Natori in Miyagi prefecture with a focus on education, communication, information and volunteerism. Abdul-Akeem Sadiq examines the relationship between disasters and development in the aftermath of the 2010 Haiti earthquake. This chapter also discusses the challenges facing Haiti's emergency management system and the opportunities at the disposal of the Government of Haiti to revamp the emergency management system. Ping Xu, Xiaoli Lu, Xi Zuo, and Huan Zhang provide an in-depth case study on the reconstruction of Shuimo town, a town that was severely destroyed in the earthquake with emphasis on how the counterpart assistance program and the sustainable development efforts have transformed Shuimo from a highly polluted town to a tourism destination.

Chapter 18
The Role of Nonprofits and Civil Society in Post-disaster Recovery and Development

Fatih Demiroz and Qian Hu

18.1 Introduction

Civil society and nonprofit organizations have played important roles throughout disaster planning, response, and post-disaster recovery and development (Eikenberry et al. 2007; Smith and Wenger 2007). Numerous nonprofit organizations provide disaster-impacted communities and residents with services that government is not able to deliver in a timely manner. While the majority of existing studies focus on the role of civil society and nonprofit organizations in disaster response and relief (Kapucu 2009; Simo and Bies 2007), this chapter focuses on the role of nonprofits in post-disaster recovery and community development.

Disaster recovery, economic development, and disaster resiliency are intertwined concepts. Disaster recovery can be considered as part of economic development policies. The effectiveness of disaster recovery and the level of economic development of a community are predictors of disaster resiliency. The process of disaster recovery determines whether a community can bounce back to pre-disaster conditions or develop stronger community capacity. Both effective recovery and economic development require participation of partnerships from public, private, and nonprofit sectors. These relationships are managed through collaborative governance practices (see chap. X/Waugh in this volume).

This chapter examines the role of nonprofits in disaster recovery and development from a collaborative emergency management perspective. Collaborative emergency management refers to collaborative governance practices (Ansell and Gash 2008; Simo and Bies 2007) in the context of disaster management. Collaborative emergency management implies the utilization of network relationships between public, private, and nonprofit sectors to achieve a common goal: effective management of

F. Demiroz (✉)
University of Central Florida, Orlando, USA
e-mail: fdemiroz@knights.ucf.edu

Q. Hu
University of Central Florida, Orlando, FL, USA
e-mail: Qian.Hu@ucf.edu

N. Kapucu, K. T. Liou (eds.), *Disaster and Development,* Environmental Hazards,
DOI 10.1007/978-3-319-04468-2_18, © Springer International Publishing Switzerland 2014

disasters and developing community resiliency. The chapter also highlights non-profits' role in building social capital, which is crucial for strengthening the fabric of communities and fostering collaboration among partners. This chapter addresses the following questions: What is the role of nonprofit organizations in post-disaster development? How do nonprofits contribute to post-disaster recovery efforts and building resilient communities? How can inter-organizational networks better work to improve post-disaster development?

The chapter is organized as follows: Section one introduces the concept of resilience and disaster recovery with a particular focus on community resiliency and post-disaster recovery and development. Section two discusses the role of nonprofits in the broader context of collaborative emergency management. Section three delves into the strengths of nonprofits in helping communities recover from disasters from social capital perspective Section four discusses how nonprofits may better play their role in collaborative post-disaster development. The last section concludes the chapter.

18.2   Post-disaster Development and Community Resilience

Disaster recovery starts in the first few hours following a disaster and may last for as long as decades. Disaster recovery is a process that is influenced by the outcomes of preparedness, mitigation, and response. It is a complex and long lasting practice that involves numerous actors from public, private, and nonprofit sectors. Philips and Neal (2007) define post-disaster recovery as a "social process in which the local government manager creates crucial partnerships to guide the affected community toward a multifaceted recovery from disaster" (p. 208). Government agencies create a networked system engaging various actors in disaster recovery.

Disaster recovery is one of the fundamental components of resiliency. The time and money spent on recovery can be used as an indicator of the level of resilience of a community. Building or rebuilding stronger and flexible economic, social, or administrative systems take place in the process of recovery after a disaster. During this process, communities get an opportunity to examine the vulnerabilities that they may have and try to address them with full support from various stakeholders. In other words, disaster recovery and resiliency can be seen as collaborative effort to engage nonprofits, civil society, local communities, and businesses in the long process of post-disaster recovery.

Disaster resiliency and recovery overlap with each other particularly in addressing vulnerabilities. Aldrich (2012a) uses Schumpeter's idea of creative destruction in disaster recovery. Originally proposed in the economics literature, creative destruction connotes the opportunities that arise after each crisis and lead to developments for a better condition. Rose and his colleagues (Rose 2009; Rose and Liao 2005) examine disasters as external shocks to economic systems. Disasters can exhibit existing weaknesses and vulnerabilities and eventually create opportunities for building better and more flexible systems. Addressing these weaknesses and vulnerabilities is part of the recovery process. During the recovery, it is crucial to pay attention to the social aspects of the community recovery, which is related with economic development practices.

In 2011, the National Disaster Recovery Framework (NDRF) was established to provide guidance and support to disaster-impacted states, tribes, and local governments to "build a more resilient nation" in the U.S. (FEMA 2011, p. 1). Building upon the concepts of the National Response Framework (NRF) Emergency Support Function # 14—Long-Term Community Recovery, the NDRF provides core principles in disaster recovery, clarifies roles and responsibilities of key stakeholders involved in recovery, defines a coordination structure for collaboration efforts, provides guidance for recovery planning, and describes the process of rebuilding a stronger and resilient communities (FEMA 2011). Throughout the entire document, the concept of resilience has been used as a measure of community recovery after disasters. According to NDRF, successful recovery should have the following characteristics:

> The community successfully overcomes the physical, emotional, and environmental impacts of the disaster.
> It reestablishes an economic and social base that instills confidence in the community members and business regarding community viability.
> It rebuilds by integrating the functional needs of all residents and reducing its vulnerability to all hazards facing it.
> The entire community demonstrates a capability to be prepared, responsive, and resilient in dealing with the consequences of disasters.
> (FEMA 2011, p. 13)

18.3   Disaster Resilience

In a broader sense, disaster resilience can be defined as a set of adaptive capacities to help communities maintain functions in the post-disaster period (Bruneau et al. 2003; Norris et al. 2008; Sherrieb et al. 2010). This definition implies adaptation to new conditions. Disaster resilience involves all four phases of emergency management–preparedness, response, mitigation, and recovery (Kapucu et al. 2013). Furthermore, building disaster resiliency involves land use planning and community development, and requires civic engagement and social capital (Aldrich 2012a, b; Kapucu et al. 2013; Paton et al. 2008). These components help communities reduce their vulnerabilities, evolve, and operate in a new environment.

Rose (2004) discusses resiliency from an economic perspective and distinguishes between inherent resilience and adaptive resilience. Inherent resilience refers to certain abilities under normal conditions. These abilities include capacities to replace the inputs that are curtailed by an external shock with new ones. Adaptive resilience, on the other hand, requires ingenuity in crisis conditions. In other words, it connotes the abilities to manage crises in the face of uncertainty.

Measuring resilience is a challenging task. A metric for measuring resilience is shaped by the question of resilience of whom and against what (Nowell and

Steelman 2013). For example, the metric for measuring economic resilience to di-sasters is different than psychological resilience. Economic resilience and psycho-logical resilience are relatively specific and easier to measure compared to com-munity resilience against disasters as a whole. Bruneau et al. (2003) use the concept of adaptive capacity for defining and measuring disaster resilience against earth-quakes. In their definition, adaptive capacity of a system depends on the level of robustness, redundancy, rapidity, and resourcefulness of its assets (Bruneau et al. 2003).

Norris and her colleagues (2008) identify four capacities for examining resil-ience, including economic development, social capital, information and communi-cation, and community competence. Norris et al. (2008) further identify five steps for enhancing community resilience. They argue that first a community should have economic resources, decrease inequalities, and reduce the social vulnerabilities that it may carry. This step is particularly important for economic development. Increas-ing economic and social inequalities pose significant threats to communities and economic development practices that aim to solve these issues (not merely focus on employment and boost of economic activities) (Blakely and Leigh 2009). During disaster recovery, it is crucial to work on eliminating economic and social vulnera-bilities rather than simply rebuilding physical infrastructure. Second, individuals in the community must have access to social capital that will allow them to participate in the phases of disaster management (particularly mitigation and preparedness). Third, preexisting organizational networks and relationships are key for mobiliz-ing resources in the aftermath of a disaster. This step is important since it includes all stakeholders of the community in the disaster response and recovery. In other words, interorganizational networks are not limited to public organizations but in-clude private organizations and nonprofits as well. Strong interorganizational social capital enables decision makers to benefit from human capital, equipment, and ser-vices that nonprofits can offer. The fourth and fifth steps touch on the importance of protecting the social support webs that individuals and families form for reducing the impact of disasters and having flexible disaster management plans that will en-able communities to manage the stress properly. These five steps are closely related to nonprofits' involvement in disaster recovery and economic development. The following sections of this chapter provide a more detailed discussion of these con-cepts within the context of collaborative emergency management and social capital.

18.4   Nonprofits in Collaborative Emergency Management

To understand the role of nonprofits in emergency management and specifically disaster recovery and development, it is important to first introduce collaborative emergency management. Collaborative emergency management implies horizon-tal and networked type of management practices. Actors from public, private, and nonprofit sectors are part of this network setting and contribute to the successful management of emergencies (Comfort et al. 2012; Kapucu 2006a).

Nonprofits serve as important service and support providers during disaster relief and post-disaster recovery. Given the intensity and frequencies of disasters, government agencies alone may not have adequate resources to help communities to prepare for, respond to, and recover from large-scale natural or man-made disasters. Private actors such as Wal-Mart and Home Depot had successful operations in response to Katrina and "their success is not surprising since supply chain management is their daily business" (Kapucu et al. 2010, p. 455). Different from the private sector, civil society and nonprofits have their advantages in responding to the needs of local communities because of their close and on-going relationships with their embedded communities (Kapucu 2006a). Due to unique organizational attributes, nonprofit organizations such as the American Red Cross, Salvation Army, and faith-based and community organizations, may provide services to local communities that government agencies cannot serve. Hence, nonprofits can complement government services or compensate weaknesses in the public sector (Kapucu 2006a; Simo and Bies 2007).

Moreover, the increasing number of disasters has prompted the rapid growth of local, national, and international nonprofit organizations to better respond to the needs of communities in disaster situations (Kapucu 2006a; Shaw and Goda 2004). In response to the 9/11 attacks, 258 new nonprofit organizations were established following the Internal Revenue Service's expedited approval procedure for tax-exempt status application (Campbell 2009). These newly created nonprofit organizations provided direct financial assistance and emotional support to families of victims of 9/11, and worked in other disaster relief areas, such as economic development (Campbell 2009). As discussed above, the role of nonprofits and civil society in disaster response has been extensively studied (Simo and Bies 2007); by contrast, studies on the role of nonprofits in disaster recovery have received relatively less attention (Chandra and Acosta 2009).

18.5   Nonprofits in Post-disaster Recovery and Development

The National Disaster Recovery Framework (NDRF) provides a post-disaster checklist for nonprofit organizations' involvement in post-disaster recovery and development (FEMA 2011). According to the NDRF, nonprofit organizations are encouraged to deliver resources and services to vulnerable individuals and communities, "provide emotional and psychological care," and "supply housing repair, reconstruction, and rehabilitation services" (FEMA 2011, p. 91). Some of the services are for short-term recovery, such as food and shelter, and other types of services belong to long-term recovery, such as reconstruction services (Stys 2011). Long-term recovery services require more systematic efforts to identify the needs, seek financial resources, and recruit sustaining volunteers (Stys 2011). To help local residents and communities recover from disasters in the long run, many nonprofits mainly focus on "helping low income and other special [needs] populations with case management services and home repair" to help disaster-impacted residents with their insurance claims and rebuild their homes (Flatt and Stys 2011, p. 9). Due to the flexible structure and multiple sources of funding, a large number of small local nonprofits, fast growing medium-sized nonprofits, and large nationwide nonprofits can fill in the service gaps that government cannot provide in a timely and effective manner. These nonprofit organizations are in a better position for engaging in long-term recovery in local communities (FEMA 2011).

Nonprofits' involvement in the preparedness, mitigation, and response phases can also contribute to disaster recovery and community disaster resiliency. The effectiveness of disaster recovery is closely related with preparedness, mitigation, and response phases of emergency management. Several studies show that nonprofits make significant contributions during these various phases of emergency management. For example, Chikoto et al. (2012) examine how nonprofit, public, and private organizations in Memphis, Tennessee differ in their mitigation and preparedness activities. They find that despite the lack of financial and human resources, nonprofit organizations adopted more mitigation and preparedness activities than private organizations. Right after the 9/11 attacks, 1,176 nonprofit organizations were involved in disaster response and relief (Kapucu 2006b). Nonprofits' engagement in other phases of emergency management can help them better understand the needs of disaster-impacted communities and individuals and build long-term collaborative relationships with other public and private organizations within networks of emergency management.

The roles and responsibilities of nonprofits are not limited to supportive tasks. Nonprofit organizations also serve as leaders and collaboration coordinators. Nonprofits take lead roles in organizing collaborative efforts to provide assistance to local residents to recover from disasters. For instance, to meet the needs of the displaced residents in New Orleans, American Red Cross led the efforts to organize more than 52 agencies from both public and nonprofit sectors to create the Greater New Orleans Disaster Recovery Partnership (Eikenberry et al. 2007). There are coordinating organizations such as the National Voluntary Organizations Active in Disaster (NVOAD) coordinating efforts of various VOADs in disaster response and recovery (Gazley 2013; Kapucu et al. 2011). To avoid service duplication through timely communication and cooperation, the NVOAD was created as a collaboration platform for member organizations to share information, resources, and coordinate efforts. Furthermore, at the local community level, coordinating mechanisms have been set up, such as the community Long-Term Planning or Recovery Committees (LTPCs, LTRCs), to coordinate efforts among nonprofit organizations, and between nonprofit organizations, government agencies, and private businesses (Gazley 2013), although this type of committee has not been widely adopted in local communities (Flatt and Stys 2011).

In addition to Voluntary Organizations Active in Disasters (VOADs) which are integrated into federal or state emergency planning documents, Gazley (2013) examined the role that Community Organizations Active in Disaster (COADs) played in disaster response and recovery. COADs differ from VOADs in a way that the roles and responsibilities of most of COADs are not clearly defined in government documents (Gazley 2013). Their missions are not specific about disaster response or recovery, but are centered on broader "community-oriented missions of social and human service provision, economic development, volunteerism or other local services" (Gazley 2013, p. 84). Nonprofits not only contribute to recovery and resiliency through their ties with government organizations at all levels, but also work as catalysts for recovery of disasters-impacted individuals. Community-level social capital helps nonprofits carry out their practices. The following section provides an overview of the concept of social capital and how social capital relates to nonprofits' engagement in disaster recovery and development.

18.6 Social Capital and Nonprofits in Post-disaster Recovery and Development

Social capital is a widely used concept in disaster research. It can be defined as everyday networks and resources that are available to people and organizations through their connections to others (Aldrich 2012a). Halpern (2007) identifies networks, norms, and sanctions as the three basic components of social capital. Social capital can be examined at individual (micro), community (meso), and national (macro) levels (Halpern 2007). In the context of disaster recovery, most of the research on social capital is conducted at the individual and community level. And the outcomes of social capital can be both positive and negative. For example, social capital might help individuals to receive assistance people in their networks. On the other hand, it might create close social networks that excludes people from different backgrounds and may lead to discrimination in the distribution of disaster relief aids (see Chap. 15/Paton et al. in this volume)

Nonprofits help build social capital in a community, which is crucial for interorganizational collaborations in post-disaster recovery. The relationships between social capital, civic engagement, and economic development have already been intensively studied. More participation in nonprofit organizations can help enhance civic health, which as a consequence can help people develop skills, enhance trust, build networks of resources (Putnam 1995, 2000). The development of nonprofit organizations can also help build positive relationships between citizens and government, which contribute to stability and development of society. All of these factors are contributors to long-term local economic development. On the other hand, nonprofit organizations serve as an important sector of economies, producing a large proportion of Gross Domestic Product (GDP) and local employment (Blackwood et al. 2012).

Individual-level social capital in post-disaster recovery can explain differences in the speed of recovery in different communities. In these cases, nonprofits play a crucial role in disseminating information, better allocating limited resources, and encouraging people to returning back to their hometowns. Organizational-level social capital pertains to more organized and collaborative efforts (mostly led by a government organization) for disaster recovery.

Research has been conducted to study how individual-level social capital contributes to effective disaster recovery and community disaster resiliency. Chamlee-Wright and Storr (2009a) analyze how a Vietnamese community in New Orleans recovered from Hurricane Katrina. Their findings show that the community was organized around the Mary Queen of Vietnamese (MQVN) Catholic Church. The church had served as a center of religious and nonreligious activities for Vietnamese and it was a religious, cultural, commercial, and political institution for the community. Chamlee-Wright and Storr (2009a) name the benefits that the church provided to its community as club goods. They argue that after the hurricane, people who left their homes because of the hurricane had to make the decision either to stay at their current location or return back to their hometown. Returning to their hometown is attractive and beneficial if there are also other people returning. Therefore there is a cost and uncertainty of returning back since a family will not know if there will be enough people returning that will be enough to revitalize the town. MQVN played a significant role in these circumstances and helped people return home (Chamlee-Wright and Storr 2009b).

During long-term post-disaster development, social capital and trust plays a crucial role in initiating and sustaining collaborations. The importance of social capital and trust in emergency management networks has been highlighted in many studies (Jaeger et al. 2007; Kapucu 2006a). Defining social capital as "a resource that is inherent in the relations between actors in a variety of locations and sectors," Kapucu (2006a, p. 209) notes that daily working relationships enable communities to function well when faced with disaster scenarios as trust can be built between public and nonprofit organizations prior to disasters.

In a few studies on community-based emergency management, social capital has been argued as a crucial factor to the effectiveness of emergency management, because the existing positive relationships among individuals, between individuals and community-based organizations, and between organizations and government authority prepare the community for emergency scenarios, and develop or enhance community resiliency to quickly recover from disasters (Jaeger et al. 2007; Murphy 2007). In a study on community-level emergency management, Murphy (2007) reviews both the positive benefits as well as the potential effects of social capital. He further proposes that the resiliency of community-based emergency management is dependent upon "the relationship between communities and other societal structural contexts such as government authorities and patterns of social status and wealth distribution" (Murphy 2007, p. 305). In other words, it is very important to build ongoing collaborative relationships between nonprofit organizations and local residents, and between nonprofit organizations and government agencies on a daily basis.

Aldrich's (2012b) retrospective study on 1923 Tokyo Earthquake shows that social capital and civil society are the most important predictors of recovery after the disaster. He identifies three mechanisms that allow areas with higher social capital recover faster. First, social ties are informal insurance for disaster victims. That helps decision makers to receive information about the post-disaster conditions of a community, which in turn helps make better and more accurate policy decisions. Second, higher political activity in a community help its members to better mobilize and overcome barriers for collective action. Third, dense social ties make 'exiting' harder for individuals and force them to use their 'voice.' These three mechanisms that Aldrich (2012a) notes are identified in several cases (Chamlee-Wright and Rothschild 2007; Olshansky et al. 2005) but most importantly in the 1923 Tokyo Earthquake. In communities that have higher social capital and trust between organizations across sectors and jurisdictions, when disasters occur, nonprofits can unite resources quickly at the local level in both disaster response and post-disaster recovery.

Organizational-level social capital also embeds individuals' relationships. It is possible to consider an interorganizational tie as a rope consisting of individual threads. The greater the number of threads, the stronger the rope will be. Each thread is a tie between individuals in different organizations. The types of relationships determine the strength of the thread and the rope. Ties can be either formal (led and maintained by formal agreements) or informal (developed and maintained by goodwill and personal relationships). Formal ties are likely to result in better outcomes than the informal ties; however, functioning formal ties are built upon trust and good reputation. These two are generated through informal ties and prior collaborations. In other words, trust and good reputation help organizations to institutionalize their informal ties and convert them into formal relationships. In most cases informal ties foster relationship building and lowers transaction costs between organizations.

The ensuing section discusses the role of nonprofits in disaster development and resiliency building through the lens of collaborative emergency management. It also presents propositions that can guide future research on this topic.

18.7   The Engagement of Nonprofit Organizations in Collaborative Post-disaster Recovery Efforts

The engagement of nonprofits in post-disaster recovery is determined by various factors. The situation may get more complex, especially when nonprofits are expected to be involved in collaborative emergency management networks. Not all nonprofit organizations are willing to participate in collaborations with government agencies or other organizations (Gazley 2010; Guo and Acar 2005, Sowa 2009). To fully engage nonprofit organizations in post-disaster development, it is worthwhile to understand the factors that may affect nonprofit organizations' decision to engage in collaborative efforts in post-disaster development. Guo and Acar (2005) examined the factors influencing nonprofit organizations' participation in collaborations with other nonprofit organizations. Based on a survey of 95 urban charitable organizations, they find that nonprofit organizations that are older, larger in budget size, receive government funding, receive funding from fewer agencies, are more likely to build collaborative linkages with other nonprofit organizations. Gazley (2010), in her study of service delivery collaborations between nonprofits and local government, notes that besides the funding factor, other factors such as "organizational capacity and age, shared professional or collaborative experience, and a respondent's ideological orientation and gender" also influence nonprofits' collaboration with government (p. 70). In another study of service delivery partnerships, Sowa (2009) assumes that resource dependency, environmental uncertainty, pressure from funders, and the development state of the organization will influence nonprofit's collaboration decisions.

Involvement in collaborative relationships and maintaining them require dedication of time, human capital, and financial resources. Organizations may or may not dedicate such resources due to several reasons. The sustainability of interpersonal and formal relationships is a function of organizational goals, network goals, types of resources shared, structure of the network and place of organizations in the structure etc. (Koliba et al. 2011). There is a strong association among organizational capacities (e.g. resources, human and social capital), network development phase, and the sustainability of a network (Gazley 2010).

The type and structure of interorganizational relationships impacts the capacity of organizations, communities, and the adaptive capacity of service delivery networks (Isett and Provan 2005). In their study of donative and commercial nonprofits, Galaskiewicz et al. (2006) find that donative nonprofits' network ties lead to a faster rate of organizational growth while commercial nonprofits showed a different result. Their results showed that donative nonprofits, with ties to urban elites, higher status, and more central positions in their network have a faster growth rate. Commercial nonprofits with greater interorganizational connectedness, on the other hand, do not have a similar pattern of organizational growth (Galaskiewicz et al. 2006).

Studies conducted by Isett and Provan (2005), Larson (1992), Lazerson (1995), and Uzzi (1996) show that relationship ties (both formal and informal) between organizations gets stronger overtime, breeds trust, and fosters multiplex relationships. From another perspective, strengthening ties between members of a network increases trust, interaction, communication, information sharing, and diffusion of innovative ideas (Bouty 2000). Organizations that are trusted and have positive reputation of collaboration are likely to be involved in new collaborative ties while the reverse situation occurs when there is a negative reputation. Additionally, past friendship ties within informal networks also promote collaborative relationships. Both past formal and informal ties help organizations operate on a trust-based relationship. This in turn reduces transaction costs and increases collaborative capacity of organizations. The conceptual framework is proposed in Fig. 18.1 below.

To summarize the factors that are important for nonprofit organizations to be more engaged in post-disaster development, this section concludes with the following propositions:

Propositions 1: Nonprofit organizations' participation in post-disaster recovery and development will be strengthened if there is a formal planning process for nonprofits to engage in post-disaster disaster recovery and development.

Propositions 2: Frequent information sharing among nonprofits and between nonprofits and government will be positively related with nonprofit organizations' involvement in post-disaster recovery and development.

Fig. 18.1 A conceptual model of nonprofits' engagement in post-disaster development

Propositions 3: Formal interorganizational management structure for long-term recovery positively correlates with nonprofit organizations' participation in post-
disaster recovery and development.

Propositions 4: In communities that have stronger social capital, nonprofit organizations' participation in post-disaster recovery and development will be more
active.

Propositions 5: Nonprofit organizations with strong organizational capacity, leadership, and positive collaboration experience, are more likely to engage in cross-
sector collaborations in post-disaster recovery and development efforts.

Conclusion

Disaster resiliency, disaster recovery, and community development are overlapping
concepts. The time and resources spent on recovery shows the level of resilience of
a community. Also, recovery efforts determine whether a community is bouncing
back to pre-disaster conditions or developing into better conditions through adaptation. Bouncing back to pre-disaster conditions repeats a community's vulnerabilities and makes it susceptible to the same threats again. Bouncing forward, on the
other hand, makes a community adapt to disaster conditions while it maintains its
vital functions (Edwards 2013). Adapting to new conditions coincides with replacing some community elements that may have failed or weakened in a disaster with
newer ones. These elements might be physical infrastructure, human capital, community competence, or communication channels (Norris et al. 2008). Additionally,
bouncing forward includes addressing social, infrastructure, and economic vulnerabilities. These are the efforts where disaster recovery sits on a common ground
with community development.

Nonprofits carry out critical roles in community development, post-disaster recovery, and disaster resilience. Nonprofits are identified as crucial partners not only in disaster recovery, but also in preparedness, mitigation, and response. Disaster recovery efforts cannot succeed without having support from community stakeholders. Including nonprofits as partners in the recovery process is crucial. In community development, nonprofits are instrumental in reducing inequalities and social vulnerabilities. Nonprofits' strengths in post-disaster recovery can be better explained using the concept of social capital. Nonprofits sit in the intersection of these three concepts and social capital explains how nonprofits function for carrying out their roles.

This chapter highlights the relationship between disaster resiliency, recovery, and community development, with a particular focus on the role of nonprofits as their conjunction point. While discussing nonprofits and social capital, we focus on individual-level and organizational-level social capital. The organizational relationships are discussed in the context of collaborative emergency management because nonprofit organizations play important roles and assume great responsibilities in collaborative post-disaster recovery.

References

Aldrich, D. P. (2012a). *Building resilience: Social capital in post-disaster recovery*. Chicago: University of Chicago Press.

Aldrich, D. P. (2012b). Social, not physical, infrastructure: the critical role of civil society after the 1923 Tokyo earthquake. *Disasters, 36*(3), 398–419.

Ansell, C., & Gash, A. (2008). Collaborative governance in theory and practice. *Journal of Public Administration Research and Theory, 18*(4), 543–571.

Blackwood, A. S., Roeger, K. L., & Pettijohn, S. L. (5 October 2012). *The nonprofit sector in brief: Public charities, giving, and volunteering*. Urban Institute Press. http://www.urban.org/publications/412674.html.

Blakely, E. J., & Leigh, N. G. (2009). *Planning local economic development: Theory and practice*. Thousand Oaks: Sage.

Bouty, I. (2000). Interpersonal and interaction influences on informal resource exchange between R & D researchers across organizational boundaries. *Academy of Management Journal, 43*(1), 50–66.

Bruneau, M., Chang, S., Eguichi, R. T., Lee, G. C., O'Rourke, T. D., Reinhorn, A. M., von Winterfeldt, D. (2003). A framework to quantitatively assess and enhance the seismic resilience of communities. *Earthquake Spectra, 19*(4), 733–752.

Campbell, D.A. (2009). Stand by me: Organization founding in the aftermath of disaster. *The American Review of Public Administration, 40*(3), 351–369.

Chamlee-Wright, E., & Rothschild, D. (2007). Disastrous uncertainty: How government disaster policy undermines community rebound. Mercatus Policy Series Comment No. 9.

Chamlee-Wright, E., & Storr, V. (2009a). Club goods and post-disaster community return. *Rationality and Society, 21*(4), 429–458.

Chamlee-Wright, E., & Storr, V. (2009b). There's no place like new orleans: Sense of place and community recovery in the ninth ward after Hurricane Katrina. *Journal of Urban Affairs, 31*(5), 615–634.

Chandra, A., & Acosta, J. (2009). *The role of nongovernmental organizations in long-term human recovery after disaster*. RAND Occasional Paper.

Chikoto, G. L., Sadiq, A. A., & Fordyce, E. (2012). Disaster mitigation and preparedness: Comparison of nonprofit, public, and private organizations. *Nonprofit and Voluntary Sector Quarterly, 42*(2), 391–410.

Comfort, L. K., Waugh, W., & Cigler, B. A. (2012). Emergency management research and practice in public administration: Emergency, evolution, expansion, and future directions. *Public Administration Review, 72*(4), 539–547.

Edwards, F. (2013). All hazards, whole community. In N. Kapucu, C. Hawkins, & F. Rivera (Eds.), *Disaster resiliency: Interdisciplinary perspectives* (pp. 21–44). New York: Routledge.

Eikenberry, A. M., Arroyave, V., & Cooper, T. (2007). Administrative failure and the International NGO response to Hurricane Katrina. *Public Administration Review, 67*(S1), 160–170.

Galaskiewicz, J., Bielefeld, W., & Dowell, M. (2006). Networks and organizational growth: A study of community based nonprofits. *Administrative Science Quarterly, 51*(3), 337–380.

Federal Emergency Management Agency (FEMA). (2011). National disaster recovery framework: Strengthening disaster recovery of the nation. http://www.fema.gov/pdf/recoveryframework/ndrf.pdf.

Flatt, V. B., & Stys, J. J. (2011). Long term recovery in disaster response and the role of nonprofits. *Oñati Socio-legal Series, 3*(2), 346–362.

Gazley, B. (2010). Why not partner with local government? Nonprofit managerial perceptions of collaborative advantage. *Nonprofit and Voluntary Sector Quarterly, 39*(1), 51–76.

Gazley, B. (2013). Building collaborative capacity for disaster resiliency. In N. Kapucu, C. Hawkins, & F. Rivera (Eds.), *Disaster resilience: Interdisciplinary perspectives* (pp. 84–98). New York: Routledge.

Guo, C., & Acar, M. (2005). Understanding collaboration among nonprofit organizations: Combining resource dependency, institutional, and network perspective. *Nonprofit and Voluntary Sector Quarterly, 34*(3), 340–361.

Halpern, D. (2007). *Social Capital*. Cambridge: Polity Press.

Isett, K. R., & Provan, K. (2005). The evolution of dyadic interorganizational relationships in a network of publicly funded nonprofit agencies. *Journal of Public Administration Research and Theory, 15*(1), 49–165.

Jaeger, P. T., Schneiderman, B., Fleischmann, K. R., Preece, J., Qu, Y., & Wu, F. F. (2007). Community response grids: E-government, social networks, and effective emergency management. *Telecommunications Policy, 31*(10–11), 592–604.

Kapucu, N. (2006a). Interagency communication networks during emergencies: Boundary spanners in multiagency Coordination. *American Review of Public Administration, 36*(2), 207–225.

Kapucu, N. (2006b). Public-nonprofit partnerships for collective action in dynamic contexts. *Public Administration: An International Quarterly, 84*(1), 205–220.

Kapucu, N. (2009). Interorganizational coordination in complex environments of disasters: The evolution of intergovernmental disaster response systems. *Journal of Homeland Security and Emergency Management, 6*(1), 1–26.

Kapucu, N., Arslan, T., & Demiroz, F. (2010). Collaborative emergency management and national emergency management network. *Disaster Prevention and Management, 19*(4), 452–468.

Kapucu, N., Yuldashev, F., & Feldheim, M.A. (2011). Nonprofit organizations in disaster response and management: A network analysis. *European Journal of Economic and Political Studies, 4*(1), 83–112.

Kapucu, N., Rivera, F., & Hawkins, C. (2013). *Disaster resilience and sustainability: Interdisciplinary perspectives*. New York: Routledge.

Koliba, C., Meek, J. W., & Zia, A. (2011). *Governance networks in public administration and public policy*. Boca Raton: CRC.

Larson, A. (1992). Network dyads in entrepreneurial settings: A study of the governance of exchange relationships. *Administrative Science Quarterly, 37*(1), 76–104.

Lazerson, M. (1995). A new Phoenix? A modern putting-out in the Modena Knit-Wear Industry. *Administrative Science Quarterly, 40*(1), 34–49.

Murphy, B. L. (2007). Locating social capital in resilient community-level emergency management. *Natural Hazards, 41*(2), 297–315.

Norris, F. H., Stevens, S. P., Pfefferbaum, B., Wyche, K. F., & Pfefferbaum, R. L. (2008). Community resilience as a metaphor, theory, set of capacities and strategy for disaster readiness. *American Journal of Community Psychology, 41*(1–2), 127–150.

Nowell, B., & Steelman, T. (2013). The role of responder network in promoting community resilience: Toward a measurement framework of network capacity. In N. Kapucu, C. Hawkins, & F. Rivera. *Disaster resilience: Interdisciplinary perspectives* (pp. 233–357). New York: Routledge.

Olshansky, R., Johnson, L. & Topping, K (2005). *Opportunity in Chaos: Rebuilding after the 1994 Northridge and 1995 Kobe Earthquakes*. Urbana: Department of Urban and Regional Planning, University of Illinois at Urbana-Champaign.

Paton, D., Parkes, B., Daly, M., Smith, L. (2008). Fighting the flu: Developing sustained community resilience and preparedness. *Health Promotion Practices, 9*(4), 45S–53S.

Philips, B., & Neal, D. (2007). Recovery. In W. L. Waugh & K. Tierney (Eds.), *Emergency management: Principles and practice for local government* (2nd ed.) (pp. 207–234). Washington D.C.: ICMA Press.

Putnam, R. D. (1995). Bowling alone: America's declining social capital. *The Journal of Democracy, 6*(1), 65–78.

Putnam, R. D. (2000). Bowling alone. The collapse and revival of American community. New York: Simon and Schuster.

Rose, A. Z. (2004). Defining and measuring economic resilience to disasters. *Disaster Prevention and Management, 13*(4), 307–314.

Rose, A. Z. (2009). A framework for analyzing the total economic impacts of terrorist attacks and natural disasters. *Journal of Homeland Security and Emergency Management, 6*(1), Article 9.

Rose, A. Z., & Liao, S. Y. (2005). Modeling regional economic resilience to disasters: A computable general equilibrium analysis of water service disruptions. *Journal of Regional Science, 45*(1), 75–112.

Shaw, R., & Goda, K. (2004). From disaster to sustainable civil society: The Kobe experience. *Disasters, 28*(1), 16–40.

Sherrieb, K., Norris, F. H., & Galea, S. (2010). Measuring capacities for community resilience. *Social Indicators Research, 99*(2), 227–247.

Simo, G., & Bies, A. (2007). The role of nonprofits in disaster response: An expanded model of cross-sector collaboration. *Public Administration Review, 67*(S1), 125–142.

Smith, G. P., & Wenger, D. (2007). Sustainable disaster recovery: Operationalizing an existing agenda. In H. Rodríguez, E. L. Quarantelli, & R. R. Dynes (Eds.), *Handbook of disaster research* (pp. 234–257). New York: Springer.

Sowa, J. E. (2009). The collaboration decision in nonprofit organizations: Views from the front line. *Nonprofit and Voluntary Sector Quarterly, 38*(6), 1003–1025.

Stys, J. J. (17 January 2011). Non-profit involvement in disaster response and recovery. http://www.law.unc.edu/documents/clear/publications/nonprofit.pdf.

Uzzi, B. (1996). The Sources and consequences of embeddedness for the economic performance of organizations: The network effect. *American Sociological Review, 61*(4), 674–698.

Chapter 19
California Seismic Safety Commission: Multi-disciplinary Collaboration for Seismic Safety Mitigation

Frances L. Edwards and Daniel C. Goodrich

19.1 Introduction

Mitigation creates the basis for community resiliency by lessening the disaster's toll on the human and physical environment. For example, the contrasting outcomes of the 2010 earthquakes in Haiti (Renoir 2012) and Chile (Kraul 2010) demonstrate the value of physical mitigation through engineering, building codes and code enforcement. Policy interventions, like building code changes, and governmental mechanisms, like tax incentives, can enhance the likelihood of mitigation steps being taken in advance of disasters, which in turn improves the resiliency of communities. Such action requires development of a consensus through multi-disciplinary collaboration. California's seismic activity in populated areas has led to the creation of mitigation measures for schools, hospitals, residential and commercial buildings, and infrastructure, which grew from multi-disciplinary collaboration, an approach which is embodied in the California Seismic Safety Commission (SSC), which oversees the state's seismic hazard mitigation plan. This chapter describes the contributions of multi-disciplinary collaboration to the development of resiliency in California as a model for other states and nations confronted with intractable natural hazards.

The chapter begins with an explanation of California's earthquake history, followed by the value of mitigation and risk reduction efforts. It then describes the history of legislative attention to seismic safety, including the development of the SSC, emphasizing its multidisciplinary collaboration structure, its political role and its risk communication challenges. It concludes with a description of its current structure and broadened focus on recovery and economic mitigation, and learning from international experiences, for example Turkey, Greece, and Taiwan (SSC 2000).

F. L. Edwards (✉) · D. C. Goodrich
Mineta Transportation Institute, San Jose State University, San Jose, USA
e-mail: kc6thm@yahoo.com

N. Kapucu, K. T. Liou (eds.), *Disaster and Development,* Environmental Hazards,   331
DOI 10.1007/978-3-319-04468-2_19, © Springer International Publishing Switzerland 2014

19.2   The Value of Mitigation

Mitigation creates the basis for community resiliency by lessening the disaster's toll on the human and physical environment. For example, the contrasting outcomes of 2010 earthquakes in Haiti and Chile demonstrate the value of physical mitigation based on geotechnical engineering, building codes, and code enforcement. In Haiti, with a 7.0 Richter event, there were no earthquake-oriented building codes, and over 316,000 people died, with $ 8 billion in damage in a country with $ 7 billion per year GDP (Renois 2012). In contrast, Chile, with an 8.8 Richter event, saw pre-1960s buildings collapse, killing 300 people, but newer buildings sustained little damage (Kraul 2010), clearly demonstrating the social and economic value of mitigation.

Policy interventions and governmental mechanisms can enhance the likelihood of mitigation steps being taken in advance of disaster, which in turn improves the resiliency of communities. Structural engineers say, "Earthquakes do not kill people; buildings kill people" (Nelson 2011). Investing in strengthening buildings and infrastructure against hazards benefits a community by protecting human life and lessening the economic losses. In 2004 a FEMA website called mitigation "the cornerstone of emergency management," noting that "every dollar spent on mitigation saves roughly $ 2 in disaster recovery costs" (Elliston 2004). A 2005 study by the Multihazard Mitigation Council (MMC) of the National Institute of Building Sciences determined that every dollar invested in mitigation saved an average of $ 4 in losses. (MMC 2005) Furthermore, FEMA's Federal Insurance and Mitigation Administration noted an average of $ 3.65 in "disaster relief costs and tax losses avoided" (FEMA 2012). Integrating mitigation into recovery "ensure[s] that communities will be physically, socially, and economically resilient to future hazard impacts" (FEMA 2012).

This same mitigation philosophy is reflected in the California Seismic Safety Commission's mission. This multidisciplinary state level commission has sponsored landmark legislation requiring seismic mitigation measures that are uniform across all California jurisdictions. The success of these measures is clear from the loss of life statistical comparisons with other jurisdictions. Within less than 5 years California experienced two significant earthquakes in densely populated urban areas with limited loss of life, while similar events in areas without strictly enforced seismic codes killed thousands, even in developed nations, as shown in Table 19.1.

History has demonstrated the value of mitigation during disasters. Chile suffered the world's largest earthquake in 1960 with 3,000 deaths. Thereafter Chile developed the strictest building codes in the world (Fleury 2008). In 2010 its loss of life was 1/10th of the 1960 losses in a very large earthquake (Kraul 2010).

19.3   "Disasters by Design"

These natural seismic hazards are intractable. However, before seismology was understood dense human habitation was developed along the Pacific Coast. Natural ports for trade also supported exploration and a gold rush that caused California's

Table 19.1 Chronology of selected major earthquakes' magnitude and loss of life. (Sources: Fleury 2008; USGS 1993; Nelson 2011; Tierney and Goltz 1997; CNN 2009; Renois 2012; Kraul 2010; Imai 2011)

Year	Richter magnitude	Earthquake name	Location	Loss of life
1960	9.5	Valdivia	Chile	3,000 *Includes deaths from tsunami in Chile, Hawaii, Japan and The Philippines
1964	9.2	Prince William Sound	Alaska	128 *plus 15 deaths from tsunami
1976	7.8	T'ang Shan Province	China	240,000
1989	6.8	Spitak	Armenia	25,000
1989	6.9	Loma Prieta	Bay Area, California	62
1994	6.8	Northridge	Los Angeles, California	78
1995	7.2	Hanshin-Awaji	Kobe, Japan	6,279
2004	9.3	Sumatra-Andaman	Ocean off Sumatra, Indonesia	*Resulting tsunami killed 230,000 in 14 countries bordering the Indian Ocean
2009	6.3	L'Aquila	Italy	300
2010	7.0	Port-au-Prince	Haiti	250,000
2010	8.8	Conception	Chile	300
2011	9.0	Great East Japan earthquake	Sendai, Japan	20,000 *Includes missing and dead from the tsunami

population to burgeon from 34,000 people in 1848 to over 100,000 people by 1850 (Gerston and Christensen 2008). Today California is the most populous state, with 38 million people (US Census 2010), most of whom live in the large coastal metropolises near the San Andreas Fault system, following a national trend for moving toward the coast (Mileti et al. 1999, p. 3). Pacific Coast cities at risk include Vancouver, Canada with 2.4 million people (Statistics Canada 2012), metropolitan Seattle with 2 million people (US Census Bureau 2013c), metropolitan Portland with 748,000 people (US Census Bureau 2013a), 7.1 million people in the San Francisco Bay area (MTC-ABAG 2010), 12.8 million people in the Los Angeles area (US Census 2010), and 3.1 million people in San Diego County (US Census Bureau 2013b).

Mileti et al. (1999) noted in *Disasters by Design* that the increasing economic losses to "natural" disasters are the result of an interaction between existing natural hazards, developing population density in those areas, and the concomitant denser built environment. Pressure for housing in desirable climates near the water has led to poor decisions about where to build. For example, in California "bulldozing steep hillsides for homes… [has] disrupted natural runoff patterns and magnified flood hazards" (p. 3). Construction on "view lots" with houses on steep slopes has led to worsened earthquake losses, such damage to homes in Malibu landslides, and the double fatality loss of a home in Sherman Oaks that slid off its view lot during the Northridge earthquake (Earthquake disaster 1994).

Scientists have noted that merely identifying the presence of faults, or the likelihood of fault rupture, fails to completely communicate the danger that the faults pose (SCEC 2013). As the fault ruptures, seismic waves are sent through the Earth's crust. Differential shaking, which causes most of the damage, is experienced based on the magnitude of the earthquake and local geological conditions, such as soils composition and instability. The Uniform California Earthquake Rupture Forecast (UCERF) predicts when and where a fault might rupture, but not the shaking intensities that will be generated. "This is an important distinction, because even areas with a low probability of fault rupture can experience shaking and damage from distant, powerful quakes" (SCEC 2013).

Earthquake predictions do, however, inform decisions about managing human habitation and the built environment. Even before reliable predictions, actual fault ruptures in California's recent history have led to legislation to change aspects of its construction codes, mitigating future damage. For example, the 1933 Long Beach earthquake with magnitude 6.4 destroyed most of the city's schools, empty at 5:54 pm (USGS 2012b), demonstrating the danger of unreinforced masonry construction, which led to the Field Act, with its strict regulation of school construction (Olson 2002). "It was the first time the state acknowledged that public steps should be taken to protect its citizens from earthquake hazards" (Wiley 2000, p. 11). The 1971 San Fernando earthquake, with 65 people killed, $ 50 million in damages and two hospitals destroyed, demonstrated the need for the first seismic construction codes specifically for hospitals (California Department of Conservation 2007).

UCERF information is also used to create legislation that limits construction in dangerous areas. Stricter building codes and requirements for geotechnical studies before construction are structural mitigation steps that have led to "The steady and major decline in fatalities from earthquake in the United States over the last half century" (Sylves 2008, p. 121). Non-structural strategies like zoning regulations can keep high density developments from inherently dangerous areas. Public education on personal preparedness, and regular training and exercises for public safety staff all help to lessen deaths and economic losses after disaster (Edwards and Goodrich 2012).

19.4   Earthquakes in California

California's San Andreas Fault system runs from the Southern California desert to the Pacific Coast just south of San Francisco, and then along the northern coast. It caused California's most deadly earthquake, the 7.9 magnitude earthquake in San Francisco in 1906 with over 3,000 deaths, and its most costly earthquake, the 6.7 magnitude earthquake in the Northridge neighborhood of Los Angeles in 1994 with $ 15 billion in losses (Table 19.2).

California's northern coast is adjacent to the "Mendocino triple junction" of the Pacific, North American and Gorda tectonic plates (University of California at Berkeley 2007). While the 1980 Eureka and 1992 Cape Mendocino quakes occurred off shore at the triple junction, resulting in no deaths and little property damage (California Department of Conservation 2007), the Gorda plate is sliding under the North American plate. Earthquakes in Chile and Alaska demonstrate the

Table 19.2 A sampling of California's largest earthquakes since 1850, by magnitude. (Source: California Department of Conservation 2007)

Magnitude	Date	Location	Comments
7.9	Jan. 9, 1857	Fort Tejon	2 killed, 220-mile surface scar
7.9	April 18, 1906	San Francisco	3,000 killed, $ 524 million in property damage, including fire damage
7.8	March 26, 1872	Owens Valley	27 killed, 3 aftershocks of 6.25+
7.5	July 21, 1952	Kern County	12 killed, 3 aftershocks of 6+
7.3	Jan. 31, 1922	West of Eureka*	37 miles offshore
7.3	Nov. 4, 1927	SW of Lompoc*	No major injuries, slight damage
7.3	June 28, 1992	Landers	1 killed, 400 injured, 6.5 aftershock
7.2	Jan. 22, 1923	Mendocino	Damaged homes in several towns
7.2	Nov. 8, 1980	West of Eureka*	Injured 6, $ 1.75 million in damage
7.2	April 25, 1992	Cape Mendocino*	6.5 and 6.6 aftershocks
7.1	Oct. 16, 1999	Ludlow (Hector Mine Quake)	Remote, so minimal damage
7.1	May 18, 1940	El Centro	9 killed, $ 6 million in damage
6.9	Oct. 17, 1989	Loma Prieta	63 killed
6.7	Jan. 17, 1994	Northridge	61 killed, $ 15 billion in damage
6.6	Feb. 9, 1971	San Fernando (Sylmar)	65 killed, $ 505 million in damage

power of subduction and the devastation that may result in future California quakes (Thompson 2012). The Sierra Nevada Fault on the state's eastern side, less active than the San Andreas, has caused large earthquakes near Reno, Nevada and the Antelope Valley (Choi 2011) and Owens Valley (Wiley 2000) in California.

The UCERF gives a 99.7% probability of a 6.7 or greater earthquake in California before 2037. The San Andreas is estimated to have a return rate of about every 150 years, making it ready for another rupture. For example, the Hayward Fault in the Bay Area has not ruptured since 1868. (USGS 2012a), while a 1988 study by USGS assigned the Crystal Springs portion of the San Andreas Fault the highest rupture probability in the Bay Area (Ward and Page 1989). The Southern San Andreas Fault traverses the Los Angeles metropolitan area with a population of 12.8 million (US Census 2010), and a 59% probability of a 6.7 or greater earthquake by 2037 (SCEC 2013).

At the Cascadia Subduction Zone, running from northern California to Vancouver Island in Canada, the Juan de Fuca plate subducts under the North American plate, and its last rupture caused a tsunami in Japan in 1700. A Cascadia rupture from end to end could generate an earthquake over 9.0 that would be felt throughout the North American west coast, as far south as San Francisco, with tsunamis traveling across the Pacific (Thompson 2012).

19.5   Risk Reduction in California

California has the largest population of any state, three of the nation's ten largest cities, two of the world's ten largest container ports, and the principal connection for the nation to the Pacific Rim global supply chain. Therefore, scientific and technical knowledge must be used to lower the risk for its citizens and economy, and the economy of the nation. One successful method for risk reduction has been the development of the Seismic Safety Commission (SSC) as a deliberative, multidisciplinary body that guides state-level legislation to enhance seismic safety through structural and non-structural mitigation.

The California SSC, which oversees the state's seismic hazard mitigation strategy, embodies a multi-disciplinary collaborative approach to seismic safety. As noted by Sylves (2008), the development of successful seismic safety policy requires the participation of multiple stakeholders. California's seismic activity in populated areas has led to the creation of mitigation measures, many of which grew from the Commission's multi-disciplinary collaboration. The commissioners include representatives from science, engineering, emergency management, local government and social services, who can apply their expertise to the evaluation of potential mitigation measures, work with the legislature for passage and funding of mitigation measures, and work with local governments for the implementation of policies and programs to create resiliency.

19.6   California Seismic Safety Commission

Mileti et al. (1999) note that mitigation must be based on an understanding of the "economic, social and political ramifications of extreme natural events" (p. 1). "The process of transforming the future requires open-minded debate" (p. 63–64), and that a "deliberative body" is needed to create consensus on acceptable losses and reasonable protective actions to enhance the safety of the interface between natural hazards and the humans living in the at-risk built environment. (Mileti et al. 1999). The California SSC was created to be that deliberative body, representing the physical and social sciences, the NGO sector and government.

19.6.1   Joint Legislative Committee and Advisory Board

In 1969 State Senator Al Alquist established the Joint Legislative Committee on Seismic Safety. The Alaska Earthquake of 1964 was a focusing event that led Stanley Scott of the Institute of Governmental Studies at the University of California, Berkeley and Professor Karl Steinbrugge of the UCB School of Architecture to raise the issue of Bay Area earthquake impacts with Senator Alquist through a paper "that suggested government could mitigate potential injuries and damage with forethought and action" (Wiley 2000, p. 12). Scott also wrote an article advocating "an earthquake policy commission" (Wiley 2000, p. 13). In June of 1969 Senator Alquist sponsored Senate Concurrent Resolution 128 that created the Joint Committee on Seismic Safety for 4 years. "The joint committee was required to 'develop seismic safety plans and policies'" and recommend legislation to limit earthquake-related damage (Wiley 2000, p. 13).

Assemblyman John Burton, chair of the Assembly Joint Rules Committee, helped get funding to start the committee. Senator Alquist and Assemblyman Paul Priolo were the co-chairs, with several legislators from each house as members. The committee developed an Advisory Board with subject matter experts, such as Chairman Henry Degenkalb as the seismic engineering expert. The Advisory Board included a political scientist, city planner, councilmember and mayor, and created advisory committees to consider specialty areas related to tasks given to the Board from the Committee (Diridon 2012). There were five subcommittees of subject matter experts, including an Advisory Group on Engineering Considerations and Earthquake Sciences.

For the first 18 months, from 1969–1971, Senator Alquist, who represented the City of San Jose in the Bay Area, funded the committee's work out of his office budget. He hired Diridon Research Corporation, a San Jose public opinion research firm, to provide staff support to the committee. One early focus of the Advisory Board was to prevent construction of buildings intended for human occupancy on the traces of active earthquake faults. The Advisory Board developed a proposal for limiting such construction, and held hearings with stakeholders to find a balance between property rights and public safety. The draft of this legislation was ready for consideration by the Legislative Counsel's Office when the Sylmar Earthquake occurred (Diridon 2012).

19.6.2 Sylmar Earthquake

On February 9, 1971 a magnitude 6.6 earthquake hit the San Fernando Valley in Los Angeles at 2:00 pm. It lasted for 60 seconds "took 65 lives, injured more than 2,000, and caused property damage estimated at $ 505 million" (USGS 2012c). One of the most surprising events was the collapse of the new seismically resistant building at the county's Olive View Hospital. (USGS 2012c), while "Older, unreinforced masonry buildings collapsed at the Veterans Administration Hospital at San Fernando, killing 49 people" (USGS 2012c). Older buildings and chimneys throughout the area were destroyed, and landslides damaged roads, utility lines, railroads and pipelines. The Lower Van Norman Dam and the Pacoima Dam were damaged, and rock falls blocked roadways, slowing emergency response (USGS 2012c).

This 'focusing event' led to funding for seismic research, and an impetus for seeking mitigation strategies that would lessen the toll of the inevitable future seismic activity in the state. The earthquake provided the impetus for the Advisory Board to craft existing ideas into 35 pieces of legislation that were introduced in 1971 and 1972. Alquist grabbed the opportunity to make progress in seismic safety in California, following his philosophy that "You only have 90 days [after an earthquake] to introduce earthquake mitigation legislation before the legislature goes back to business as usual" (Alquist 1994). The impact of the loss of life through building damage and landslides from earthquake shaking overcame the legislature's usual reluctance to interfere in private property rights. Gael Douglas, Senator Alquist's chief of staff, led the effort to create bills that were reviewed by the Board, and in turn referred to the Legislative Counsel's office. Hearings on these bills were held by the Joint Legislative Committee on Seismic Safety, who approved the bills and sent them to the authorizing committees in the Assembly and Senate. (Diridon 2012)

19.6.3 Post-Sylmar Seismic Safety Laws

Several pieces of land use legislation were among the Joint Commission's proposals that were passed by the legislature. Government Code, Sections 65000 et seq. empowers cities and counties to pass zoning and subdivisions laws, and create specific plans for areas of a community. The Joint Committee on Seismic Safety and its Advisory Board and groups determined that seismic safety should be a consideration in the development of communities near the state's known fault zones. SB 351 Seismic Safety General Plan Element (Chapter 150, Statutes of 1971) required city and county general plans to include a seismic-safety element. Government Code, Section 65302(b) requires the Seismic Safety Element to identify geologic hazards like fault rupture zones, ground failure potential (mudslides, landslides and slope instability), and tsunami zones (City of Berkeley n.d.), leading to zoning regulations that limit construction in dangerous areas.

On December 22, 1972 another land use measure based on the work of the Joint Committee and its multidisciplinary Advisory Board and groups was passed, the Alquist-Priolo Earthquake Fault Zoning Act (SB 520) (Diridon 2012). Section 661 of the Act also changed the composition of the oversight organization, the State Mining and Geology Board, from 9 to 11 members, and specified the members across a variety of professions: "two should be mining geologists, mining engineers, or mineral economists, one should be a structural engineer, one should be a geophysicist, one should be an urban or regional planner, one should be a soils engineer, two should be geologists, one should be a representative of county government, and at least two shall be members of the public having an interest in and knowledge of the environment" (Statutes of 1972, Chapter 1354, section 661). This revision of the board's composition reflected the breadth of knowledge needed to oversee the fault zone mapping program. The act required the State Mining and Geology Board to create maps that defined the Special Studies Zones along the faults in California, "to prohibit the location of developments and structures for human occupancy across the trace of active faults as defined by this board" (California Government Code, 1972 as amended, 2621.5). The Alquist-Priolo maps are regularly updated, with additions and revisions issued in 2012 (California Department of Conservation 2012).

The damage to two major hospitals during the Sylmar Earthquake led Senator Alquist to propose the Alquist Hospital Safety Act (SB 519, Chapter 1130, Statutes of 1972). The county's Olive View hospital had been engineered for seismic resistance, but it suffered catastrophic failures, including the separation of wings from the main building and the collapse of three stairwells (USGS 2012c). Of the 65 people who died in the earthquake, 49 died at the Veterans Administration Hospital site where a wall fell off the buildings. Severely disabled patients who were dependent on medical gasses for respiration died when the gas lines were severed by the shaking (Johnson 1995).

In the 1970s medical technology was evolving rapidly, requiring new facilities to be built to house the machinery and meet electricity demands. The Alquist Hospital Safety Act required that all newly constructed hospitals comply with strict construction standards based on the Field Act requirements for schools (Holmes n.d.). In addition to requiring state oversight of construction and design by structural engineers, the law required that buildings be designed to withstand seismic shaking, include geotechnical studies for the site, and "meet design requirements for non-structural elements" (Holmes n.d., p. 2). Furthermore, the law created a multi-disciplinary Building Safety Board to oversee the implementation of the law, including "design professionals and facility experts" (Holmes n.d., p. 2). Considering that hospitals house people with limited mobility, and are an essential community resource during disaster response, their design and construction had to "resist, insofar as practical, forces generated by earthquakes, gravity, and wind" (Holmes n.d., p. 2). Ultimately the slow progress of building replacement led to amendments that required seismic retrofits to existing buildings.

Governor Ronald Reagan also recognized the importance of enhancing seismic safety in California. In January, 1972 he created the Governor's Earthquake Council to study the San Fernando Earthquake and make recommendations for enhancing the state's seismic safety.

19.6.4   Creation of the Seismic Safety Commission

The Advisory Board was an adjunct to the Joint Committee, whose legislated lifespan was up in 1974. The Joint Legislative Committee's final report, *Meeting the Earthquake Challenge*, encouraged the legislature to create the California Commission on Seismic Safety to focus on the development and implementation of earthquake safety programs (Wiley 2000, p. 25). The concept was also supported by the Governor's Earthquake Council.

The value of the work done by the Advisory Board was clear to Senator Alquist and Assemblyman Priolo, so in the 1973–74 legislative term they developed legislation to create a permanent SSC, independent of the legislature, but acting as an advisory body to the governor and the legislature (Diridon 2012). SB 1729, signed by Governor Reagan in September of 1974, established a 17-member SSC for 2 years. Fifteen members were to be appointed by the governor, from technical and non-technical professions, and Governor Jerry Brown made the first appointments in May of 1975. The other two members represented the State Senate and Assembly. The commission was advisory to the governor and legislature, conducting studies and sponsoring and reviewing legislation (Wiley 2000).

On August 1, 1975 a 5.7 magnitude earthquake damaged the Oroville Dam, giving the commission its first role as an investigative body, a pattern that has been followed in every earthquake since. Senator Alquist extended the life of the commission to January 1, 1981 (SB 1340), and the law allowed the representatives of the State Senate and Assembly to appoint alternates to attend meetings on their behalf. In 1979 Senator Alquist used his position on the Senate Finance Committee to extend the commission to 1986, and following the 1983 Coalinga Earthquake, SB 1782 (Alquist) eliminated the sunset clause for the SSC (Wiley 2000).

19.6.5   Multidisciplinary Collaboration

Recognizing that earthquakes affect the entire community, the framers of the SSC's structure created a multidisciplinary body of 15 members who would bring multiple perspectives and knowledge bases to the evaluation of policies and legislative proposals, and creation of commission reports. Technical representatives included seismology, soils engineering, structural engineering, mechanical engineering and geology. Practice-based representatives included emergency services, fire service, and architecture and planning. Community-based representatives included insurance, social services and utilities. There are four local government—city and county—representatives, of whom one must be a local building official. These 15 members are appointed by the governor for four-year terms. The State Senate and the Assembly each appoint one member. As investigations are conducted of various earthquakes and public policies, multiple perspectives are used in the discussions and analysis.

The 15 members representing different communities of interest in the state bring a wealth of resources beyond their own expertise. This multidisciplinary approach "provides a window into the professions," (Diez 2012) since the commissioners can refer to their professional organizations for additional information collection. For example, when the commission recommended the development of a statewide emergency management system after the Loma Prieta earthquake (SSC 1991b, p. 23), the emergency management representative shared that idea with his colleagues at a Southern California Emergency Services Association meeting, engendering interest among his colleagues and support for the development process for the Standardized Emergency Management System (SEMS) when it began in 1992. Likewise, the recommended enhancements to communications (SSC 1991b, p. 24) resulted in the adoption of a satellite-based phone system by the Governor's Office of Emergency Services that was supported by amateur radio operators, fire service personnel and emergency managers who had first heard about the proposal from the commissioner representing emergency services.

The commission's reports reflected the multidisciplinary character of both the commission itself and its advisory committees. This was reflected in their decision to include disaster recovery in their mandate after the Coalinga Earthquake of 1983. Kathleen Tierney, a sociologist, wrote the portion of the Coalinga Earthquake report that described the community and economic impact of the event, which caused no loss of life but damaged the adjacent oil fields, the 12 block business district and much of the community's housing stock. It also revealed the lack of preparedness for the earthquake within the local government structure. Evaluating issues like legal challenges to the demolition of damaged buildings, funding for long term repairs, and decisions about rebuilding plans incorporated the knowledge of technical, practice-based and community commissioners in developing recommendations for future research and capability enhancements. In a later example of multidisciplinary research, Dr. Donald Cheu's report on the hospitals' response to the Northridge Earthquake grew out of his connection to the commission through the social services commissioner, Patricia Snyder, who was a Red Cross volunteer and nurse.

Early in its existence the commission began issuing studies of seismic safety challenges like dam safety, highway retrofit needs and school building safety. In the 1979–1980 legislative session AB 2202 (Vicencia) refocused the Commission's mandate on "earthquake hazard mitigation" (Wiley 2000, p. 24). In 1980 it issued the first comprehensive evaluation of California's overall seismic risk and earthquake hazard mitigation potential, *Earthquake Hazards Management: An Action Plan for California and the Executive Summary, A report to the Governor & the Legislature* (SSC 1982). The recommendations in the plan created a blueprint for areas the commission would study for the purpose of developing strategies to enhance seismic safety for individuals and communities. The first edition of *California at Risk: Reducing Earthquake Hazards 1987–1992, Vol. I and II* (SSC 1986) created a comprehensive workplan for the commission and its committees. The post-Loma Prieta Earthquake report, *California at Risk: Reducing Earthquake Hazards 1992–1996,* included the looming challenge of financing the mitigation programs already in place, and the issue of recovery, including financial support (SSC 1991a) providing the basis for the commission's evolving focus on economic recovery.

Post-earthquake field reconnaissance work by the commissioners enriched the understanding of the earthquake's political, social, economic and technical lessons. For example, the Loma Prieta Earthquake was the first in California to strike a major urban area with modern building codes (Wiley 2000). The commissioners conducted field reconnaissance to understand community impacts from the event, issuing the report, *Loma Prieta's Call to Action* (SSC 1991b). The influence of the multiple professions was clear in both the summary sections of the report and the recommendations that it made. For example, while earthquake hazard mitigation based on technical information was recommended, other recommendation categories included improving response, enhancing the statewide emergency management system, enhancing post-disaster communications capabilities, improvements in mass care and shelters, medical services, and training evaluation and enhancements. (SSC 1991a) The perspectives of local government, emergency responders (including voluntary agencies) and community representatives like utilities greatly expanded the concept of earthquake mitigation embodied in this report from physical retrofitting to a system-wide network of improvements.

After the Northridge Earthquake the commissioners determined that simply recognizing the earthquake risk was inadequate. The action plan, issued every 5 years, was renamed *California Earthquake Loss Reduction Plan,* and served as the state's federally mandated earthquake hazard mitigation plan (SSC 2002). The plans have been issued for 1997–2001, 2002–2006, and 2007–2011, focusing on 11 "building blocks" of seismic safety: geosciences, research and technology, education and information, economics, land use, existing buildings, new buildings, utilities and transportation, preparedness, emergency response and recovery, again reflecting the multidisciplinary character of the commission's membership.

Commission report topics range from technical reports on structural engineering issues distributed by the Applied Technology Council to the *Homeowners Guide to Earthquake Safety* that is distributed by realtors to their clients purchasing homes built before 1960. The *Homeowners Guide*, which was developed after the Loma

Prieta Earthquake to support legislation that required sellers of pre-1960 homes to disclose all seismic weaknesses in the structure (AB 2959), is an excellent example of the benefits of using the multidisciplinary approach to appointing commissioners. As the staff developed the guide, and its companion *Commercial Property Owner's Guide* (AB 1968), the commissioners reviewed its contents with their professional colleagues. The technical commissioners ensured that drawings and specifications were all clear and correct, while the practice-based members reviewed the content for ease of use by non-specialist community members, and the community representatives and local government representatives considered the implementation steps for the retrofits. Review by professional colleagues in emergency management resulted in suggestions for rewording instructions for retrofits that could be done by homeowners, like water heater strapping, and adding text to suggest actions that should only be done by construction professionals. The *Guide* was also translated into Spanish, California's second most widely spoken language.

Perhaps the clearest demonstration of the multidisciplinary approach is the report *Public Safety Issues from the Northridge Earthquake of January 17, 1994* (SSC 1995). Each commissioner wrote a brief statement on the issues most closely related to his or her profession. The information was developed from both the field reconnaissance performed by the commissioners right after the earthquake, and the testimony of first responders and community members at hearings held by the commission in February and March of 1994. Some statements are co-authored to offer insights from two professions on the same topic. For example, the planner and emergency services representative developed the statement on health care facility recommendations, and the planner and building official developed the recommendations on essential services buildings improvements.

19.6.6   Seismic Safety Commission's Political Role

The commission's mechanism for engendering change was primarily political. The 15 non-legislative members were governor's appointees, who generally came from his party, sometimes on the recommendation of professional organizations like the Structural Engineers Association of California. Because commissioners might be appointed at any point during a governor's term, and the commission terms are for 4 years, there was sometimes overlap between administrations that resulted in both Republican and Democratic commissioners on the commission at the same time. However, the orientation of individual commissioners was generally toward their professional roles rather than as political representatives. They approached debates and legislative review from the perspective of applying professional expertise to evaluate what was best for the state rather than through a political lens (Diez 2013).

Generally the commissioners were recommended to the governor by other elected officials who were members of the governor's political party, and the commissioners were generally politically active, or were in the family of politically active people. This meant that they had several lines of communication into the legislative process. Some had access to lobbyists for professional organizations (insurance industry and AFL-CIO fire fighters union, for example), or for cities and counties. Others had access to the state's Public Utilities Commission members (also appointees of the governor) or local political leaders with an interest in seismic safety. Through these informal channels of communication they were able to suggest topics for bills that embodied the recommendations created by the commission.

The commission itself also had a political role. One of its tasks was to review proposed legislation and recommend a position to the legislature as a whole and the governor. The staff analyzed the bills initially and passed their suggestions on to the commission for discussion and action in their monthly public meetings. The commission could support, support if amended or oppose legislation as submitted to the legislature (Diez 2012). The legislation related to earthquakes in Table 19.3 demonstrates the breadth of topics that commissioners discussed.

The multiple professional perspectives enabled the commission to take a balanced view of legislation, recognize the likely economic and social impacts of the bill's implementation, and evaluate likely community response to new mandates. For example, while the technical members of the commission supported strengthening of buildings codes and other mandates to enhance structural seismic safety, the commissioners representing communities and local governments could articulate the challenges for the local governments in implementation.

Perhaps the most significant outcome of the commission's Coalinga (1983) earthquake recovery study was the adoption of AB2865 that removed earthquakes from the covered named perils under the homeowners' insurance policy, and required insurance companies selling residential insurance to offer a separate earthquake policy to their customers (Reeves 1984). The insurance challenge would arise again after Northridge.

During the Loma Prieta Earthquake (1989), soft story construction homes (first floor garage, second and third floor living space) pancaked into the garages in the San Francisco Marina District, causing no loss of life but destroying the homes. When new regulations were considered, commissioners expressed a concern that the cost of retrofitting could be prohibitive for owners, resulting in abandoned buildings and leaving behind "ghost towns" of abandoned areas that would further damage property values and lower property tax income (Diez 2012). Thus the technical knowledge of some commissioners about physical retrofitting for enhanced seismic resistance was tempered with the political insights of others, and no soft story retrofit mandates were proposed. The issue came back in 1994 after the deaths of 17 people in the Northridge Meadows Apartments, a soft story construction design, during the Northridge Earthquake, but again the political realities tempered the technical knowledge.

The discussion of legislation often resulted in a recommendation that was different from the initial staff review, often offering amendments that strengthened the bill. For example, the Field Act that governed school construction standards frequently came under attack from special interests that would benefit from loosening the requirements for building plan review. Strong opposition from the commission helped to balance the construction community's desire to use design/build approaches in the development of new community college facilities, fending off

Table 19.3 Historic California earthquakes and significant legislation. (Source of data: Wiley 2000, p. 91–93)

Earthquakes	Legislation
February 9, 1971—San Fernando Valley	Seismic Safety General Plan Element (SB 519), Hospital Seismic Safety Act (SB 352), Alquist-Priolo Earthquake Fault Zoning Act (SB 520), Strong Motion Instrumentation Program (SB 1374), State Capitol Seismic Evaluation (SCR 84), Dam Safety Act (SB 896)
August 1, 1975—Oroville	No liability for earthquake predictions (SB 1950), School inspection program revised (AB 2122)
August 13, 1978—Santa Barbara	Earthquake hazard reduction program (SB1279), Registration of construction inspectors (SB 1367)
August 6, 1979—Gilroy/Hollister	Lower building standards for pre-1933 buildings (SB 445), So. CA Earthquake Preparedness Project (AB 2202), Seismic safe gas valves (AB 2438) and Study of mobile home bracing (ACR 96)
November 8, 1980—Eureka	Mobile home bracing devices (SB 360), Earthquake Education (SB 843), Hospital inspections (SB 961)
May 2, 1983—Coalinga	Coalinga investigation (SB 612), Removal of SSC sunset date (SB 1782), Earthquake Education (SB 1893), Earthquake preparedness (AB 2662), and Earthquake insurance (AB 2865)
April 24, 1984—Morgan Hill	Essential buildings (SB 239), Unreinforced masonry building program (SB 547), Hazardous buildings (SB 548), and Private school construction (AB 3249)
October 1, 1987—Whittier	Funding for Strong Motion Instrumentation Program (SB 593), and Earthquake insurance study (AB 1885)
October 17, 1989—Loma Prieta	Levels of risk (SB 920), Earthquake bond (SB 1250), Hospital anchorage (SB 2453), Mobile home bracing (AB 631), So. CA Earthquake Preparedness Project (AB 725), Safety elements (AB 890), Water heaters (AB 1890), Homeowner's guide (AB 2959), Retrofit guidelines for state buildings (AB 3313), Seismic hazard mapping program (AB 3897), Liability for volunteers, (SBX1 46), Essential buildings (SB 122), Seismic retrofit guidelines (SB 597), Permits for mobile home bracing (SB 1716), Retrofit not new construction (AB 43), Earthquake preparedness guide distribution (AB 200), Liquefaction in safety elements (AB 908), Local bonds for seismic safety (AB 1001), Commercial owner's guide (AB 1968), Hospital seismic requirements (SB 1953)
January 17, 1994—Northridge	Transportation bond act (SB 146), Earthquake research (SB 1864), CA Earthquake Authority (AB 13 and SB 1993)

changes that would have removed the community colleges' Field Act protections. In 1999 the commission held hearings on strengthening and updating the Field Act that included post-earthquake school tagging authority for the Office of State Architect, and support for cost-benefit studies for Field Act requirements (Wiley 2000).

The commission also proposed strengthening existing programs like the Unreinforced Masonry (URM) Building posting requirements, which were strongly opposed by the Building Owners and Managers Association and real estate interests. Because the California Constitution (Article XIII B, Section 6; and Gov. C. Section 17514) forbids the state from creating unfunded mandates, and the state budget has been in distress since the passage of Proposition 13 (1978, which limits the growth of property tax revenues), the legislature is reluctant to pass programs that create mandates for which it may become financially liable. However, commissioners were able to support stronger programs at the local level that benefitted from information developed through SSC hearings and published reports. A draft model ordinance (90–01) and recommended model ordinance (91–03) were published to assist local governments that wanted to take action on URM. Annual status reports informed legislators and the public of the progress of the URM program. Long Beach, Los Angeles and San Jose provide early examples of local URM programs that led to retrofits and other mitigation, and soft story mitigation programs in Berkeley and Oakland benefitted from the SSC's discussion of state-wide action.

19.7   The Commission, Public Knowledge and Risk Communication Challenges

One challenge to the commission's efforts to enhance seismic safety is the relative lack of knowledge about seismic risk among the members of the public. The commission has created the *Homeowner's Guide* and *Commercial Property Owner's Guide* mentioned above, but those are generally only given to people at the time of a real estate transaction. USGS has issued the earthquake probabilities described earlier, but the general public often finds the interpretation of the information difficult. Although the commission's reports are available at little cost, and now often free on-line, earthquake safety is not a salient topic on the public agenda except during the very brief window of opportunity after a major earthquake that is local or somehow relates to the community member, what Birkland calls a "focusing event" (Birkland 1997). It has been almost 20 years since the last big urban California earthquake (although smaller quakes in more remote areas have occurred with regularity), so the public has lost interest in spending public or private funds for retrofits and seismic upgrades (Diez 2013).

The understanding of the functional definition of "earthquake resistant" among the public is poor. People believe that if the building, bridge or road is "built to code" it is "earthquake resistant" and therefore needs no additional attention. It has been difficult to get the general public, the legislature, and building owners in particular to understand that the codes are intended for life safety not building functionality. Commission Executive Director Dick McCarthy notes that if you can walk out of a building after an earthquake it has done its job. The code is designed to ensure that the building does not kill the occupants. "If they have to bulldoze the building the next day, that has met the intent of the code" (Diez 2013).

Technical members of the commission understand that a small increase in the cost of construction can create significant mitigation against earthquake damage for the building. "Value engineering", a concept that allows the owner to determine what level of post-earthquake functionality he wants to purchase, has been suggested in commission reports. The problem is that owners think the code will leave them with a functional building, so they refuse to invest in additional resistance elements (Diez 2013). Only school and hospital building codes are designed to provide a level of post-earthquake functionality; all others are life safety only.

The commission even worked with the legislature to get seismic retrofits exempted from real estate tax enhancements (SCA 33/ Prop 127; AB 43; AB 1291). The challenge is that the fire code and Americans with Disabilities Act (ADA) requirements are generally triggered by the cost of the seismic upgrades. While local governments can exempt buildings within their jurisdiction from fire code value triggers (usually 50 % or more of the value of the building), ADA is a federal mandate that cannot be exempted. Having to pay the cost of ADA retrofitting has discouraged many owners from seismic retrofitting (Diez 2012).

While defense of the Field Act remains a priority, the challenge of retrofitting all of California's hospitals has proven daunting. Hospitals' revenue for patient care has not kept up with costs, so few hospitals can afford the replacement or retrofit of seismically unsafe buildings. Closing the only hospital in a community that delivers day to day critical care because it might perform poorly in an earthquake is not politically acceptable. There are 900 hospitals that have not been retrofitted even though the first deadlines have passed. Unless a bond measure can be passed to retrofit the 300 most dangerous buildings, it is unlikely that further progress will be made on hospital retrofits (Diez 2012).

The commission is focusing on using social media to communicate brief, targeted messages on Facebook, Twitter and the Internet. They are concentrating on creating one page briefings for legislators, focusing on what the element of seismic safety "means for their constituents, why this is of value" (Diez 2013). As political scientist Larry Gerston has noted, "Term limits push out [state legislators]. Since their days are numbered from the minute they take office, elected officials operate with one eye on current issues and the other on their next possible office. …why dedicate yourself to detailed and potentially painful, long term policy initiatives…if you won't be around long enough to see it through? " (Gerston 2012, pp. 199–200). Without focusing events in California, it is difficult to develop legislators' interest in seismic safety when there are so many competing social problems that have more immediacy with constituents (Diez 2013).

19.8 The Current Commission Structure and Focus

California legislative term limits were imposed in 1990 through Proposition 140, limiting state officials to three two-year terms in the Assembly and two four-year terms in the State Senate. Senator Al Alquist was forced out of office by term limits in 1996 at 88 years old. He had served for 33 years in the state legislature (Aleman 2007), becoming the chair of the Senate Budget Committee, a key position for ensuring the SSC's continued budgetary support (Wiley 2000). With the loss of the "father of the Seismic Safety Commission," and the challenge of delivering inherently political evaluations of legislation, the commission's future was uncertain. Unfortunately, as a result of the legislative review mandate, the commission developed opponents in the legislature whose bills had been rated "oppose." When budgetary challenges arose after the dot.com bust economic downturn in the early days of the twenty-first century, opponents began proposing that the commission be abolished. Governor Gray Davis moved the commission off the General Fund budget and onto a new funding source, a 12 cents per property insurance policy surcharge. Initially this generated about $ 1.3 million per year, but the income declined during the foreclosure crisis (Diez 2013).

In 2006, SB 1278 revised the 1975 Seismic Safety Act. The commission was renamed the Alfred E. Alquist Seismic Safety Commission and moved from the Office of the Governor to the California State and Consumer Services Agency, part of the state's bureaucracy. Three members were added to the commission, representing three other agencies in the executive branch: the Building Standards Commission, the State Architect's Office, and the Emergency Management Agency (Alfred E. Alquist SSC 2007). As part of the executive branch the commission no longer comments on bills, rather acting in an advisory capacity to legislative staff members who are developing ideas for bills related to seismic safety. Thus the commission tries to provide advice and "explanations on technical issues or policy issues" as they are developing (Diez 2012).

The commission provides an annual report to the attorney general on their research program and products. In 2011 their $ 500,000 investment of state funds "leveraged an additional $ 1,020,000 in matching funds from participating partners" (McCarthy 2012). The 2011 projects included "Pilot program for evaluation of most seismically vulnerable California Public School Facilities," a template benefiting public school organizations that need to set priorities for public school seismic upgrades. The second project, conducted in partnership with the California Emergency Management Agency, the California Earthquake Authority, and the American Insurance Association, developed "Public education through public television" "on how to take action to reduce injuries and damage from earthquakes" (McCarthy 2012). Previous years' projects have included tsunami risk, residential retrofits, standards for future tall buildings and a Field Act building performance study (McCarthy 2012). These projects all draw from the multiple disciplines on the commission for input and evaluation.

The commission has shifted to a stronger focus on recovery after a damaging earthquake. Studying recent events in New Zealand, Italy and Japan the commission has considered what businesses and industries are most at risk from economic damage following an earthquake. Their studies have reinforced earlier post-Kobe research on the impact of loss of component manufacturing capabilities on the global supply chain, including production on other continents. Loss of Toyota parts from the Sendei area stopped Ford assembly lines in the United States, and Italy's biomedical manufacturing facility's damage is likely to have significant impact on their economy. Italy also lost $ 500 million in cheese that was on racks when the earthquake struck. This fact is being used to try to encourage agri-business to consider how it might be impacted by an earthquake. While farmers consider themselves self-sufficient, they are dependent on water and power that may be disrupted by an earthquake elsewhere in the state. The commission is using its technical, practice-based and community knowledge to collaborate with the Governor's Office of Economic Development to learn about economic recovery from earthquakes in other nations, with the goal of keeping jobs in California and restarting the economy after the next big earthquake (Diez 2012).

19.9   Conclusion

The Alfred E. Alquist Seismic Safety Commission has evolved from the 1969 Joint Legislative Committee on Seismic Safety with a focus on structural hazard mitigation to an internationally-focused organization learning lessons about economic recovery, job retention, and global supply chain resiliency from earthquakes in other nations. The presence of technical specialists, practice-based professionals and community representatives, along with state and local elected officials, enables the commission to collaborate across many disciplines to meet the challenge of earthquake-based global supply chain disruption. They have become a collaborator and resource to the governor's economic development staff and to legislators considering seismic safety projects and programs. Leveraging their research funds with public and private partners enhances their scope of work, and generates products with immediate benefit to California's residents.

References

Aleman, K. (2007). Guide to the Alfred E. Alquist Papers, Biogrphical history. ScholarWorks, San Jose State University. http://scholarworks.sjsu.edu/cgi/viewcontent.cgi?article=100 2&context=speccoll_archives. Accessed 6 Dec 2012.

Alfred E. Alquist Seismic Safety Commission. (2007). History and responsibilities. http://www.seismic.ca.gov/about.html. Accessed 12 Dec 2012.

Alquist, Senator Alfred. (11 May 1994). Talk given at the seismic safety commission legislative day, State Capital, Sacramento, CA.

Alquist-Priolo Earthquake Fault Zoning Act. (1972). Statutes of 1972, Chapter 1354, Section 661.

Birkland, T. (1997). *After disaster: Agenda setting, public policy and focusing events*. Washington, DC: Georgetown University Press.

California Department of Conservation. (2007). California's big earthquakes. http://www.conservation.ca.gov/index/earthquakes/pages/qh_earthquakes_calbigones.aspx. Accessed 18 Nov 2012.

California Department of Conservation. (2012). California geological Survey-Alquist-Priolo earthquake fault zones. http://www.consrv.ca.gov/cgs/rghm/ap/Pages/index.aspx. Accessed 3 Dec 2012.

California Public Resources Code. (1972, as amended). Alquist-Priolo Special Studies Zones Act of 1972. PRC, Division 2. Chapter 7.5. Special studies zones. Geology, mines and mining, 2621–2630.

Choi, C.Q. (19 October 2011). Sierra Nevada faults pose major quake risk, study finds. Science on NBC News.com. http://www.msnbc.msn.com/id/44960429/ns/technology_and_science-science/t/sierra-nevada-faults-pose-major-quake-risk-study-finds/. Accessed 4 Dec 2012.

City of Berkeley, CA. (n.d.). Seismic safety/safety element. Department of planning and building. http://www.ci.berkeley.ca.us/contentdisplay.aspx?id=514. Accessed 4 Dec 2012.

CNN. (2009). At least 150 killed in Italian quake, officials say. http://www.cnn.com/2009/WORLD/europe/04/06/italy.quake/index.html. Accessed 18 Nov 2012.

Diez, M. (2012). Interview of Richard McCarthy, Executive Director, Alfred P. Alquist Seismic Safety Commission. Sacramento, CA.

Diridon, The Honorable Rod. (3 December 2012). Personal communication.

Earthquake: disaster before dawn, 6.6 in the southland. (18 Jan 1994). Los Angeles Times (Pre-1997 Fulltext). Retrieved from ProQuest Database.

Edwards, F. L., & Goodrich, D. C. (2012). Non-FEMA mitigation: Local government actions. In A. Jerolleman & J. Kiefer (Eds.), *Natural hazard mitigation*. Boca Raton: CRC Press.

Elliston, J. (2004, September 15). FEMA-AGENCY PROFILE: Disaster In The Making. *San Francisco Bay Guardian.*

FEMA. Federal Insurance and Mitigation Administration. (2012). *Mitigation's value to society: Building stronger and safer*. Washington, DC: FEMA.

Fleury, M. (2008). World's strongest earthquake. http://suite101.com/article/worlds-strongest-earthquake-a51011. Accessed 27 Nov 2012.

Gerston, L. (2012). *No so golden after all: The rise and fall of California*. Boca Raton: CRC Press.

Gerston, L., & Christensen, T. (2008). *California government and politics*. Boston: Wadsworth.

Holmes, W. (n.d.). Background and history of the seismic hospital program in California. Oakland: Rutherford and Chekene. http://host.uniroma3.it/dipartimenti/dis/ricerca/Hospitals/Articoli_pdf/h.01.eng.pdf. Accessed 4 Dec 2012.

Imai, T. (2011). *Local government responses to the Great East Japan Earthquake*. Sendai: General Affairs and Planning Bureau.

Johnson, J. (1995). Personal communication.

Kraul, C. (2010). Chile reels from 8.8 earthquake. LA Times. http://www.latimes.com/news/la-fg-chile-earthquake28-2010feb28,0,7096321.story. Accessed 28 Feb 10.

McCarthy, R. (14 January 2012). Status Report on California research and assistance funds expended in 2011. Sacramento, CA: Seismic Safety Commission.

Mileti, D., et al. (1999). *Disasters by Design*. Washington, DC: John Henry Press.

MTC-ABAG. (2010). Bay area census: Population by county, 1860-2010. Metropolitan Transportation Commission-Association of Bay Area Governments. http://www.bayareacensus.ca.gov/historical/copop18602000.htm. Accessed 27 Nov 2012.

Multihazard Mitigation Council, National Institute of Building Sciences. (2005). *Natural hazard mitigation saves: An independent study to assess the future savings from mitigation activities* (Vols 1–2). Washington, DC: National Institute of Building Sciences.

Nelson, S. A. (2011). Earthquake hazards and risks. http://www.tulane.edu/~sanelson/geol204/eqhazards&risks.htm. Accessed 27 April 2012.

Olson, R. A. (2002). Legislative politics and seismic safety: California's early years and the "Field Act," 1925–1933. Earthquake Engineering Research Institute, Mitigation Center. http://mitigation.eeri.org/files/Olson_Legislative_Politics_CA.pdf. Accessed 14 Oct 2012.

Reeves, B. (17 December 1984). Earthquake Insurance-AB2865. Lodi Sentinel. http://news.google.com/newspapers?nid=2245&dat=19841217&id=LZwzAAAAIBAJ&sjid=zDIHAAAIBAJ&pg=2207,6289488. Accessed 12 Dec 2012.

Renois, C. (2012). Still in ruins: Haiti marks two years after quake. http://news.yahoo.com/haiti-quake-victims-stuck-time-warp-211733034.html. Accessed 12 Jan 2012.

SCEC. (2013). Uniform California earthquake rupture forecast. http://www.scec.org/ucerf/. Accessed 4 Jan 2013.

SSC. (1982). *Earthquake hazards management: An action plan for California and the executive summary*. A report to the Governor & the Legislature. Sacramento, CA: Seismic Safety Commission.

SSC. (1991a). *California at risk: Reducing earthquake hazards 1992–1996*. Sacramento, CA: Seismic Safety Commission.

SSC. (1991b). *Loma Prieta's call to action*. Sacramento, CA: Seismic Safety Commission.

SSC. (1995). *Public safety issues from the Northridge earthquake of January 17, 1994: A compendium of issue statements by the commissioners of the Seismic Safety Commission*. Sacramento, CA: Seismic Safety Commission.

SSC. (2000). *Seismic safety commission findings: A report to the governor and the legislature on lessons learned from recent earthquakes in Turkey, Greece, and Taiwan*. (Report 2000–2003). Sacramento, CA: Seismic Safety Commission.

SSC. (2002). *California earthquake loss reduction plan*. Sacramento, CA: Seismic Safety Commission.

Statistics Canada. (7 March 2012). Population of census metropolitan areas. http://www.statcan.gc.ca/tables-tableaux/sum-som/l01/cst01/demo05a-eng.htm. Accessed 1 Dec 2012.

Sylves, R. (2008). *Disaster policy and politics*. Washington, DC: CQ Press.

Thompson, J. (13 March 2012). The giant, underestimated earthquake threat to North America. Discovery Magazine. http://discovermagazine.com/2012/extreme-earth/01-big-one-earthquake-could-devastate-pacific-northwest. Accessed 27 Nov 2012.

Tierney, K., & Goltz, J. (1997). *Emergency response: Lessons learned from the Kobe Earthquake*. Newark: Disaster Research Center, University of Delaware.

University of California at Berkeley. (2007). Where the San Andreas Ends. Seismological laboratory. http://seismo.berkeley.edu/blog/seismoblog.php/2008/10/27/where-the-san-andreas-fault-ends. Accessed 18 Dec 2012.

US Census, 2010. (2010). Census 2010 P.L. 94-171 Profile-Los Angeles-Long Beach-Santa Ana, CA Metro Area (US0002091). http://proximityone.com/cen2010/pl/plUS0002091.htm. Accessed 27 Nov 2012.

US Census Bureau. (10 January 2013a). State and county quick facts: Multnomah County, Oregon. http://quickfacts.census.gov/qfd/states/41/41051.html. Accessed 12 Jan 2013.

US Census Bureau. (10 January 2013b). State and county quick facts: San Diego County, California. http://quickfacts.census.gov/qfd/states/06/06073.html. Accessed 12 Jan 2013.

US Census Bureau. (10 January 2013c). State and county quick facts: King County, Washington. http://quickfacts.census.gov/qfd/states/53/53033.html. Accessed 12 Jan 2013.

USGS. (1993). Historic earthquakes: Prince William Sound, Alaska. http://earthquake.usgs.gov/earthquakes/states/events/1964_03_28.php. Accessed 27 April 2013.

USGS. (18 July 2012a). 2008 bay area earthquake probabilities. USGS earthquake hazards program. http://earthquake.usgs.gov/regional/nca/ucerf/. Accessed 1 Dec 2012.

USGS. (1 November 2012b). Historic earthquakes: Long Beach, CA. http://earthquake.usgs.gov/earthquakes/states/events/1933_03_11.php. Accessed 18 Dec 2012.

USGS. (1 November 2012c). Historic earthquakes: San Fernando, California. http://earthquake.usgs.gov/earthquakes/states/events/1971_02_09.php. Accessed 18 Dec 2012.

Ward, P. L., & Page, R. A. (1989). *The Loma Prieta earthquake of October 17, 1989; what happened, what is expected, and what can be done*. Washington, DC: United States Government Printing Office.

Wiley, K. (2000). *Living where the earth shakes: A history of the California Seismic Safety Commission*. Sacramento: California Senate Office of Research.

Chapter 20
Multi-sector Partnerships in Disaster Housing Recovery: An Examination of Housing Development Approaches in India

Simon A. Andrew and Sudha Arlikatti

20.1 Introduction

Responding to disasters is no longer the sole prerogative of the concerned governments. As evidenced by the response to recent disasters, disaster management is a collaborative effort purposely designed by government agencies, international humanitarian agencies, as well as private corporations and local and international nonprofit agencies in order to jointly provide relief and support to disaster survivors (Tierney et al. 2001; Özerdem and Jacoby 2006). Disasters elicit complex collective actions by a wide array of organizations, irrespective of their ideologies and backgrounds, and are sometimes effective but often fraught with challenges (Kapucu 2006; Kapucu and Van Wart 2006; Waugh and Streib 2006). As noted by Comfort (2005), intense coordination is required between multi-sector agencies in order to achieve common goals of effective response and recovery. Post disaster housing repair and rebuilding is one such challenge faced by public, private, and nonprofit agencies, as they try to balance a quick return to normalcy with sustainable development goals that are intertwined with physical, psychological, and livelihoods recovery of disaster survivors (Lindell et al. 2006).

The pursuit of sustainable housing recovery has gathered momentum in recent years (Berke and Beatley 1997; Smith and Wenger 2006) with the essential focus on achieving post disaster recovery that emphasizes not only physical recovery but psycho-social and economic recovery through the active involvement of disaster survivors, policy makers, and private and nonprofit organizations. The assumption is that communities can recover faster and more equitably when projects incorporate disaster risk reduction strategies through a participatory approach, and pursues overall development goals such as livelihood recovery, maintains social networks and is culturally sensitivity (Arlikatti and Andrew 2012). Within the context of

S. A. Andrew (✉) · S. Arlikatti
University of North Texas, Denton, USA
e-mail: sandrew@unt.edu

S. Arlikatti
e-mail: Sudha.Arlikatti@unt.edu

N. Kapucu, K. T. Liou (eds.), *Disaster and Development,* Environmental Hazards,
DOI 10.1007/978-3-319-04468-2_20, © Springer International Publishing Switzerland 2014

sustainable housing recovery, the partnership arrangements developed by public agencies with nonprofits, private enterprises and donor agencies is a complex process requiring time and preparation, especially when seeking citizens involvement. Without clear philosophical underpinnings, appropriate deliberative process, and representation, there is a danger of marginalizing local housing culture, social infrastructure, and households' livelihoods (Barakat 2003). Disaster scholars have argued that, although temporary housing assistance in terms of financial or material aid may help disaster survivors' perform basic household functions, in the long run, the ability to transform such available assistance varies by gender and vulnerable marginalized populations (Enarson 2010; Peacock et al. 1997; Oxfam 2005). Much of the current research on social vulnerability compels scholars to posit that, in order to minimize the differential impact of disasters, permanent housing recovery must be "supplemented with higher levels of recovery resources" particularly in areas that are largely occupied by low-income and minority households (Peacock et al. 1997, p. 265).

Housing recovery for displaced populations is yet another challenge. Poorer households, when displaced, are likely to be removed from their familiar environment and social support structures. The displacement makes them markedly disadvantaged if their "livelihoods were destroyed and assets lost" (Doocy et al. 2006, p. 278).

Displaced households also face a complex set of issues, ranging from finding suitable lands and building materials to securing property rights and rebuilding their social infrastructure (Peacock et al. 2006). Settling in new physical environments affects their capability to be self-reliant and thus, a cause for concern if employment opportunities are not forthcoming. Not surprisingly, most displaced and subsequently relocated and resettled households often continue to have a strong attachment to their old villages and may seek to go back despite the threats from future disasters (UNDRO 1982; Coburn et al. 1984; Partridge 1989).

Organizational and management challenges are also created due to the socioeconomic, language, cultural, and livelihood differences that create issues of environmental justice and equity. While the roles of these three sectors are crucial to effective disaster response and recovery, local community participation is also underscored (Alexander 1993; Arlikatti and Andrew 2012; Barenstein and Leeman 2012). Scholars have consistently noted that, wherever donor agencies and government entities have worked in conjunction with disaster survivors, these households are able to better adapt to their changing environments and recover faster.

This was aptly demonstrated by the success of numerous cash for work (CFW) interventions instituted after the Asian tsunami in Aceh, Indonesia, and Sri Lanka to serve thousands of displaced households whose livelihoods had been destroyed. CFW programs were operated by multiple international non-governmental organizations (INGOs) and the United Nations Development Program (UNDP) often in partnership with local non-governmental organizations (NGOs) and public sector agencies (Doocy et al. 2006).

We argue that post-disaster housing recovery packages can serve as a window of opportunity for community development, especially when spearheaded by public sector agencies that can establish collaborative arrangements and partnerships, and implement timely quality control and monitoring mechanisms. From a normative perspective, the formation of partnerships is of practical importance because they tend to be created with a broad set of economic development goals for the greater good (see Chap. 17 in this Volume). While some observers welcome these public-private partnerships in housing recovery, others criticize the approach as a strategy employed by government agencies to shirk their social responsibilities and pass it on to the private sector (Mahadevia 2001, p. 3670). Moreover, in the immediate aftermath of a disaster, government agencies are pressed for time while trying to address infrastructure repair, population protection, search and rescue, and temporary sheltering requirements, and may rush into these agreements without much forethought or planning. This may lead to inadequate attention while crafting appropriate partnership agreements, specifying built-in quality control, and monitoring mechanisms, which often have long-term consequences on disaster recovery trajectories.

To investigate these conflicting observations and to better understand how housing programs are initiated as an integral part of development, we review different housing policies and program approaches focusing on three disasters in the largest world democracy—India. We examine the housing policy changes in India through the lens of post-disaster housing recovery programs instituted in the aftermath of the Killari earthquake in the Latur district of Maharashtra in 1993, the Bhuj earthquake in Gujarat in 2001, and the Great Tsunami of 2004 in Nagapattinam district of Tamil Nadu. The objectives are to highlight: (1) the various approaches adopted by the state governments to initiate public-private-nonprofit partnerships in jumpstarting disaster housing recovery, (2) to highlight the challenges of village adoption and matching programs initiated in the disaster impacted communities, and (3) to provide lessons learned and suggest improvements to foster synergistic partnerships for housing recovery that are self-sustaining and intertwined with overall economic development objectives. Emphasis is laid on the challenges in implementing public-private partnerships that are responsive to local politics and caste differentials, allowing for representation and participation by recipients of the housing redevelopment programs.

The following sections start with a discussion of the housing policy changes that have occurred in India—from a government institutionalized strategy to a more market-driven strategy—where private and nonprofit organizations have started playing a key role in the implementation of urban and rural housing programs. The Central government of India's gradual shift in the nation's housing policy and its relevance to the production and provision of permanent housing during the recovery phase of disasters is then discussed. Subsequent sections examine the various housing approaches adopted through public-private partnerships by three state governments in response to two earthquakes and a tsunami. The chapter is concluded with lessons learned in order to ensure the delivery of equitable housing recovery programs that meet sustainable development goals.

20.2   Housing Policy in India

With an estimated 1.2 billion citizens (July 2012), India is second only to China in population making it the largest democracy in the world (The World Factbook 2012). In 2011, there were 53 urban agglomerations in India with a population of 1 million or more, as against 35 in 2001 (Census of India 2011). India consists of 28 states and 7 union territories, and is the seventh largest country in the world by geographical area. The economy of India is the tenth-largest in the world by nominal GDP and the third largest by purchasing power parity (PPP). On a per capita income basis, India ranked 140th by nominal GDP and 129th in 2011 by GDP (PPP), according to the International Monetary Fund (in comparison, the US ranks 6th on a per capita income basis and the highest by PPP). A recent World Bank report (2010) found that 32.7 % of the Indian population lives below the international poverty line of US\$ 1.25 per day (PPP) while 68.7 % lives on less than US\$ 2 per day.

The traditional approach to housing reconstruction in disaster-impacted areas in India generally emphasized retrofitting and improved building codes and design to combat building failures, thus overlooking housing units as part of the subunits of the built environment. Housing reconstruction and rehabilitation projects put less emphasis on community level design, collective spaces, social development activities, and education. More recently, influenced much by multilateral agencies and international approaches and mandates, the country's recovery activities have tended to include community development projects, education and skill training programs, and access to medical and technical assistance. There is also a general shift to treat housing as a living system consisting of social infrastructure as well as surrounding physical environment rather than only as a structural unit.

Of special relevance to the study of post-disaster housing programs in India initiated through multi-sector partnerships, are the political and socio-economic circumstances. Most housing recovery programs are linked to community rehabilitation and poverty alleviation programs funded largely by international aid agencies and implemented by private and nonprofit organizations. Yet, in the area of housing recovery involving private contractors selected by NGOs and international aid agencies, the notion of public-private partnerships are often a poorly understood. In a world of uncertainties—where parties involved in collaborative efforts may attempt to default from their shared responsibilities and withdraw from their initial commitment—officials working for government agencies must decide on the best way to minimize political outrage over poor implementation of public housing programs. That is, the problems associated with delegating the task of housing program implementation to other organizations.

The production and provision of public housing in India has largely been based on policy objectives set by the Central government through its Five-Year Plans. Often referred to as a top-down approach, state governments are mostly responsible for the implementation of urban and rural housing programs (Barenstein 2010; Mahadeva 2006; Pandey and Sundaram 1998). Multilateral agencies have also influenced the shaping of disaster recovery programs in India. For example, the "*Global*

Strategy for Shelter to the Year 2000," the "*Millennium Declaration 2000*," and the "*Hyogo Framework for Action*" have redirected the Indian government's attention to initiating housing programs that are tied to poverty alleviation and disaster mitigation, so as to build disaster resilient communities.

20.2.1   Housing Policy in the 1960–1970s

In the late 1960's and early 1970's the aim of the housing policy in India was to increase the country's housing stock through the development of vacant lands, clearing of slums, relocation/resettlement programs, and upgradation of existing housing stock with basic amenities such as potable water, electricity, and sewage systems. The traditional housing policy implemented during this period relied on the supply of low interest rate loans and subsidies for housing redevelopment, which Mukhija (2001) refers to as the "institutionalized government strategy" (p. 214). This strategy assumed the security of land tenure as an important determinant of the quality of housing including settlement programs in rural areas (Friedman et al. 1988; Kaliappan 1991; Keivani and Werna 2001). However, most of the components of the approach received criticism for being inflexible and unable to meet the increasing demands for urban housing (Sengupta 2006), and the financial burden placed on poor households (Mukhija 2001; Singh and Das 1995; Wadhwa 1988). The problems related to bureaucratic red-tape preventing equitable distribution of housing resources were also a major concern. Investments to improve existing housing stock were equally lacking (Godbole 1999).

20.2.2   Housing Policy in the 1980–2000's

Subsequently, over the next three decades, there was a general shift in the Central government's approach. Some scholars suggest that the central tenet of housing policy during this period emphasized a "market enabling strategy," which relied on the logic of a public good market (Keivani and Werna 2001, p. 192). Depending on the types of housing programs, the strategy redefined and, to some extent, broadened the roles of state governments. The objective was to create a favorable housing market and incentive structure that could simulate infrastructure investments and liberate the financial sector, thus increasing the supply of housing and improving the quality of private dwellings (Sengupta 2006). To some extent, the approach also encouraged the involvement of private and nonprofit organizations in production and provision of public housing (Sen 1998). This reduced the administrative burdens placed on state governments by spreading investment risks associated with public housing and encouraged specialization through the contributions of the private sector. Such an approach also recognized innovative solutions to specific community's housing concerns through development proposals made by private housing developers.

At the national level, the Indian government's National Housing Policy (in the early 1990s) and the New Economic Policy (in the mid 1990s) shaped much of the funding, types of housing development, and resettlement projects (Singh and Das 1995). The National Housing Policy of 1994 heralded improvements on several key housing indicators (Mahadeva 2006). Depending on state governments' funding streams, housing programs were implemented by state-level Housing Board's or quasi-governmental corporations, through "sites and services" projects, by identifying and clearing vacant land (i.e., the site) and providing much needed services like water and sewage connectivity and civic amenities (i.e., services). The implementation of low cost housing projects, for the most part, relied on public-private partnerships, where a household could purchase a plot of land with low interest loans and housing subsidies during normal times and contract out the construction.

20.2.3 Influence of Global Risk Reduction Strategies on Housing Policies

The housing policy and disaster recovery programs were also influenced by development principles introduced by international humanitarian agencies. For example, the World Bank's structural adjustment programs and its project-based approach (in the early 1980s), and the adoption of the *Global Strategy for Shelter to the Year 2000* approved in New Delhi in 1988 provided a framework for the Indian government to initiate housing programs that were tied to poverty alleviation. A stronger emphasis was also placed on social vulnerability, which was advanced by the United Nations General Assembly when it declared the 1990s as the *International Decade for Natural Disaster Reduction* (UNDP 1994). Further, the *Millennium Declaration 2000* also played a large part in directing multilateral agencies' emergency relief and development strategies towards reducing poverty and mitigating the devastating effects of natural disasters rather than limited response activities. Similarly, the major international agreement "*The Hyogo Framework for Action*" on disaster risk reduction, proposed at the World Conference on Disaster Reduction in Kobe, Japan (January 2005), heightened the need to reduce disaster vulnerability and build resilient communities (UNISDR 2005).

While at first glance it appears as if the Indian national government has responded positively to the aforementioned international strategies, it is difficult to determine whether the concepts have filtered down successfully, and are an integral part of the housing development programs implemented across the various states of India. This is key as individual state governments can choose to work alone or partner with private corporations, or international and/or local nonprofit agencies (Pandey and Sundaram 1998, p. 87). Further, there are housing schemes adopted by various state governments tailored specifically to funding availability, and community characteristics and requirements. Especially, following major disasters the housing reconstruction and rehabilitation projects often adhere to the requirements stipulated by donor agencies such as the United Nations Development Program, the World Bank etc. The details, when translated into entitlement packages, vary depending on the scope of devastation, the traditional settlement designs and local architecture style of a region, and state governments' pre-existing housing schemes for urban and rural areas.

20.3 Post-Disaster Housing Development Cases

India is no stranger to major natural disasters like earthquakes, cyclones, floods, droughts and tsunami (Arya et al. 2006; Nath et al. 2008). Traditionally, the approach to post-disaster housing reconstruction and rehabilitation centered on compensation and relocation programs. Only in the last three decades, have public, nonprofit and private sector partnerships proliferated. Following a disaster, the Central government plays a coordinating role by providing financial assistance and mobilizing key personnel, while the state governments supplement this effort. However, the actual implementation of the emergency response and recovery efforts fall squarely on the shoulders of district level agencies that are subunits of each state. The District Collector's office is charged with coordinating the repairs and rebuilding activities of private and nonprofit organizations at the local level (Prater et al. 2006; Arlikatti et al. 2010). Depending on the scale of the disaster, specialized agencies have also been established by state governments to coordinate and oversee disaster housing recovery programs (Barenstein 2006) as demonstrated in the following three case studies discussed.

20.4 The 1993 Killari Earthquake, Latur District, Maharashtra State

On September 30th 1993, at approximately 3:56 a.m. local time, a 6.4 magnitude earthquake on the Richter scale struck western India. The earthquake devastated the districts of Latur and Osmanabad in the state of Maharashtra. A total of 52 villages were demolished and approximately 20,000 people died, while another 30,000 sustained injuries.

Characteristics of the Public, private, nonprofit partnerships: The involvement of private and nongovernmental organizations was exemplified by various strategies adopted by the Maharashtra state government during the reconstruction and rehabilitation phases following the 1993 Killari earthquake. Under the *Donor-driven Reconstruction (DDR)* program, the state government of Maharashtra invited international and national NGOs, private enterprises, community based organizations, and faith-based charitable institutions to assist affected villages with housing reconstruction. These donor agencies and their engineering consultants and private contractors, were invited to present housing designs, technical bids, and construction plans to the local village councils.

The selection of donor-agencies was based on their constitution or organizational charter, past and current organizational experiences related to housing reconstruction activities, and funds available with the donor-agencies, and construction teams established to undertake the reconstruction work (Parasuraman et al. 1995). Donor-agencies selected by the state government were then matched with villages depending on the needs and their capacities, while faculty from the reputable Tata Institute of Social Sciences (TISS), a private University in Mumbai, India were charged with mediating the process. The teams worked closely with affected villagers during the project implementation stage by identifying beneficiaries, selecting sites, planning village layouts, housing design, and allocation of plots and permanent housing. In some villages, the certificate of participation, issued by TISS, was an important element of the public, private, nonprofit partnership before construction began. In terms of oversight, the state government appointed technical audit and quality assurance consultants to monitor the quality of housing construction.

Damaged homes were to be retrofitted and all new reconstruction was required to adhere to earthquake resistant building codes. Consistent with the community participatory approach to development, the donor agencies were required to work in consultation with the affected communities. Under the DDR program, funds were allocated by the Chief Minister's Earthquake Relief Fund to donor-agencies to conduct their work. Although there were no specific guiding documents, the scope of reconstruction was characterized by housing designs and layout plans prepared by the Maharashtra Housing and Area Development Agency and approved by the Indian Institute of Technology—Mumbai, Central Building Research Institute (CBRI), or the Roorkee University's Department of Earth-Science (Parasuraman et al. 1995).

Outcomes: Public-private partnerships in executing DDR programs post Latur earthquake had both positive and negative outcomes. Since consultations with villages took longer than expected before housing reconstruction could begin, issues related to temporary resettlement and housing arrangements were carefully considered. In earthquake prone areas, planning for resettlement of whole communities was emphasized as a crucial development strategy rather than focusing on individual damaged homes. While the process involving local villages and marginalized communities was not easy, a flexible time-frame was developed to ensure trust and adjustment to emerging community perceptions, especially in affected villages with households from diverse religious, economic, and social backgrounds. Concerted efforts were made to involve affected households in rebuilding (free-labor and work contributions) in order to avoid the misconception that donor-agencies were contractors of government agencies rather than development partners with the local communities (Parasuraman et al. 1995).

Yet, scholars and independent observers have been critical of the government's top-down agency driven DDR policy for its overemphasis on physical improvements of the structure, while missing the broader development oriented strategies, thereby reinforcing local community vulnerabilities (Salazar 1998; Vatsa 2010). Concerns related to the inadequacy of strategies adopted by donor agencies to achieve meaningful consultations with affected villages and traditionally marginalized groups have also been reported (Narasimhan 2003). Most importantly, the long-term negative impacts of adopting a resettlement policy that depended on outside contractors rather than utilizing the expertise of the affected people for in situ repairs was pointed out. The DDR approach was also blamed as being a traumatic experience, impacting livelihoods and leading to social disarticulation of the recipients (Barenstein and Leeman, 2012).

20.5   The 2001 Earthquake in Kutch, Gujarat State

On January 26th, 2001, at approximately 8:46 a.m. local time, a 7.7 magnitude earthquake on the Richter scale, occurred in western India, where around 20 million people lived and worked. The earthquake devastated the entire state of Gujarat, one of the most economically well established states in India, with the death toll reaching over 20,000 and 167,000 people sustaining severe injuries. Over 90% of the deaths were reported in Gujarat's largest and most vulnerable district of Kutch. An estimated one million homes were damaged or destroyed. Cities like Anjar and Bachau were completely destroyed, damages reported in 7,633 villages and towns and over 300 villages nearly flattened (Shaw et al. 2001).

Characteristics of the Public, private, nonprofit partnerships: Two weeks after the earthquake, on February 2001, the Gujarat state government established the Gujarat State Disaster Management Authority (GSDMA) to coordinate a comprehensive rehabilitation program targeted at the housing, agriculture, tourism, industrial and arts and crafts services, as well as rural and special populations capacity building efforts (Nakagawa and Shaw, 2004; Shaw and Sinha, 2003). The proposal was the relocation of the most affected villages, assistance for in situ reconstruction of severely affected villages and assistance for home repairs in less damaged locales. However, a systematic public consultation with survivors from 450 of the most severely affected villages carried out by the NGO network *Kutch Nav Nirman Abhiyan*, found that 90% of those consulted, were not amenable to relocating (Barenstein and Pettet 2012). This led to the abandonment of the state government's relocation plans and the adoption of the Owner-Driven Reconstruction (ODR) approach at such a large scale for the first time, and ratified by the State Government as its official reconstruction policy. Reconstruction initiatives fostered the involvement of a large number of national and international NGOs and private corporations, who were free to support housing reconstruction through material or cash assistance, without having to formally adopt whole villages, resulting in a 50–50 government-private partnership (Sheth et al. 2004; Shaw and Sinha 2003).

This resulted in five different approaches adopted by NGOs and private corporations, clearly enumerated by Barenstein and Pettet (2012): Owner Driven Reconstruction (ODR) approach where households either renounced NGO assistance in favor of the government's housing reconstruction program or received housing support from the government or NGO or both in the form of financial or material assistance; a Subsidiary Housing Reconstruction Approach (SHA) where NGOs specifically targeted households in remote hamlets, thatwere not entitled to any government compensation as they were not officially registered; a Participatory Housing Reconstruction Approach (PHA) where NGOs assumed a leading role in reconstruction but consulted with communities in the finalization of proposed housing designs; a Contractor Driven Reconstruction Village In Situ (CODIS) where the task of housing reconstruction was contracted out to a professional construction company who imported construction materials and expertise from outside the target community; and finally the Contractor Driven Reconstruction Approach *Ex-Nihilio* (CODEN) where the full village is relocated and reconstructed on a new cleared site.

The GSDMA's work gained international recognition and was awarded the 2003 UN Sasakawa Award for its outstanding work in managing post-earthquake reconstruction and disaster management and risk reduction (Barenstein and Pettet 2012, p. 75). The ODR approach found more favor with the communities with about 82 % of the housing reconstruction in the affected regions adopting this approach (Sheth et al. 2004). The public-private partnership during the recovery phase was described as a model that emphasized "village adoption."

The major component of the approach "matched"[1] up a registered donor agency that was classified as an "adopter" of a particular devastated village, who then deposited funds into that village's reconstruction committee's bank account. Under the arrangement, the state government of Gujarat contracted out the reconstruction of houses and in-situ rehabilitation activities to private/corporate entities and NGOs as potential donors (Mahadevia 2001). This arrangement was to ensure that donors conducted an assessment of the devastated villages and estimated their ability to provide assistance based on the scale of destruction and expected recovery funding; and then, made a commitment. Households in villages that were not adopted by any NGO or private corporation received cash compensation from the Gujarat government depending on the extent of damage.

Alternative housing recovery packages tailored to suit different geographic areas, extent of damage, and structural types, were made available to rural households enhancing owner-driven reconstruction efforts in collaboration with public, private and local or international NGOs partners (Nakagawa and Shaw 2004, p. 20). Such success was not immediately evident in urban areas due to time spent by public sector agencies, in meeting the rezoning requirements set by the Gujarat State Disaster Management Authority, conducting site suitability analysis, and acquiring land

---

[1] The arrangement allowed the donors to choose their prospective "adopted villages" by inspecting or touring the affected areas, and if the village reconstruction committee agreed, the planning and reconstruction processes would begin with the blessing of the government.

for resettlement and housing redevelopment. Perhaps one of the most important lessons learned during the recovery phase was the important role of government and quasi-government agencies in coordinating multiple stakeholders' activities. For example, the Bhuj Development Council (BDC) was involved in information dissemination and encouraging community participation, and mediating meetings between disaster survivors, government officials and private consultants. BDC also "engaged in the rehabilitation of slums and the informal sector" (Nakagawa and Shaw 2004, p. 22).

Outcomes: In response to the 2001 Gujarat earthquake, disaster management was shifted from the Ministry of Agriculture to the Ministry of Home, and the National Institute of Disaster Management center was set up. Community-based disaster risk management programs were set up in 17 states. Furthermore, hazards and disaster management topics were introduced in high school curriculum in the state of Gujarat. However, the challenges in implementing housing reconstruction and rehabilitation programs through public-private-nonprofit partnerships were "intricately intertwined with existing social processes" (Mahadevia 2001). According to Mehta (2001), NGOs preferred to work in small homogenous villages rather than large villages with multi-caste populations due to fear of caste conflicts and village politics. Some villages were not willing to work with NGOs citing that these NGOs sought out only the village elites with vested interests rather than the more egalitarian village committee, thereby compounding caste divisions and village politics (Barenstein and Pettet 2012). Similar observations have been reported elsewhere in the state of Andhra Pradesh, where NGOs generally preferred to operate in relatively homogeneous single-caste villages, leaving other multi-caste villages without adequate assistance (Bosher 2007).

The subsidiary housing reconstruction approach (SHA), the owner-driven reconstruction approach (ODR), and the participatory housing reconstruction approach (PHA), in that order received the highest ratings on satisfaction and effectiveness by their recipients, and touted as successful and exemplary. The SHA approach enabled the most vulnerable and marginalized to obtain goods and services and mitigated some of the risks of the ODR approach. Not surprisingly, the contractor driven approaches in-situ and ex-nihilio (relocated), although favored by a large number of international NGOs who are unfamiliar with local culture and construction techniques, were least satisfying to the communities for numerous reasons. Homeowners complained about the poor quality of construction, inability to closely monitor or supervise the contractors as houses were distributed randomly, only after construction was completed. Homeowners expressed distrust as they perceived fraudulent alliances between contractors and village elites who then received preferential treatment and more houses. Concerns about the exogenous building materials and designs brought in, that were unsuitable for the climate and lifestyle of the villagers were also noted. As a result, participation of homeowners and local artisans and craftsmen was extremely limited (Barenstein and Pettet 2012).

20.6   The 2004 Indian Ocean Tsunami, Nagapattinam district, Tamil Nadu State

On 26 December 2004, an underwater earthquake of 9.1 magnitude on the Richter scale struck Indonesia triggering a series of devastating tsunamis along the coasts of most landmasses bordering the Indian Ocean, killing more than 300,000 people in eleven countries (Arlikatti et al. 2010). The Nagapattinam District of Tamil Nadu in South India, which has a population of 1.5 million comprising mostly of rural fishing and agricultural households was the hardest hit. Over 6,000 people were killed, 196,000 displaced, and 28,000 sheltered in temporary relief camps (Prater et al. 2006). The Nagapattinam District Collectorate's office (a *subunit of the s*tate) identified 81 villages as "impacted by the tsunami" and requiring relief aid (Prater et al. 2006).

Characteristics of the Public, private, nonprofit partnerships: In response to the 2004 tsunami, a different model of public-private partnership was adopted by the southern state of Tamil Nadu in India. The state government, together with various multilateral agencies developed a comprehensive Emergency Tsunami Reconstruction Project (ETRP)—a collaborative governance structure purposely designed to develop and coordinate acceptable strategies for repairing and rebuilding damaged homes. Drawing upon the concept of a participatory approach, state agencies worked jointly with domestic and international NGOs and local stakeholders to develop a set of policy guidelines related to the implementation and construction of disaster-resistant homes. The ETRP's response and recovery activities not only provided financial aid and building materials but also rehabilitation activities such as outreach and educational programs—awareness that is critical in establishing resilient coastal communities. The strategy aimed at encompassing activities associated with rebuilding of public infrastructure as well as reviving the livelihood of the affected communities.

The Nagapattinam District Collector set up an NGO Coordination and Resource Center (NCRC) at the *Collectorate*, which allowed NGOs to be grouped to deal with several issues including housing reconstruction. NGOs and donor agencies were required to register and declare how many homes they could rebuild and then sign a Memorandum of Understanding (MOU) committing to this task. These agencies were then allocated the village by the collector's office (Prater et al. 2006). This was to provide equitable housing to all, especially in multi-caste and far flung remote villages reflecting lessons learned from the Gujarat experience.

The public, private, nonprofit partnerships adopted during the 2004 Tsunami recovery also emphasized consultation with village level committees. As part of the rebuilding process, NGOs and private corporations consulted with the local village committee leaders (the *Panchayat*) on the design of their new homes before a formal agreement was reached. A tripartite agreement was then signed between the District Collector, the private contractor, and the beneficiary before the construction was authorized. Once the new homes were constructed, the legal title was assigned as a joint ownership between husband and wife. The task for the payments of compensation was delegated to a District Level Negotiation Committee to purchase or acquire lands from private land-owners, religious institutions, and vacant forest reserved lands (G.O M.S No. 774). And then, through the administrative process, the rebuilding and reconstruction of new homes was assigned to domestic and international NGOs.

However, these processes were complicated by existing land-use regulations. Under the Environmental Protection Act (1986), the Coastal Regulation Zone (CRZ) Notification of 1991 declared the vast coastal stretches (7,516 km) of the Indian peninsula as ecologically sensitive areas and vulnerable to natural hazards, requiring various degrees of protection from residential development. Following the tsunami, the state government of Tamil Nadu (GoTN) was mandated to consider the provisions of this CRZ notification while relocating and resettling the tsunami survivors. The GoTN also adopted building practices that incorporated disaster resistant technologies as stipulated by the Bureau of Indian Standards Codes. Subsequently, in compliance with these requirements, on 30 March 2005, the G.O Ms. 172 was issued stating that all government, NGO and private corporations sponsored houses would be constructed (with appropriate compensations) beyond 200 m of the high-tide line (HTL) off the coast of the Indian Ocean.

The most ambiguous part of the G.O Ms. 172 was the relocation of affected communities whose homes were located closest to the coastline, less than 200 m from the HTL. While the CRZ Notification was clear on new constructions permitted within a specific category of the zoning law, in 2005, immediately after the tsunami, the state government was not sure on what stance to take regarding the rebuilding of hutments and traditional homes that did not have the necessary permissions or authorization records prior to the tsunami but had been there for generations. Further, while relocation was optional for owners whose permanent homes were located within 200 and 500 m of the HTL, homeowners located beyond the 500 m radius did not have to move, but could apply for assistance to repair homes if damaged.

Outcomes: Integrating disaster recovery programs and development was not without its challenges following the 2004 tsunami. It was observed that multi-sector partnerships in housing recovery programs have been hampered by the ambiguities of government regulations and resettlement policies, the complexities of the land tenure system, and the building of disaster-resistant housing within government stipulated and enforced setback lines (Gokhale 2005; Sridhar 2005; Srinivasan and Nagaraj 2006; Arlikatti and Andrew 2012).

The relocation programs embedded within recovery packages took some households away from their livelihoods, forcing them to spend considerable amount of financial resources "to get to the sea and then back home" (Kumaran and Negi 2006, p. 381). Although the World Bank implicitly stated in its policy guidelines that it does not support projects that would lead to involuntary resettlement[2], given the stated land-use regulations in India, the financial incentives, and requirements

---

[2] World Bank Operational Manual (December 2001), Operational Policies: Draft OP 4.12 Involuntary Resettlement. The Operational Policy statement was updated in March 2007 consequent to the issuance of OP/BP 8.00, Rapid Response to Crises and Emergencies.

for receiving housing assistance, the relocation of households were mostly indirect. Kumaran and Negi (2006) noted, for instance, those involved in these projects were "generally frustrated and disappointed with political leaders and NGOs, especially on failures to use available funds for rebuilding strong, well-looked after communities" (p. 384) The implementation of some housing reconstruction programs by private or non-profit organizations have also been reported as being insensitive to the housing culture and rituals of local fishing communities thus making them difficult for the populace to get acclimatized to (Barenstein 2010; Arlikatti and Andrew 2012).

20.7   Discussions

The three case studies illustrate that, although addressing the immediate housing needs of disaster impacted communities is crucial to their return to normal routines, there are potential risks involved if problems related to local political institutions, social, and economic aspects are not interwoven into the post-disaster housing recovery packages.

20.7.1   What Hinders Development and Recovery

In the context of recovery efforts in India, our review of the cases demonstrate that major disasters sometimes reinforce pre-existing social and economic imbalances. Reports of marginalized and rural households generally not trusting private contractors employed by the government, INGOs or NGOs to carry out housing reconstruction projects persist.

20.7.2   Facilitators of Development and Recovery

The success of these programs necessitate the active participation of civil society/ disaster survivors in the public-private-nonprofit partnership arrangements, especially in the planning and housing reconstruction phase of recovery.

1. Partnership arrangements between local community, private contractors, and specialized agencies need to be part of the process of empowering local communities in building disaster-resistant houses. Given a choice, members of the community are interested in owner driven reconstruction (ODR) approaches with technical and monetary assistance from the government, NGOs or private entities. They are eager to rebuild stronger, but using local building materials and traditional design practices as demonstrated in Gujarat.

2. Design experts and engineers need to understand this preference and provide innovative design alternatives that use traditional eco-friendly materials in hazard resistant stronger design alternatives that are more cost effective, easily repaired or replaceable seasonally and can be executed by the owners themselves or by local tradesmen and artisans. There is also an urgency to develop products that are easy to build by the locals without over dependence on expensive materials and technology. For example, Tiwari (2001) argues that the entire philosophy in housing reconstruction in India should focus on locally available materials, citing the construction of *pucca* (strong) houses made up of bricks, cements steel, and timber as too expensive for the majority of households. The use of mud blocks and roofing tiles as building materials, or other indigenous based construction technologies is often not in the minds of planners, architects, and home owners, and this needs to change.

3. More importantly, in order for building materials to be climatically appropriate and meet the needs of the local communities, reconstruction of mass housing must consider locally available materials that are cheap and in abundance with low energy input (Harrison and Sinha, 1995). As suggested by Hirway (1987), rural housing programs should be "socially and culturally valid and should be suitable to the climate as well as to the economic conditions of the poor" (p. 1456). Barenstein and Pittet (2012) have also emphasized that housing in India, especially in the coastal Tamil Nadu area "is culturally sensitive and a highly ritualized process," (p. 122) suggesting a need to have reconstruction programs that are sensitive to affected communities' built environment and housing culture.

4. The frequency of financial and technical assistance received from multilateral agencies thus becomes crucial in explaining the type of public-private-nonprofit partnerships. Such an arrangement emphasizes the importance of community development and participation. One of the lessons learned is that, instead of focusing on the allocation and disbursement of emergency funds for immediate public safety, the general approach in housing reconstruction and rehabilitation programs needs to continually pay special attention not only to the immediate recovery of infrastructure, and the provision of temporary sheltering but also emphasizing sustainable mitigation efforts in housing recovery that takes into account psychological, economic, and social recovery.

20.8 Conclusion

Under the flagship of the housing reform movement since the early 1990s, the adoption of public, private and nonprofit partnerships has gained favor in India (Sengupta 2006, p. 449). The state governments approaches to housing reconstruction and rehabilitation has tended to center around compensation and relocation programs as well as third-party contracting with private and non-governmental organizations.

In situations where local villages have been involved in the planning, site selection, and design of housing units, the public-private-nonprofit partnerships take on a healthy form. The role of the government is generally to facilitate the working relationships of contractors and the local community. The involvement of multilateral agencies, guided much by their organizational missions and philosophical beliefs, has also influenced a state government's tendency to adopt certain types of partnership arrangements.

Building a sustainable housing recovery program must start with prevention and mitigation such as building education and awareness and blending traditional building techniques with modern technologies. As such, a collaborative effort in relief and rehabilitation highlights the interactions of multiple service providers in pursuit of a common purpose. Embedded in these networks are collective actions, local empowerment, and capacity building efforts. Also present, but not obvious, are the pressures under which these agencies work, which includes but is not limited to pressures from political parties and potential donors or funders. The expectations of the people and beneficiaries also matter. Without effective monitoring mechanisms in place, there is the threat that nonprofit and private sector involvement may fail to address the recipients' needs and not allow for feedback from the beneficiaries leading to dissatisfaction and mistrust and slow development and recovery trajectories.

If designed and implemented effectively, housing recovery programs can not only improve the physical living conditions of poor, marginalized households but also help affected communities rebuild their livelihoods and return to normalcy expeditiously. Multisector partnerships can influence a wide array of household recovery processes, from the most basic household functions such as accessibility to sanitation, clean water and food, to substantial aspects of recovery such as access to schooling and medical health, employment opportunities, and social infrastructure. As argued by Kapucu (2006), "public-nonprofit partnership is not simply an administrative matter" (p. 215). Such partnerships encourage 556,557, action, capacity building, effective decision-making and community involvement. When multiple agencies from different sectors and backgrounds collaborate for a common purpose, the key to understanding context specific issues is to listen to the front-line workers who have more knowledge because of their regular and close communications with the public. Each partnering agency learns from the other and the shared resources make it more effective in relief provision. There can nevertheless be differences among the agencies in terms of communication, and pressures from funders and politicians.

Yet, when such multi-sector partnership approaches are applied to a country like India, known for its social and cultural diversity, large and overwhelming bureaucracy, it requires prompt decisions with a confluence of national, international, and local nongovernmental agencies seeking to provide emergency relief and rehabilitation services. Challenges, beyond reaching a consensual agreement, include an understanding of the local issues, empowering the local people, building trust among the agencies as well as with the local people, and managing and coordinating the activities of multiple agencies.

References

Alexander, D. (1993). *Natural disasters*. New York: Chapman and Hill.

Andrew, S. A., Arlikatti, S., Long, L., & Kendra, J. (2012). The effect of housing assistance arrangements on household recovery: An empirical test of donor-assisted and owner-driven approaches. *Journal of Housing and the Built Environment* (Early view), 1–18. doi:10.1007/s10901-012-9266-9.

Arlikatti, S., Bezboruah, K., & Long, L. (2012). Role of Voluntary Sector Organizations in Post-tsunami Relief: Compensatory or Complementary? *Journal of Social Development Issues, 34*(3), 64–80.

Arlikatti, S., & Andrew, S. A. (2012). Housing design and long-term recovery processes in the aftermath of the 2004 Indian ocean tsunami. *Natural Hazards Review, 13*(1), 34–44.

Arlikatti, S., Peacock, W. G., Prater, C. S., Grover, H., & Sekar, A. S. (2010). Assessing the impact of the Indian ocean tsunami on households: The modified domestic assets approach. *Disasters, 34*(3), 705–31.

Arya, A. S., Mandal, G. S., & Muley, E. V. (2006). Some aspects of tsunami impact and recovery in India. *Disaster Prevention and Management, 15*(1), 51–66.

Barakat, S. (2003). Housing reconstruction after conflict and disaster. Humanitarian practice network paper, No. 43 (December). London: Overseas Development Institute (ODI).

Barenstein, J. D. (2010). Who governs reconstruction? Changes and continuity in policies, practices, and outcomes. In G. Lizarralde, C. Johnson, & C. Davidson (Eds.), *Rebuilding after disasters: From emergency to sustainability* (pp. 149–176). New York: Spon Press.

Barenstein, J. D. (2006). *Housing reconstruction approaches in post-earthquake Gujarat: A comparative analysis*. Humanitarian practice network paper, No. 54. London: Overseas Development Institute.

Barenstein, J. D., & Leeman, E. (Eds.). (2012). Post-disaster reconstruction and change: Communities' perspectives. Boca Raton: CRC Press (Taylor and Francis Group).

Barenstein, J. D., & Pittet, D. (2012). A social and environmental assessment of pre- and post-tsunami housing and building practices in Tamil Nadu. In J. D. Barenstein & E. Leeman (Eds.), *Post-disaster reconstruction and change: communities' perspectives* (pp. 119–136). CRC Press (Taylor and Francis Group): Boca Raton.

Berke, P., & Beatley, T. (1997). *After the hurricane: Linking recovery to sustainable development in the Caribbean*. Baltimore: Johns Hopkins University Press.

Barenstein, J. D., & Leeman, E. (Eds.). (2012). *Post-disaster reconstruction and change: Communities' perspectives*. Boca Raton: CRC Press (Taylor and Francis Group).

Bosher, L. (2007). A case of inappropriately targeted vulnerability reduction initiatives in Andhra Pradesh, India? *International Journal of Social Economics, 34*(10), 754–771.

Census of India. (2011). Government of India, Ministry of Home affairs, Office of Registrar General and Census Commissioner. http://ancsdaap.org/cencon2011/Papers/India/India_slides_2011.pdf. Accessed 23 Nov 2012.

Coburn, A., Hughes, R., Illi, D., Nash, D., & Spence, R. (1984). The construction and vulnerability to earthquakes of some building types in the northern areas of Pakistan. In K. J. Miller (Ed.), *The International Karakoram project* (Vol. 2, pp. 226–252). Cambridge: Cambridge University Press.

Comfort, L. K. (2005). Fragility in disaster response: Hurricane Katrina 29 August 2005. *The Forum, 3*(3), 1–8. http://www.bepress.com/forum/vol3/iss3/art1. Accessed 17 Aug 2007.

Doocy, S., Gabriel, M., Collins, S., Robinson, C., & Stevenson, P. (2006). Implementing cash for work programmes in post-tsunami aceh: Experiences and lessons learned. *Disasters, 30*(3), 277–296.

Enarson, E. (2010). "Gender." In B. D. Phillips, D. S. K. Thomas, A. Fothergill, & L. Blinn-Pike (Eds.), *Social vulnerability to disasters* (pp. 123–154). Boca Raton: CRC Press.

Friedman J., Jimenez E., & Mayo S. (1988). The demand for tenure security in developing countries. *Journal of Development Economics, 29*, 185–198.

Gokhale, V.A. (2005). Analytical study of living environment in the tsunami-affected areas of Tamil Nadu, India. *ISET Journal of Earthquake Technology, Technical Note, 42*(4), 219–225.

Godbole, M. (1999). Outdated rent laws and investment in housing. *Economic and Political Weekly, 34*(7), 387–390.

Harrison, S. W., & Sinha, B. P. (1995). A study of alternative building materials and technologies for housing in Bangalore, India. *Construction and Building Materials, 9*(4), 211–217.

Hirway, I. (1987). Housing for the rural poor. *Economic and Political Weekly, 22*, 1455–1460.

Kaliappan, T. P. (1991). Shelter programme in Tamil Nadu. *Building and Environment, 26*(3), 277–287.

Kapucu, N. (2006). Public-nonprofit partnerships for collective action in dynamic contexts of emergencies. *Public Administration, 84*(1), 205–220.

Kapucu, N, & Van Wart, M. (2006). The evolving role of the public sector in managing catastrophic disasters: Lessons learned. *Administration and Society, 38*(3), 279–308.

Keivani, R., & Werna, E. (2001). Refocusing the housing debate in developing countries from a pluralist perspective. *Habitat International, 25*, 191–208.

Kumaran, T. V., & Negi, E. (2006). Experiences of rural and urban communities in Tamil Nadu in the aftermath of the 2004 tsunami. *Built Environment, 32*(4), 375–386.

Lindell, M., Prater, C. & Perry, R. (2006). Fundamentals of emergency management. http://training.fema.gov/EMIWeb/edu/fem.asp. Accessed 19 July 2011.

Mahadevia, D. (2001). Privatizing earthquake rehabilitation. *Economic and Political Weekly, 36*(39), 3670–3673.

Mahadeva, M. (2006). Reforms in housing sector in India: Impact on housing development and housing amenities. *Habitat International*, 30(3), 412–422.

Mehta, L. (2001). Reflections on the Kutch earthquake. *Economic and Political Weekly, 36*(31), 2931–2936.

Mukhija, V. (2001). Upgrading housing settlements in developing countries: The impact of existing physical conditions. *Cities, 18*(4), 213–222.

Nakagawa, Y., & Shaw, R. (2004). Social capital: A missing link to disaster recovery. *International Journal of Mass Emergencies and Disasters*, 22(1), 5–34.

Nath, S. K., Roy, D., & Thingbaijam, K. K. S. (2008). Disaster mitigation and management for West Bengal, India—an appraisal. *Current Science, 94*(7), 858–864.

Narasimhan, S. (2003). Lesson from Latur: A decade after the earthquake. *Economic and Political Weekly, 38*(45), 4730–4732.

Oxfam. (2005). The tsunami's impact on women. http://www.oxfam.org.uk/what_we_do/issues/conflict_disasters/downloads/bn_tsunamiwomen.pdf. Accessed 20 April 2010.

Özerdem, A., & Jacoby, T. (2006). *Disaster management and civil society: Earthquake relief in Japan, Turkey and India*. London: I.B. Tauris.

Partridge, W. (1989). Involuntary resettlement in development projects. *Journal of Refugee Studies, 2*, 373–384.

Pandey, R., & Sundaram, P. S. A. (1998). Volume and composition of housing subsidies in India through the central Government. *Habitat International, 22*(2), 87–95.

Parasuraman, S., Ramble, V., & Menezes, K. (1995). People's involvement in the donor housing programme: The situation in two villages of Latur District. In Organization and administration of relief and rehabilitation following Marathwada earthquake, 1993. Center for Research on the Epidemiology of Disasters—Tata Institute of Social Sciences Bombay, India.

Peacock, W. G., Morrow, H.B., & Gladwin, H. (1997). *Hurricane Andrew: Ethnicity, gender, and the sociology of disaster*. London: Routledge.

Peacock, W. G., Dash, N., & Zhang, Y. (2006). Shelter and housing recovery following disaster. In H. Rodriguez, E. L. Quarantelli, & R. Dynes (Eds.), *The handbook of disaster research* (pp. 258–74). New York: Springer.

Prater, C. S., Peacock, W. G., Arlikatti, S., & Grover, H. (2006). Social capacity in Nagapattinam, Tamil Nadu after the December 2004 Great Sumatra earthquake and tsunami. *Earthquake Spectra, 22*, (Special Issue III), 715–730.

Radhakrishnan, J. (2005). India and tsunami recovery: Government officials think back. Disaster risk management. http://go.worldbank.org/M54F0YTWF0. Accessed 8 Sept 2009.

Salazar, A. (1998). Disasters, the World Bank and participation: Relocation housing after the 1993 earthquake in Maharashtra, India. *Third World Planning Review, 12*(1), 83–101.

Sen, S. (1998). On the origins and reasons behind nonprofit involvement and noninvolvement in low income housing in urban India. *Cities, 15*(4), 257–268.

Sengupta, U. (2006). Government intervention and public–private partnerships in housing delivery in Kolkata. *Habitat International, 30*, 448–461.

Shaw, R., & Sinha, R. (2003). Towards sustainable recovery: Future challenges after Gujarat earthquake. *Risk Management, 5*(2), 35–51.

Shaw, J., Mulligan, M., Nadarajah, Y., Mercer, D., & Ahmed, I. (2010). Lesson from tsunami recovery in Sri Lanka and India: Community, livelihoods, tourism, and housing. Report 1. Globalism Research Centre, RMIT University, Melbourne. www.rmit.edu.au/globalism/publications/reports. Accessed 8 Aug 2010.

Sheth, A., Jain, S., & Thiruppugazh, V. (2004, August). *Earthquake capacity building and risk reduction measures in Gujarat post Bhuj 2001 earthquake.* Paper presented at the 13th World conference on earthquake engineering. Vancouver, B.C., Canada.

Singh, G., & Das, P. K. (1995). Building castles in air: Housing scheme for Bombay's slum-dwellers. *Economic and Political Weekly, 30*(40), 2477–2481.

Smith, G. P., & Wenger, D. (2006). Sustainable disaster recovery: Operationalizing an existing agenda. In H. Rodriguez, E. L. Quarantelli, & R. Dynes (Eds.), The *handbook of disaster research* (pp. 234–257). New York: Springer.

Sridhar, A. (2005, March). Statement on the CRZ notification and post-tsunami rehabilitation in Tamil, Nadu. (New Delhi, India: UNDP). http://www.dakshin.org/DOWNLOADS/Statement-percent20onpercent20CRZpercent20Complete.pdf. Accessed 30 July 2010.

Srinivasan, K., & Nagaraj, V. K. (2006). The state and civil society in disaster response: Post-tsunami experiences in Tamil Nadu. *Journal of Social Work in Disability and Rehabilitation, 5*(3/4), 57–80.

The World Factbook. (2012). Washington DC: Central intelligence agency. https://www.cia.gov/library/publications/the-world-factbook/index.html. Accessed 21 Jan 2013.

Tierney, K., Lindell, M. K., & Perry, R. W. (2001). *Facing the unexpected: Disaster preparedness and response in the United States.* Washington DC: Joseph Henry Press.

Tiwari, P. (2001). Housing and development objectives in India. *Habitat International, 25*, 229–253.

United Nations Development Program. (UNDP). (1994). Human development report. http://hdr.undp.org/en/reports/global/hdr1994/. Accessed 23 Nov 2012.

United Nations Disaster Relief Organization—UNDRO. (1982). *Shelter after disaster: Guidelines for assistance.* New York: UNDRO.

UNISDR. (2005). Hyogo framework for 2005–2015: Building the resilience of the nations and communities to disasters (Electronic Version). www.unisdr.org/wcdr/intergover/official-docs/Hyogo-framework-action-english.pdf. Accessed 4 Jan 2007.

Vatsa, K. S. (2010). Earthquake reconstruction in Maharashtra: Impacts on assets, income and equity. In S. B. Patel & A. Revi (Eds.), *Recovering from earthquakes: Response, reconstruction and impact mitigation in India.* New Delhi: Routledge.

Wadhwa, K. (1988). Housing programmes for urban poor: Shifting priorities. *Economic and Political Weekly, 23*(34), 1762–1767.

Waugh, W. L., Jr., & Streib, G. (2006). Collaboration and leadership for effective emergency management. *Public Administration Review, 66*(S1), 131–40.

World Bank. (2010). The World Bank annual report 2010. http://siteresources.worldbank.org/EX-TANNREP2010/Resources/WorldBank-AnnualReport2010.pdf. Accessed 23 Nov 2013.

Chapter 21
Re-Development, Recovery and Mitigation After the 2010 Catastrophic Floods: The Pakistani Experience

Sana Khosa

21.1 Introduction

Pakistan has experienced many natural disasters in recent years such as floods, landslides, earthquakes, droughts and cyclones. However, what it experienced in the summer of 2010 was of unimaginable and unprecedented scale since its creation in 1947. The 2010 floods affected over 78 districts (compared to a total of 141 districts) (NDMA 2011; UN 2011) that cover 100,000 km^2 of the country (ADB 2010) and impacted 20 million people (out of a total population of nearly170 million people) of which 14 million required immediate assistance (Kronstadt et al. 2010; IEG 2010), 8 million required urgent health care (UN 2011), and 3.5 million were children (NDMA 2011). Due to breeched levees, water flowed to rural flood-plains destroying agricultural land and resulting in mass destruction of houses and causing a high internal displacement of people. Alongside many roads, bridges and transportation routes were destroyed, causing havoc to the overall infrastructure in many regions across the country. The 2010 Pakistan Floods were referred to as the worst disaster in the history of the country (ADB 2010). The United Nations (UN) Secretary-General, Ban Ki-moon upon visiting the country declared that this disaster was larger than the accumulated impact of major disasters such as the 2004 Asian Tsunami, the 2005 Kashmir Earthquake, the 2008 Nargis Cyclone and the 2010 Haiti Earthquake (Solberg 2010). The Secretary-General of UN also referred to as the floods as a slow-motion tsunami (UN 2011).

The 2010 catastrophic floods were considered a national tragedy and required comprehensive relief, recovery, and reconstruction efforts. Relief and immediate recovery were led by the government of Pakistan, the Armed forces of Pakistan, and the international community being led by UN agencies. Long-term recovery and reconstruction is going to take years since the scope of the disaster is enormous and has caused massive destruction of infrastructure, massive internal displacements of houses, and livelihoods in the country. Also the fact that Pakistan is a country

S. Khosa (✉)
University of Central Florida, Orlando, USA
e-mail: Sana.Khosa@ucf.edu

N. Kapucu, K. T. Liou (eds.), *Disaster and Development,* Environmental Hazards,
DOI 10.1007/978-3-319-04468-2_21, © Springer International Publishing Switzerland 2014

marred with political and social unrest impedes wholehearted attempts at long-term recovery and sustainable development. This makes the terrain of recovery and re-development more convoluted and complex.

This chapter provides an outline of the country's existing disaster management structure and key information on the impact of the floods. At a second step an over-view of the recovery and reconstruction system and efforts taken after the floods that included reconstructing homes, redeveloping community infrastructure (i.e. schools, roads, and bridges), addressing problems of massive internal displace-ments and the loss of livelihoods (i.e. the destruction of agricultural land). Third, the chapter provides important initiatives aimed at disaster recovery and reconstruction that have been taken by the Pakistani government, UN agencies and other local and international non-governmental and humanitarian organizations.

The purpose of this chapter is not to list down all the recovery and rehabilitation projects that were underway after the floods of 2010, but to highlight the variety of recovery efforts due to the nature of the disaster and identify the need for sustain-able and improved development in impacted regions. The chapter also describes the importance of collaboration as a tool for achieving better recovery and restoration and the role of the international humanitarian community in such collaborative ini-tiatives and partnerships.

21.2   Hazards and Vulnerabilities in Pakistan

Pakistan is a country that faces great threats posed by manmade and natural disas-ters. Natural disasters such as floods, earthquakes, landslides, avalanches, drought and cyclones along with threats caused by civil conflicts, terrorism, health epidem-ics, oil spills, urban fires, challenges associated with a high number of internally displaced populations (IDPs), etc occur quite often in the country. Pakistan lies on a seismic belt and thus experiences earthquakes pretty often but of small magnitude. Pakistan's long coastline also increases its risks to potential tsunamis and cyclones (Khan and Khan 2008). Pakistan is also one of the top ten countries in Asia that will suffer due to climate change (Amir n.d.).

Khan and Khan (2008) suggest that floods are one major hazard 'against which an effective protection network of dykes and flood water regulatory infrastructure has been built over the years' (p. 9). Prior to the large-scale, unprecedented devas-tation caused in the 2010 floods, flood events of 1950, 1992, and 1998 have been massively destructive as well. According to Khan and Khan (2008) floods have hit all provinces ranging from riverine flooding to flash floods and landslides in both mountainous northern areas and flat areas in the provinces of Sindh and Punjab. In Pakistan floods happen regularly in the monsoon months from July to September. These floods originate from the Bay of Bengal and pass through lower central India into the northern parts of Pakistan. The mountain ranges in the north of Pakistan help as a recurrent source providing inflow to rivers (Khan and Khan 2008). Given threats and hazards the country faces, one may expect that by now Pakistan should have developed a comprehensive disaster management and mitigation system. Unfortunately, the disaster management and risk reduction framework is still under development in the country and what has been created so far remains to be tested. The next section lays out the specifics of the existing system.

21.3 Disaster Management System in Pakistan

In Pakistan the approach to managing disasters is reactive. The long-term vision of managing and mitigating disasters is not in place but rather a management style dealing with quick-fixes is applied (Khan and Khan 2008). After every disaster resources are utilized for relief and recovery efforts. The disaster management system and the structure in Pakistan is three tiered—divided into federal/national, provincial/state, and district/local levels.

21.3.1 National-Level Disaster Management

In October 2005 Pakistan experienced one of its worst natural disasters—a 7.6 magnitude earthquake that resulted in the death of 80,000 people and loss of 3.5 million of people's dwellings (Khan and Khan 2008). Prior to the 2005 Earthquake in Pakistan, no single central organization was responsible for overlooking and mobilizing disaster response and relief efforts. By far the most important and pivotal institutional change came about after the massive destruction caused by the October 2005 Earthquake. The National Disaster Management Authority (NDMA) was created in 2006 (Zaidi 2012). Thus, the 2005 earthquake served as a 'focusing event' for Pakistan and led to new legislation and the creation of a central body to manage disasters. Ordinance No XL of 2006 was issued by the government which set up a body for oversight and developing disaster management policies and plans called National Disaster Commission. This body was chaired by the Prime Minister. Alongside, this ordinance also set up the NDMA and its provincial branches to implement plans and policies and coordinate disaster management and response efforts (Young et al. 2007)..

The NDMA in Pakistan is the central body responsible for leading and coordinating disaster preparedness, response, and recovery efforts by different organizations which include different government departments, international agencies and donors, and the military. In addition, the Economic Affairs Division (EAD) is tasked to coordinate humanitarian donations (Kronstadt et al. 2010). There is also Federal Flood Commission (FFC) which is responsible for flood risk management and for developing and implementing a National Flood Protection Plan. The policy creation organization for risk management is the National Disaster Management Commission (NDMC) (ADB 2010).

21.3.2   Provincial and Local-Level Disaster Management

Just like the disaster management structure in the United States has state and local level emergency management organizations, in Pakistan there are Provincial
Disaster Management Authorities (PDMAs) at the state/provincial level and there
are some District Disaster Management Authorities (DDMAs) at the district/local.
All provinces in the country have established PDMAs, but not all districts in the
country have DDMAs. Only those districts have established DDMAs that have local capacities and capabilities to develop and operate them (ADB 2010).

21.3.3   Disaster Management Plans and Policies

Important legislation and existing structures include the West Pakistan National
Calamities Act 1958 that focuses on organizing relief and response operations. An
Emergency Relief Cell (ERC), a cabinet division cell was created in 1971 to coordinate and monitor disaster response at the federal level and also provide financial
resources to provincial governments during a disaster and also to foreign countries
experiencing major disasters (NDMA 2010).

In 1974 the ERC developed a national disaster plan which outlined responding
agencies and the procedures for relief operations. However, this plan was never
activated and put into action (Zaidi 2012). A National Disaster Risk Management
Framework was published in 2007. This framework was created to guide the development of disaster management plans and policies along with strengthening and
building the capacity of existing disaster management institutions in the next five
years. This framework lists the UN under other key stakeholders and explicitly
states that the UN agencies have to work closely with the NDMA and work in accordance with the policies set out by NDMA (Young et al. 2007).

Just before the devastating floods of 2010, NDMA had created and published another plan called the National Disaster Response Plan (NDRP). However this plan
was new and its execution was not possible with the existing apparatus and capacity
of disaster management organizations at the national, provincial and district levels
(Zaidi 2012). The 2010 NDRP aimed at solving the issue of coordination difficulties in large-scale disaster response and at involving all major stakeholders to the
process of developing policies and plans in their respective areas of jurisdiction
(Dorosh et al. 2010). The NDRP document outlines the roles and responsibilities
of government bodies and other partnering agencies at every level of operation according to their respective areas of jurisdictions. Standard operating procedures for
various relief functions and responding agencies is also described in the document
(NDMA 2010; Zaidi 2012). The institutional framework set out in NDRP is shown
in Fig. 21.1.

The framework reflects that NDMA is the focal organizations responsible for
disaster management. The NDMC is the planning and policy making body while
NDMA is the implementation body. NDMA works is close coordination with line

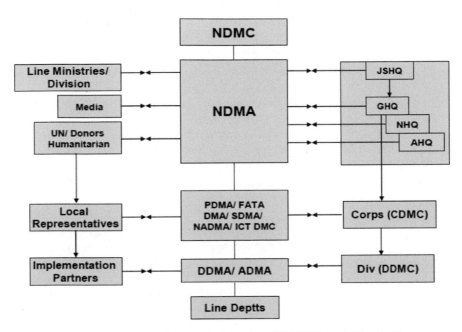

Fig. 21.1 Disaster management framework. Acronyms: *NDMC* National Disaster Management Commission, *NDMA* National Disaster Management Authority, *ADMC* Army Disaster Management Cell, *CDMC* Corps Disaster Management Cell, *DDMC* Division Disaster Management Cell, *PDMA* Provincial Disaster Management Authority, *SDMA* State Disaster Management Authority, *NADMA* Northern Area Disaster Management Authority, *ICTDMA* Islamabad Capital Territory Disaster Management Authority, and *FATA DMA* Federal Administered Tribal Area Disaster Management Authority. (Source: NDMA 2010)

ministries and divisions at the federal level along with donor organizations. At the provincial level the PDMAs are responsible for developing regional and state level risk reduction plans and implementing them in accordance with the national level plans and policies. They are also responsible for ensuring that district level plans have been made and are being implemented.

Although the new response plan and the recent experience with disasters has helped to develop national and provincial disaster management authorities and offices there is need to develop the capacity of these institutions (Dorosh et al. 2010). Therefore in order to enhance implementation of legislation and plans on disaster management and to increase the capacity of responding agencies and offices at the district, provincial and federal level financial resources must be provided (Zaidi 2012).

The response capabilities to handle catastrophic disasters such as the 2005 Kashmir Earthquake and the 2010 Floods are weak in Pakistan. The country relies heavily on the Army and the humanitarian community for support and relief operations (Amir n.d.). The Pakistan Army plays a major role in providing immediate response through sear and rescue and evacuation operations (Khan and Khan 2008). It helps by providing relief good to calamity-struck areas and regions that can be reached via helicopters and choppers only. The Flood Commission also plays an integral role in monitoring the threat of floods by evaluating the water levels at dams and barrages and by communicating closely with all provincial governments in case of irregular and unusual discharge of water level. It also maintains contacts during and after the floods (Khan and Khan 2008). For the last few years the United Nations Office for the Coordination of Humanitarian Affairs (UN OCHA) has proved to be a strong partner and leader in many respects in managing disasters in Pakistan and in advocating the United Nations' Cluster system. UN OCHA helps in carrying out situation and needs assessments, plays a strong broker role in coordinating with different agencies working in national, local and international capacities, and helps to mobilize resources (Young et al. 2007). UN OCHA is a part of the UN Secretariat and is responsible for coordinating humanitarian response in emergencies.

The United Nations' 'cluster system' is a popular response, relief and immediate-recovery approach that is used in huge disasters through which relief agencies coordinate their efforts in clusters and sectors. The main goal of the cluster system is to provide timely and coordinated response in disasters. Each cluster is guided by a specific humanitarian service and is assigned a lead agency to oversee and coordinate efforts and also individuals that are referred to as cluster coordinators. The 'cluster approach' is a top-down, UN centered initiative that aims to offer timely and effective response and improve coordination between various responding agencies and actors in huge disasters (Thomas and Rendon 2010). It is right to say that currently there are two approaches that exist in managing disasters in Pakistan. One approach has been applied several times while the other one has yet to be tried and tested. The newly developed NDRP by NDMA is yet to be fully implemented in response to disasters. However, the other approach, lead by the United Nations has been tried and tested within Pakistan.

The cluster system works to organize relief according to functional operations within different sectors with a predefined and predetermined leadership. This approach was developed and implemented with the aim to improve and increase overall efficiency and effectiveness in a number of areas such as: global capacity for responding to existing and future crises; predictable leadership at both the global and local levels; strong partnerships between responding agencies such as UN agencies, international NGOs and local agencies; accountability and transparency in relief operations; and strategic prioritization and coordination in implementing various clusters during emergencies (UNOCHA 2007). This approach was first implemented and tested in the South Asian Earthquake/Kashmir Earthquake in October 2005, then in the 2007 Sindh and Balochistan Floods and then in the 2010 Pakistan Floods.

The cluster system was piloted in the 2005 Pakistan earthquake. During this time the system tremendously helped in improving relief coordination. But it is important to understand that this is still a system in transition (Young et al. 2007). The issue has been that the UN cluster system and the NDMA itself are still in its development phases and are trying to understand new mandates and operational procedures, etc. In the past NDMA-UN relations were strained due to various reasons. Some of these reason are: "the lack of a shared agreement as to the objectives and strategy of the whole operation; misunderstanding of each other's mandates, roles and responsibilities; lack of systematic data-gathering and, from NDMA's perspective, sharing of information on international relief efforts and capabilities; a fundamental difference of approach between overtly centralized, on the one hand, and the more participative and consultative approach of the humanitarian community on the other. This has produced unrealized expectations and disappointment on both sides" (Young et al. 2007, p. 18).

The most recent policy approved by the National Disaster Management Commission has been the National Disaster Risk Reduction (DRR) Policy in 2013. This policy aims to strengthen the institutional framework of NDMA as it will be the lead agency for implementing this policy. This policy has also been an important step in meeting the goals of the UN Hyogo Framework for Action (HFA) 2005–2015.

Recovery and reconstruction is a completely different ball game compared to initial response and relief. The organizations and agencies assigned relief and response activities may not necessarily be equally assigned the tasks of reconstruction. According to the National Flood Reconstruction Plan of 2010, the monitoring and oversight tasks for reconstruction projects and activities was assigned to federal and provincial agencies, while the program execution and implementation was assigned to relevant local agencies, relevant departments and even private agencies (Planning Commission 2011). At the federal level, the Council of Common Interest (CCI) is the policy making body providing oversight and direction to different levels of the government to ensure proper allocation of reconstruction funds and resources and also resolve conflicts regarding interprovincial issues concerning the reconstruction phase. National Oversight Disaster Management Council (NODMC) was developed directly as a result of the floods in September 2010 to coordinate recovery and reconstruction efforts and programs, along with monitoring funds and donations and ensuring that progress is made on various projects and plans (Planning Commission 2011).

Alongside the Flood Reconstruction Unit (FRU) was created in the Planning Commission to develop a flood reconstruction plan, to finalize reconstruction programs and facilitate funds from development budgets and donors, etc. In addition to these federal level agencies and units, federal ministries and departments were responsible for reconstruction projects that fell under their mandates and sectors. At the provincial and local level, the provincial planning and development departments (P & DDs) are tasked with the coordination of reconstruction projects and ensure compliance with set out federal level plans and strategies (Planning Commission 2011).

Although a comprehensive apparatus for response and recovery has been outlined in the recent plans, the challenge is to build the capacity for its implementation at the district and local levels. Currently existing disaster and relief departments across the country, and at different jurisdictional levels lack the capacity or the training for disaster management (Khan and Khan 2008). The federal government and the NDMA should work towards building community-level capacity by offering trainings for community education and outreach for disaster mitigation and response and disaster management related certification to personnel working in disaster management cells and offices.

21.4   The 2010 Pakistan Floods

Pakistan experienced the worst floods since its creation in 1947. Flooding began in the northern regions of Pakistan and within days it reached the Arabian Sea, which lies at the southern part of Pakistan. Within a matter of days the entire Indus Valley and surrounding regions were flooded (Webster et al. 2011). The 2010 monsoon rains stood out as a period of above average rainfall in northern parts of Pakistan (Houze et al. 2011) compared to the 1998–2010 period. A year earlier in 2009 the monsoon rains were sparser and let to deforestation which helped to exacerbate the flash floods and their run off in the mountainous regions of north (Webster et al. 2011).

The 2010 floods were ranked as the worst natural disasters in Pakistan in terms of total population impacted and economic loss suffered (NDMA 2011). It was a profound humanitarian disaster (Houze et al. 2011) since coping with the destruction was not possible for any national government alone. According to NDMA (2011) when the number of affected population, total area impacted and households damaged are all taken into consideration it can be claimed that this disaster was bigger than the combined impact of five major disasters in the last ten years which are: the 2004 Indian Ocean Tsunami, Hurricane Katrina in 2005, the 2005 Earthquake in Pakistan, the 2008 Nargis Cyclone in Myanmar and the Haiti Earthquake in 2010. Usually the international community views a disaster as catastrophic due to the deaths and injuries it causes. This disaster is often compared with the Haiti Earthquake and rightly so because both disasters took place in the same year but had fairly different dynamics. The death toll was fairly high in the Haiti Earthquake compared to the death toll of around 2000 in the floods, but the area impacted and total population impacted was far more. The impacted area was around 20 times more than Haiti and the total displacements were 13 times more than the displacements in the Haiti Earthquake (Malik 2011; Webster et al. 2011).

The floods impacted different parts of the country in a dissimilar fashion. The flashfloods in KPK and Balochistan were very intense due to the mountainous terrain of the regions. However, the Punjab and northern Sindh areas are flatter and the riverine flooding had a slow pace but affected massive areas of cultivation and densely populated regions as well (WFP 2010). The biggest challenge was to attend to the massive displacements and to provide the displaced survival goods and services such as safe drinking water, sanitation, basic food, medical and health facilities and temporary shelter (UN 2011).

According to a study conducted by Kirsch et al. (2012) out of the families affected by the floods, 90 % belonged to rural areas. This implies that 90 % of the families required substantial help to support their survival and provide them relief services. Their study clearly suggests a disproportionate impact on the rural households and communities. Disasters of such nature have a cascading effect. Within weeks of the disaster there was huge threat of malnutrition amongst the survivors. Most of the people impacted by the floods were unskilled laborers or farmers. These people live either below the poverty line, on the poverty line or just barely above it. 60 % of these survivors had lost access to their livelihoods and around 3/4th of the affected had limited access to the supply of food (WFP 2010).

Along with killing 2,000 people, and injuring around 3,000 people, the floods also killed several thousand livestock (20,000 cattle drowned) (Webster et al. 2011) and many standing crops (around 2 million ha) as it wiped areas of cultivated land (NDMA 2011; WFP 2010). The flash flooding resulted in a huge agricultural crisis which will take years to recover (Webster et al. 2011).The irrigation sector struggled a great deal after the floods as many systems were destroyed and the plantation and sowing of many crops were delayed. The agricultural costs were believed to exceed 500 million US$ (Webster et al. 2011).

Moreover, floods water and heavy downpour destroyed many roads and homes, public buildings and offices, electricity grids and stations and around 2.4 million ha of land that it cultivated every year (UN 2011). According to the UN report "over 1.6 million homes, over 430 health facilities, and an estimated 10,000 schools were damaged or destroyed" (UN 2011, p. 19). Standing water in many regions weeks after the floods started have not only resulted in massive areas of uncultivable land but has also resulted in the spread of water-borne and skin diseases amongst the affected population (Malik 2011).

21.4.1 Disaster Response and Recovery

The response to the floods was initially slow and very challenging due to the havoc caused by destroyed or blocked roads, bridges and transportation routes in affected areas. Security concerns in northern areas of the country also hindered flood response (Kronstadt et al. 2010; Webster et al. 2011). According to reports the major relief effort was lead by the Pakistani government. The United Nations along with International NGOs and NDMA also helped in relief and response stages of the disasters. Due to the international and national efforts, 1.5 million people were rescued by the 20,000 army/military troops deployed by November 2010. It is also believed that despite the slow onset on response and many relief challenges, search and rescue operations and timely distribution of food and medical assistance were overall successful in saving many lives and handling the breakout of deadly water-borne diseases (Oxfam 2011).

The Pakistan Army led the rescue and evacuation efforts in the KP province while humanitarian agencies began providing relief goods to displaced people in August. On August 1st the government realizes the scale and scope of the disaster and announces that the floods have impacted 1 million people only to realize 2 weeks later that the actual impact affects 15 million people. It took a while to realize the extent of the disaster for both the government and the international community. The international community became more active after the UN launched an initial floods emergency response appeal at $ 459.7 million on August 11 (UN 2011). After the UN Secretary General Ban Ki Moon's visit to Pakistan on the 15th of August (around 3 weeks after the floods began), on the 18th of August a special session of the General Assembly was conducted to urge the international community to support relief efforts in Pakistan (UN 2011).

During the floods, the NDMA, being a constitutionally mandated agency, was expected to coordinate the overall response efforts between federal, provincial and district governments along with both local and international NGOs (Malik 2011). However, due to its lack of experience in coordinating such a huge disaster and leading the response efforts on its own, NDMA partnered closely with the United Nations resident coordinator to come up with a response framework. Moreover, all international organizations had to seek the permission of the government of Pakistan before providing any relief operations.

Moreover, despite a newly developed NDRP in March 2010 that outlines the role of federal, provincial and district level disaster management offices (NDMA 2010), the different levels of government were unclear about their roles their local level representatives could play to manage the floods. Moreover, there was also rarely any preparedness efforts, evacuation plans and manuals at the district levels (Malik 2011).

On the 30th of July, the government of Pakistan formally asked the Pakistan Army to carry out search and rescue operations while collaborating closely with the NDMA. Overall, the Pakistan Military helped to evacuate and rescue around 1.4 million people while deploying 20,000 troops who used either helicopters or boats. The Military also distributed essential survival items such as water and food to the affected population. The Military also set up camps for the displaced population and worked closely with the NDMA and PDMAs (UN 2011).

The UN played the most important support role in the floods via its cluster approach. Recent catastrophic disasters in Pakistan have resulted in an increased familiarity with the cluster system since it has been implemented a number of times already. The UN also played a very important role to pledge donor support, create awareness about the scale and scope of the disaster, and urge the international and humanitarian community to respond to the disaster (UN 2011).

The scope and scale was such that no government could have managed it on its own. The government of Pakistan urged the international humanitarian community to help and support relief and response efforts. Scaling-up the response by the international community and International non-governmental organizations (INGOs) also met enormous challenges since many of their resources, financial and non-financial, were being utilized in the Haiti Earthquake that took place few months earlier than the floods. However some UN agencies with a strong presence in the country had already developed a network of partners and garnered resources to scale-up in a short time (UN 2011).

Overall, one can conclude that the role of the international community in managing disasters in Pakistan is very important. Without international organizations such as the United Nations organizations and other INGOs it is not possible to deal with such massive internal displacements and provide relief services such as food, shelter, health and medical facilities, and temporary housing (UN 2011).

The Government of Pakistan officially launched the early recovery phase led by the NDMA with the collaboration of United Nations and other humanitarian agencies. This was in the form of a plan called the 'Pakistan Flood Relief and Early Recovery Response Plan' that was launched on the 5th of November 2010; around

Table 21.1 Sector-wise damage and needs assessment summary. (Source: ADB 2010)

Sector	Damage (Rs. millions)	Damage (US$ millions)	%
Transportation and communication	112,911	1,323	13.2
Irrigation	23,600	278	2.76
Energy	26,300	309	3.08
Agriculture	428,805	5,045	50.2
Education	26,464	311	3.1
Health	4,222	50	0.49
Water and sanitation	9,306	109	1.09
Environment	992	12	0.12
Government	5,976	70	0.7
Housing	135,014	1,588	15.8
Private sector	23,932	282	2.8
Finance	57,251	674	6.7
Total	*854,773*	*10,056*	*100*

3 months after the floods had started. The initial recovery phase according to the plan was expected to finish in 12 months and the recovery plan was going to cost 1.9 billion Rs. Reconstruction on the other hand, was a far ambitious plan, and according to a Damage and Needs Assessment survey carried out by ADB and WB the estimated total reconstruction costs ranged from Rs. 578 billion to Rs. 758 billion (Planning Commission 2011).The government also developed a National Flood Reconstruction Plan in 2011 to ensure that the reconstruction phase was well planned and orderly.

The Table 21.1 below shows the preliminary damage estimates according to the damage and needs assessment conducted by ADB and WB.

According to the damage and needs assessment carried out by WB and ADB, the agricultural sector suffered the greatest in terms of damages and costs (50 % of the total damages), while the housing sector suffered 16 % of the total losses. Short-term recovery has focused on rebuilding homes of people who lost their homes and restoring livelihoods of people who had lost their jobs and livelihoods (UNOCHA 2011). The strategy on reviving livelihoods has focused on providing compensation and income support primarily through the cash compensation via Watan Cards.

The main recovery strategy that the government focused on was the Watan Card Scheme, a cash-based compensation scheme for those who had suffered losses in the floods (Jillani 2010). The Watan cash compensation program is an innovative tool that helps to provide immediate relief and cash for quick recovery for victims of disasters (UNHCR 2011). This card operates as a regular ATM card (IOM 2011).This cash compensation program was to provide immediate relief to the most vulnerable of the 20 million people impacted by the floods. In the second phase, the compensation was given to houses that suffered damages and destruction. Alongside widows and disabled heads of families were also given the second phase of compensation irrespective of whether their houses were damaged or not (UNCHR 2011).

The government, through its National Database and Registration Authority (NADRA) started issuing Watan Cards to the impacted population. During the first phase of short-term recovery they received Rs. 20,000 from NADRA counters. However, the government also promised that it will give additional Rs. 100,000 for reconstructing homes to people (Jillani 16 September 2010). NADRA issued these cards, based on verification detailed and needs assessments provided by Provincial Disaster Management Authorities (PDMAs) (IOM 2011).

Alongside the government, international agencies were also involved in the implementation of the initiative. Transparency was ensured in the process as the Asian Development Bank and the World Bank were monitoring the distribution of these cards. UNHCR was also involved in the implementation of the program (Jillani 2010). IOM is providing technical help and support in the initiative by building capacity of watan card centers and staff working in the centers to improve beneficiary-specific communications. These supportive activities are funded through UK's Department for International Development (DFID) (IOM 2011).

Although this scheme was beneficial to many, there were some challenges attached to its implementation. There was a threat of registering twice for compensation. There were some people that were registered in two relief camps and thus could receive compensation twice as well. For instance, if the families of two brothers lived together, they might receive separate compensations but it will go to a single household. Also families and individuals had to register to the scheme themselves and sometimes had to travel great distances to reach the NADRA branch (Jillani 2010). There was also the need to ensure that this scheme is accessible to all the vulnerable and poor people impacted and is not limited to those residing in the large cities (UNCHR 2010). Moreover, this scheme also requires a computerized national identity card (CNIC) that is verified when registration is made to receive compensation (Jillani 16 September 2010).

A year after the floods 1.6 million Watan cards were distributed to survivors of the floods. Around 30 billion Rs. were disbursed to flood survivors in the first phase of the scheme (Pakistan Today 2011), while the government had distributed around 37 billion to over 950,000 beneficiaries in the second phase. A total of 71 billion Rs. have been disbursed in both phase 1 and 2 combined (The Express Tribune 2012). A total of 580 million $ have been invested in this program that is a collaboration between the GoP and international agencies such as the World Bank, International Organization for Migration UNCHR, and USAID (The Express Tribune 1 November 2012).

Other recovery projects across the country focused on immediate solutions for destroyed schools, critical infrastructures and agricultural and irrigated land. Although the response stage of disasters is primarily led by government agencies and the military in the country, many recovery initiatives are taken by private companies and international NGOS, nonprofits and humanitarian agencies (APF 2010).

A partnership between the American Pakistan Foundation (APF), Save the Children and the International Rescue Committee was developed to rehabilitate flood-damaged schools in the Khyber-Paktunkhwa province and in Balochistan. The rehabilitation of schools was focused on replacing damaged classroom furniture and supplies along with teacher training and health and hygiene classes. A total of 17 flood-damaged schools were successfully rehabilitated and a total of 4, 220 children were provided better quality education. Moreover, Save the Children initiated Back-to-School Campaign in the regions which managed to increase the student enrollment in targeted communities by 23 % (APF 2010).

APF was also involved in recovery projects to improve livelihoods for communities in the KPK province. Through this project lactating goats were distributed to rural families that had lost their livestock in the floods and beneficiaries were also provided livestock management training. 765 households were selected and identified through the help of Relief International and local community agencies (APF 2010).

There is still a lot that needs to be done. Although the support of the humanitarian community and the resilience of the people of Pakistan has helped a great deal in meeting the challenges of recovery. Reconstruction and reconstruction geared towards sustainable practices will take more effort and resources. This requires institutional restructuring and alterations in the existing tax administration (Planning Commission 2011). Currently, the country suffers from a high rate of tax evasion and low tax literacy and tax incentives (Kleven and Waseem 2011).

The Damage and Needs Assessment carried out by the WB and ADB chalked out three reconstruction options. The first Option cost 578 billion Rs., and was focused on reconstructing the damaged and destroyed infrastructure. Option 2 cost Rs. 630 billion since the emphasis was on building the lost infrastructure but building better used improved methods and technologies. They referred to this option as Build Back Better (BBB). The third option cost the most at Rs. 758 billion and integrated provisions for future flood protection (Planning Commission 2011). Quite obviously, the third Option should have been chosen since Pakistan is a country prone to the threat of flooding. However, the choice was not as simple as it seems. The government chose Option 1 due to the budgetary and financial constraints.

One key part of the reconstruction strategy has been focused on rebuilding of destroyed housing and reconstruction of settlements to accommodate displaced persons. The focus is to move beyond temporary housing and to rebuild better. The government of Punjab has initiated the development of model villages and settlements. The Turkish Housing Development Authority is collaborating with the Government of Punjab for the construction of 2,120 modern homes in the affected regions (Planning Commission 2011).

The sector-wise summary of the number of reconstruction projects and their expenditures are identified in Table 21.2. The summary shows that the most projects are related to restoring and improving irrigation, drainage and flood management. There is a lot that has been recommended to follow in terms of improving the existing irrigation system changes such as reviewing maintenance standards for flood systems, increasing the capacity of barrages for floods, improving forecasting and warning systems (Planning Commission 2011).The highest expenditures are identified for the transport and communication sector. This is because, firstly, transport infrastructure projects are costly, and secondly, the transport and communication sector suffered a huge blow since many roads and bridges on key routes were destroyed across the country and requires immediate reconstruction and re-development.

Table 21.2 Flood reconstruction projects sector-wise summary. (Source: Planning Commission 2011)

Sector	No. of projects	Cost (Rs. millions)	Cost (US$ millions)
Agriculture	16	12,877.36	150.68
Education	13	32,953.41	385.58
Energy/Power	11	3,642.60	42.62
Environment	1	15.87	0.19
Governance	19	26,999.17	315.91
Health	7	6,555.98	76.71
Irrigation, drainage and flood management	113	73,492.12	859.92
Transport and communication	58	188,773.04	1,565.26
Waterand sanitation	7	1,881.73	22.02
Total	*245*	*292,191.28*	*3,418.88*

The agricultural sector suffered the largest blow—Many standing Kharif crops (domesticated crops sown in the monsoon region in the Asian subcontinent) that were destroyed such as cotton, sugarcane, maize, vegetables and fruits. Much of fertile land was also damaged for the coming seasons. Steps taken were to assist farmers and subsidize fertilizers, farm credit and machinery and also provide compensation to those who had lost agriculture related livelihoods and also their livestock. The education sector also suffered huge losses as 10,348 schools, 23 colleges and 21 vocational institutions were damaged. Temporary tent schools were developed and initiated by the UN agencies, nonprofit agencies, and efforts were funded by much philanthropy. Strategies for reconstruction include: combining low enrollment schools and making more co-ed institutions (Planning Commission 2011).

By the sector-wise summary provided, one can easily gauge that reconstruction is not possible without the help and support of the international and humanitarian community for a country like Pakistan. Many countries such as China, Japan, Abu Dhabi, Saudi Arabia made commitments to support the reconstruction costs. Additional international agencies such as the WB, ABD, etc also pledged to finance part of the reconstruction process (Planning Commission 2011). The role of the international and humanitarian community can never be underestimated during disasters in a country like Pakistan. In 2010 the Haiti Earthquake and the Pakistan Floods stood out as catastrophic events and were referred to as mega-disasters as around 95% of funding by international agencies in 2010 went to these two events alone (Ferris and Petz 2011).

21.4.1.1 Reconstruction, Mitigation, and Development

Disasters have a direct relationship with economic and social development in countries that are inflicted with catastrophic disasters. This is especially true for developing countries such as Pakistan. Prior to the 2010 floods, Pakistan was undergoing financial problems and was part of the IMF stabilization program since 2008. The financial and economic situation worsened after the 2010 floods as huge expenditures were incurred in rehabilitation, recovery and reconstruction (Planning Commission 2011).

The existing fiscal structure in the country was altered to support additional reconstruction and rehabilitation in the country. This was done by shifting resources from "low-priority current and development spending and additional resources would be mobilized through restructuring of tax structure and through tariff rationalization of the energy sector" (Planning Commission 2011, p. i).

Prior to the floods, Pakistan was already suffering from slow social and economic development due to the low social development indicators, high population rate and high poverty. The law and order situation in the country is also bleak and impacts the economy directly since cities are dealing with high internally displaced populations due to civilian unrest and terrorist attacks. This has the tremendously slowed the path to investment and growth in the country (Planning Commission 2011).

Due to flood recovery, rehab and reconstruction projects, the fiscal deficit has also increased to over 5 % of GDP. The floods have enormously contributed to decreasing budget revenues and deteriorating the economic and financial health of the country. The large reconstruction projects have also shifted spending expenditures to more pressing needs of victims (Planning Commission 2011).

21.5   Collaboration as a Tool for Sustainable Recovery and Redevelopment

The need for partnerships and collaboration is imperative for effective recovery efforts. Most efforts require enormous funding which was made possible by international agencies, foundations and philanthropies. However, these agencies lack the local knowledge and without the help of national level agencies and local agencies are unable to achieve much. For instance, the American Pakistan Foundation worked closely with local non-profit agencies to provide necessary facilities to damaged schools, provided micro-finance through household grants and livestock programs to provide livelihood recovery for affected populations, and invest in building micro-infrastructure for irrigation (APF 2010). And although their projects were funded by international philanthropies and private agencies, the need for developing close links with local communities and the government were considered essential for implementation and capacity building.

Community involvement and empowerment is very important in successful and effective recovery projects. According to the APF the projects on relief and recovery initiated by them had a community involvement and empowerment component where in the first stage of the project, the foundations' local partners mobilized the community through social events. This helped to develop the interest and support for recovery projects at the local level and increased the sustainability and accountability components of the projects as well (APF 2010). Identifying community agencies as partners for externally funded recovery efforts is essential to ensure the long-term sustainability of the project and the development and recovery the project aims for.

Another exemplar program for reconstruction and development after the floods was initiated by the United Nations Development Programme (UNDP) and was called the Post Floods Early Recovery Programme. UNDP's implementation partners were 70 local partners along with the NDMA and PDMAs. The funding partners were the government of Japan, EU, government of US, Australian government, Italian government, State of Kuwait and COFRA Foundation. The aim of the program was the build back better in the worst impacted regions by investing in long-term mitigation measures and developing resilience against the floods. Alongside developing the destroyed community infrastructure, creating jobs and livelihood opportunities for women and men was promoted under the programme. Another area for development was identified as the restoration of local government capacities for long-term sustainable development and disaster response and relief (UNDP 2013). Some highlights and accomplishments of the program include: Apart from restoring and reconstructing 120 government buildings, these buildings were also provided digitization equipment and technical support; Over 11,000 people especially females were offered cash grants to start their own small businesses and income generating activities; More than 10,000 community organizations along with some citizen community boards were established in many affected regions; 115,000 farmers were provided agriculture kits; and Flood control measures were applied such as stabilization of slopes and biotechnical soil engineering.

Another major partnership was initiated in 2011 by United Nations Educational, Scientific and Cultural Organization (UNESCO) with the Government of Japan and the Government of Pakistan to improve the existing flood forecasting apparatus in the country and conduct risk mapping of floodplains (UNESCO 2011).

21.6   Conclusion

The government of Pakistan has to realize that due to its geographical location and seismic risks floods are going to continue to occur and so complete mitigation and control is not possible. There is a dire need to focus on changing the reactive disaster management style to one that is proactive and emphasizes mitigation.

One major concern after the 2010 floods was to gauge whether these floods were predictable (Webster et al. 2011), and if so why wasn't the government prepared to curtail the flooding or deal with its after effects in a better way. A study by Webster et al. (2011) shows that heavy rainfall could have been predicted a week in advance of the floods. And if they were predicted in time, the government and water management authorities and irrigation departments could have taken proactive measures to release water before flash flooding.

Provinces such as the Sindh province are nowhere near being prepared for dealing with catastrophic disasters. Despite many measures taken for essential flood control after 2010 Floods, one year later the southern part of the country was flooded again. That event mostly affected poor people and large families and caused massive destruction of homes and farmlands (Roopanarine 2012). There are many lessons that can be drawn from the 2010 floods. The key lessons learned are:

the existing risk analysis apparatus is weak and spreads across multiple agencies; the forecasting and warning system for floods is also outdated and needs technical updating; the need for identifying vulnerable points and high risk regions to put up better flood defenses and also relocate residents permanently from those regions (Roopanarine 2012).

The main reason for the unprecedented scale of the 2010 Floods can be directly attributed to the fact that poor disaster risk management and mitigation is being followed in the country. According to a research study conducted by the Disasters Emergency Committee (DEC), a group of United Kingdom (UK) based NGOs, the Pakistani government has been unable to introduce effective measures for disaster risk reduction. Although they have plans on paper the country has not paid enough attention to mitigation and the country also lacks the required funds and political support for effective disaster risk reduction.

The first step in order to get started on risk mitigation measures is to carry out hazard and risk mapping throughout the country. Another important step is to invest in improving the flood early warning system. There is also an essential need to invest in promoting community based disaster risk management. Overall, there is a need to develop better land use plans and enforce building regulations, and also develop improved storm water drainage plans for cities.

Disaster risk reduction and mitigation seem to be the ignored part of disaster management policies and plans in Pakistan. Even leading national agencies do not have enough capacity to invest in developing better interoperable systems and conducting trainings and exercises to improve the capabilities of personnel working within these offices and agencies. Khan and Khan (2008) suggest that the disaster management agencies suffer from "a dearth of knowledge and information about hazard identification, risk assessment and management" (p. 11). They are certainly also suffering from a lack of financial resources. Therefore, although the disaster management system and existing structure reflect that many organizations are involved in flood management, improvements in interaction and coordination of tasks between different entities is required. Also there is a need to delineate clear roles and responsibilities of agencies to ensure there is no duplication or overlap in them (ADB 2010). Although experience with frequent floods has improved the flood control and management system in Pakistan enormously, there is a lot that still needs to be done. The Pakistan Floods of 2010 were an eye-opener for Pakistan.

References

American Pakistan Foundation (APF). (2010). Disaster recovery. http://americanpakistan.org/disasterrecovery. Accessed 3 Jan 2013.

Amir, P. (n.d.). Climate change: Disaster management in Pakistan. International Union for Conservation of Nature (IUCN). http://cmsdata.iucn.org/downloads/pk_cc_dm_vul.pdf. Accessed 20 Oct 2012.

Asian Development Bank (ADB). (2010). Pakistan 2010 floods: Preliminary damage and needs assessment. http://www.gfdrr.org/gfdrr/sites/gfdrr.org/files/publication/Pakistan_DNA.pdf. Accessed 2 Feb 2010.

Dorosh, P., Malik, S., & Krausova, M. (2010). *Rehabilitating agriculture and promoting food security following the 2010 Pakistan floods: Insights from South Asian experience. IFPRI discussion paper 01028*. Washington, DC: International Food Policy Research Institute (IFPRI).

Ferris, E., & Petz, D. (2011). A year of living dangerously: A review of natural disasters in 2010. The Brookings institution-London school of economics project on internal displacement. http://reliefweb.int/sites/reliefweb.int/files/resources/04_nd_living_dangerously.pdf. Accessed 10 April 2011.

Houze, R. A., Rasmussen, K. L., Medina, S., Brodzik, S. R., & Romatschke, U. (2011). Anomalous atmospheric events leading to the summer 2010 floods in Pakistan. *Bulletin American Meteorological Society, 92,* 291–298.

Independent Evaluation Group (IEG). (2010). Response to Pakistan's floods: Evaluative lessons and opportunities. The World Bank Group. http://siteresources.worldbank.org/EXTDIRGEN/Resources/ieg_pakistan_note.pdf. Accessed 2 Feb 2010.

International Organization for Migration [IOM]. (15 September 2011). Second phase of IOM-supported Watan Card Project launched in Khyber Pakhtunkhwa. http://reliefweb.int/report/pakistan/second-phase-iom-supported-watan-card-project-launched-khyber-pakhtunkhwa. Accessed 5 Dec 2012.

Jillani, S. (16 September 2010). Watan Cards launched to compensate flood survivors. The Express Tribune. http://tribune.com.pk/story/50486/watan-cards-launched-to-compensate-flood-survivors/. Accessed 10 Nov 2012.

Khan, H., & Khan, A. (2008). Natural hazards and disaster management in Pakistan. http://mpra.ub.uni-muenchen.de/11052/1/MPRA_paper_11052.pdf. Accessed 20 July 2012.

Kirsch, T. D., Wadhwani, C., Sauer, L., Doocy, S., & Catlett, C. (22 August 2012). Impact of the 2010 Pakistan floods on rural and urban populations at six months. *PLOS Currents Disasters.* doi:10.1371/4fdfb212d2432.

Kleven, H., & Waseem, M. (2011).Tax notches in Pakistan: Tax evasion, real responses, and income shifting. LSE working paper, Mimeo, London School of Economics.

Kronstadt, K. A., Sheikh, P. A., & Vaughn, B. (2010). *Flooding in Pakistan: Overview and issues for Congress*. Washington, DC: CRS Report for Congress.

Malik, A. A. (2011). The Pakistan floods 2010: Public policy lessons. *Policy brief series, 1*(1),1–6. (International Policy and Leadership Institute).

National Disaster Management Authority (NDMA). (2010). National disaster response plan. Islamabad: NDMA.

National Disaster Management Authority (NDMA). (2011). Annual report 2010. Islamabad: NDMA.

Oxfam. (2011). Six months into the flood: Resetting Pakistan's priorities through reconstruction.144 Oxfam briefing paper. http://www.oxfam.org.uk/oxfam_in_action/emergencies/downloads/oxfam_pakistan_6mth_briefing_note.pdf. Accessed 2 Feb 2010.

Pakistan Today. (7 June 2011). Watan Cards: 2nd phase likely to start soon. http://www.pakistantoday.com.pk/2011/06/07/city/karachi/watan-cards-2nd-phase-likely-to-start-soon/. Accessed 5 Dec 2012.

Planning Commission. (2011).National flood reconstruction plan 2010. Government of Pakistan, planning commission flood reconstruction unit. http://www.pc.gov.pk/hot%20links/Nation%20Reconstruction%20Plan%202010/National%20Flood%20Reconstruction%20Plan%202010.pdf. Accessed 5 Jan 2013.

Roopanarine,L.(3August2012).Pakistannowherenearpreparedforanothermajordisaster.TheGuardian. http://www.quardian.co.uk/globaldevelopment/2012/aug/03/pakistan-nowhere-preapred-major-disaster. Accessed 3 Jan 2013.

Solberg, K. (2010). Worst floods in living memory leave Pakistan in paralysis. *Lancet, 376,* 1039–1040.

The Express Tribune. (1 November 2012). Over Rs71bn disbursed among flood victims since 2010: NADRA. The Express Tribune website http://tribune.com.pk/story/459329/over-rs71bn-disbursed-among-flood-victims-since-2010-nadra/. Accessed 2 January 2013.

Thomas, A., & Rendon, R. (2010). Confronting climate displacement: Learning from Pakistan's floods. Refugees international November 2010. http://www.refugeesinternational.org/policy/in-depth-report/confronting-climate-displacemen. Accessed 4 June 2011.

UNHCR. (2011). The WATAN scheme for flood relief: Protection highlights 2010-2011. http://floods2010.pakresponse.info/LinkClick.aspx?fileticket=_SpKC9jJClY%3D&tabid=206&mid=1604. Accessed 4 Dec 2012.

United Nations. (2011). Pakistan floods: One year on 2011. The United Nations in Pakistan: Islamabad. http://unportal.un.org.pk/sites/UNPakistan/Floods%20One%20Year%20On%202011/One%20Year%20On%202011.pdf. Accessed 13 Sep 2012.

United Nations Development Programme (UNDP). (2013).Project brief: Post floods early recovery programme. http://www.undp.org/content/dam/brussels/docs/Other/Post%20Floods%20Early%20Recovery%20Pakistan.pdf. Accessed 5 July 2013.

United Nations Educational, Scientific, and Cultural Organization (UNESCO). (2011). UNESCO launches a comprehensive project to strengthen flood forecasting and management capacity in Pakistan. http://www.unesco.org/new/en/no_cache/unesco/themes/pcpd/dynamic-content-single-view/news/unesco_launches_a_comprehensive_project_to_strengthen_flood_forecasting_and_management_capacity_in_pakistan/. Accessed 5 July 2013.

United Nations Office for the Coordination of Humanitarian Affairs (UNOCHA) (2007). Appeal for building global humanitarian response capacity. Office for the Coordination of Humanitarian Affairs, New York, NY. http://ochaonline.un.org/cap2006/webpage.asp?Page=1566. Accessed 15 April 2012.

United Nations Office for the Coordination of Humanitarian Affairs (UNOCHA). (17 September 2010). Pakistan floods: Timeline of events (as of 17 Sep 2010). http://documents.wfp.org/stellent/groups/public/documents/ena/wfp225987.pdf. Accessed 10 Nov 2012.

Young, M., Khattak, S. G., Bengali, K., & Elmi, L. (2007) *IASC inter-agency real time evaluation of the Pakistan floods/cyclone Yemyin*. Islamabad: IASC.

Webster, P. J., Toma, V. E., & Kim, H. M. (2011). Were the 2010 Pakistan floods predictable? *Geophysical Research Letters, 38,* L04806, doi:10.1029/2010GL046346.

World Food Programme (2010). Pakistan Floods Impact Assessment.

Zaidi, M. (23 April 2012). Poor disaster management. Dawn Newspaper. http://dawn.com/2012/04/23/poor-disaster-management/. Accessed 8 Aug 2012.

Chapter 22
Community-Based Recovery and Development in Tohoku, Japan

Rajib Shaw

22.1 Introduction

On a Friday afternoon, at 14:46 of 11th of March 2011, a large earthquake of magnitude 9.0 in the Richter Scale occurred around 250 km off-coast of the northern Japan, resulting a massive tsunami with differential height, and destroyed several hundred kilometers of the coastline. The Japan Meteorological Agency (JMA) named this earthquake "The 2011 off the Pacific coast of Tohoku Earthquake" (JMA 2011), however the mass media termed this as the "Great East Japan Earthquake and Tsunami" or the "East Japan Earthquake and Tsunami (EJET)." As of 9th January 2013 (NPA 2013), the total number of casualty was 15,879, and missing persons were 2,700. The disaster struck Japan, when the country is going through an economic recession, political instability, and lack of leadership. Earthquake and tsunami were not new to the region, which experienced several past disasters like 1960 Chile Tsunami, 1933 Showa Sanriku Tsunami, 1896 Meiji Sanriku Tsunami, etc. Thus, the local governments, people, and the institutions made significant attempts to develop the tsunami scenario, identifying the evacuation places, evacuation routes and conducted regular drills with the schools and communities. In spite of all these preparations, there was unprecedented devastation, which posed serious questions on the disaster risk reduction approaches of Japan.

At the aftermath, first few days were the time of chaos and confusion. One of the key reasons was the lack of information. The affected areas consist of smaller towns and villages with relatively limited capacities in terms of necessary infrastructures and human resources. When this limited system is affected and hit directly by the tsunami, it was difficult to return back to operation in first few days. Also, the level of devastation was so high that the local governments, people and institutions were overwhelmed. Due to the collapse in the governance system in many of these local governments, the initial few days were full of confusion. An additional collateral hazard was the nuclear meltdown, which added to the already chaotic situation. The

R. Shaw (✉)
Graduate School of Global Environmental Studies of Kyoto University, Kyoto, Japan
e-mail: shaw.rajib.5u@kyoto-u.ac.jp

N. Kapucu, K. T. Liou (eds.), *Disaster and Development,* Environmental Hazards,
DOI 10.1007/978-3-319-04468-2_22, © Springer International Publishing Switzerland 2014

impacts were deep reaching, also affecting significantly the basic services in the Tokyo and the surrounding metropolis areas.

It is common knowledge that the people at the community level have more to lose because they are the ones directly hit by disasters, whether it is a major or a minor one (Shaw 2012). They are the first ones to become vulnerable to the effects of such hazardous events. The community therefore has a lot to lose if they do not address their own vulnerability. On the other hand, they have the most to gain if they can reduce the impact of disasters on their community. The concept of putting the communities at the forefront gave rise to the idea of community-based disaster management (CBDM). At the heart of the CBDM is the principle of participation. Through the CBDM, the people's capacity to respond to emergencies is increased by providing them more access and control over resources and basic social services Using a community-based approach to managing disasters certainly has its advantages (Kapucu 2008).

Community based disaster related activities are termed differently over time. Over more than 100 years ago, before the existence of most of the states, people or communities were taking care of themselves through collective actions during the disasters. After the formation of state, government-based disaster risk reduction program started, which failed to serve the needs of the people and communities. Over last 20–30 years, we are now again talking on the need of community involvement (Shaw 2012).. Thus, community-based approach is not new. Rather, we are going back to the old and traditional approaches of risk reduction. CBDM has been a popular term in later 1980s and 1990s, which gradually evolved to CBDRM (Community Based Disaster Risk Management), and then to CBDRR (Community Based Disaster Risk Reduction). CBDRM and CBDRR are often used with similar meaning, with enhanced focus on "risk," however there still exists a thin line of distinction. While CBDRR focuses more on pre-disaster activities for risk reduction by the communities, CBDRM focuses a broader perspective of risk reduction related activities by communities; both during, before and after the disaster.

The recovery process depends on the people's power, its networking, neighborhood tie, and resilience (Shaw and Takeuchi 2012a). People of Japan are known for its resilience and to cope with the natural disasters. This disaster recovery will also need people's power through strengthening the resilience of the affected people. A complete recovery needs time and resources. A well-coordinated, planned and decisive recovery policy with well-thought participation of different stakeholders is useful and efficient. This chapter focuses on three recovery issues: temporary housing, local information sharing, and education sector. In all these three issues, community participation is a key common element. Identifying some of the major issues of the disaster in the following section, the chapter presents the key initial findings on the three issues described above.

Table 22.1 Major aftershocks on the day of the disaster

Date	Aftershock magnitude	Occur time	Region/Prefecture
11th March	7.0	15:06	Sanriku Oki
	7.4	15:15	Ibaraki-ken Oki
	7.2	15:26	Sanriku Oki
	6.1	15:57	Ibaraki-ken Oki
	6.8	16:15	Fukushima-ken Oki
	6.6	16:29	Sanriku Oki
	6.7	17:19	Ibaraki-ken Oki
	6.0	17:47	Fukushima-ken Oki
	6.4	20:37	Iwate-ken Oki
	6.1	21:13	Miyagi-ken Oki
	6.0	21:16	Iwate-ken Oki

22.2 East Japan Earthquake and Tsunami

At the aftermath of the disaster, there were several challenging issues, which made significant barriers in terms of time, space, human resources and other related complexity. The following part describes some of the issues, which emerged to be critical after the disaster.

Once in 1,000 Years Event The Tohoku earthquake was the fourth mega earthquake known to date; the other three were the Chili earthquake in 1960, the Alaska earthquake in 1964 and Sumatra earthquake in 2004. After the earthquake, a massive tsunami swept across the coast of Tohoku (Shaw and Takeuchi 2012a). Similar mega-earthquake was observed during Jogan era (869 AD), when three seismic sources broke together to generate a mega earthquake. Paleo-seismic analysis also pointed out similar record of mega-earthquake and tsunami almost 3,000 and 5,000 years back, however the population and the built environment was different. The tsunami first reached the Japanese mainland 20 min after the earthquake and ultimately affected a 2,000 km stretch of Japan's Pacific coast. The tsunami height, when it struck the coastline was already 10 or 15 m in some cases, and reached up to 40 m, when then tsunami water entered into the narrow gorges within the river basin. For the Sendai plain, the water entered into more than 6–7 km inland from the coast region.

Large Aftershocks The main event was followed by a series of aftershocks of large magnitudes. Table 22.1 below shows the major aftershocks on 11th of March 2011 itself, the day of the disaster. It is to be noted that the earthquake happened at 14:46, and the within 1 h, there were three major aftershocks of magnitude more than 7. This posed a serious problem in the immediate evacuation after the disaster. Personal interviews with the affected people by the author revealed that the first shock of the earthquake was too strong, and it was very difficult for the people to stand during the disaster. Especially, for the aged people, it was a challenge to move to the safer place immediately after the earthquake. This time loss of first 10–15 min was very crucial. After that when people started evacuating, three consecutive aftershocks of more than magnitude struck the region, which affected the evacuation behavior of the people very much.

Differential Topographic Characteristic The affected area of Tohoku has two very distinct characters, which has significant consequence of the people's evacuation behavior, and nature and magnitude of damage. First, in the Sanriku coast, the coastline is saw-toothed, and the mountain is very near to the coast region. This has the significance in the sense that people could immediately evacuate to the higher ground, back of their houses. Thus, the lead-time of evacuation was rather short. Second, in the Sendai plain, the nearest higher ground (or mountain range) is more than 10–12 km away in most cases (Shaw and Takeuchi 2012b). This has a significant meaning in the sense that community facilities like schools and community halls become more important in this type of situation. The character of this nature of the coastline also affects the community relationship and decision making in the recovery process. While in the Sanriku area, a community is often restricted to one valley, the border of the community in the Sendai plain does not restrict to the topographic nature. Therefore, the decision making for the recovery process needs more time and discussion the adjacent communities. This is especially important for the decisions on the development of infrastructures like dykes, or the school buildings as potential community centers.

Damages of Local Governments and Key Facilities In several cities and municipalities, the city offices and other critical facilities (like emergency management center) were damaged. If government building is survived, the government can start response and recovery works earlier. In addition, the government can spare spaces for community. If local government had data of community, they can start recovery works earlier. In the cities or towns where the government buildings were affected, it was difficult to start the post disaster response and recovery processes. Many local governments did not prepare backup data and temporally offices.

Large Amount of Debris The disaster produced a huge amount of debris, which posed significant barriers to the post disaster immediate search and rescue as well as recovery issues. An estimated of 25 million t of debris was produced. However, there were significant other amounts of debris, which was under sea, and no estimation was made for that part. At the aftermath, space was a major problem for the debris clearance, and its processing. Ownership of some of the asset related debris (cars, bike, etc.) was another problem, since several of the owners either passed away or were missing. Due to the nature of the disaster, the debris were rather mixed, and many of the memory or emotion related issues like photos, certificates etc. need to be separated from the other debris. This also took significant time.

Nuclear Meltdown as Collateral Hazard Needless to say that the nuclear meltdown within the first week of the earthquake had changed the disaster landscape with complexity of developing internally displaced people. This was the first time in the post-world war II, Japan had internally displaced people in different provinces. The meltdown issue also drew significant attention, criticism, policy chaos, and public demonstration at different levels. Thus, the aftermath of the disaster should be considered with the backdrop of the nuclear meltdown event.

The above issues are not the exhaustive one, there are many more issues which are not discussed here. This shows the complexity of the event and the need for a long term and comprehensive plan. The government's recovery policy focuses very strongly the community participation. Some of the areas like infrastructure recovery need strong government support. However when it comes to the housing or social infrastructure and communication related issues, there needs to be a strong community participation. In this context, education, communication, information and volunteerism are the four key elements, found to be important. The common thread is the community, and its behavior. These elements are equally important for pre, during, and post disaster phases. After the disaster, the national government has asked the city or town governments to make its own recovery plans through community consultation. This was a challenging process for many local governments, since many of them lost their staffs and had limited resources. However, this is the time when the community came in, and community-local government linkages became stronger based on mutual understanding and communication.

Community's role in disaster management has been widely argued, debated and accepted (Kapuc 2008; Shaw 2012) (see Chaps. 18/Demiroz/Hu and 10/Waugh/ Chen in this volume). The Kobe earthquake pointed out that the community social capital plays an important role in disaster recovery. Comparing the case of Kobe and Gujarat, India, Nakagawa and Shaw (2004) demonstrated that the role of leadership and community cohesion is extremely important for decision making in the recovery phase. The bridging and bonding of social capital are equally important, where the bonding keeps the cohesion of the community, and the bridging links the community to external resources.

The Gujarat experience (Shaw et al. 2003) also pointed out a unique community based recovery experiences. It is important for the outside actors to establish a sense of trust and mutual understanding with the community. In case of Patanka village in Gujarat, this trust was established through constructing a house for the neediest people in the village, an aged widow who lost her properties in the disaster. Community based recovery or owner-driver house takes time. While at the aftermath of the disaster, there is tremendous pressure on the government for a faster recovery, there should not be any compromise with the quality of the process. Several past events pointed out that where the community involvement is higher, the quality of the recovery is also higher. Thus, the government's role is to establish a policy environment and framework, which should provide spaces of community consultation, involvement, participation, and ownership.

Similar experiences were also observed in Aceh in Indonesia after the Indian Ocean tsunami of 2004. The World Bank's housing recovery project had a strong component of community involvement, which was embedded in the Kechamatan (village) Development Program (Shaw 2008). The study by Ochiai and Shaw (2009) revealed that in a devastated community, where several people left their hometown, community building during housing reconstruction was important. The role of housing facilitators (who are the external people with professional expertise in engineering, architecture and social science) has been found to be effective in the community development process. A comparative analysis of

Aceh and Yogjakarta disaster (2005) shows that Yogjakarta had the unique community participation culture (locally called Gotang Royang or mutual help), which was very effective in the recovery process. Thus, the role of housing facilitators was different in case of Yogjakarata (housing facilitators were used in technical problem solving) than that of Aceh (where housing facilitator were used for social problem solving). Thus, the community character also defines the level of community involvement and participation.

With this context, the following section provides examples of community involvement in three different sectors: temporary housing in Kesennuma, emergency radio in Natori and school recovery in Kamaishi. All these three cities had high rate of damages of lives, livelihoods, houses and necessary infrastructures. All the three cities were proactive to open to external assistance (especially in terms of technical knowhow), and made community based recovery as the core of the recovery planning.

22.3 Temporary Housing and Community Involvement in Kesennuma

Kesennuma city is located in the North East end of Miyagi Prefecture along the coast of Pacific Ocean and had population of 70,000 in an area of 333 km^2. The main industries of the city were fishery and tourism. The city has been one of the largest catch of tuna, bonito, and sanma (Pacific saury) in Japan, and developed seafood processing industry complex from early times. The area north of the center of Kesennuma on the coast of Karakuwa Peninsula forms Ria coast. Current Kesennuma city comprises of three administrative districts, former Kesennuma city, Karakuwa town, and Motoyoshi town. After completion of consolidation of these three municipalities in September 2009, one administrative office is placed in each area in three districts.

Community building process in Kesennuma city is characterized with the food culture produced from the rich natural environment including mountains, rivers, and the ocean. In 2006, a slogan was announced, "Kesennuma Slow Food" as well as "Fish-Eating-Healthy-City" to promote health by eating fish. This way, Kesennuma citizens live in a close relationship to ocean, and fishery is considered as a main industry by the municipality for city planning. On the other hand, Sanriku area including Kesennuma city is prone to tsunamis and had been hit by large earthquakes several times in the past: "Meiji Sanriku Tsunami" in 1896 with 27,122 deaths, "Showa Dai Tsunami" in 1933 with 3,008 deaths, and "Chile Earthquake Tsunami" in 1960 with 119 deaths (KCC 2003).

The tragedy of Kesennuma is that in addition to the damage caused by tsunami, the central part of the city where many seafood processing companies were located, was completely destroyed by the massive fire that spread from an oil tank located at the bay entrance. More than 1,316 died or were missing, with 15,611 losing their homes—affecting 9,500 households (as of 29th May 2012). The city had 26,622 households in 2010—this means that more than one third of the population was affected. As mentioned above, the area is prone to tsunami disasters and had been hit by large earthquakes several times in the past. With these tsunami disaster experiences, the coastal area was equipped with coastal levees to prevent tsunami disasters. However, some experts mentioned the construction of such protection might be negatively affected, which led to more casualties since many residents did not take necessary action to evacuate at the time of tsunami. Also, according to Kesennuma city officials, out of 4,102 registered business institution, 3,314 business or 80 % of the total were considered to be affected and local economy was devastated, and led to a huge number of unemployed people.

Most of affected people moved to temporary housings prepared by the government. Due to lack of construction sites, building temporary housing sites prolonged than expected. As each site completed, people moved in. Construction of the sites went on all over the city until the end of December, eight months after the completion of the first one, which completed in April 2011. After all, they have built 93 sites consists of 3,503 housings. To be fair to everyone, residents were selected by lots. Consequently, many of them had to move from the place with few acquaintances where they lived before the earthquake (SEEDS Asia 2012).

In Japanese context of post disaster response/recovery, temporary housing play an important role. Immediately after the disaster, people take shelter in the evacuation center in public facilities like schools or public halls. Depending on the nature of the disaster, it may vary for 3–4 days [for typhoon or rainfall related disasters] to 3–6 months (for earthquake or tsunami). As per the government regulation, the local governments (city and/or prefecture) need to provide temporary housing for the affected people. The duration of the temporary housing is up to 3 years, however, in the Great Hanshin Awaji Earthquake (commonly known as Kobe earthquake), people lived 5 years in the temporary housing. There are several issues reported in temporary housing in Kobe due to its nature of alignment (closely placed grid pattern, which affect privacy of people), heat and sound insulation problems (which has both physical and mental effects), new neighbors (which prohibits people to come out of their rooms) etc. (UNCRD 2003; Shaw and Goda 2004). Volunteers played a very important role in socialization process in the temporary housing by organizing different events among different age groups. Thus, in post Kobe disasters, special care has been taken to reduce the risk of isolation in the temporary housing through volunteer mobilization.

One of the key characteristics of the temporary housing in Kesennuma was that due to lack of availability of land, the sites were spread over 96 different locations with different numbers of units (varying from 5 or 10 in some places to around 250 units for others). Therefore, volunteer coordination has been a challenging issue in temporary housing. To overcome this problem, the city government has outsourced a newly established local non-profit group named Kesennuma Reconstruction Association (KRA). With some funding from the city government (especially targeting to generate local employment for the people who lost their jobs in the disaster), KRA has employed around 100 people to look after these temporary housing. A Kobe based professional NGO called SEEDS Asia (with experiences of disaster risk

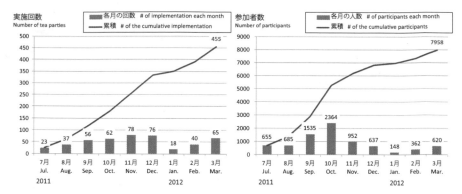

Fig. 22.1 Summary of tea party activities: *Left*: number of tea parties organized, and *Right*: number of participants in the tea parties

reduction in Asian region) tied up with KRA to provide technical support, drawing from the lessons of past disasters with Japan. Thus, SEEDS Asia and KRA formed a team to support the temporary housing through different activities, including holding regular tea parties, health check events, children's events, craft making events, and temporary housing leader's meeting.

There are two major objectives of tea party at temporary housing sites. First is to have a look over people such as elderly live-alones for prevention of their isolation. Second is to build communities among residents of temporary housing sites and promoting their community activities. The first tea party at temporary housing site by SEEDS Asia in collaboration with KRA was held on 30 June 2011, then the activities have been spread to all of temporary housing sites in the city. Tea parties were held at least once a month at each temporary housing site: twice a day, morning and evening, each for 2 h in different places in Kesennuma city. Figure 22.1 depics the number of parties and number of participants in the fiscal year of 2011 (SEEDS Asia 2012). There has been an increase in the number of participants till October (due to several other associated events along with the tea-parties).

One of the objectives of tea parties is to build relationship to know each other in the sites and lead establishment of neighborhood association. For example, during conversation in tea parties KRA staff put words to encourage the establishment such as "If someone keeps the keys to common rooms in the site, it makes easier to hold tea party." "Neighborhood associations will help aid organizations operate more easily." Elders in Kesennuma city use specific dialect, which can only be understood by the locals. For that, listening activities by the locals from KRA were very effective.

After establishment of neighborhood association, arrangement of tea parties was done thorough the leader of the association. During autumn and cold weather, tea parties were held at common rooms inside temporary housing sites. In order to encourage people who stayed inside temporary housing and were not interacting with other people to come outside for interaction, KRA was assisted by SEEDS Asia on planning/conducting recreational events in temporary housing sites. Winter is severe in Tohoku and there were some worrying voices from tea party participants on how to cope with cold in temporary housings. This was one example of concerns that leaders of temporary housings were facing to make living environment better. Some leaders of neighborhood associations individually requested city office for possible measures, but no collective action had been taken. Therefore, exchange meetings for leaders of temporary housing sites in Kesennuma area was held, as an occasion to figure out solutions together for problems at each site.

One of the key issues was to know the lessons from the past disasters. In this regard, SEEDS Asia arranged several exchange visits and seminar with KRA team to bring people and experts from other disaster hit areas in Japan [like Kobe, Chuetsu]. These were found to be very fruitful and practical in terms of solving several problems in the temporary housing. In summary, a few highlights of the community building through tea parties can be summarized as follow:

- A few months after construction of temporary housing sites many residents tended to participate in tea parties therefore it was very effective as a space for residents to communicate and know each other.
- Number of participants to tea parties decreased and was fixed with the increasing number of tea parties. In addition to tea parties, another approach such as door-to-door visit was necessary for prevention of isolation at temporary housings.
- Not only tea parties but also various community events were required in order to promote participation of residents in community activities at temporary housings. However, it is also found that tea parties, which provide opportunities for rest or conversation with others can be easily held in combination with other various community events.

22.3.1 From Emergency FM Radio to Community FM in Natori

Community radio can play significant role in disaster management through promoting preparedness, warning and rehabilitation programs before, during and after any disaster in the coastal areas (Ideta et al. 2012). In the pre-disaster phase, community radio can promote different pre-disaster activities and also local development activities for the local communities. Community radio can disseminate the messages of warning signals, indigenous coping mechanism i.e. behavior of animal, local weather situations, on first aid, emergency food, evacuation, gathering in shelter in particular space and on sanitation practice and facilities during emergency and before or afterwards. Moreover, community radio can reflect on plan for emergency responses. During the disaster community radio can standby and function in warning dissemination in local language/accents. Local young people can form a volunteer group as community radio listeners club and can work using radio messages for effective publicity for readiness. As different other communication ways are disrupted during the disaster phase, community radio can continue linkage with central government control room and liaison with the local governments. In post-disaster phase, community radio can broadcast programs and messages on identification of dead, disposal of waste, restoration of safe water supply and basic sanitation, promotion of alternative livelihood options, appropriate technology for restoration of communications, At a later stage, it can help on need assessment through pre-selected volunteers, organize and coordinate relief operations by mobilizing local resources, and prioritize primary health care services. For the effective role of community radio in disaster management, the CR staff should be properly motivated and radio programs should be designed with much emphasis on the disaster preparedness.

In Kobe earthquake of 1995, the community FM played a crucial role. It broadcasted earthquake information in Korean, Japanese, and many other foreign languages. Due to interruption of other communication system, the FM radio was able to provide local daily information, which is required for the daily living of the affected people. It also aired music at different intervals, which help the affected people to listen and relax.

In case of East Japan Earthquake and Tsunami, many local radio networks keep broadcasting relief operations (some do in multi-language), yet localized vital information is not reaching the most needed due to lack of information receivers. While internet-based information is accelerating the overall emergency response within/ from outside, affected aged community are still not in this loop. As availability of radio (band frequency of Japanese radio stations: FM 76—108 MHz (or 90 MHz)/ AM 530—1,600 KHz) is getting very limited, we are collecting second-hand radio within Japan and compatible band frequency radio from other parts of the world. FMYY estimates 30–40 thousands of receivers are needed for the affected areas in addition to the radios that are already collected for relief response by the Japanese government (including in-kind donations from leading electronic manufacturers such as Sony and Panasonic). In order to collect ground information and coordinate among local radio network, a representative of FYMM departed today and will report from the affected Tohoku area (Kyoto University 2012).

Not all emergency radio will continue as long-term information provider as Community radio. However, the emergency radio has made immense attempts to reduce the gap of information. In several cities, where bousai musen (emergency communication) did not work, the radio provided useful information on warning during the first few days of the aftershocks. While the mass media provides overall damage information, the affected people needs local information, like which road is accessible, which convenience or gasoline stand or supermarket store is opened, when and where the local food distribution (including takidashi, making hot food) is given, which bus route is opened etc. This set of information is especially relevant for the old people, who do not have access to the other media social media in the Internet. The nature of information changes over time, and gradually, the community radio needs to focus more on the community development issues. It is a major challenge though, since the sustainability of the community radio depends on the funding issues as well. During the emergency phase, there are funding supports, volunteer supports etc., however, over time, it gradually decreases. Therefore, community contribution is very much important to sustain these efforts. Following parts provide specific example of emergency radio in Natori city.

Natori city is located south of Sendai, the biggest city of Miyagi prefecture. The worst damaged area in the city was Yuriage district, a coastal area, with a long history. The area east of the speedway of east Sendai and Joban express were mostly flooded by the tsunami. The Yuriage area, which had a fishery harbor city central, with a population of 5,000, was entirely destroyed. 52 % of paddy fields or 2,200 ha were flooded. And because there is no upland in this area, many people found shelter at the speedway of east Sendai. The wall was also heavily damaged. However, the city central district of Natori, located upland further inland, only sustained little damage. The total death toll was 911.

According to an NPO report, just soon after the earthquake, in the coastal areas in Miyagi Prefecture, landline phones and mobile phones did not work well because of regulation and blackout (Chunichi Newspaper 18 July 2011). Natori City had 20 community wireless systems, which communicate emergency information to residents during a disaster and a disaster prevention radio that gives out evacuation warnings and instructions but, of these, seven stations were damaged by the tsunami. However, actually, the report shows that the community wireless system dealt with blackout using battery for few hours and was able to give people information about the disaster and evacuation order. The questionnaire survey shows that 100 % of people listened to the information from the community wireless system. Despite this, the damage was extensive. This means that the people did not actually believe the information from the community wireless system. At the temporary housing, there were no information system and they use megaphone to inform residents about urgent matters.

In this context, temporary emergency FM 'Natoraji' was launched as a temporary emergency radio station in Natori City, Miyagi Prefecture with the assistance of Set World Creation broadcast technology company. It was launched in order to provide disaster information to people living in Natori City. Natori City Hall applied for a license from the Ministry of Internal Affairs and Communications, Regional Bureau of Telecommunications of Tohoku on April 10, 2011. According to a staff of the Policy Planning Division of Natori City Hall, until now, Natori City still has no radio station, which has been expected to be a valuable source of life information. Likewise, Natori City Hall, which launched Natoraji thinks that it does not cover such a wide range of information, like detailed information such as confirming the safely of victims, the restoration of lifeline, and the route of the water tank truck.

A survey with the residents of the temporary housing and non-affected areas was done in October-November 2011 to understand people's perception on Natoraji (Ideta 2012; Ideta et al. 2012). While at the initial stage, the perception on Natoraji was higher only with the affected people (who took shelter in the temporary evacuation places like schools, and got several daily information for their daily lives), gradually Natoraji became popular with the local communities, irrespective of the age group. This was done through the volunteer activities from the local universities and other groups. Figure 22.2 depicts a distinct increase in the number of listeners irrespective of the affected or non-affected communities or whether they live in temporary housing or not. The key reason of this increase in the number of listeners is due to local information by Natoraji, which is not broadcasted by the mass

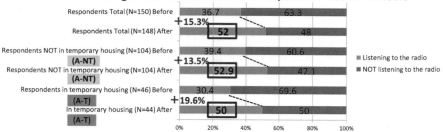

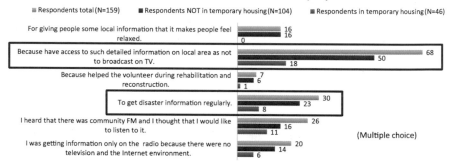

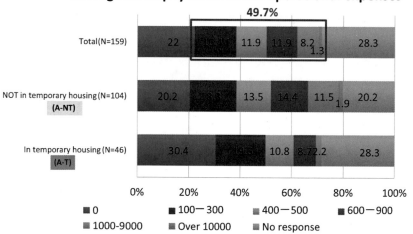

Fig. 22.2 Community survey in Natori: *Top*: Changes in the listeners of FM radio before and after the disaster, *Middle*: the reasons for listening to radio, and *Bottom*: Willingness to pay fund to continue the operation of FM radio

media. It seems that they need the radio as a tool to get disaster-related information in their dwelling area to prepare them for disasters in daily life and also to be able to respond immediately to a disaster when it occurs. The survey also shows that more than 50 % of the people are willing to pay to continue the radio, even after the completion of the emergency radio phase.

Based on the above results, a new model of radio is proposed as an evolution model of Natoraji. There are four issues in terms of the transformation process: (i) money for operation of the radio station, (ii) manpower, (iii) reliability of information, and (iv) airing or broadcast area. If the holder of the radio license is relegated from the government to a civil corporation, it is difficult to anticipate stable income. As a result, it is impossible to secure highly professional staff. Therefore, the ability of broadcasting would decrease and the reliability to the listeners would also decrease. Also, the frequency of the radio decreases, which is provided by Ministry of Internal affairs and Communications (MIC). As a result, the airing area reduces. Therefore, some areas cannot listen the broadcast. For the future of the Natoraji, a dual system of community FM radio is proposed, which should ideally be a community FM, and can be turned into emergency FM in case of a disaster. The Natoraji should be based on the strong local community involvement. A workshop with the local residents and radio experts was held in January 2012, with cooperation with the city government, and this new model of community radio was approved for future adoption (Kyoto University 2012). Bringing this to future, the Natoraji was enrolled as a community radio in July 2012 with the help of the local community, and made a new model of community radio, with a balanced approach of "on-air" (broadcasted activities) and "off-air" (community activities) activities.

22.3.2   School Centered Community Recovery in Kamaishi

At the aftermath of the disaster, on 11 October 2011, Ministry of Education Sports Culture Science and Technology (MEXT) Minister Masaharu Nakagawa sent a notification to 15 prefectures and one city of (East Japan Earthquake and Tsunami) affected region to propose the concept of "School-Centered Community Building" for school recovery. The concept has three main pillars: (1). Ensuring safety and security of school, (2) Provision to improve DRR and eco-friendly features and (3) Combining school with other public facilities to make it a central public facility. Although the main purpose of this concept is to rehabilitate or newly construct safe schools, the significance of this concept lies in the new idea of making the new school to become a multi-functional facility by combining such facilities as children's daycare center, public library, community center and DRR facilities in which the whole community can utilize and benefit from. Placing the school at the center of recovery and community building is believed to be a feasible option for school recovery because schools in Japan are usually considered to be a symbolic facility that the most community members are familiar with, especially in rural cities. In addition, If properly implemented, schools will not only as educational facility, but can also function as a place for community integration among different age groups, allowing the entire community to look over their children and take part in nurturing them (Matsuura and Shaw 2013).

This part describes the experiences of Kamaishi, a small city of 37,000 people in Iwate prefecture of Tohoku region. The city is known for its active role in disaster education. Several schools conducted regular disaster drills with the school and the communities, which resulted in "zero casualty" in the school buildings, although several of the schools were severely destroyed. In the secondary schools, the children evacuated along with the elementary school children. The role of teachers in implementing disaster education in schools needs to be highlighted. This is often termed as "Kamaishi Miracle" by media, and the credit goes to the proactive model of disaster education, the city adopted several years before. There were several school buildings destroyed in the disaster. Among these, the city government took an early decision of school centered community recovery, as presented by the MEXT, Government of Japan for two specific locations: Unosumai in the north and Toni in the south. This part focuses on the Toni area, where a decision has been taken to combine the elementary and junior high school along with the other community facilities.

Toni town is located in deeply-indented coastline along the Pacific Ocean. A mountainous terrain accounts for more than 90 % of the town. Based on data from the Kamaishi city government, population is 2,106 and the number of households is 956 in Toni-Cho. Recently, the town exhibits an aging population with very few children. Toni town's major industry is fishery particularly scallop, kelp and trout and the cultivation of brown seaweed. Most of the young people, however, work in offices in the central part of Kamaishi city or in another city. The central part of Kamaishi city is a 20 min travel by car from Toni town, so many find it convenient to commute. The town has seven neighborhood associations (jichikai) namely Kojirahama, Katakawa, Yamaya, Hongo, Kerobe, Oishi and Arakawa (Suda et al. 2012). Each Jichikai conducts a disaster drill on March 3rd every year. March 3rd is the date that the Meiji-Sanriku tsunamiof 1896. This drill was conducted in increments by neighborhood association (Jichikai) in the morning and people evacuate to a designated evacuation site. However, the participation rate of the disaster drill varies according to the Jichikai. According to head of the Jichikai, participation rate in Kojirahama is low. On the other hand, most of the community people in Kerobe participate in the disaster drill. The reason for the low participation in Kojirahama is that the population of commuters out of Toni town is more than the other Jichikais. Since they have to leave early in the morning, they cannot participate in the drill. In Kerobe and Oishi, however, people are bounded by solidarity because they have lived in the area for years and population is smaller than Kojirahama. There are books made by the people living in Toni town about past tsunami and its hazards. Students and adults used these books as a tsunami hazard prevention tool. Therefore, the area has a strong community bond, as well as high level of awareness and risk reduction activities.

The author conducted interviews with the principals of Toni elementary and junior high schools (Takeuchi and Shaw 2012). Both the schools are located in

Sanriku mountain area. Characteristic of Sanriku mountain area is narrow flat area and steep slope. 11 m concrete dyke was developed along the coast, and Toni ES was located near this dyke. This school was established in 1982 in this area. Before that, Toni elementary school was together with Toni junior high school. The school had 14 teacher/staffs and 68 students on 11th March 2011. At 14:46, every student was taking classes in the building. After the shaking stopped, students gathered in the ground and moved to shrine in the nearby mountain. Some voluntary fire fighters came to school and helped to evacuate. This area received large tsunami waves three times in this disaster. After the 2nd wave, school principle thought, "next one may be larger", and then school students and teacher/staff moved to even higher area (National route 45). After evacuation to the national route 45, they moved to the community hall and stayed one night. School facility received strong damage by tsunami. Tsunami reached up to the 3rd floor in school building. Consequently, the school facility could not be used to restart the class. Toni ES got some temporary space in Heita ES. After restart school, students attended by school bus from temporary houses. After 5 months from the disaster, education board of Kamaishi city made joint plan with Toni ES, Toni JHS and other public community facility. School principal agreed to this plan, but had worry about difference of ES and JHS, and safety issue. The Toni ES is currently undergoing in the temporary school facility built in the Toni JHS premise.

Unlike the Toni elementary school, the junior high school (JHS) was located in higher ground. Therefore, it was not directly affected by tsunami. But, since the building was quite old, it had significant earthquake damages, and could not be used for the evacuation center. When the earthquake happened, 47 students, 11 teachers were inside the school. Students were taking classes and immediately evacuated to the school ground. They went to higher ground on Route 45 fearing that the tsunami may arrive. The school principal had the knowledge that there would be a time lag of it would take 30 min between the earthquake tsunami arrivals. Therefore, he instructed the students to help the aged population to evacuate together to Route 45 after the event. The students spent the night-time together with the local community in a construction site office near the Route 45, and then the school started handing over the children to their parents on the next day. Since the Route 45 was broken and affected in many places, there was no immediate connection to the outside people, and the school and the community had to depend on the local resources available with the local people. After 1 month, when the school restarted, the graduation and welcome ceremony were held in the gymnasium. The school also re-started in the gymnasium, and from January 2012, the school started in the temporary building built in the playground. One of the key lessons was the proper judgment of the school principal, who had to take different responsibilities, apart from the usual education programs.

Kamaishi BoE (Board of Education) established the School Construction Consultation Committee (headed by the author) in December 2011, in order to discuss with the communities on the ways to implement and realize the recovery plans through their community leaders. The other members include school principals, local resident associations, local shrine head, fishing community head and PTA

(Parent teacher association). A detailed survey was conducted with the residents to understand the priority and needs in the new facility (as a combined school and community center). Through a unique consulting process, the committee had a regular meeting to discuss and decide the location of the new school, the need and priorities, the possibilities of different facilities to combine with the new school etc. However, issues such as budget availability, finalization of land use planning, consensus building of the community and the sequence of these factors actually have been delaying the implementation process. As recovery and community building issues go beyond the schools and education, which are conventional responsibilities of BoE, related local government departments must place more efforts to coordinate and at the same time, conduct surveys and analysis to investigate the social situation and needs of the communities.

Thus, the school-centered recovery is a unique participatory process, which goes beyond the school construction. Ideally, the construction of these types of facilities is the responsibility of the local government, with very little participation from the community. However, in the post disaster situation, participatory decision-making is important, and through active involvement in the construction committee, the communities feel the ownership of the process. It is also suggested that during the school construction period (which will be around 1.5–2 years), the school children should also have opportunity to visit the sites, to interview the construction workers and thus be part of the process. There are several literatures on the importance of school, and school community linkage (Shaw et al. 2011). This unique live experiment in Toni town of Kamaishi city shows the complexity as well as importance of the process. The national government's role here was to provide a policy framework, the local government's role was to provide the implementation opportunity, and the community's role was to provide local knowledge and resources in decision making.

22.4   Conclusion

Shaw and Takeuchi (2012c) from the earlier study concluded the importance of six different elements in the recovery process: community, information, education, evacuation, communication and voluntarism. Among these, this chapter deals with three of them: community, information, and education. The link of education and communication can be enhanced by drills, disaster maps etc., as experienced in Kamaishi case. The link between communication and information can be local and community owned media (community FM radio) with a strong link of on-air and off-air activities. The ultimate goal is to develop a "*Culture of Safety*," which is linked to post disaster recovery, and long-term risk reduction measures. Shaw and Takeuchi (2012c) also suggested the policy and resource linkages, in term of a "*Cone of response, recovery and risk reduction*".

Tohoku area is in the third year of the recovery process. The recovery planning in most of the cities have been completed through a long community consultation process. Budgets are calculated and some priority construction works have been started. As it is said that the physical part of Kobe has been recovered within 5 years after the disaster, however, the social and human aspects of Kobe is still recovering. Community-based recovery issues will take time, and it is a dynamic process. It is important to understand, analyze and document this process very carefully, and provide proper feedback both to the local communities and local governments. Community recovery is not a myth, it is a reality, and can happen over a longer period of time. The examples of Kesennuma, Natori and Kamaishi provide the human dimensions of the recovery process, which are linked to social infrastructures recovery. Different change agents are required at different timeframe and in different context. While non-profit, non-government organizations played important role in Kesennuma, it was the local governments which played strong role in the beginning of emergency FM radio, and gradually transferring the ownership to the local community in the form of community radio. In case of Kamaishi, although school construction is a public responsibility, the process involves communities, and then only, it can become a truly vital community facility to serve the community during the disaster. The Tohoku area is prone to tsunami and other disasters. Thus, the community involvement would be essential for the resource poor local governments, most of which have higher aged population.

At the aftermath of a disaster, there has always been pressure to quickly restore support systems, livelihood and repair damages. In most of the cases, this undermines the quality of relief, reconstruction and rehabilitation works. The pressure of time and other constrains such as the difficulties in communication and transport in the post-disaster environment make it difficult to restore and lives and livelihoods with enhanced resilience. However, recovery is a balance between speed and quality. The speed is higher when it is done in a centralized way, by single agency. However, when it comes to cooperation and collaboration among different stakeholders, departments and agencies, the process becomes slow. However, the question is do we always need a fast recovery? A community based consultation and decision-making needs time, it needs to be linked to different culture and local situation. Thus, the recovery should not be measured just on the speed. The quality becomes equally important. Thus, based on the above observations and discussions, following specific recommendations can be made to link disaster and development issues in terms of sustainability.

Local DRM Needs to be Linked to Development Issues like Health, Education It is always a challenge to continue the DRM activities at the local level for a longer period of time. This sustainability issue is often discussed and analyzed. There are several models of sustainability of local DRM, linking it to the local communities, or providing ownership to the communities, or depending on the local leadership etc. After the disaster, community based DRM becomes popular, and after a few years, the initiative gradually diminishes. Thus, to sustain the local DRM over a longer period of time, it is required to link it to the daily needs, like welfare, health, environment etc. The nature of the community or local context defines the specific entry point for sustainability of DRM activities. For Kamaishi, the education is

found to the entry point of long-term recovery and sustainability. In Natori and Kesennuma, it is the link of community radio and tea parties in temporary shelter, where
information sharing for health and welfare issues for the aged population becomes
important. This ensures the sustainability of the process.

Resource Commitments of the Local Governments are Required One of the key
points of local DRM is the policy provision for utilizing appropriate budget and
funding. Preferably, a certain percentage of local development fund needs to be
allocated for local DRM in long run. Specific activities like regular training, capacity building programs or awareness-raising programs need to be budgeted properly
in the local level.

Regular Updating and Testing of Local DRM Initiatives is Essential Community is
dynamic, and there are changes in the communities. Thus, the local DRM needs to
be dynamic as well. The approaches may be same, but needs to be customized based
on the local changes and context. The technology and governance system needs to
be adjusted accordingly. Also, the traditional knowledge needs to be transferred
over generation and context.

Linkage with Local Resource Knowledge Institutions Needs to be Strengthened The
link with the local institutions is important for sustaining the knowledge base of
local DRM. The local non-government or volunteer agencies, the local universities
can be the resource organization to provide technical expertise to the local governments and communities, and thereby making a sustainable knowledge management system. It is often argued that the data acquisition and data quality at the
local level is rather poor. This can be upgraded with the link to the education and
research agency at the local level, and providing support to capacity building of
those agencies.

Acknowledgements The author acknowledges the support from the city governments of Kesennuma, Kamaishi and Natori and the affected communities for their help and cooperation. Funding
support from CWS Asia Pacific and MERCY Malaysia is highly acknowledged.

References

Chunichi Newspaper. (2011, July 18). In Japanese, Chunichi Shimbun, 18 July, 2011.
Ideta, A. (2012). Role of FM radio in East Japan earthquake and tsunami. Master Thesis, Kyoto
    University.
Ideta, A., Shaw, R., & Takeuchi, Y. (2012). *Post disaster communication and role of FM radio:
    Case of Natori. In East Japan earthquake and tsunami: Evacuation, communication, education
    and volunteerism* (pp. 73–108). Singapore: Research Publishing.
Japan   Meteorological   Agency   (JMA).   (2011).   http://www.jma.go.jp/en/tsunami/fo-
    cus_04_20110311145000.html. Accessed 18 March 2011.
Kapucu, N. (2008). Collaborative emergency management: Better community organizing, bet-
    ter public preparedness and response. *Disasters: The Journal of Disaster Studies, Policy, and
    Management, 32*(2), 239–262.
KCC. (2003). *Kesennuma Chamber of Commerce "Marukajiri Kesennuma Guidebook"* (p. 59).

Kyoto University. (2012). Road map of Natori emergency FM radio "Natoraji", 33 pages., Kyoto, Japan.

Matsuura, S., & Shaw, R. (2013). Overview of school centered community recovery. In R. Shaw (Ed.), *Disaster recovery: Used or misused development opportunity (pp. 165–196)*. London: Springer Publication.

Nakagawa, Y., & Shaw, R. (2004). Social capital: A missing link to disaster recovery. *International Journal of Mass Emergency and Disaster, 22*(1), 5–34.

NPA. (2013). National police agency: Status of 2011 earthquake and tsunami. http://www.npa.go.jp/archive/keibi/biki/higaijokyo.pdf. Accessed 14 Jan 2013.

Ochiai, C., & Shaw, R. (2009). Participatory urban housing reconstruction study in Aceh, Indonesia. In: R. Shaw, H. Srinivas, & A. Sharma (Eds.), *Urban risk: An Asian perspective* (pp. 233–254). UK: Emerald Publication.

SEEDS Asia. (2012). Community recovery of the Great East Japan earthquake and tsunmai: Kesennuma experiences (25 p.). Published by SEEDS Asia, Kobe, Japan.

Shaw, R. (2008). Kechamatan development program, Indonesia, case study in building resilient communities, the World Bank, pp. 243–267.

Shaw, R. (2012). Community based disaster risk reduction. UK: Emerald Publisher 402 p.

Shaw, R., & Goda, K. (2004). From disaster to sustainable community planning and development: The Kobe experiences. *In Disaster, 28*(1), 16–40.

Shaw, R., Gupta, M., & Sharma, A. (2003). Community recovery and its sustainability: Lessons from Gujarat earthquake of India. *Australian Journal of Emergency Management, 18*(2), 28–34.

Shaw, R., Shiwaku, K., & Takeuchi, Y. (2011). *Disaster education*. UK: Emerald Publisher, 162 p.

Shaw, R., & Takeuchi, Y. (2012a). *East Japan earthquake and tsunami: Evacuation, communication, education and voluntarism*. Research Publisher, Singapore. 280 p.

Shaw, R., & Takeuchi, Y. (2012b). *Introduction and overview of East Japan earthquake and tsunami. In East Japan Earthquake and Tsunami: Evacuation, Communication, Education and Volunteerism (pp. 1–14)*. Singapore: Research Publishing.

Shaw, R., & Takeuchi, Y. (2012c). Towards long term recovery after the mega disaster. In *East Japan Earthquake and Tsunami: Evacuation, Communication, Education and Volunteerism* (pp. 255–269). Research Publishing, Singapore.

Suda, Y., Shaw, R., & Takeuchi, Y. (2012). *Evacuation behavior and its implication: Case of Kamaishi. In East Japan earthquake and tsunami: Evacuation, communication, education and volunteerism (pp. 25–57)*. Singapore: Research Publishing.

Takeuchi, Y., & Shaw, R. (2012). *Damages to education sector and its recovery. In East Japan earthquake and tsunami: Evacuation, communication, education and volunteerism (pp. 143–164)*. Singapore: Research Publishing.

UNCRD. (2003). From disaster to community development: The Kobe Experiences. 109 p., Kobe, Japan.

Chapter 23
The Impact of the 2010 Haiti Earthquake on Disaster Policies and Development

Abdul Akeem Sadiq

> *"Decision-makers who ignore ... relationships between disasters and development do a disservice to the people who place their trust in them."*
>
> —Stephenson (1994 p. 11)

23.1 Introduction

The Haiti earthquake of January 12, 2010 caused the death of hundreds of thousands of people and caused billions of dollars in economic damage. After more than 3 years since the earthquake, it is important to examine the factors that contribute to Haiti's vulnerability to disasters in general and the reasons for the huge devastation caused by the 2010 earthquake. By understanding Haiti's vulnerabilities and the reasons why the 2010 earthquake caused huge destruction, we will have a better understanding of the challenges facing Haiti's long road to recovery and redevelopment. Before discussing the particulars of disaster and development in Haiti, it is important that the concepts of disaster and development are first defined.

A disaster is defined as a "... serious disruption of the functioning of a community or a society causing widespread human, material, economic or environmental losses which exceed the ability of the affected community or society to cope using its own resources" (United Nations International Strategy for Disaster Reduction (UNISDR) 2004). In defining development, the author would like to emphasize the importance of sustainability in development. According to the United Nations (UN) (1987) sustainable development is development which "meets the needs of the present without compromising the ability of future generations to meet their own needs" (p. 15). The author revisits these two concepts later in the chapter.

A. A. Sadiq (✉)
Indiana University Purdue University Indianapolis (IUPUI),
Indianapolis, USA
e-mail: asadiq@iupui.edu

N. Kapucu, K. T. Liou (eds.), *Disaster and Development,* Environmental Hazards,
DOI 10.1007/978-3-319-04468-2_23, © Springer International Publishing Switzerland 2014

This chapter begins with a background on Haiti and the Haiti earthquake and takes a look at the vulnerabilities of Haiti to disasters in general. Next, it discusses the response of the international community to the 2010 earthquake and the steps taken to help Haiti recover from this disaster. This chapter then discusses the challenges facing Haiti's emergency management system and the opportunities at the disposal of the Government of the Republic of Haiti (GoH) to revamp its emergency management system. Next, this chapter examines the key differences between Haiti and the United States with respect to disasters and development. Finally, this chapter concludes by suggesting some recommendations based on lessons learned from past disasters in both developed and developing countries.

23.2 Background on Haiti and the Haiti Earthquake

Haiti is the world's oldest black republic, gaining its independence from France in 1804 (Telegraph 2010). Haitians speak mainly French and/or Creole and about 95 % of the population is black, with 5 % being mulatto and white (Central Intelligence Agency (CIA) 2012). Haiti is bounded by the North Atlantic Ocean to the north, the Dominican Republic to the east, and the Caribbean Sea to the south and west. In terms of geographical size, Haiti is a little smaller than the state of Maryland with an area of 27,750 km² (CIA 2012). Haiti is a country with a long history of political instability, violence, and devastating disasters (Brattberg and Sundelius 2011). Haiti also has the worst malnutrition and income disparities in the hemisphere and half of Haiti's school-age kids do not attend school (Kaufman 2012). In addition, Haiti is one of the most densely populated countries in the Western Hemisphere and has an unemployment rate of 40 % (CIA 2012; Klose and Webersik 2010).

On January 12, 2010, at about 4:53 p.m., a 7.0 magnitude earthquake hit about 15 miles south west of Port-au-Prince, Haiti's capital (United States Geological Survey (USGS) 2011). The intense shaking, which lasted for 35 s and was extremely violent (McEntire et al. 2012) killed 316,000 people (including 17 % of the GoH work force) (Archibold 2011; Bellerive and Clinton 2010). In addition to the death toll, the earthquake injured 300,000 people, totally destroyed about 105,000 houses, damaged about 208,000 houses, and displaced 1.3 million people (GoH 2010; USGS 2010). The earthquake also left behind 10 million m³ of rubble (Kaufman 2012). Experts have put the overall damage estimates caused by the 2010 earthquake at $ 7.2–13.2 billion (Lacey 2010). Based on the aforementioned statistics, it is not surprising that the Haiti earthquake has been described as "the worst recorded natural disaster in the Western Hemisphere" (Weisenfeld 2011, p. 1097). The reasons for the huge destruction include, but are not limited to, poor building practices, absence of building codes, and a lack of awareness about earthquake risk and planning (Kapucu and Ozerdem 2013) (Fig. 23.1).

Fig. 23.1 Houses destroyed by the 2010 Haiti earthquake. (Source: Author)

23.3   Vulnerability of Haiti to Disasters

Haiti is vulnerable to myriads of disasters like hurricanes, floods, earthquakes, and climate change (CIA 2012; Slagle and Rubenstein 2012). For example, in 2008, four hurricanes—Ike, Fay, Hanna, and Gustav—killed over 1,000 people, damaged countless properties, and destroyed 60 % of Haiti's agricultural crops (Slagle and Rubenstein 2012). Flooding, which is typically brought about by hurricanes and torrential rainfalls, is also a common hazard in Haiti (Fordyce et al. 2012). For example, Hurricane Gordon in 1994 killed over 1,000 people and caused extensive flooding (Poncelet 1997). When it comes to vulnerability to climate change, Haiti is ranked number one out of 200 nations (Slagle and Rubenstein 2012).

Haiti's vulnerability to disasters is due to factors such as geographical location, human factors, and poverty. Haiti is located on the island of Hispaniola (along with its neighbor the Dominican Republic), an area right in the center of a hurricane belt (CIA 2012) and on the Enriquillo-Plaintain Garden Fault between the Caribbean and North American tectonic plates (USGS 2010). In addition, human factors, including but not limited to environmental degradation, exacerbate the destruction and damage from hurricanes, floods, and earthquakes. For instance, Haiti has only 1 % of forest cover (compared to its neighbor, the Dominican Republic with 28 %) (Mainka and McNeely 2011). In fact, some experts (e.g., Wdowinsky et al. 2010) believe that deforestation may have contributed to the devastation caused by the 2010 earthquake. And in the aftermath of the earthquake, the already scanty forest cover was further depleted as a result of Haitians constructing makeshift tents made from tree branches, as well as due to the high demand for charcoal for cooking. Finally, Haiti is the poorest country in the Western Hemisphere (Weisenfeld 2011). With 80 % of its population living below the poverty line (Brattberg and Sundelius 2011) and a per capita income of US $ 1,338 (in 2009) (Klose and Webersik 2010), poverty is at the root of Haiti's vulnerability to disasters (Pantelic 1991).

For example, it is common practice for Haitians to cut down trees and use them as fuel for cooking. The removal of forest cover, as discussed above increases Haiti's vulnerabilities to disasters.

23.4 The Ubiquity of Non-Governmental Organization (NGO) in Haiti

Before the 2010 Haiti earthquake, Haiti had the highest per capita NGO presence in the hemisphere, with more than 8,000 NGOs in various humanitarian and development sectors in the country (Lawry 2010). In fact, Haiti's development sector is run by International Non-Governmental Organizations (INGOs) and their donors (Lawry 2010) to such an extent that Haiti is sometimes referred to as the "Republic of NGOs" (Kristoff 2010). As a result of the prevalence of NGOs and the over reliance of Haiti on NGOs for service provisions, including public safety (Brattberg and Sundelius 2011), NGOs can be regarded as an important part of Haiti's emergency management system (Fordyce et al. 2012). Despite the prevalence of INGOs in Haiti, the response to the 2010 Haiti earthquake was ineffective. A lack of coordination among the myriad INGOs and countries that responded to the earthquake is an often cited reason (e.g., Kristoff 2010; Weisenfeld 2011). Coordination, which can be defined as the cooperative efforts among responding agencies to work toward the same goals in the aftermath of disasters (McEntire 2009) is important for an effective response. In the case of the Haiti earthquake, common goals included rescuing trapped victims, providing food, water, and medical services, removing debris and engaging in long-term development programs.

The UN recognizes the importance of coordination after a disaster and as a result implemented the "Cluster System" in Haiti (Weisenfeld 2011). The Cluster System was developed in 2005 to improve coordination of humanitarian response to disasters by specifying lead agencies and support agencies in different need areas like nutrition, protection, and health (United Nations Office for the Coordination of Humanitarian Affairs (OCHA) 2012). The first application of the Cluster System was to the Pakistan earthquake of 2005. Since then, the Cluster System has been used in more than 30 major disasters (OCHA 2012). Despite the use of the Cluster System, the coordination of humanitarian efforts in Haiti was not as effective as expected due to factors like inconsistencies in methodology and the tools used to gather information on needs assessment (Weisenfeld 2011). The lack of coordination after the Haiti earthquake led to duplication of efforts and inefficient utilization of resources, among other coordination-related problems (Weisenfeld 2011). To improve international coordination to future disasters, Weisenfeld (2011) recommends establishing procedures that define the roles and responsibilities of responding organizations when a major surge is needed.

Fig. 23.2 A United States
Army base in Haiti. (Source:
Author)

23.5   Response to the Haiti Earthquake

Although, the main focus of this chapter is on development and not on response, I will briefly talk about the response phase and the key participants in the response to give an idea to the reader about the outpour of support from the international community. The scale of destruction caused by the 2010 earthquake, which can be described as unimaginable, was met by an unprecedented and commensurate international response. The outpour of support in the form of personnel and supplies came from 129 countries (Weisenfeld 2011) including the United States, France, England, and organizations like the UN, the European Union (EU), and the Caribbean Disaster Emergency Management Agency (CDEMA). While the author recognizes the important roles played by these and other players, the focus will be on the contributions made by the US in response to the Haiti earthquake.

The United States Agency for International Development (USAID) led the US response to the earthquake, which also included participation by Federal Emergency Management Agency (FEMA), Center for Disease Control and Prevention (CDC), and the US Department of Defense (DOD) (Weisenfeld 2011), to mention a few. USAID's Disaster Assistance Response Team was on the ground in Haiti within 24 hours of the quake and provided logistical and material support to Haiti (Weisenfeld 2011). The US military deployed about 13,000 troops and equipment; including a total of 264 military aircrafts and 23 Navy ships (Brattberg and Sundelius 2011). As of January 2011, the US had spent $ 1.1 billion in relief assistance to Haiti (United States Departments of State (USDS) 2011). This money was spent on a wide range of services and materials like search and rescue, medical supplies and treatment, food, water, shelter, building camps, and immunizations (USDS 2011; Fig. 23.2).

23.6   Haiti's Emergency Management System

There are several challenges confronting Haiti's emergency management system. First, Haiti has a weak emergency management system that is over reliant on regional (e.g., CDEMA) and international bodies (e.g., INGOs) to mitigate, prepare for, respond to, and recover from disasters (Fordyce et al. 2012). By relying on INGOs, the GoH did not invest in the development of effective mitigation, preparedness, response, and recovery capabilities. Perhaps, it was due to this realization that the World Bank was in the process of revamping Haiti's emergency management system shortly before the 2010 earthquake struck (Kapucu and Ozerdem 2013).

The poor response to the Haiti earthquake by INGOs and other humanitarian bodies, as discussed above, should serve as an impetus for the GoH to overhaul its emergency management system. Second, political instability has diminished Haiti's economic status (Lundahl 1989) and Haiti's capability to mitigate, prepare for, respond to, and recover from disasters. Without a stable government, the political commitment to a better emergency management system is bound to be feeble and unstable at best (Fordyce et al. 2012). Finally, another challenge confronting Haiti's emergency management system is illiteracy. As said earlier, a little over half of Haiti's population, age 15 and older, is considered literate (CIA 2012). Illiteracy has a negative impact on Haiti's economy because without proper education, Haitians may not be able to gain employment that is necessary to contribute to their economy. In addition, it will be difficult to entrench a culture of disaster preparedness (World Bank 2012) among Haitians if Haitians do not have the basic education needed to understand risk information presented during community awareness campaigns and school education programs on mitigation and preparedness measures (Fordyce et al. 2012).

The aforementioned challenges notwithstanding, there are some important opportunities that the GoH can leverage to improve its emergency management system. First, NGOs, CDEMA, UN, EU, etc., can provide the necessary resources to improve Haiti's emergency management system. An important first step in establishing a good emergency management system is to establish emergency management offices at the local, regional, and federal levels, and staff the offices with individuals knowledgeable about emergency management (e.g., individuals who are certified emergency managers). The US government along with other humanitarian bodies can provide the necessary training for individuals to become certified emergency managers and also teach them how to develop and maintain emergency plans. In addition, it is important for GoH to use the donations and expertise from the international community wisely—provide sufficient funding for emergency management offices so that these offices have the resources to mitigate, prepare for, respond to, and recover from disasters. Second, the new democratically elected president, President Michel Martelly, who came into office in October 2011 (Kaufman 2012), provides an immense opportunity for the GoH to establish a politically stable platform on which to build a better emergency management system and implement sustainable development programs like tree planting, drainage system expansion projects, and housing redevelopment. Lastly, the window of opportunity opened by the tragic earthquake—a focusing event (Birkland 1997)—allows the GoH to roll out programs that focus on educating Haitians about disaster risks and the appropriate mitigation and preparedness measures to adopt for disasters. In addition, the GoH should use this opportunity to include disaster management in school curricula with a view to building a culture of prevention in Haiti.

23.7   Disasters and Development

In this section, this chapter revisits the concepts of disaster and development and examines the relationship between disasters and development. As evident in the definition of a disaster presented in the introduction, when disasters occur, they have the tendency not only to cause loss of lives; they can invariably lead to the destruction of the built environment. That is to say disasters can have tremendous impact on development programs in disaster-stricken communities (Stephenson 1994). Similarly, inherent in the definition of sustainable development is the notion of intergenerational equity as well as maintaining a balance among three spheres—environmental protection, social development, and economic development. From this definition, it is clear that any sustainable developmental program should ensure fairness between the current and future generations. In other words, "all development has to take into account its impact on the opportunities for future generations" (United Nations Economic Commission for Europe (UNECE) n. d.). In addition, any sustainable development program should meet the tripartite goals of environmental protection, social development, and economic development. To illustrate this last point, let us take the construction of low-cost housing as an example. To make sure that the housing project is sustainable, it should meet the goals of social development (e.g., provide housing at affordable prices to the poor), environmental protection (e.g., built from materials like recycled bamboo and recycled rubble), and economic development (e.g., provide construction jobs for local residents).

Researchers have emphasized the importance of the relationship between disasters and development. For example, Collins (2009) argues that "disaster reduction is fundamentally a development issue ... development is also about learning and adapting to the risk of disaster" (p. 251). I discuss the relationship between disasters and development, a historically neglected area of study (Collins 2009) using Stephenson's (1994) four themes (see Fig. 23.3). First, disasters can have an adverse impact on development. For example, the 9/11 terrorists attack against the US destroyed the North and South World Trade Center buildings. Second, disasters can create an opportunity to initiate sustainable development programs. For example, after over 90% of the City of Greensburg, Kansas structure was destroyed by an EF-5 tornado in May 2007 (Berkebile and Hardy 2010), the City of Greensburg is embarking on various sustainable development programs (e.g., some newly constructed buildings are Leadership in Energy and Environmental Design (LEED) Platinum certified). In the absence of disasters, it is relatively difficult to

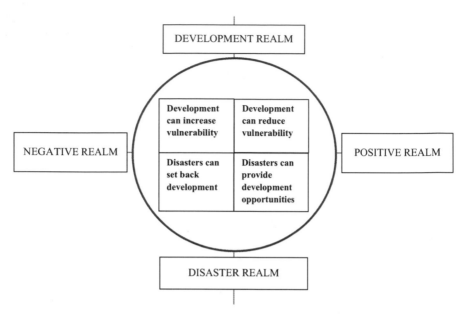

Fig. 23.3 The four themes of disaster and development. (Source: Stephenson 1994)

incorporate sustainable development programs due to many factors: (i) people do not like change, (ii) resources have already been invested in current development, and (iii) the opportunity cost of a new sustainable development program. As a result of these obstacles to implementing sustainable development in communities, those in charge of the development effort have to make sure that members of the community are aware of the benefits inherent in a proposed sustainable development program. Third, development programs can increase community vulnerability to disasters. If a development program is not properly designed, implemented, and maintained, it can make a community more susceptible to disasters. For example, a lack of maintenance is one of the reasons that the system of levees protecting New Orleans failed during Hurricane Katrina. In this third theme, the notion of *disaster risk management* is very applicable. Disaster risk management "is an approach which can be used in development planning … in order to reduce development-induced risks and vulnerability" (UNISDR 2004). And fourth, development programs can decrease community vulnerability to disasters. For example, constructing good drainage systems in flood prone areas of a community might help that community to be less susceptible to floods. As a result of the aforementioned relationships between disasters and development, it is imperative that any plan to redevelop Haiti should incorporate the need to embark on sustainable development projects as a way to reduce future vulnerabilities to a wide range of disasters.

23.8   Disasters and Development in Haiti and the United States

There are many differences between the US and Haiti when it comes to disasters and development. For example, in the case of the US, there is documented land ownership, political stability, capacity to respond to disasters, disaster planning, and a fairly good emergency management system. Conversely, undocumented land ownership, political instability, a lack of capacity to respond to disasters, inadequate disaster planning, and a weak emergency management system make Haiti more vulnerable to disasters, unable to respond and recover quickly from disasters. Taking the land tenure issue as an example; in the United States, land ownership is documented so that after a disaster, a home owner can rebuild on his or her land. In Haiti, however, there is no national land registration system (Olshansky and Etienne 2011). As a result, after the 2010 earthquake, there were disputes over land ownership and subsequent delays in redevelopment projects like reconstruction of homes. The newly sworn-in democratic government should help to resolve on-going and future land ownership-related problems (Kaufman 2012).

Another example that shows a stark contrast between the United States and Haiti is the differences in their emergency management capabilities and structures. Let us take a quick look at how both countries carried out mass fatality management (MFM) in the aftermath of the Haiti earthquake. The MFM effort of the GoH after the earthquake was very ineffective due to many challenges, including but not limited to, a lack of resources (Sadiq and McEntire 2012). McEntire et al. (2012) compared the MFM process of US citizens that died in the Haiti earthquake with Haitians and concluded that the MFM of the former was well coordinated while that of the latter was nonexistent because many of the dead were buried in mass graves following the directive issued by the GoH (McEntire et al. 2012). In the case of US citizens, the Disaster Mortuary Operations Response Team (DMORT) was able to recover, identify, and return the corpses of US citizens to their families for burial. It is important to note here that the number of US citizens that perished in the Haiti earthquake is nowhere near the number of Haitians killed by this earthquake. Nevertheless, if the US suffers the same number of casualty from a disaster as we saw after the 2010 Haiti earthquake, it is unlikely that the US government would issue a directive for mass burial. Finally, the US has in place many disaster-related plans and frameworks at the federal, state, and local levels that communities can use in planning for disasters. Examples of these include, but are not limited to, Local Emergency Operations Plans, State Comprehensive Emergency Management Plans, the National Response Framework, and the National Disaster Recovery Framework. Conversely, before the earthquake, Haiti had no disaster risk reduction plans. It is noteworthy that with the help of United Nations Development Program, Haiti is now putting together a national level Seismic Risk Reduction Plan (Homeland Security News Wire 2012; Fig. 23.4).

Fig. 23.4 The mass burial
site in Titanyen, Haiti.
(Source: Author)

23.9 Lessons Learned from Past Disasters in Developed and Developing Countries

When disasters occur, they can have severe impact on communities. For instance, Japan's triple disasters—earthquake, tsunami, and nuclear accident—killed 17,500 people, injured another 6,109, damaged over a million buildings, and caused $ 210 billion in economic loss (World Bank 2012). As tragic as this and other disasters can be, disasters provide communities—both affected and unaffected—the opportunity to learn some lessons. Such lessons can provide valuable insight into which risk-reducing measures worked and which ones failed. A thorough assessment of the performance of risk-reducing measures would enable a community to make necessary improvements in their risk-reducing measures with the goal of reducing their vulnerabilities to future disasters. After the assessment, communities can share information about measures that worked with other communities. This section discusses and applies some of the lessons learned from past disasters in both developed and developing countries to Haiti.

Any sustainable development strategy for Haiti should incorporate the basic principle of "building back better" (Lawry 2010). In its Action Plan for National Recovery and Development, the GoH (2010) also recognizes this principle by noting that, "Rebuilding Haiti does not mean returning to the situation that prevailed before the earthquake" (p. 5). The following recommendations can help to build Haiti better and make it less vulnerable to future disasters. First, long-term reconstruction should consider the design, implementation, and enforcement of new building codes for both existing and new construction (Mainka and McNeely 2011; Pantelic 1991). For example, the presence and strict enforcement of Japan's building codes were credited for minimizing loss of lives and economic damage from the 2011 earthquake (many of the recorded deaths were from the tsunami)

(World Bank 2012). In the case of Haiti, building codes for hurricanes and earthquakes can make buildings more resilient to these natural hazards. For example, zoning ordinances can be enacted to make sure that residential units are not built close to fault lines or near coastal areas. In recognition of the importance of rebuilding stronger, the GoH implemented a voucher program for homeowners who needed to repair their homes. As part of the program, homeowners were asked to use government certified construction companies that would make sure homes were built to the required hurricane and earthquake standards (Kaufman 2012).

Second, Haitians (including Haitians abroad) must be the vanguard of the long-term development strategy, supported by other stakeholders like donors, INGOs, UN, US, the EU, etc. The idea that Haitians should take over the reconstruction was echoed by Thomas Adams, the U.S. special coordinator for Haiti. He said, "We do want this to be a Haitian-led reconstruction … and sometimes we just have to wait for the Haitians to lead because it's their country. We can help, but they have to really make the key decisions …" (Kaufman 2012). The Interim Haiti Recovery Commission, whose purpose, among others, is to empower the GoH (Olshansky and Etienne 2011), also supports this idea. There are many benefits to letting Haiti take the lead with support from other stakeholders. One such benefit is that Haiti will have the opportunity to include its social and cultural values in the redevelopment process and enjoy the "invigorating experience of rebuilding" (Pantelic 1991). Another benefit is that Haitians' participation will legitimize the redevelopment process and increase the likelihood that the redevelopment process will be successful. Finally, by letting Haitians take the lead, there is likely to be infusion of local knowledge and work experiences that are necessary for a successful redevelopment process (Pantelic 1991). Haiti can learn from the recovery planning efforts of Greensburg, Kansas after the EF-5 tornado. Greensburg organized a public recovery planning workshop to afford its residents the opportunity to contribute to the recovery of their community by developing a "unified community vision, goals, and a strategy to move from devastation to renewal" (Berkebile and Hardy 2010). The GoH should establish this type of community mobilization effort in order to build a solid foundation on which to roll out sustainable development programs.

Third, and along the same line as the previous recommendation, is for the GoH to reach out to the Haiti diaspora. In the United States alone there are more than 800,000 Haitians that have migrated to areas like New York, Florida, and Boston (Carlson et al. 2011). Further, it has been reported that the diaspora sends 1.8 billion USD to Haiti annually and in 2010, the donations accounted for 20 % of Haiti's GDP (Carlson et al. 2011). Moreover, a poll done by New America Media (2010) reports that 78 % of Haitian adults in the US have sent money to Haiti and that they are both willing to contribute more since the earthquake and contribute in new ways, for example adopting or fostering Haitian orphans. The willingness of the diaspora to contribute to the country can be a tremendous resource for the recovery and redevelopment of Haiti. These resources of the Haiti diaspora should be leveraged to aid in building Haiti back better.

Fourth, the international community must help Haiti build capacity so that Haiti can be more resilient to future disasters (see Chap. 10 in this volume). Before the earthquake, Haiti's capacity to respond to disasters was virtually absent due in part to the over reliance on INGOs. As part of this process, the GoH would need to hire new workers to replace those lost to the earthquake and train new workers in emergency management procedures like urban search and rescue. The international community can aid in this process. The importance of having search and rescue team cannot be over emphasized. For example, after the 1971 Sylmar earthquake, which affected parts of the San Fernando Valley and the 1987 Whittier earthquake in eastern Los Angeles County, search and rescue teams did not have the necessary tools to extricate and remove trapped victims. As a result of this deficiency, FEMA established 28 urban search and rescue task forces that are capable of responding to disasters anywhere in the United States within hours of a disaster (Collins 2002). The training of Haitians by GoH and INGO's on how to assess buildings and the training of Haitians in engineering and construction by the USAID are steps in the right direction (Olshansky and Etienne 2011).

Fifth, one of the most important needs of Haitians according to recent polls is jobs (Kaufman 2012). A way to fulfill this need is to make sure that sustainable development projects are designed in such a way that they provide jobs for Haitians. After the 1985 earthquakes that hit Mexico City, Mexico, Renovacion Habitacional Popular created 115,000 jobs and filled these jobs with local people during the reconstruction process and as a result helped to strengthen the local economy (Pantelic 1991). In the case of Haiti, job-creating development projects are beginning to emerge. A good example of job creation for Haitians during the reconstruction phase is the establishment of Caracol Industrial Park, which is capable of providing 65,000 jobs to Haitians (Kaufman 2012). While there are some problems associated with the establishment of Caracol Industrial Park—low wages, location in an environmentally sensitive area, and deforestation—these problems can be ameliorated (Sontag 2012). Some potential avenues for fostering the success of the park include building affordable and environmentally-friendly homes near the park and developing sustainable initiatives such as planting trees and the recycling of wastes from Caracol Industrial Park. One objective that should be made clear in the establishment of jobs, such as at Caracol, is that the GoH and the international community must ensure that the jobs created during the redevelopment process are being filled by Haitians and Haitians earn good wages for their labor.

Sixth, the GoH has to depopulate Port-au-Prince. Prior to the earthquake, more than half of the 9.8 million Haitians lived in urban areas. As many as 2.1 million lived in Port-au-Prince (CIA 2012), a city initially built for only 50,000 residents (Joseph 2010). The overpopulation created an unprecedented demand for natural resources like firewood and as a result increased Haiti's vulnerability to earthquakes and other disasters, as witnessed in 2010. The GoH should adopt land-use planning "to mitigate disasters and reduce risks by discouraging high-density settlements and construction of key installations in hazard-prone areas, control of population density and expansion ..." (UNISDR 2004). A complementary strategy to land-use planning is to provide job opportunities and infrastructure—affordable housing, roads, schools, communication systems, water, power, etc.—in other less populated but safe areas of Haiti.

Lastly, it is necessary to rehabilitate the environment as part of an effective sustainable development program. For instance, trees should be planted to serve as buffer zones for hurricane-force winds (Mainka and McNeely 2011) and wetlands should be revitalized so that they can help to soak up hurricane storm surges. Additionally, land use planning should be used to relocate homes in vulnerable areas to safer grounds and to restrict construction in environmentally sensitive areas of Haiti, such as wetlands and along slopes (Pantelic 1991).

23.10   Conclusion

There is no gainsaying the fact that the 2010 Haiti earthquake is one of the most catastrophic natural incidents in recent history. The impact of this earthquake was exacerbated due to the inherent vulnerabilities of Haiti—geographical location, environmental degradation, and poverty. The colossal devastation caused by this disaster was met by a commensurate humanitarian response from the US, UN, EU, INGOs, CDEMA, etc. The window of opportunity created by the earthquake and the inauguration of a democratically elected President can provide the needed impetus to lift Haiti up from the doldrums. Before this can happen, Haiti has to overcome some challenges—over reliance on INGOs and its weak emergency management system. After overcoming these challenges, Haiti should follow the aforementioned recommendations, which are based on past disasters in developed and developing countries, to build back better and become more resilient to future disasters, both natural and man-made.

References

Archibold, R. C. (2011). Haiti: Quake's toll rises to 316,000. New York Times. http://www.nytimes.com/2011/01/14/world/americas/14briefs-Haiti.html. Accessed 28 Nov 2012.

Bellerive, J.-M., & Clinton, B. (2010). Finishing Haiti's unfinished work. New York Times. http://www.nytimes.com/2010/07/12/opinion/12clinton-1.html. Accessed 28 Nov 2012.

Berkebile, R., & Hardy, S. (2010). Moving beyond recovery sustainability in rural America. *National Civic Review, 99,* 36–40.

Birkland, T. A. (1997). *After disaster: Agenda setting, public policy, and focusing events.* Washington, D.C.: Georgetown University Press.

Brattberg, E., & Sundelius, B. (2011). Mobilizing for international disaster relief: Comparing U.S. and EU approaches to the 2010 Haiti earthquake. *Journal of Homeland Security and Emergency Management, 8*(1), 1547–7355.

Carlson, W. L., Desir, A., Goetz, S., Hong, S., Jones, S., & White, J. (2011). The Haitian diaspora and education reform in Haiti: Challenges and recommendations. http://sipa.columbia.edu/academics/workshops/documents/SSRC.BureauofHaiti_Capstone.pdf. Accessed 28 Nov 2012.

Central Intelligence Agency (CIA). (2012). The world factbook: Haiti. https://www.cia.gov/library/publications/the-world-factbook/geos/ha.html. Accessed 28 Nov 2012.

Collins, L. (2002). Eight years later: Lessons from the Northridge earthquake. Fire Engineering. http://www.fireengineering.com/articles/print/volume-155/issue-4/departments/the-rescue-company/eight-years-later-lessons-from-the-northridge-earthquake.html. Accessed 28 Nov 2012.

Collins, A. E. (2009). *Disaster and development*. New York: Routledge.

Fordyce, E., Sadiq, A. A., & Chikoto, G. (2012). Haiti's emergency management: A case of regional support, challenges, opportunities, and recommendations for the future. In D. McEntire (Ed.), *Comparative emergency management: Understanding disaster policies, organizations, and initiatives from around the world*. http://www.training.fema.gov/EMIWeb/edu/CompEmMgmtBookProject.asp. Accessed 28 Nov 2012.

Government of the Republic of Haiti (GoH). (2010). Action plan for national recovery and development of Haiti: Immediate key initiatives for the future. http://haiti.org/files/Haiti_Action_Plan.pdf. Accessed 28 Nov 2012.

Homeland Security News Wire. (2012). Haiti implements national plan for disaster risk reduction. http://www.homelandsecuritynewswire.com/srdisasters20120328-haiti-implements-national-plan-for-disaster-risk-reduction. Accessed 28 Nov 2012.

Joseph, R. (2010). We're going to have some order in Haiti. PBS NEWSHOUR. http://www.pbs.org/newshour/bb/weather/jan-june10/haiti5_01-15.html. Accessed 28 Nov 2012.

Kapucu, N., & Ozerdem, A. (2013). *Managing emergencies and crises*. Burlington: Jones and Bartlett Learning.

Kaufman, S. (2012). "New traction" in place for Haiti's quake recovery. http://iipdigital.usembassy.gov/st/english/article/2012/01/20120110175117nehpets0.2364923.html#axzz2BusTPYY7. Accessed 28 Nov 2012.

Klose, C. D., & Webersik, C. (2010). Long-term impacts of tropical storms and earthquakes on human population growth in Haiti and the Dominican Republic. Nature Precedings. http://dx.doi.org/10.1038/npre.2010.4737.1. Accessed 28 Nov 2012.

Kristoff, M. (2010). Haiti: A republic of NGOs? PEACEBRIEF. http://www.usip.org/files/resources/PB%2023%20Haiti%20a%20Republic%20of%20NGOs.pdf. Accessed 28 Nov 2012.

Lacey, M. (2010). Estimates of quake damage in Haiti increase by billions. New York Times. http://www.nytimes.com/2010/02/17/world/americas/17haiti.html. Accessed 28 Nov 2012.

Lawry, S. (2010). *Building back better: Revisiting the roles of government, donors, and INGOs in Haiti's reconstruction. Humanitarian & development NGOs domain. Connecting scholars and practitioners. Catalyzing reflection and exchange*. Boston: Hauser Center for Nonprofit Organizations: Harvard University. http://hausercenter.org/iha/2010/01/31/building-back-better-revisiting-the-roles-of-government-donors-and-ingos-in-haitis-reconstruction/. Accessed 28 Nov 2012.

Lundahl, M. (1989). History as an obstacle to change: The case of Haiti. *Journal of Interamerican Studies and World Affairs, 31*(1/2), 1–21.

Mainka, S. A., & McNeely, J. (2011). Ecosystem considerations for post-disaster recovery: Lessons from China, Pakistan, and elsewhere for recovery planning in Haiti. *Ecology and Society, 16*(1), 13.

McEntire, D. (2009). *Introduction to homeland security: Understanding terrorism with an emergency management perspective*. Hoboken: Wiley.

McEntire, D., Sadiq, A. A., & Gupta, K. (2012). Unidentified bodies and mass-fatality management in Haiti: A case study of the January 2010 earthquake with a cross-cultural comparison. *International Journal of Mass Emergencies and Disasters, 30*(3), 301–327.

New America Media. (2010). Poll of the Haitian diaspora on the earthquake. http://news.newamericamedia.org/news/view_article.html?article_id=3b09ad4d30ff54b0d4ad8f72c5c5ea44. Accessed 28 Nov 2012.

Olshansky, R. B., & Etienne, H. F. (2011). Setting the stage for long-term recovery in Haiti. *Earthquake Spectra, 27*, S463–S486.

Pantelic, J. (1991). The link between reconstruction and development. *Land Use Policy, 8*, 343–347.

Poncelet, J. L. (1997). Disaster management in the Caribbean. *Disasters, 21*(3), 267–279.

Sadiq, A. A., & McEntire, D. (2012). Mass fatality management challenges: A case study of the 2010 Haiti earthquake. *Journal of Emergency Management, 10*(6), 459–471.

Slagle, T., & Rubenstein, M. (2012). Climate change in Haiti. State of the Planet. http://blogs. ei.columbia.edu/2012/02/01/climate-change-in-haiti/. Accessed 28 Nov 2012.

Sontag, D. (2012). Earthquake relief where Haiti wasn't broken. New York Times. http://www. nytimes.com/2012/07/06/world/americas/earthquake-relief-where-haiti-wasnt-broken. html?pagewanted=all. Accessed 28 Nov 2012.

Stephenson, R. S. (1994). Disaster and development. http://iaemeuropa.terapad.com/resources/8959/ assets/documents/UN%20DMTP%20-%20Disaster%20&%20Development.pdf. Accessed 28 Nov 2012.

The Telegraph. (2010). Haiti cholera outbreak: Country factfile. http://www.telegraph.co.uk/news/ worldnews/centralamericaandthecaribbean/haiti/8140452/Haiti-cholera-outbreak-country-factfile.html. Accessed 28 Nov 2012.

United Nations. (1987). Report of the world commission on environment and development: Our common future. http://conspect.nl/pdf/Our_Common_Future-Brundtland_Report_1987.pdf. Accessed 28 Nov 2012.

United Nations Economic Commission for Europe (UNECE). (n. d.). Sustainable development—concept and action. http://www.unece.org/oes/nutshell/2004-2005/focus_sustainable_development.html. Accessed 28 Nov 2012.

United Nations International Strategy for Disaster Reduction (UNISDR). (2004). Terminology: Basic terms of disaster risk reduction. http://www.unisdr.org/files/7817_7819isdrterminolo gy11.pdf. Accessed 28 Nov 2012.

United Nations Office for the Coordination of Humanitarian Affairs (OCHA). (2012). The cluster approach. http://ochanet.unocha.org/p/Documents/120320_OOM-ClusterApproach_eng.pdf. Accessed 28 Nov 2012.

United States Department of State (USDS). (2011). Fast fact on U.S. government's work in Haiti: Funding. http://www.state.gov/p/wha/rls/fs/2011/154143.htm. Accessed 28 Nov 2012.

United States Geological Survey (USGS). (2010). "Poster of the Seismicity of the Caribbean Plate and Vicinity." http://earthquake.usgs.gov/earthquakes/eqarchives/poster/regions/caribbean. php. Accessed 24 Feb 2012.

United States Geological Survey (USGS). (2011). Magnitude 7.0—HAITI region 2010 January 12 21:53:10 UTC. http://earthquake.usgs.gov/earthquakes/recenteqsww/Quakes/us2010rja6. php#summary. Accessed 28 Nov 2012.

Wdowinsky, S., Tsukanov, I., Hong, S., & Amelung, F. (2010). Triggering of the 2010 Haiti earthquake by hurricanes and possibly deforestation. *American Geophysical Union*, Fall Meeting.

Weisenfeld, P. E. (2011). Successes and challenges of the Haiti earthquake response: The experience of USAID. *Emory International Law Review, 25*, 1097.

World Bank. (2012). The great East Japan earthquake: Learning from megadisasters. The World Bank. http://www.gfdrr.org/gfdrr/sites/gfdrr.org/files/publication/KNs_Learning_from_Megadisasters. pdf. Accessed 28 Nov 2012.

Chapter 24
Post-Wenchuan Earthquake Reconstruction and Development in China

Ping Xu, Xiaoli Lu, Kelvin Zuo and Huan Zhang

24.1 Introduction

At 2:28 p.m. on May 12, 2008, a deadly earthquake that measured at 8.0-magnitude hit Wenchuan county, Sichuan province, in Western China. The earthquake and numerous aftershocks caused secondary disasters such as landslides, avalanches, debris flow, and formation of barrier lakes (Kapucu 2011; Shi et al. 2010). As one of the most damaging catastrophes in contemporary China, the earthquake resulted in 69,227 deaths, 374,643 injuries, 17,923 missing, and an estimated direct economic loss of 845.2 billion RMB (State Council of the PRC 2008b). It affected lives of more than 120 million people, left 5 million people homeless, and caused extensive damages to the economy and local critical infrastructures.[1]

---

[1] As a result of the earthquakes, large areas of farmland, natural forests and wildlife habitat were lost. Approximately 19.5 million cubic meters of timber were lost as a direct result of the earthquakes (Yang 2008). The direct loss was equivalent to 200 billion RMB in Sichuan, 2 billion RMB in Gansu, and 1.63 billion RMB in Shaanxi (Chen et al. 2008). The disaster area in Sichuan had an industrial base for chemical engineering, which suffered from a direct economic loss of 12 billion RMB as well as over 4,000 casualties and injuries among the industry workers (Shaanxi Provincial Development and Reform Commission 2008).

P. Xu (✉)
Department of Political Science, Gender and Women's Studies,
University of Rhode Island, Kingston, USA
e-mail: pingxu@mail.uti.edu

X. Lu
Center for Crisis Management Research, School of Public Policy and Management,
Tsinghua University, Beijing, China

K. Zuo
Department of Civil and Environmental Engineering, University of Auckland,
Auckland, New Zealand

H. Zhang
School of Social Development and Public Policy, Beijing Normal University, Beijing, China

N. Kapucu, K. T. Liou (eds.), *Disaster and Development,* Environmental Hazards,
DOI 10.1007/978-3-319-04468-2_24, © Springer International Publishing Switzerland 2014

The most severely damaged zones covered 51 townships in Sichuan, Gansu, and Shaanxi Provinces. Some counties near the epicenter, such as Beichuan and Wenchuan, were literally flattened. The affected communities were spread throughout more than 400 cities and towns, totaling 500,000 km^2 (State Council of the PRC 2008a, b). After the earthquake, it was estimated that more than 6 million houses in rural areas and 102 million square meters of apartments in urban areas required reinforcement or rebuilding (State Council of the PRC 2008b). In the meantime, critical infrastructures and public facilities, such as public transportation, power grids, telecommunications, and water supply systems, were also in need of recovery.

Such deadly losses were the result of both the natural disasters and the long-existing vulnerability in the earthquake stricken communities. The region was relatively underdeveloped and remote, and the previous developmental models were unsustainable due to the pollution caused by unregulated industrial factories. All these factors directly increased vulnerabilities of the region in terms of its low preparedness and high environmental risks.

After the earthquake, the Chinese government issued a series of policies aimed at re-building and re-developing the earthquake stricken areas. Post-disaster reconstruction was concluded in less than 3 years, producing positive economic and social outcomes (Dunford and Li 2011; The United Nations Office for Disaster Risk Reduction (UNISDR) 2010). According to the State Council (2012), a total of 1.7 trillion RMB was invested to achieve six major goals of reconstruction: housing for every family, job stability for at least one family member, basic welfare, economic, infrastructure, and ecological improvements in the affected areas. How did the Chinese government achieve these ambitious goals in such a short period of time after a highly destructive natural disaster? In this chapter, we attempt to answer this question by analyzing reconstruction and developmental policies and strategies, and studying the implementation of reconstruction efforts in the case of Shuimo town.

The chapter starts with an introduction of the pre-disaster conditions of the region. We contend that poverty and imprudent developmental models directly resulted in vulnerabilities to natural disasters in the region. In Sect. 2, we examine the major post-earthquake developmental strategies adopted by the Chinese government. We maintain that both the counterpart assistance program and the emphasis of sustainable development result in positive outcomes of the reconstruction. In Sect. 3, we use the post-disaster development of Shuimo town as a case study to illustrate how these two strategies were utilized to shape a sustainable development model in Shuimo. In the last section, we provide concluding remarks, critics, and recommendations.

24.2   Underdevelopment and Vulnerabilities to Disasters

Previous literature of emergency management pointed out that the impact of a disaster results from not only the magnitude of the natural hazard event, but also the vulnerability of the disaster hit area (see Dunford and Li 2011, p. 999; Strömberg

2007; Wisner et al. 2004). Indeed, underdevelopment and imprudent developmental models had caused high levels of vulnerability in the stricken region prior to the earthquake (see Chap. 2 in this volume).

24.2.1   Underdevelopment of the Region

Beginning in 1978, Deng Xiaoping initiated the "open-up reform" by introducing market economy to the country. The East coast region of China has enjoyed special economic development policies and experienced rapid economic development. However, the West region[2], which the Wenchuan earthquake heavily impacted, has remained relatively poor, undeveloped, and stayed as a piece of "forgotten land." In most regions, agriculture has been the largest sector of the local economy. For example, in some counties in Sichuan Province, the percentage of agriculture related labor exceeded 73 % of the total labor force (Sichuan Provincial Bureau of Statistics 2006). As of 2008, the West region occupied 56.8 % of the total land area of China, but its GDP only accounted for 14.3 %, with its per capita GDP approximately equivalent to only 40 % of that in the East (Jin 2008). It is estimated that West China fell behind the economic development of the East Coast by at least 20 years. In 2008, the annual per capita net income of rural residents in Sichuan was only $ 593, and the annual per capita disposable income of the urbanites was $ 1,819 (ChinaView 2008; Xinhua English News 2006). The GDP per capita of Sichuan, Shaanxi, and Gansu ranked 24th, 18th, and 29th, respectively, out of China's 31 provinces and municipalities (China National Bureau of Statistics 2011).

Underdevelopment in this region directly contributed to the vulnerability of communities to natural disasters. The remote mountain areas lacked basic transportation infrastructures to connect to adjacent regions. Numerous residential houses and school buildings failed to sufficiently adopt safety protocols and standards. The whole region did not prepare enough emergency supplies such as tents, medicine, first aid items, and did not have clear contingency plans for potential disruptions to essential utilities such as electricity, water, and telecommunication systems.

24.2.2   Unsustainable Developmental Models

While experiencing rapid economic development throughout the past three decades, China's East coastal region also witnessed heavy pollution and increased labor costs (Zhang 2009).[3] As a result of these negative consequences, many high-energy-

---

[2] According to the definition of the Chinese government, West China covers six provinces: Gansu, Guizhou, Qinghai, Shaanxi, Sichuan, and Yunnan; one municipality: Chongqing; and six autonomous regions: Ningxia, Tibet, Guangxi, Inner Mongolia and Xinjiang; parts of Hunan and Hubei provinces. The earthquake stricken area all belonged to West China.

[3] For instance, Guangdong province reported 5,490 cases of occupational diseases from 1989 to 2004, including 2,418 cases of lung disease caused by air pollution. The cost of lung diseases caused by dust air was up to 200 million RMB (Xian Zhang et al. 2009).

consuming and high-polluting enterprises were relocated from the East coastal region to the relatively underdeveloped West region (Zhang et al. 2009). For example, in 2008, Guangdong province in the Eastern coastal region, encouraged hundreds of high-energy-consuming and high-polluting enterprises to relocate out of Guangdong (Zhang et al. 2009).

Since the West had only experienced a relatively short period of development, local governments have not fully realized the importance of a sustainable development model.[4] In China, an important criterion for local officials' career advancement is the performance of local economic development. Such a criterion was largely based on the number of projects and enterprises established and the growth rate of local GDP figures (Landry 2008). Therefore, even though some of the developmental projects were visionless and imprudent, many local governments still competed to attract more investment, build more factories, and construct more buildings (Landry 2008; Yin 2001). When the East was eager to relocate their high-energy-consuming and high-polluting enterprises out of their own region, the West was more than happy to host them.

Due to such imprudent developmental models, the economy in the West was not only underdeveloped, but also poorly structured. Heavy industry, especially high-energy-consuming and high-polluting enterprises, dominated the economy in many regions. For example, the iron and steel industry in Sichuan was ranked second among all provinces in China (China Guide 2012; China Perspective 2011). In 2009, Shaanxi ranked third in China for its production of coal, natural gas and crude oil (China Perspective 2012b). In Gansu, most of the province's economy was based on mining and mineral extraction (China Perspective 2012a). One *Industrial Demonstration Zone* in Aba Prefecture (where the earthquake heavily impacted) in Sichuan had 63 high-energy-consuming and high-polluting enterprises, all of which discharged wastewater and emissions to the environment.

When the earthquake hit the region, the highly concentrated polluting industry caused more damage to the environment. For instance, pipelines in a number of chemical plants broke during the earthquake releasing poisonous gases like liquid chlorine and ammonia (Jin 2008). In addition, high-energy-consuming and high-polluting enterprises in this region released wastewater, emissions and other pollutant and made the ecological system extremely fragile (He and Jiang 2008; Liu 2008; Liu 2003).

---

[4] It was not until 2000 when the State Council led by then Premier Zhu Rongji decided to develop the economy in the West in order to catch up with the East coast. The State Council established a leadership group in charge of the "Western Development Campaign" Which was written in law in 2004. Before 2008, several significant projects were completed to help develop the education, transportation and energy industries in the West. For example, in 2005, the Chinese government exempted the tuition and fees for students in their 9-year compulsory education. In 2006, a railway from Qinghai to Tibet was built.

24.3 Post-Wenchuan Reconstruction and Development Policies and Strategies

Even though post-Wenchuan reconstruction was full of challenges, it was completed within 3 years and generated positive social and economic outcomes in many regions. Using post-disaster reconstruction as an opportunity, many towns and cities set their development on a much more sustainable path. Our analysis of why this could happen in China in such a short period of time led us to two important factors—the counterpart assistance program and the emphasis of sustainable development models.

Shortly after the earthquake, the Chinese central government collected suggestions from experts in various fields and drafted guidelines for the reconstruction.[5] These guidelines were finished and posted on all major medium outlets (i.e., major TV stations, web sites, etc.) on August 12, 2008. After receiving public suggestions and comments, the State Council publicized the final version of the guidelines, the *Post-Wenchuan Earthquake Recovery and Reconstruction Master Plan* (referred to as the *Master Plan* aforementioned) on September 19, 2008. This *Master Plan* served as an ultimate blueprint for all stakeholders involved in the reconstruction process. It included general principles and instructions on rebuilding infrastructure, residential houses, historical sites, public service facilities, industries, and spiritual homes; as well as policies in relating to ecological and environmental protection and disaster mitigation.

This *Master Plan* clearly stated the timeline and goals of the reconstruction, which included six major goals to be achieved within 3 years. The six major goals are: (1) rebuilding a house or apartment for every family; (2) ensuring the job stability for at least one member of each family, with annual disposable personal income exceeding the pre-disaster level; (3) providing basic social welfare for disaster survivors, i.e., 9-year free public education, public health and basic medical care, social welfare and other basic public services; (4) restoring and upgrading public facilities and infrastructures; (5) further developing the economy of earthquake stricken area; (6) improving the ecological environment with enhancements in ecology, environment, disaster mitigation and preparedness capacities (State Council of the PRC 2008b; Zuo 2010). Two characteristics of the *Master Plan* stood out in facilitating the development of the disaster hit region, the counterpart assistance program and the emphasis of sustainable development.

---

[5] China has a top-down unitary government system; therefore, the Chinese central government has the ultimate power of directing sub-national governments. Since the earthquake impacted multiple provinces and the reconstruction required participation by stakeholders from governments at different levels, it naturally became the central government's call to make the reconstruction guidelines.

24.3.1   The Counterpart-Assistance Program

Counterpart assistance is a resource allocation mechanism that promotes fast development in one area or one policy domain by devoting resources from other areas or policy domains. It existed in China before the Wenchuan Earthquake and had been used in three manners: assisting the development of border areas and minority areas, supporting key infrastructure projects (such as the Three Gorges Dam), and supporting disaster response and reconstruction (Wang and Dong 2010).

The use of the counterpart assistance program in disaster reconstruction started in the 1950s and became salient in responding to the 1978 drought disaster in Hubei Province (Zhong 2011).[6] In the 1980s, central ministries and local disaster stricken areas were paired up in the counterpart assistance practice. For instance, after the 1991 Tai Lake flood, the central government instructed ministries to counterpart-assist disaster stricken areas in Anhui province. In 2007, the administrative procedure of using counterpart assistance in disasters was specified in the *Emergency Response Law*. When an area is affected by an emergent incident and support in rehabilitation and reconstruction from the higher level is needed, the local government can submit a request to the government at the next higher level for support. According to the losses suffered by the affected area and its actual condition, the governing authority at the next higher level shall provide financial and material support, technical guidance, and mobilize other regions to provide support in terms of external funding, materials, and human resource (Emergency Response Law of the People's Republic of China 2007).

After the 2008 Wenchuan Earthquake, the central government decided to mobilize the counterpart assistance program (2006). The details of the practice were explained in the *Master Plan* and *Post-Wenchuan Earthquake Recovery and Reconstruction Counterpart Assistance Program*. In this practice, 20 economically developed provinces, unaffected by the earthquake, were assigned to assist 18 counties and severely afflicted areas in Gansu and Shaanxi Provinces. Table 24.1 shows the list of donor counterpart provinces and their paired-up disaster hit counties. For example, Shandong province was paired up with Beichuan County, and Guangdong province was paired up with Wenchuan County.

In the counterpart assistance practice, the donor provinces were required to spend at least 1 % of the governmental revenue from the previous year to assist their paired-up disaster-hit areas for three consecutive years. The donor provinces should provide assistance in a full range of services, which include: (1) planning and designing, construction, expert consulting, project construction, inspection, (2) renovation of residential communities, (3) restoring and upgrading public facilities and infrastructures, such as schools, hospitals, roads, water, gas, disposal and

---

[6] In 1950s, counterpart assistance in disaster reconstruction was on a small scale (scattered mutual assistance occurred between military and communities, and between different areas). In the 1978 drought disaster in Hubei Province, counterpart assistance was applied on a much larger scale (major state-owned enterprises and non-disasters stricken areas provided counterpart assistance to disaster stricken areas).

Table 24.1 The counterpart assistance programs. (Source: State Council 2008a)

Disaster Hit counties	Counterpart provinces	Disaster Hit counties	Counterpart provinces
Beichuan county	Shandong	Maoxian county	Shanxi
Wenchuan county	Guangdong	Lixian county	Hunan
Qingchuan	Zhejiang	Heishui county	Jilin
Mianzhu city	Jiangsu	Songpan county	Anhui
Shifang city	Beijing	Xiaojin county	Jiangxi
Dujiangyan city	Shanghai	Hanyuan county	Hubei
Pingwu county	Hebei	Chongzhou city	Chongqing
Anxian county	Liaoning	Jiange county	Heilongjiang
Jiangyou city	Henan	Severely afflicted area in Gansu province	Guangdong (mainly Shenzhen)
Pengzhou city	Fujian	Severely afflicted area in Shaanxi province	Tianjin

sewage, social welfare, and other public services, (4) selecting and sending doctors and teachers to assist hospitals and schools in the disaster hit areas, and providing equipment, tools and facilities, construction materials to help the reconstruction, (5) providing training services, education and schooling to students who cannot go back to school at home, and (6) restoring basic infrastructure for trading and operational centers (State Council of the PRC 2008a).

Counterpart assistance served as one of the key financial sources for post catastrophe reconstruction and recovery in China.[7] During the reconstruction after the Wenchuan Earthquake, donor provinces provided over 80 billion RMB via counterpart assistance service by the end of September in 2011. Funds from assisting counterparts were used in 3,668 reconstruction projects, 3,662 of which were completed by 2011 (Sichuan Provincial Government 2011). After the Wenchuan Earthquake, the counterpart assistance in catastrophe reconstruction has been institutionalized as a reconstruction assistance model and was written in the revision of the *Law of the People's Republic of China on Protecting against and Mitigating Earthquake Disasters* in 2008. After 2008, counterpart assistance in disasters was used in other major catastrophe reconstruction in China.[8]

The benefits of a counterpart assistance program lie in the following aspects. First of all, counterpart assistance could mobilize resources quickly, and expedites the process of recovery and reconstructions in targeted disaster-impacted areas. Secondly, the command and coordination from the higher level (i.e., the central government in the Wenchuan case) under the counterpart assistance mechanism could balance available resources out evenly to the disaster region, and therefore reduce the resource convergence in some disaster-impacted areas with more public

---

[7] The other channels of funding for recovery and reconstruction included budget allocation of local government, social donations, domestic bank loans, capital market financing, foreign emergency loans, urban and rural self-possessed and self-collected funds, self-possessed and self-collected funds of enterprises, and innovation financing.

[8] Counterpart assistance in disasters was soon applied in the reconstruction of Yushu Earthquake in Qinghai Province in 2010.

attention. Thirdly, counterpart assistance can potentially help bring in advanced technology and quality services from the economically developed provinces to the affected areas. For instance, in the counterpart assistance between Beijing and Shifang, education institutions in Beijing signed an agreement with 35 schools in Shifang to help improve teaching and research quality (Sichuan Government Website 2009). Fourthly, in order to sustain local development after the 3-year counterpart assistance, some donor provinces created job opportunities for disaster survivors in the disaster-affected areas. For instance, Shanxi Province provided 4,000 jobs for Maoxian County and helped 482 local residents to work in other parts of the country through the counterpart assistance program in the first year after the earthquake (Zhang 2009). Last but not the least, in practice, the counterpart assistance program extended beyond the disaster recovery time period. Some donor provinces even built up long-term cooperation relationships with the recipient local governments regarding future local development. For instance, Guangdong province signed a long-term agreement with Wenchuan county, named *"Guangdong Wenchuan Long Term Cooperation Framework Agreement,"* to assist Wenchuan's long-term development after the recovery and reconstruction time period. Under the framework, Guangdong will continue to assist Wenchuan county in areas such as labor migration and tourism development (Li 2010).

Nonetheless, the counterpart assistance program has some flaws in its design and could be problematic during its implementation. First of all, the central government's regulation that donor provinces must allocate 1 % of last year's revenue to disaster-hit areas was at odds with the *Budget Law*. Article 13 of the *Budget Law* specifies that the budget can only be approved by People's Congress at the corresponding level, yet there was no specification in the budget law to allocate funding for disaster assistance in other areas (Wang and Dong 2010).[9] In its implementation, the absolute allocation of 1 % revenue exerted financial burdens to some relatively underdeveloped donor provinces. Moreover, the political mobilization in counterpart assistance programs could easily trigger competition between donor provinces/cities, sometimes causing unnecessary wasteful investment. Thirdly, the donor counterparts could sometimes dominate the reconstruction process because they have more financial and human resources, as well as expertise. The disaster-hit regions, however, often times lose most of its capacities during the earthquake. Therefore, it could be a common phenomenon that reconstruction designs lack considerations of local conditions.

---

[9] Article 32 of the *Budget Law* specifies: "Reserve funds in government budgets at various levels shall be established at a ratio of 1–3 % of the budgetary expenditures at the corresponding level for coping with the relief for natural calamities and other unexpected expenditures in the implementation of the current year's budgets." This budget for disaster relief did not mention that it could be used in other areas. Article 31 writes on funds for assisting other areas: "Necessary funds shall be arranged in the central and relevant local budgets to assist the developing areas such as areas of regional national autonomy, old revolutionary bases and outlying and poverty-stricken areas, in developing undertakings of economy and culture." However, it did not mention that these funds could be used to assist disaster reconstruction in other areas.

24.3.2   Emphasis of Sustainable Development

The other positive element of the *Master Plan* was its emphasis of sustainable development models in various phases of the reconstruction process. First and foremost, the *Master Plan* required the reconstruction teams to consider the sustainability of the industry as a top priority to town/cities reconstructions when they design reconstruction plans. As mentioned, historically, many highly energy-consuming and highly polluting enterprises were relocated to West China from East Coast area. The *Master Plan,* therefore, encouraged towns/cities under reconstruction to phase out the highly polluting and high-energy-consumption industry and to eliminate mineral mining enterprises that are not in compliance with safety requirements as well as the enterprises that contaminate drinking water sources. Considering that the slow recovery of the agricultural sector due to the destruction of large areas of farmland, the *Master Plan* required the government to rearrange and optimize the industry structure. This shifted the focus from agriculture to other industries, such as manufacturing, education, and tourism (Disaster Relief Expert Panel 2008).

Secondly, the *Master Plan* emphasized environmental protection and preservation of historical and cultural heritages. Based on the *Master Plan,* the reconstruction team was required to "respect" nature, to fully consider the environmental and resource capacity, as well as exposure to potential hazards and other risks when choosing reconstruction sites and considering their functions (i.e., population distribution, industry structure and productivity). The *Master Plan* also required stakeholders involved in the reconstruction process to protect ethnic and traditional cultures and heritages, protect valuable historical and ethnic architectures, and preserve traditional appearances of cities, towns and villages. For instance, reconstruction sites should be selected away from all natural reserves, historical and cultural sites, and water source protection sites. Stakeholders involved in reconstruction processes should be strictly required to protect farmland, forests, and be environmental friendly. Simultaneously, reconstruction units should establish environment protection facilities as soon as possible. At the same time, the *Master Plan* highly encourages enterprises to recycle and reuse various construction materials, industrial solid waste, and develop new environmental friendly wall materials. It also called for everyone to save resources and avoid extravagant and wasteful development.

Thirdly, the disaster stricken areas were encouraged to lead their own development, instead of relying on external assistance in the long run. The *Master Plan* prioritized the restoration of capabilities of local government agencies so that they could actively participate in the reconstruction process. It also encouraged the local governments and residents to take an active role in the decision-making process, to determine the distribution of resources, as well as to suggest solutions to problems encountered in the process. Local governments and residents were also encouraged to be self-reliant. In contrast, over reliance on superior government agencies or external help was highly discouraged.

Last but not the least, in order to build up a more resilient system, local governments were required to develop their capacities in identifying and coping with future potential hazards. Natural hazard observing centers, disaster prediction and information centers, disaster reduction and mitigation centers, shelters and disaster education centers were required to be set up in the disaster areas.

Even though sustainable development models were emphasized throughout the *Master Plan*, they were not always taken into consideration in the reconstruction process due to the time constraints. The government set 3 years as the time frame to finish the reconstruction, which evolved into a political task. All reconstruction teams were trying to rush through and meet up with the time requirement set by the *Master Plan*. As a result, the goal simply shifted to building up a new town or city that can quickly resume its own economic production. It was very typical for reconstruction teams in different areas to develop one standard model, and reconstruct different towns with different geographic features and historical backgrounds into one uniform style. The designs were often detached from the region's original cultural and social backgrounds and broke up the continuity of the regional development (Jin 2008). The counterpart assistance program some-times a counter factor of sustainable development. Since the donor counterparts were from a more developed region, they had more developmental experience and could dominate the reconstruction process. They did not always fully understand local conditions, and therefore were often reluctant to incorporate suggestions from local residents and governments.

24.4   The Post-earthquake Reconstruction of Shuimo Township: A Case Study

The reconstruction of Shuimo Town is a relatively successful post-disaster development case. The post-disaster reconstruction has transformed Shuimo into a modern artistic and sustainable town. According to a former United Nations Environment Program official, Shuimo "highlighted ecologically-friendly and low-carbon concepts," and featured "reconstruction projects that were well-integrated with people's lives" (Xinhua News 2012). In 2011, Shuimo Town received the "Best Global Implementation of Post-Disaster Reconstruction" award from the Sixth Global Forum on Human Settlements of the United Nations (Xinhua English News 2011). This section explores how the counterpart assistance program and the emphasis of sustainable development have contributed to such positive outcomes of the post-disaster reconstruction.

Before the earthquake, Shuimo Town was a mountainous small town with a population of approximately 15,000. It had the only high-energy-consumption industrial zone in the Aba Tibetan and Qiang Autonomous Prefecture, with 63 high-energy-consuming and high-polluting enterprises discharging wastewater and emissions year round. Due to the fact that the township boasted abundant hydropower resources that provided low-price electricity, numerous family-run or village-run small businesses mushroomed in service of these large enterprises, exacerbating the pollution. Seriously polluted air and water in Shuimo not only disrupted local agricultural production, but also impacted health conditions of local residents.[10]

Shuimo was only about 10 km from the epicenter, Yingxiu Town in Wenchuan County. During the earthquake, 20 % of the local residential houses collapsed and another 55 % were severely damaged. Most local public service facilities and infrastructures were wiped out (Liu 2011, p. 170). The earthquake also ruined critical transportation infrastructure -destroying the only highway and leaving the partially damaged Shuimo Bridge as the only connection to the outside world. After the earthquake, traffic jams often stretched as long as 30–40 km on the bridge (Liu 2011, p. 15).[11]

24.4.1   The Counterpart Assistance Program in Shuimo

Under the counterpart-assistance program, Guangdong province, paired up with Wenchuan County, further broke down the assistance task and paired its 13 prefecture cities up with 13 townships in Wenchuan County. In this practice, Shuimo Township was paired up with Foshan City[12]. Assistance provided by Foshan was crucial to Shuimo's successful transformation in the reconstruction process. Foshan City not only helped raising a total of 3 billion RMB for Shuimo's reconstruction, but also actively participated in the actual design and reconstruction process. The assistance team from Foshan City had a wealth of urban development experience. Its leader, Dr. Liu Hongbao, former Director and Party Secretary of the Foshan Development and Reform Bureau, holds a doctorate degree in engineering and has extensive experience in regional economic development and transition.

First and foremost, the counterpart assistance program played a crucial role in raising funds for the reconstruction of Shuimo. Initially, Foshan City itself made a preliminary pledge to commit RMB 640 million to the reconstruction of Shuimo Town, a figure exceeding Shuimo's 60 years'annual revenue before the year 2008 (Liu 2011, p. 33). During the process of reconstruction, the reconstruction team realized that 640 million was far from enough so that they decided to raise funds from other sources. They soon found out that their home province, Guangdong Province, had set up a separate provincial assistance fund to support the reconstruction of key townships. By presenting their well-designed reconstruction plan to Guangdong province, Foshan reconstruction team successfully made Shuimo a third key

---

[10] According to relevant surveys conducted by the Foshan medical assistance team, more than 300 children of the small township with a population of less than 20,000 suffer from congenital heart diseases (Liu 2011, p. 18).

[11] Shuimo town was connected to the outside world only through a secluded mountainous road and a 5-meter-wide bridge called "Shuimo Bridge."

[12] Foshan City, located in central Guangdong, was the third largest industrial and manufacturing base in the Pearl River Delta. It benefited tremendously from the open-up policy and enjoyed a per capital GDP of 80,579 RMB (equivalent of about $ 13.000) in 2009.

township and was awarded around 200 million RMB provincial assistance.[13] In addition, Foshan City also raised nearly 200 million RMB of public donations from various sources in Foshan. The overall assistance funding contributed by Foshan towards Shuimo's reconstruction reached 1.07 billion RMB.

Secondly, Foshan assistance team helped bring resources and ideas to create a reconstruction plan. The goal of the reconstruction was to shift Shuimo from a high-energy-consuming and polluting industrial little town to an ecological town with a cultural identity. The Foshan assistance team attached great importance to a suitable reconstruction plan and design. However, when they first arrived at Shuimo, they found that many of the local residents and cadres still lived in tents or temporary housing, and local students yearned to get back to school as soon as possible. Therefore, local people were eager to kick off the reconstruction process. The Foshan assistance team had contradictory opinions with local cadres on another matter. While the local cadres believed that Shuimo should orient its development toward the center of the Wenchuan County, and therefore adopt a strategy "to grow northward," Foshan assistance team suggested a different developmental strategy to "grow southward." Their argument was that among all townships in Wenchuan, Shuimo is the furthest from the center of Wenchuan County, but the closest to the provincial capital Chengdu and a UNESCO World Heritage Site Dujiangyan. By growing southward, Shuimo could be easily integrated into the Chengdu Economic Circle. The Foshan assistance team made great efforts to persuade local residents and cadres that priority should be given to a strategic reconstruction plan and also persuaded that "growing southward" could make Shuimo unique from other towns in Wenchuan (Liu 2011, p. 32).

Experts and professionals brought by the Foshan assistance team designed an initial reconstruction plan. However, they soon found that the plan lacked its own characteristics and resembled other townships' plans. Additionally, they realized that the plan was supposed to function best on relatively flat terrain, yet Shuimo Township was a mountainous town. The Foshan assistance team chose to conduct more research and eventually decided to introduce a bid for a better reconstruction plan of Shuimo. Among numerous proposals sent for the bid, the proposal led by Professor Chen Keshi, Director of Chinese Urban Design Research Center of Peking University and another prestigious urban planning expert, won out (Liu 2011, p. 37).[14] Different from the original plan, Professor Chen's plan was characterized as "One Lake, Two Shores and Four Zones", with the lake as the township center and the bridge as its framework (Liu 2011, p. 172).

This awarded bid outweighed the original plan in many ways. It not only innovatively enriched the historical and cultural nature of Shuimo, but also respected

---

[13] On March 17, 2009, in the "Outstanding Achievements in Planning and Design During Post-Earthquake Reconstruction in Wenchuan" contest organized by the Guangdong Province Assistance Team and Wenchuan County government, the Foshan assistance team won 6 awards including an award in "urban planning for post-earthquake reconstruction" (Liu 2011, p. 49).

[14] The team led by Prof. Chen is affiliated with Chinese Urban Design Research Center, Zhongying Urban and Architectural Design Center and China Southwest Architectural Design & Research Institute.

and improved its ecological environment. In addition, it optimized its industrial structure and created long-term employment opportunities in the tourism industry for local residents. It also combined residential settlements in the short term with the concept of sustainable development in the long run (Chen et al. 2011). The fact that the Foshan assistance team prioritized planning and spent time in selecting a most suitable plan rather than rushing through the reconstruction process provided a solid foundation for the reconstruction.

Lastly, Foshan assistance team actively participated in the actual rebuilding process. The reconstruction of Shuimo relied heavily on the strenuous and painstaking efforts of external assistance, led by the Foshan assistance team. For instance, even though local residents were encouraged to participate in renovating their own residential houses, Foshan assistance team and outside workers were responsible for renovating the street-facing part within 5 m from the edge of the street.

24.4.2 Sustainability and Transforming the Development Models

The success of Shuimo's transformation was not merely due to the counterpart assistance program, but also the consistently emphasized concept of sustainable development. From the design to the actual reconstruction process, sustainability was a core concept of Shuimo's reconstruction. While designing the plan, the reconstruction team intentionally designed Shuimo into a modern ecological town with a sustainable tourism industry. The reconstruction team understood that maintaining a good environment was essential to the sustainable development of the town. Therefore, they made efforts to persuade local cadres and residents on this matter.

After the earthquake, most local high-polluting enterprises soon managed to resume their operation, which posed challenges to the goal of building a sustainable town set by the assistance team. Keeping these enterprises in the town, the reconstruction would only have restored Shuimo to be what it was before the disaster. If the reconstruction team decided to get rid of the polluting enterprises, they faced even bigger questions: how could Shuimo relocate or remove all those enterprises, and what new industries could sustain the local economy and employment? While local residents and cadres were terrified to lose all the jobs in the highly polluting industries, Foshan assistance team persuaded them that Shuimo should develop a more sustainable tourism-oriented economy and eliminate the polluting enterprises. They organized local residents and cadres to tour two successful tourism towns, Lijiang in Yunnan and Luodai in Sichuan. Lijiang was successfully reconstructed into a tourist destination after a 7.0-magnitude earthquake in 1996. These trips deeply impressed representatives of local residents and officials, who could visualize a similar future for Shuimo. Through continuous efforts, the reconstruction team obtained consent of every household in the town (Liu 2011, pp. 60–64). All stakeholders in Shuimo's reconstruction finally agreed that the priority should be given to optimizeing the local industrial structure and transforming the local economic model from a combination of high-energy-consuming industry and family-based agriculture to a modern service-based industry and urban ecological agriculture (Liu 2011, p. 172).

The leader of the reconstruction team, Dr. Liu, paid a visit to the party secretary of the Aba Prefecture in October 2008 and convinced him of a more sustainable development plan for Shuimo. The party secretary was impressed with the plan and eventually decided to officially fund the relocation of the polluting enterprises (Liu 2011, pp. 29–30).

After the removal of the high polluting enterprises, the reconstruction team discovered that the township enjoyed an advantageous geographical location. To its east lies Dujiangyan, a UNESCO World Heritage Site; to its southeast is the famous Qingcheng Mountain, a scenic spot of historic and cultural significance; to its west and north lies respectively the Wolong National Nature Reserve well known for its pandas, and Yingxiu Township, the epicenter of the Wenchuan earthquake, which attracted global attention after the earthquake. Therefore, they believed that Shuimo Township is well positioned to link all those neighboring scenic attractions and form a high-end tourist route. After studying the local historical records, they discovered that the township had been honored as a "longevity village" as early as the Han Dynasty and long hailed as a Shangri-Lain western Sichuan Province (Liu 2011, p. 21).

The reconstruction team also decided that Shuimo would promote the Qiang minority culture instead of Tibetan culture. Even though Tibetans were the largest minority group in Shuimo and Qiang only accounted for 4.19% of the local population, the reconstruction team found that numerous other townships in Wenchuan chose to promote Tibetan culture in the reconstruction plan. The reconstruction team realized that the value of Shuimo lay only in embracing a different culture, and that the Qiang culture should be celebrated as the soul of the township. Besides, the local existence of Qiang ethnic minority can be traced back a few thousand years. In drawing on its rich Qiang history and culture, Shuimo Township would emerge as a unique tourist destination lying in close proximity to Chengdu, the regional metropolis (Liu 2011, p. 30).

Based on numerous surveys and constant reflections on natural, geographic, historic and traditional features of Shuimo, the reconstruction team decided to transform Shuimo into "an ecological town featuring a famous Qiang culture in West China" (Liu 2011, pp. 29–30). The designing proposal that was finally adopted successfully integrated developing tourism as a more sustainable sector, creating employment opportunities, and improving residential settlements all together. For instance, in designing housing projects, this plan used the model of "store front and residence in the back" or "residence upstairs with store downstairs", integrating residential settlement, livelihood, and family business for the tourism together.

Drawing upon lessons learned from the Dujiangyan City, the design also utilized natural watercourses and created a lake called "Shouxi Lake."[15] Strategically, Shuimo would be made a lake-centered mountainous tourist township in Wenchuan County. The design featured local architecture with an extensive use of sloped roofs and traditional Tibetan and Qiang hues using cement, straw and iron circles.

---

[15] Dujiangyan City in Sichuan province used a similar strategy to dredge the sand deeper and build the dam lower, so that the lake could create a tourism site.

These designs not only promoted the Qiang and Tibetan cultures, but also created a sun-proof, water-resistant and durable mud-like effect on the façades of buildings. Qiang artists were also invited to make collages of all kinds of Qiang ethnical patterns to decorate the buildings, with the Qiang tradition and regional culture fully exploited and the traditional Qiang art brilliantly expressed in a modern way (Chen et al. 2011).

The reconstruction team also endeavored to introduce the educational industry to Shuimo. Initially, the Foshan assistance team had difficulties finding schools willing to relocate to Shuimo because of its remote location and heavy pollution. The Foshan assistance team decided to lobby Aba Normal College first, because its ethnic dancing programs, especially their Qiang dancing program, would be a good fit to Shuimo's Qiang cultural identity. The university administration was convinced by Shuimo's visionary development plan. Ma Hongjiang, President of Aba Normal College, remarked that, "Shuimo Town is located in close proximity to Chengdu, and lies in the key area wedged between two UNESCO World Heritage Sites. Besides, the new town has a strong Qiang cultural atmosphere and is therefore ideal for our school to grow" (Liu 2011, p. 53). After Aba Normal College decided to relocate to Shuimo, Sichuan Conservatory of Music soon followed, moving its Research Institute of Tibetan and Qiang Culture to Wenchuan and its School of Tibetan and Qiang Art to Shuimo. The two colleges were almost the only two high education institutions researching Qiang culture. Hosting them made Shuimo well positioned to become a "stronghold of Qiang culture" in West China.

The design of Shuimo's reconstruction plan took 7 months, during which reconstruction operations in other townships had already achieved significant progress.[16] In March 2009, after 7 months in design, the physical aspect of Shuimo's reconstruction finally kicked off and was completed in less than 2 years. Shuimo's reconstruction turned out to be a successful case. The reconstruction of Shuimo Town involved demolition and relocation of 722 households, with 2,520 people affected. This difficult task was nonetheless rapidly accomplished with minimal complaints. The newly built township is now the home for a few colleges and schools, namely, the Wenchuan No.2 Kindergarten, the Bayi Primary School and the Shuimo Middle School; as well as two tertiary education institutes, Aba Normal College and Sichuan Conservatory of Music. Shuimo is now recognized as an important cultural and education center in the Aba Prefecture. Its water-based scenic spots, such as the Shouxi Lake, have effectively restored the township's ecological system. The goal of Shuimo's future development is to achieve sustainable and harmonious coexistence of population, resource and the environment. The living conditions of local residents have improved as well: Shuimo has achieved a per capita annual income of $ 3,000 in 2010 and has become a wealthy township in western Sichuan. Within only 2 years, by taking advantage of external assistance, Shuimo has achieved the development level that might have previously taken 50 years.

---

[16] For instance, the Huizhou assistance team almost completed their reconstruction of the Sanjiang Township.

24.5   Conclusion

The success of Shuimo Town's reconstruction not only lies in the renewed small town's beautiful architecture, state-of-the-art infrastructures, and improved living conditions of local residents, but more importantly, in the sustainable development model that the town has embarked on. The two important pieces of experience that Shuimo can share with the rest of the world in post-disaster development are first of all, a good design bearing "sustainable development," and second, the counterpart assistance program that brought in a wealth of financial and human resources, as well as development expertise and experiences from developed regions.

Foshan City has contributed enormously to the reconstruction of Shuimo Township as a counterpart/donor. However, the post-disaster reconstruction of Shuimo would not be successful unless local residents, officials, and external aids also participated in the process. Foshan and the Guandong Province alone devoted 1.07 billion RMB to the reconstruction of Shuimo, and assisted a total of 78 projects. Instead of allocating billions of dollars directly to Shuimo, the Foshan assistance team integrated their own development experience with Shuimo's local culture and situations. They made sure that local residents and officials agreed upon the reconstruction plan and participated in the planning and reconstruction process. Local residents did not easily accept the new development ideas at first, because it changed their familiar ways of living. The Foshan assistance team's efforts to incorporate local residents in the process, i.e., organizing tours to the benchmark towns and encouraging residents to build up parts of their houses, turned out to be effective. The success of the reconstruction program also lies in recognizing limitations of the official assistance, using the assistance as a leverage to mobilize all possible resources, and involving different stakeholders into the reconstruction.

Many lessons could be drawn from the Shuimo case. First of all, development models pursuing economic growth at the expense of environment and health damages are not sustainable, and therefore should not be encouraged. Sustainability should be a crucial factor in future development. Secondly, a forward-looking, strategic plan that fits the region's cultural identity is important, and it should be agreed upon before any reconstruction starts. Even though post-disaster reconstruction is often under time pressure, the practice of simply copying other cities' planning without understanding one's own historical and cultural heritage should be avoided. Thirdly, the disaster-hit region should take advantage of global attention following major disasters to consolidate all possible external resources and facilitate an open and multi-players involved reconstruction campaign. Efforts should be made to establish governmental collaboration and seek help from more developed areas to the disaster zone (Kapucu 2011).

One should also bear in mind that the Shuimo reconstruction case is unique and not replicable in many ways. For instance, the amount of the planned assistance fund that had been injected into the Shuimo reconstruction was over 3 billion RMB, which is unprecedented and rare among all the 700 affected townships that received counterpart assistance during the post-Wenchuan reconstruction. In addition, Shuimo's reconstruction process also benefited from ambitious, responsible, and competent assistance team leaders and an experienced and talented chief urban designer. Lastly, the counterpart assistance program was a unique Chinese disaster relief strategy, not necessarily universally replicable. In addition, the fact that the Chinese central government could potentially pair provincial governments with disaster-affected counties directly on such a large scale is not applicable to all regimes either.

Of course, the post-Wenchuan reconstruction is a large-scale and highly complex project. Shuimo was only one tip of the iceberg. Many cities and towns experienced dramatically different reconstruction experiences from that of Shuimo. Even though Shuimo's reconstruction has lent us successful post-disaster experience, lessons can also be drawn from other regions. For instance, since the *Master Plan* did not clearly specify the exact responsibilities of various stakeholders who had different interests and agendas to follow, a main dilemma was prioritizing interests and coordinating between different players. Lacking timely and accurate communications between different players became a common problem. Since the *Master Plan* did not stress the equity issue, inter-regional in equality turned out to be another problem when counties and towns tried to compare the allocated resources among themselves. Lastly and most importantly, as we mentioned, due to the time constraints set by the *Master Plan,* the post-Wenchuan reconstruction evolved into a political task. Reconstruction in many cities/towns was simply rushed through without enough deliberation on the designing plan. Many projects used one standard model to mold areas with different geographic and historical features into a uniform style.

We hope the post-Wenchuan reconstruction model in China could lend experience to post-disaster development in other regions of the world. At the same time, we hope that the lessons drawn from the Chinese experience could help prevent post-disaster recovery and development in other countries from falling into similar straps.

References

Chen, X., Han, Q., & Xiao, J. (2008). Post-disaster industry recovery—An opportunity to rearrange and reprioritize the industry structure (in Chinese). Retrieved 18 June, 2008.

Chen, K., Zhou, J., & Jiang, W. (2011). Exploring City Designing from the Reconstruction of Shuimo Town of Wenchuan, China. Architectural Journal. 2011 (4) 11–15.

China Guide. (2012). An introduction to Sichuan province. http://www.china-guide.de/english/a_profile__of_china/sichuan_province/index.html. Accessed 15 Dec 2012.

China National Bureau of Statistics. (2011). *China statistics yearbook* (in Chinese). Beijing: China Statistics Press.

China Perspective. (2011). Sichuan province: Economic news and statistics for Sichuan's economy. http://thechinaperspective.com/topics/province/sichuan-province/. Accessed Dec 2012.

China Perspective. (2012a). Gansu economic news and data. Retrieved Dec 15, 2012.

China Perspective. (2012b). Shaanxi province: Economic news and statistics for Shaanxi's economy. http://www.thechinaperspective.com/topics/province/shaanxi-province/. Accessed 15 Dec 2012.

ChinaView. (2008). Counting the economic costs of China's earthquake. http://news.xinhuanet.com/english/2008-05/29/content_8277443.htm. Accessed Dec 2012.

Disaster Relief Expert Panel, M. o. S. a. T., NDRC. (2008). Chapter 14: Post-disaster industry recovery and development. In Z. Haiyan, P. Han, & J. Li (Eds.), *Wenchuan earthquake social management policy analysis*. Beijing: Science Publishing House.

Dunford, M., & Li, L. (2011). Earthquake reconstruction in Wenchuan: Assessing the state overall plan and addressing the "Forgotten Phase." *Applied Geography, 31,* 998–1009.

Emergency Response Law of the People's Republic of China. (2007). Order of the President of the People's Republic of China (No. 69) C.F.R. (2007).

He, T., & Jiang, Y. (2008). Yinghua town: The way forward for the rich mountainous town that over-relying on exploiting natural resources (in Chinese), *Guangdong Daily, A9.*

Jin, L. (2008). Research on the key subjects of the comprehensive development of urban and rural calamity reduction in China: With special interests to the post-disaster reconstruction of Wenchuan 5.12 earthquake (in Chinese). *South Architecture, 2008*(6), 04–08. doi:1000-0232(2008)06-0004-05.

Kapucu, N. (2011). Collaborative governance in international disasters: Nargis cyclone in Myanmar and Sichuan earthquake in China cases. *International Journal of Emergency Management, 8*(1), 1–25.

Landry, P. F. (2008). *Decentralized authoritarianism in China: The communist party's control of local elites in the post-Mao era.* Cambridge University Press.

Li, Q. (2010). Guangdong counterpart assistance program to Wenchuan is completed (in Chinese). http://news.ifeng.com/gundong/detail_2010_10/09/2726032_0.shtml. Accessed 10 Dec 2012.

Liu, T. (29. August 2003). Sichuan Yanbian: A booming factory and a declining city (in Chinese). *China Broadcasting Online.*

Liu, J. (24. December 2008). Survived the earthquakes but not the pollution (in Chinese). *Chinese Economic Times.*

Liu, H. (2011). *Transition: Personal experience and reflection of an assistance leading cadre on reconstruction operations in Wenchuan earthquake-affected areas* (in Chinese). Guangzhou: Yangcheng Evening News Publishing House.

Shaanxi Provincial Development and Reform Commission. (2008). The significant 'aftershocks' to the industry structure of Sichuan province caused by Wenchuan earthquakes (in Chinese). Retrieved June 18, 2008.

Shi, P., Wang, M., Wang, J., Xu, W., Tang, D., Chen, W., & Shuai, J. (2010). *Criteria for a catastrophe and catastrophe insurance based on the cases of the Wenchuan Earthquake and the Southern Snowstorm.* Paper presented at the National Forum on Disaster Prevention and Reduction and Sustainable Development, Beijing.

Sichuan Government Website. (2009). 35 Schools of Shifang to join hands with 25 schools of Beijing in twins. http://www.sc.gov.cn/zt_sczt/zhcjmhxjy/cjjy/kjcj/200912/t20091217_871603.shtml. Accessed 10 Dec 2012.

Sichuan Provincial Bureau of Statistics. (2006). *Sichuan statistics yearbook 2006*. Chengdu: China Statistics Press.

Sichuan Provincial Government. (2011). Official update on Wenchuan earthquakes disaster reconstruction and recovery situations (in Chinese). http://news.sina.com.cn/c/2011-10-14/095123302670.shtml. Accessed 10 Dec 2012.

State Council of the PRC. (2008a). *Post-Wenchuan earthquake restoration and reconstruction counterpart provinces supporting program [2008] 53*. ([2008] 53). Beijing: State Council General Office.

State Council of the PRC. (2008b). *Post-Wenchuan earthquake restoration and reconstruction Master Plan [2008] 31*. ([2008] 31). Beijing.

State Council of the PRC. (2012). Press conference on Wenchuan earthquake recovery and reconstruction have been fully completed (in Chinese). http://www.scio.gov.cn/xwfbh/xwbfbh/wqfbh/2012/0224/. Accessed 19 Jan 2013.

Strömberg, D. (2007). Natural disasters, economic development, and humanitarian aid. *Journal of Economic Perspectives, 21*(3 Summer), 199–222.

The United Nations Office for Disaster Risk Reduction (UNISDR). (2010). Wenchuan earthquake 2008: Recovery and reconstruction in Sichuan province.

Wang, Y., & Dong, L. (2010). A preliminary study of post disaster counter-part assistance model in China. *Contemporary World & Socialism, 1,* 131–136.

Wisner, B., Blaikie, P., Cannon, T., & Davis, I. (2004). *At risk: Natural hazards, people's vulnerability and disasters* (2nd ed.). London: Routledge.

Xinhua English News. (2006). Sichuan posts per capita income rise in 2005. http://news.xinhuanet.com/english/2006-02/02/content_4128534.htm. Accessed Dec 2012.

Xinhua English News. (2011). Shuimo township an "unbelievable" model for post-disaster reconstruction. http://english.cntv.cn/20110426/105354.shtml. Accessed 20 April 2013.

Xinhua News. (2012). Shuimo in Wenchuan: The rebirth of an old town (in Chinese). http://news.xinhuanet.com/city/2012-12/14/c_124096981.htm. Accessed 20 April 2013.

Yang, S. (2008). Wenchuan Earthquakes damaged 4.93 million mu forest in Sichuan, reduced the forest coverage by 0.5 % (in Chinese). http://unn.people.com.cn/GB/7398673.html. Accessed 15 Dec 2012.

Yin, W. (2001). Causes of China's local market fragmentation. *China & World Economy, 6.*

Zhang, X. (2009). Shanxi province in 2009 to provide 4000 jobs to disaster-affected areas in Maoxian county.

Zhang, X., Wu, H., & Li, S. (2009). Several issues on post-Wenchuan industry reconstruction and sustainable development. *Financial and Economic Science, 257*(8/2009).

Zhong, K. (2011). Counter-partner assistance program in disaster reconstruction: Origin and development. *Comparative Economic & Social Systems, 6,* 140–146.

Zuo, K. (2010). *Procurement and contractual arrangements for post-disaster reconstruction.* (PhD in Civil Engineering). Auckland: The University of Auckland.

Chapter 25
Disasters and Development: Lessons and Implications from Global Cases and Issue

Kuotsai Tom Liou and Naim Kapucu

The purpose of this book is to examine the dynamic relationship between disaster and development for resilient and sustainable communities. The topic of disaster and development is important because in recent years we have observed increasing disaster risks and hazards that are associated with physical, social, economic, and environmental vulnerabilities. The environmental vulnerabilities are, directly or indirectly, related to development conditions of individual communities and nations. Scholars and practitioners of emergency management have recognized the need to examine the complex relationship between disasters and development and provided important theoretical explanations and arguments as well as individual case studies and findings (e.g., Collins 2009; Fordham 2007; Kapucu et al. 2013; Manyena 2012; McEntire 2004).

To support the study of disaster and development, researchers have emphasized an integrated approach to consider research from different disciplines, which include, for example emergency management, development studies, environmental management, physical planning, and social and behavior studies (e.g., Collins 2009; Fordham 2007). As indicated in the Hyogo Framework for Action 2005–2015, policymakers have also indicated that we need to adopt a systematically integrated approach to connect disaster management policies with sustainable development policies and we have to build the necessary capacities at the community and national levels to achieve the goals of sustainable development, poverty reduction, good governance, and disaster risk reduction (United Nations 2005).

While recognizing the importance of relationships between disaster and development, previous studies have not examined this topic and related issues from different disaster cases from global perspectives. To address the need for a comprehensive study of the disaster and development relation, this book has collected chapters

K. T. Liou (✉) · N. Kapucu
School of Public Administration, University of Central Florida, Orlando, USA
e-mail: kliou@ucf.edu

N. Kapucu
e-mail: kapucu@ucf.edu

N. Kapucu, K. T. Liou (eds.), *Disaster and Development,* Environmental Hazards,   447
DOI 10.1007/978-3-319-04468-2_25, © Springer International Publishing Switzerland 2014

to provide a comprehensive study about policy and managerial issues that are related to disaster management and sustainable development in different countries and communities. The chapters in the book offer both conceptual and theoretical arguments and specific case investigations from recent disasters in developed and developing countries. The conceptual issues discussed in the chapters include, for example, arguments and suggestions about the comprehensive vulnerability management, the whole community approach, the measure of disaster resilience, and the role of nonprofit and civil society organizations. The disaster cases investigated consist of such disasters as earthquakes, wild fire, floods, hurricanes and storms in about 20 countries in the North, Central and South America, Middle East, East, South and Central Asia, Europe, Australia, and New Zealand.

This conclusion chapter provides a brief review of chapter issues and findings and introduces important lessons and implications from these studies. Based on their major arguments, the chapters are arranged into three related sections, including Risk Reduction and Policy Learning, Disaster Resilience and Capacity Building, and Recovery, Development, and Collaborative Emergency Management. We first provide brief summary of the findings of chapters in these sections and then identify major lessons from the studies.

25.1 Findings of Chapter Issues

The introduction chapter, by Naim Kapucu and Tom Liou, provides a review of theoretical concepts about the background of disaster and development and introduces a framework of integrated model of disaster and development for sustainability and disaster resilience. The model presents the close relationship between community-based solutions, inter-agency and inter-organizational collaboration civic engagement, and community and organizational capacity, as well as their influence to the effective disaster policy and governance, community resilience and sustainable recovery and development.

The first part, entitled Risk Reduction and Policy Learning, consists of eight chapters to discuss emergency management policy changes and risk reduction and development in different communities and countries. Rejina Manandhar and Davie McEntire explained the complex relationship between disaster and development and promoted a comprehensive vulnerability management model to emphasize vulnerability reduction through reducing risk, reducing susceptibility, increasing resistance and resilience. The chapter by Thomas Birkland and Megan Warnement analyzed earthquake cases in Nicaragua, China, and Haiti as focusing events for these governments to seek opportunities for policy change and learning. Emel Ganapati examined disaster management reforms after 1999 Marmara Earthquake in Turkey and suggested the importance of development plans to integrate social vulnerability at the national level and the participation of local governments and community-based organizations to reduce vulnerability. Vener Garayev studied the 2010 Kura river flood case to examine the emergency management in Azerbaijan. While the system was effective in eliminating disasters, in terms of short-term structural response and recovery measures, the management system does not address long-term policies and non-structural measures to link mitigation with community planning. John Kiefer and Alessandra Jerolleman used the 2005 Hurrican Katrina case in New Orleans, Louisiana to explain the progress in social, technical, and political systems to address major vulnerabilities in communities and to promote community resilience and sustainable development after the catastrophic disaster. Christopher Hawkins and Claire Knox also used the focusing event arguments to examine policy change and learning about disaster planning and response after the 1992 Hurrican Andrew in Florida and compared the Lewis Report recommendations with the results of after action reports from three major hurricanes (Charley, Frances, and Jeanne) in three counties of Central Florida in 2004. In the review of Lebanon's disaster management challenges and sustainable development issues, Thomas Haase suggested that local communities and civil society organizations might promote risk resilience in the absence of support and guidance from the central government. D.K. Yoon reviewed the development and change of disaster management system and policy in Korea and emphasized the importance of improving disaster management to build a disaster resilient community. He discussed the cause and effect relationship between disasters and development through the case of the 2011 Seoul flood.

The second part of the book, entitled Disaster Resilience and Capacity Building, also includes eight chapters to discuss issues and efforts in building managerial capacity to strengthen disaster resilience among communities and organizations. Willaim Waugh and Cathy Liu introduced an integrated model to assess community characteristics and recovery-related capacities to include social vulnerability, the structure of the local economy, business diversity, nonprofit density, and government capacity. They pointed out the engagement of the whole community, the building of social capital and the need of effective leadership as important factor in the relationship between disaster recovery and sustainable development. Steve Scheinert and Louise Comfort proposed a measure of resilience that is based in a complex adaptive system approach to include four components: presence on the ground, internal coordinator, external coordination, and adaptation. They examined the effectiveness of the resilience measure from data collected in the field study of the UN intervention cases in Bosnia-Herzegovina and in Haiti. Samuel Brody argued that natural features of the landscaper (e.g., wetlands, riparian areas, and soils) can support community-level resiliency with respect to flooding and flood impacts. Based on evidences in Texas and Florida, he suggested the need to evaluate the impact of development patterns and to measure the effectiveness of specific land use planning strategies to inform communities. Melanie Gall and Brian Gerber studied the intersection of mitigation and resilience and their relationship with local and national economic development practices by analyzing the case of coastal and interior vulnerability to the flood hazard among the Netherlands, Great Britain, and Germany. Daniel Nohrstedt and Charles Parker investigated two storms (Gudrun in 2005 and Per in 2007) in Sweden to understand policy learning and changes for building resilience to extreme events and emphasized the importance of combining policy intervention, organizational change, and experiential learning as a basis for resilience. Douglas Paton and his colleagues examined the relationship between resilience, recovery and development by studying the 2009 Victoria, Australia wildfires and the 2011 Christchurch earthquake in New Zealand. They developed a model to explain factors affecting disaster recovery and development, which include responses to adaptive demands over time and the adaptive capacities and interdependencies at personal, community, cultural, and institutional and environmental levels. Emphasizing the women's roles in disaster resiliency efforts in Ghana, Kiki Caruson and colleagues noticed that the local residents perceive their community (or the people) as the greatest asset in the context of disaster and use personal networks to enhance communication, disseminating information, and building leadership; but they found that local women represent an under-utilized resources for enhancing resiliency and community development. Focusing on the evolution of disaster management system in the Philippines, Ralph Brower and colleagues explained changes of related laws and challenges to their implementation (e.g., limitations of the new technology and resistance from the above and below). And pointed out that disaster management is synonymous with development, not an unfortunate interruption of it.

The last part, entitled Recovery, Development, and Collaborative Emergency Management, collects seven chapters to explain inter-governmental and inter-sectorial collaboration activities among organizations and communities to promote disaster recovery, development, and the effectiveness of our emergency management system. Faith Demiroz and Qian Hu provided conceptual discussions about the role of nonprofit organizations in the collaborative emergency management and the nonprofits' contribution to community resiliency and post-disaster recovery and development. Frances Edwards and Dan Goodrich used the case of California Seismic safety Commission to support the contributions of multi-disciplinary collaboration (e.g., engineering, science, emergency management, and social services) to the development of resilience in California as a model for others confronted with intractable natural hazards. Simon Andrew and Sudha Arlikatti emphasized the importance of multi-sector partnerships in disaster housing recovery by reviewing reforms and challenges of the Indian housing policy changes after the Killari earthquake in the Latur district of Maharashtra in 1993, the Bhuj earthquake in Gujarat in 2001, and the Great Tsunami of 2004 in the Nagapattinam district of Tamil Nadu. Sana Khosa studied the redevelopment, recovery, and mitigation activities in Pakistan after the 2010 catastrophic floods in Pakistan and explained collaborative efforts by the Pakistani government, international agencies, and international and local non-governmental organizations. Using the East Japan Earthquake and Tsunami in 2011, Rajib Shaw analyzed community-based recovery from three cities in Iwate and Miyagi prefectures and he maintained the need to link disaster risk management to development issues (e.g., education and health), the recourse commitments from local governments, the importance of regular updating and testing local DRM initiatives, and the connection to local resources knowledge institutions. Based on the 2010 Haiti Earthquake case, Abdul-Akeem Sadiq examined the relationship between disasters and development and provided suggestions for future sustainable development in Haiti, which are the incorporation of the principle of building back better, the long-term development, the contribution of the Haiti diaspora, the support from the international community to build capacity, the focus on job opportunities and change of the population and the environment programs. Ping Xu and colleagues reviewed the reconstruction and development experience in China after the 2008 Wenchuan earthquake and emphasized the importance of the counterpart assistance program and the sustainable development approach in the transform of Shuimo town from a highly polluted town to a tourism destination.

25.2   Lessons and Implications

From the review of key issues of the book chapters, we want to identify four major lessons about the study of disaster and development. The four lessons are: vulnerability and resilience/sustainability concerns; collaboration practices and integrated approaches; policy learning and management adaptation, and individual characteristics and common practices.

A. Vulnerability, Resiliency, and Sustainability Concern The first lesson we learned is the recognition of two important terms, vulnerability and sustainability, which affect the relationship between disaster management and economic development. As emphasized by McEntire (2004) and in the chapter by Manandhar and McEntire and another chapter by Sadiq, community vulnerability is related to the interaction between disaster and development (e.g., development reduces or increase vulnerabilities to disasters). From the disaster management perspective, vulnerability concern is one of the important factors as many theories and policies have been developed to reduce risks and increase resilience that are related to vulnerability problems before and after the occurrence of disasters. From the development perspective, less developed communities or nations are more vulnerable than developed communities and nations in dealing with disasters because of the lack of resources, knowledge and techniques to deal with challenges of disasters. For developed communities and nations, they also have vulnerability challenges because these communities and nations need to face negative side problems of development, such as different types of pollutions and changes of physical environment and climate. While the development communities and nations may have the resources to address vulnerability issues, they have to develop policies and methods to consider methods and approaches to achieve the balance between economic development and environmental protection.

Sustainability concern is another important term for both disaster and development studies. Again, many scholars have recognized the importance of sustainable development in their studies of disaster reduction or hazard mitigation (e.g., Berke 1995; Mileti and Gailus 2005). Sustainable development has also been emphasized by researchers of local economic development as one of the important development goals (e.g., Greenwood and Holt 2010; Liou 2009). In this book, the chapters by Manandhar and McEntire and the chapter by Waugh and Liu also maintained that disasters may set back development or provide development opportunities, especially in the case of sustainable development. While agreeing on the relationship between risk/hazard reduction and sustainable development, Tobin (1999) and Manyena (2012) have warned us about difficulties and challenges in connecting disaster reduction and sustainable development if we ignore important implementation factors, such as local context, political interests, social activities, and economic concerns. Indeed, many disaster management practices tend to be reactive and short-term oriented, which do not consider long-term effects and consequences. The reactive approach of emergency management may not include sustainable development goals and strategies in the community's comprehensive planning and misses the opportunity for the long-term development. Similarly, many economic development policies (e.g., business incentive policies) only consider jobs, growth and related economic changes from a short-term perspective. This short-term and narrow approach is related to political interests of policymakers to get quick results or improvements of local economic conditions (Platt 1999; Mileti 1999). They did not consider other social and environmental consequences of economic development, which include social equity and environmental protection issues. Focusing on the proactive and long-term perspective, sustainable development thus has become another importance concern for studies of disasters and development.

B. Collaboration Practices and Integrated Approaches Beside vulnerability and sustainability concerns, we also noticed the emphasis on the collaboration and integration approach in the study of disaster and development relation. The collaboration approach has been identified as one of the major goals or strategies in the literature of public administration (e.g., Ansell and Gash 2008). The importance of collaboration or partnership has also been stressed by studies of emergency management (e.g., Waugh and Streib 2006) and of sustainable development (Gray 2007). The collaboration approach refers to the need of establishing a collaborative emergency management system, which includes inter-government, inter-agency, inter-organization, and inter-sector collaboration in disaster reduction, recovery, and development. The collaboration approach covers the relationship between central and local governments, the horizontal and vertical connection among departments and agencies, the involvement of nonprofit and community-based civic organizations, and the collaboration between public and private sectors (Kapucu 2008). These different types of collaboration and related issues in different communities and nations have been examined in this book, especially the chapters by Demiroz and Hu, Edwards and Goodrich, Andrew and Arlikatt, and Xu et al.

Closely related to the collaboration idea, the integrated approach has also been emphasized in the study of disasters and development. For example, the integration approach was explained in the UN's Hyogo Framework for Action (2005), addressed in the study of vulnerability (McEntire et al. 2010) and emphasized by the FEMA's the whole community approach to emergency management (2011). In this book, the integrated approach was especially emphasized in many chapters and by many authors. First, the introduction chapter introduced an integrated framework about sustainable disaster recovery and development for disaster resilient communities. The integrated framework considers such factors of community-based solutions, inter-agency and inter-organizational collaboration civic engagement, community and organizational capacity, and the causation relationship from effective disaster policy and governance to community resilience and to sustainable recovery and development. The integrated approach has also been emphasized in the chapter of the comprehensive vulnerability management model to reduce risk and susceptibility and increase resistance and resilience. The chapter by Waugh and Liu also introduced the integrated model of community resilience, which consists of five dimensions—social vulnerability index, economic structure, business diversity, nonprofit density, and government capacity.

C. Policy Learning and Management Learning Adaptation The lesson of policy learning and management adaptation is related to governance issues of the disaster and development studies. From the policy perspective, researchers are interested in studying the implementation and effectiveness of various disaster reduction and development policies and programs. For example, to understand the implementation of Hyogo Framework for Action, UNISDR (2011) in their mid-term review noticed that, despite some progress achieved (e.g., the passing of national legislation), there are still many problems and concerns of the implementation, which include, for example, the integration of disaster risk reduction into sustainable development policies and planning at national and international level, and the insufficient level of implementation of the Hyogo Framework for Action at the local level. In this book, the chapter by Birkland and Warnement and by the chapter by Hawkins and Knox applied the principles of focusing events to examine policy changes, resilience and learning. The chapter by Yoon also used the event of disasters to introduce policy changes and to assess the effectiveness of their management system. These changes may include new policies and ideas, structural arrangements, intergovernmental relations and politics.

On the management side, we have learned the importance of building and strengthening capacity among local governments, emergency management institutions, nonprofit and community organizations. As indicated in the UN'S Hyogo Framework for Action (2005), one of the strategic goals to build the resilience of nations and communities to disaster is the development and strengthening of institutions and capabilities at all levels. Haigh and Amaratunga (2009) also emphasized the importance of capacity building for post disaster infrastructure development and management. In this book, the chapter by Waugh and Liu indicated capacity building is not only the means to resilience for disaster relief, but also the project for local sustainable development. The chapter by Scheinert and Comfort addressed the issue of resilience measure in the improvement of emergency management. Other chapters also identified capacity building as one of major suggestions in the management learning, which include, but not limited to, better planning, training, preparation, implementation, and physical and technical improvements in community risk reduction and resilience as well as in post-disaster recovery and redevelopment activities.

D. Individual Characteristics and Common Practices The last lesson of common practices and individual characteristics reflects the nature about findings of issue and case studies in this book. In previous literature, we noticed studies of individual case of economic development and disaster reconstruction after Katrina (Waugh and Smith 2006) and the relationship between disaster and development from the case in Peru (Schilderman 1993). In this book, we have provided studies of individual disaster cases among different communities and countries. For example, the chapter by Ping Xu and colleagues introduced the counterpart assistance program in the reconstruction and development case after the Wenchuan earthquake. The counterpart assistance program involves the collaboration between different provinces and cities and is unique in China's disaster recovering policies. From these findings, we learned that the study of disaster and development needs to consider individual characteristics of various types of natural and man-made disasters and unique political, economic, social and technological environments and supporting infrastructures. To support the goal of disaster recovery and sustainable development, we believe it is to develop an effective emergency management system to consider unique environmental factors of local communities and the quality of the supporting infrastructure as these characteristics will affect the scope, the timing, and the outcome of disaster recovery and sustainable development.

Recognizing individual different characteristics in these studies, we have also leaned many common practices from the findings of these chapters. For example, many chapters provided similar literature reviews to develop their arguments and concluded with similar findings and suggestions. The previous three lessons discussed here can be considered as common practices learned from these studies. All countries are concerned about their community vulnerability and sustainable development. Scholars and practitioners recognized the importance of collaborative and integrated approach to develop an effective disaster mitigation recovery and development system. It is important to learn from disaster cases and to use the disaster experience, positive or negative, to promote policy reforms and strengthen the managerial capacity of government agencies and related public, nonprofit and community organization.

25.3   Conclusion

This book has developed a comprehensive report about the complex and dynamic relationship between disaster and development and contributed to the literature about disaster management, disaster resilience, and sustainable development. The book collects 23 chapters to study theoretical issues and investigate practical cases about policy, governance and managerial lessons of in dealing with different types of disasters (e.g., earthquakes, floods, and hurricanes) among 20 countries and communities in the world. The introduction chapter provides theoretical conceptual framework for the understanding of major issues emphasized in this book. The following chapters address theoretical, policy and managerial issues in the three sections of: risk reduction and policy learning, disaster resilience and capacity building, and recovery, development and collaborative emergency management. This conclusion chapter provides a brief review of major issues and findings of all chapters and also identifies four lessons of case studies, which include vulnerability and sustainability concern, collaboration and integration approach, policy and management learning, and individual characteristics and common practices.

The findings of this book are useful for future studies of the complex relationship between disaster and development. Future studies may continue the investigation of this topic from two areas. First, future studies may want to emphasize accountability issues regarding the implementation of disaster risk reduction and sustainable development and the performance and effectiveness of related policies and programs. As indicated in the UNISDR's mid-term review (2011), we need to develop a broadly representative mechanism to include accountability measures to measure progress and to ensure coordinate and coherent action at all aspects of disaster risk reduction action across different sectors of government. The findings of many chapters support the importance of accountability studies about the implementation and performance because they recognized the weakness of general policies and guidelines and the need for specific standards to assure the quality in policy implementation and performance outcome.

Besides the accountability measures of implementation and performance, we also suggest future studies to include different approaches to improve our study of disaster and development. As reported in this book, most of the chapters provided either theoretical discussions or case studies to investigate the topic issues. Researchers of emergency management have provided good conceptual review and theoretical model to address different concerns of the relationship between disaster and development in terms risk reduction, disaster resilience and recovery, and sustainable development. But for the analysis of disaster events, we noticed different approaches about case studies of these disasters with inconsistent arguments and findings. To improve the quality of our studies, we may want to develop a good framework to guide future case analysis in terms of theoretical review, case background, and critical analysis of specific disaster issues and lessons. In addition, as reported in Scheinert and Comfort's chapter about resilience measure, we also welcome more empirical studies to provide data-driven and evidence-based investigations of many important topics in risk reduction, community resilience, disaster recovery, and sustainability development and the interaction among these topics. The application of theoretical discussions, case studies, and empirical analyses will be valuable to enhance our full understanding of the relationship between disaster and development.

Finally, even though the chapters in this book contribute significantly to disaster and development from single case perspective (with some comparative perspectives), disasters do not respect borders. Especially with globalization and new developments in technology, transboundary disasters have become more common and impactful requiring jurisdictional collaborative solutions. Dealing this level of inter-jurisdictional, cross-sector, and interorganizational coordination is a significant challenge for policy makers and disaster managers. Future research should focus more on policies and adaptive governance structure dealing this very nature and challenge of disasters/crises. Future research should also include elements of economic growth and sustainable development while recovering from catastrophic disasters. We hope that the book provided some initial steps in this respect as well.

Reference

Ansell, C., & Gash, A. (2008). Collaborative governance in theory and practice. *Journal of Public Administration Research and Theory, 18*(4), 543–571.

Berke, P. R. (1995). Natural-hazard reduction and sustainable development: A global assessment. *Journal of Planning Literature, 9*(4), 370–382.

Collins, A. E. (2009). *Disaster and development*. New York: Routledge.

Federal Emergency Management Agency (FEMA). (2011). *A whole community approach to emergency management: Principles, themes, and pathways for action*. Washington, D.C.: FEMA, FDOC-008-1, December.

Fordham, M. (2007). Disaster and development research and practice: A necessary eclecticism? In H. Rodríguez, E. L. Quarantelli, & R. R. Dynes (Eds.), *Handbook of disaster research* (pp. 335–346). New York: Springer.

Gray, B. (2007). The process of partnership construction: Anticipating obstacles and enhancing the likelihood of successful partnerships for sustainable development. In P. Glasbergen, F. Biermann, & A. P. J. Mol (Eds.), *Partnerships, governance and sustainable development: Reflections on theory and practice* (pp. 29–48). Northampton: Edward Elgar.

Greenwood, D. T., & Holt, R. P. F. (2010). *Local economic development in the 21st century: Quality of life and sustainability*. Armonk: M.E. Sharpe, Inc.

Haigh, R., & Amaratunga, D. (2009). Capacity building for post disaster infrastructure development and management. *International Journal of Strategic Property Management, 13*(2), 83–86.

Kapucu, N. (2008). Collaborative emergency management: Better community organizing, better public preparedness and response. *Disasters: The Journal of Disaster Studies, Policy, and Management, 32*(2), 239–262.

Kapucu, N., Hawkins, C., & Rivera, F. (Eds.). (2013). *Disaster resiliency: Interdisciplinary perspectives*. New York: Routledge.

Liou, K. T. (2009). Local economic development in China and the United States: Strategies and issues. *Public Administration Review, 69*(S1), 29–37.

Manyena, S. B. (2012). Disaster and development paradigms: Too close for comfort? *Development Policy Review, 30*(3), 327–345.

McEntire, D. A. (2004). Development, disasters and vulnerability: A discussion of divergent theories and the need for their integration. *Disaster Prevention and Management, 13*(3), 193–198.

McEntire, D. A., Crocker, C. G., & Peters, E. (2010). Addressing vulnerability through an integrated approach. *International Journal of Disaster Resilience in the Built Environment, 1*(1), 50–64.

Mileti, D. S. (1999). *Disasters by design: A reassessment of natural hazards in the United States*. Washington, D.C.: Joseph Henry Press.

Mileti, D., & Gailus, J. (2005). Sustainable development and hazards mitigation in the United States: Disasters by design revisited. In C. Haque (Ed.), *Mitigation of natural hazards and disasters: International perspectives* (pp. 159–172). Springer Netherlands.

Platt, R. (1999). *Disasters and democracy: The politics of extreme natural events*. Washington, D.C.: Island Press.

Schilderman, T. (1993). Disasters and development: A case study from Peru. *Journal of International Development, 5*(4), 415–423.

Tobin, G. A. (1999). Sustainability and community resilience: The holy grail of hazards planning? *Global Environmental Change Part B: Environmental Hazards, 1*(1), 13–25.

United Nations. (2005). *Hyogo framework for action 2005–2015: Building the resilience of nations and communities to disasters*. World Conference on Disaster Reduction, Kobe, Hyogo, Japan, 18–22 January. United Nations, A/CONF.206/6.

United Nations International Strategy for Disaster Reduction (UNISDR). (2011). *Hyogo framework for action 2005–2015: Building the resilience of nations and communities to disasters, mid-term review, 2010–2011*. www.unisdr.org/preventionweb/files/18197_midterm.pdf. Accessed March 2011.

Waugh, W. L. Jr., & Smith, R. B. (2006). Economic development and reconstruction for the Gulf after Katrina. *Economic Development Quarterly, 20*(3), 211–218.

Waugh, W. L. Jr., & Streib, G. (2006). Collaboration and leadership for effective emergency management. *Public Administration Review, 66*(1S), 131–140.

Index

A
Aakre, S., 226
Abell, R.A., 112
Abel, N., 255
Acar, M., 325
Acemoglu, D., 291, 307, 309, 311
Acosta, J., 321
Acreman, M., 205
Adaptive capacity, 237, 256, 320, 326
Adger, N., 237
Adger, W.N., 237, 255
Aerts, J.C.J.H., 217, 224
Afedzie, R., 25, 26
Afrifa, E.K.A., 274
Agsaoay-Saño, E., 299, 300, 301, 302
Ahn, J.-Y., 158
Ahrens, J., 80, 82, 90
Akcar, S., 276
Aldrich, D.P., 172, 173, 176, 318, 319, 323, 324
Aleman, K., 347
Alesch, D., 172, 174
Alexander, D., 20, 21, 22, 26, 29, 291, 352
Alexander, D.A., 151
Alhassan, O., 279
Alibe, T., 306
Altintas, K.H., 69
Amaratunga, D., 28, 453
American Red Cross (ARC), 101, 321, 322
Amir, P., 372, 375
Anderies, J., 237
Anderson, M.B., 61, 75
Andrew, S.A., 351, 352, 363, 364
Andries, J.M., 255
Ansell, C., 317, 452
Antoun, R., 140, 143
Antwi, V.S., 274
Appley, R., 260
Archibold, R.C., 412

Arellano, A.L.V., 223
Arlikatti, S., 351, 352, 357, 362, 363, 364
Armah, F.A., 274
Arnold, C.L., 206
Arquilla, J., 182, 184
Arroyave, V., 317
Arslan, T., 321
Arya, A.S., 357
Atherfold, C., 266
Athukorala, P., 19
Augustin, M.E., 170
Axelrod, R., 184, 185, 189
Aydogdu, S., 69
Azerbaijan, 79, 80, 82, 83, 84, 85, 86, 90, 91, 448

B
Bach, R.L., 266
Bailey, S., 176
Balamir, M., 61, 71
Bannister, S., 257
Barabasi, A.-L., 186
Barakat, S., 61, 352
Barbee, D.G., 82
Barenstein, J.D., 168, 173, 352, 354, 357, 359, 360, 361, 364, 365
Barenstein, J.E.D., 173, 176
Batt, J., 266
Baumgartner, F., 239
Baumgartner, F.R., 113
Baxter, V., 95
Beatley, T., 114, 202, 208, 351
Beatly, T., 208
Begum, S., 222
Bellerive, J.-M., 412
Bengali, S., 104
Berkebile, R., 417, 421
Berke, P., 114, 351
Berke, P.R., 28, 30, 32, 33, 451

N. Kapucu, K. T. Liou (eds.), *Disaster and Development,* Environmental Hazards, 459
DOI 10.1007/978-3-319-04468-2, © Springer International Publishing Switzerland 2014

460   Index

Made in the USA
Las Vegas, NV
10 February 2022